VARIABILITY AND MOTOR CONTROL

Karl M. Newell, PhD
University of Illinois at Urbana-Champaign

Daniel M. Corcos, PhD
University of Illinois at Chicago

Editors

Human Kinetics Publishers

Library of Congress Cataloging-in-Publication Data

Variability and motor control / Karl M. Newell, Daniel M. Corcos, editors.

    p. cm.

    "This book is a product of a conference on variability and motor control held in Chicago, April 19-21, 1991''--Pref.

    ISBN 0-87322-424-8

    1. Human mechanics--Congresses.   2. Motor learning--Congresses. I. Newell, Karl M., 1945. II. Corcos, Daniel M., 1954- III. Conference on variability and motor control (1991 : Chicago, Ill.)

QP303.V38   1993

612'.044--dc20

                                       92-33077

                                          CIP

ISBN: 0-87322-424-8

The articles in this book were presented at a conference on variability and motor control held April 19 to April 21, 1991, in Chicago, Illinois.

Printed in the United States of America

10 9 8 7 6 5 4 3 2 1

Human Kinetics Publishers
Box 5076, Champaign, IL 61825-5076
1-800-747-4457

Canada:
Human Kinetics Publishers, P.O. Box 2503, Windsor, ON N8Y 4S2
1-800-465-7301 (in Canada only)

Europe:
Human Kinetics Publishers (Europe) Ltd., P.O. Box IW14, Leeds LS16 6TR, England
0532-781708

Australia:
Human Kinetics Publishers, P.O. Box 80, Kingswood 5062, South Australia
618-374-0433

New Zealand:
Human Kinetics Publishers, P.O. Box 105-231, Auckland 1
(09) 309-2259

Contents

Preface

This book is the product of a conference on variability and motor control held in Chicago, April 19-21, 1991. For a number of reasons it seemed timely to hold such a conference. First, variability is an issue that touches many theoretical lines of enquiry in the study of motor control and skill acquisition. Second, no matter what the theoretical orientation, variability of some movement parameter is nearly always estimated in empirical studies of motor control. Third, and perhaps most importantly, there was a suggestion from recent developments in the field that a new theoretical perspective on variability and motor control was in the offering.

Traditionally, movement variability was, and largely still is, interpreted as a problem for system control. In this view, movement variability is associated with noise—an artifact that compromises at a variety of levels of analysis the deterministic relation between input and output. Furthermore, studies of the movement speed-accuracy relation have suggested that the noise of the system is enhanced when greater effort (i.e., more speed, more force) is required in realizing a particular outcome. This traditional interpretation of noise or variability in motor control is negative in connotation in that it views variability as a problem to be eliminated or minimized. This view of variability is prevalent in a range of theoretical lines of enquiry in motor control, from motor programs to feedback, and studies of the kinetics and kinematics of movement variability.

Dynamical systems approaches to movement control, including connectionist orientations, suggest a different interpretation for variability. Here noise, within certain bounds, can be viewed as a positive factor in a range of issues for system control. This positive interpretation is even suggested in the changing nature of the accompanying nomenclature. Variability is seen as an index of movement fluctuations rather than a reflection of movement error! This contrasting view to the interpretation of variability was a major focus of the conference discussion and is a polemic that is likely to be a central source of debate in future work on variability and motor control.

It should be noted that the conference did not cover all aspects of variability and motor control. Limited emphasis was given to variability in perception or the input of information and how this relates to variability of the output. There were no discussions of the relation between variability in particular biological subsystems and variability in the resultant movement dynamics. Our approach was simply to invite as speakers those who had worked directly on the variability and motor control issue in the last few years.

The initial plans for the conference meeting were relatively local and modest in that it was set to simply have a meeting on variability and motor control for interested parties from the two campuses of the University of Illinois. However, as funding for the meeting became available we were able to expand our horizons and invite additional speakers. We would like to acknowledge and thank the University of Illinois president's office and Human Kinetics Publishers for their financial support of the meeting. Additional support was provided by the College of Kinesiology at the University of Illinois at Chicago and the Department of Kinesiology at the University of Illinois at Urbana-Champaign. Many individuals from the two campuses helped in both arranging and running the meeting, particularly members of the Motor Control Laboratory, Rush Presbyterian Medical School, Chicago, and the Coordination, Control, and Skill Group of the Department of Kinesiology at the University of Illinois at Urbana-Champaign. A special thanks to Carol Farmer for her many organizational efforts on behalf of the conference and this book.

Karl M. Newell
Daniel M. Corcos

Chapter 1

Issues in Variability and Motor Control

Karl M. Newell
University of Illinois at Urbana-Champaign

Daniel M. Corcos
University of Illinois at Chicago

Variability is inherent within and between all biological systems. The considerable differences that may be discerned among the motor abilities of humans is strong testament to this observation, as is the fact that it seems impossible for a given individual to generate identical movement patterns on successive attempts at performing the same task. The focus of this book is the within-subject variability evident in executing solutions to motor tasks in a variety of contexts, with less emphasis given to between-subject performance variability.

Human movement is an emergent property that arises from the harnessing of the many degrees of freedom of the sensorimotor system in the service of realizing a given task goal. The total number of system degrees of freedom that are coordinated and controlled in the execution of human movement is very large. Indeed, there is a progressive increase in the number of degrees of freedom as one shifts observation to a more micro level of analysis of the sensorimotor system, such as from joints, to muscles, to motor units, to cells. This large number of degrees of freedom, which requires coordination within and between levels of analysis, naturally affords variability in human movement and, more generally, in biological motion. Furthermore, the considerable redundancy of the sensorimotor system also leads to the variability evident in repetitive attempts to realize a solution to a given task demand. These potential types and sources of variability contribute to variations in the output of the motor system, or in what Sherrington (1906) described as the final common pathway.

The role that variability plays in the coordination and control of the sensorimotor system is a central issue for the study of motor control. Indeed, all theoretical accounts of motor control either implicitly or explicitly consider variability on one dimension or another. Several types of variability have been identified in motor control, in part because variability has been considered at several levels of analysis or frames of reference. Although variability is a pervasive phenomenon in motor control, it has not been a unifying or encompassing theoretical concept, due in large part to the considerable number of sources and types of variability available for study.

In this chapter we briefly outline some of the persistent issues in variability and motor control. This introductory overview is not exhaustive and is largely confined to the issues on variability and motor control that arise as central to the contributing chapters of the book. We discuss issues in variability and motor control through a consideration of the concepts of noise, stability, and skill. Some of the types and sources of variability in motor control are also introduced. To begin, however, we need to address the issue of the measurement of variability to provide some common ground for understanding *what* is actually being measured in examinations of variability in motor control, together with the limitations that naturally arise from the use of such measures.

MEASURES OF VARIABILITY

The variability of the sensorimotor system is usually operationalized by considering the standard deviation of a given system parameter that arises from repeated measures of that parameter over successive attempts at satisfying a particular movement task. The concept of variability, therefore, is often based on a *single* statistic that is drawn from a distribution of measures of a given system parameter. If the distribution of the parameter measure is normal, then the mean and standard deviations of the distribution are sufficient to describe the distribution. Thus, even with a normal distribution of a given movement parameter, one cannot consider the standard deviation measure, or the variability, in isolation from the mean of the distribution. When the distribution of the system parameter is not normal, then higher order moments are required to fully describe the distribution of the system parameter and aid in veridical interpretation of the changes in the standard deviation or variability measure (Newell & Hancock, 1984).

These elementary principles of descriptive statistics are emphasized here because there is a tendency for the operational emphasis on variability in motor control to mask the basic point that the standard deviation is only one measure of a statistical distribution. The interpretation of the variability, therefore, cannot be considered without reference to the full set of statistics that are required to adequately characterize the distribution. The standard deviation also does not provide any index of the structure in the trial-to-trial (or moment-to-moment) relations that may or may not be present in the variability. The standard deviation informs only about the *degree* of variability of a given system parameter.

Many sensorimotor system parameters can be recorded from a variety of frames of reference. For example, the accuracy of movement can often be recorded in different planes of motion, leading to different estimates of system variability. Variability of movement can also be measured relative to some external frame of reference, or it can be recorded relative to the performer or the mean of the distribution of the system measures (the coefficient of variation). The variability of system parameters can be examined at more microlevels of analysis than the limb and torso dynamics, as in the various mechanisms of neuromuscular activity. The mechanisms of neuronal transmission and muscular contraction have been widely studied in the neurophysiology of motor control, but considerably less attention has been directed to the variability of measures that arise from these respective studies (cf. Hoyle,

1983). The relation of variability *within* the various levels of analysis to variability *between* these levels of analysis is an essential element in the development of a general theory of motor control. However, most theories of motor control are confined to a given level of analysis of the sensorimotor system, and thus the issue of variability in motor control tends to follow suit.

The choice of the frame of reference for determining variability in motor control is sometimes arbitrary—that is, it merely follows a tradition that has been initiated in a certain experimental protocol. The use of different frames of reference can lead to different descriptions and interpretations of the role of variability of the sensorimotor system. The determination of the frame of reference for measurement should be motivated a priori by a theoretical perspective and not merely the operational traditions in motor control.

The term *variability* is used in motor control as both a label for a theoretical construct and an operational measure. A clear distinction between these uses of the term *variability* is essential to understanding the role of variability in motor control. For example, a traditional and still-common approach in motor control is to consider variability to be an index of noise in the sensorimotor system. This issue will be taken up more thoroughly in a later section of this chapter, but we need to reemphasize at this juncture the measurement point that, operationally, variability is only the standard deviation. Or, to put it another way, operational variability is only operational variability. Variability of a given system parameter or task outcome cannot *necessarily* be equated with noise. One reason for this limitation is that recent developments in nonlinear dynamics and chaos theory demonstrate that deterministic inputs to a system can have random consequences on the output. Noise, therefore, is a theoretical inference that may be based in part upon an operational variability measure, but caution needs to be observed in conceptualizing the relative deterministic and random (noise) contributions to the resultant variability observed in motor control.

The introduction of nonlinear dynamics and chaos theory to the study of biological systems has opened the door to new ways to conceptualize variability (Glass & Mackey, 1988). This approach has also led to new ways to operationalize variability beyond the traditional use of the standard deviation of some system parameter. A particular difference is the consideration of the initial conditions to the dynamics and the resultant topological principles governing the dynamical trajectories in phase space.

The general measurement point is that the distinction between construct and observation in variability and motor control is sometimes finessed or forgotten. As the motor ouput is a reflection of the final common pathway of the system, there are many potential sources of the observed variability. This inherent feature of biological systems increases the importance of the basic measurement requirement of content validity to the selected indices of movement variability. That is, one needs to insure that the operational measure of movement variability and motor control reflects the to-be-studied construct of variability.

TYPES AND SOURCES OF VARIABILITY

Although variability is inherent at every level of analysis or frame of reference for the measurement of movement, it has not been a unifying concept in theories of motor control. As a consequence, different types of variability have been identified in motor control, and various sources of this variability have been proposed in somewhat isolated or disjointed fashions. The theoretical link between the different types and sources of variability is not well articulated in the general motor control literature. A number of chapters in this book address this issue and provide many examples of particular types and sources of variability in motor control.

The predominant theoretical and operational emphasis in the motor control literature has been on variability in the outcome or product of an action—that is, on variability of the outcome of the movement sequence as defined by the task criterion. A related but different type of variability is that arising from the dynamics of the individual torso and joint biomechanical degrees of freedom in the control of movement. The link between these two types of variability in motor control is not well understood, even though they are observed at the same so-called behavioral level of analysis.

It has long been recognized that there is a sensorimotor equivalence in the conduct of action. That is, a variety of coordination solutions often exist to the realization of a given task constraint (Bernstein, 1967; Hebb, 1949; Lashley, 1929; Turvey, Shaw, & Mace, 1978). Indeed, it is the inherent sensorimotor equivalence of the system that affords the adaptability and flexibility of human action. Various potential sources of this sensorimotor equivalence have been identified, including anatomical, neural, and mechanical sources. The distinction between the systematic and the random sources of this sensorimotor equivalence is of central concern to theories of motor control.

The theoretical and empirical emphasis given through the years to this sensorimotor equivalence problem has focused on the presumed noise or random sources of movement variability. This emphasis has contributed to an underlying negative perspective on the role of variability in system control, without a sufficient consideration of the potential positive aspects of noise or variability in biological motion. This latter perspective is, however, an emerging viewpoint in motor control, due largely to the introduction of dynamical systems approaches to the field of study; Mpitsos and Soinila (chapter 10), Kelso and Ding (chapter 11), Riccio (chapter 12), and Turvey, Schmidt, and Beek (chapter 14) provide examples of some of the issues emerging from this orientation in the study of variability.

VARIABILITY AND NOISE

Noise is presumed to be random fluctuations and is evident in all the various neuromuscular mechanisms at all levels of analysis. Indeed, random variation is an inherent element of biological systems. A central challenge is to understand how order and regularity arise in the coordination and control of movement with noise as an inherent component to the system. This issue gives direct recognition to the fact that the so-called degrees-of-freedom problem (Bernstein, 1967) is a between- as well as within-levels-of-analysis problem. The relations among the variabilities observed at different levels of analysis in motor control are an enduring theoretical and empirical problem and one that is rarely tackled directly (although see Kugler & Turvey, 1987, for a recent effort).

The prevailing theoretical approach to variability and motor control is to treat the within-subject variability as a reflection of noise in the sensorimotor system. The greater the variability as operationalized by a standard deviation of a given system variable, the greater the level of noise in the system. For example, this theoretical-operational relation is central to the impulse variability theory of the movement speed-accuracy trade-off proposed by Schmidt, Zelaznik, Hawkins, Frank, and Quinn (1979). In this particular theory, the noise is postulated to be peripheral in neuromuscular mechanisms, although there have been no direct tests of a central-peripheral distinction to variability in this discrete-movement experimental protocol. Some years ago Wing and Kristofferson (1973) proposed a method to partition the variance due to peripheral and central sources in repetitive timing tasks, and this approach to variability is utilized by Turvey, Schmidt, and Beek (chapter 14), and Ivry and Corcos (chapter 15).

In many extant accounts of motor control, the sensorimotor system output is presumed to be deterministic, with these various unwanted sources of noise superimposed. The models in this framework are essentially linear, with noise added to certain system parameters. In this view, system noise is interpreted as a limiting factor in system control and as such should be eliminated or, at worst, minimized. The information processing approach to motor control, with its black-box depiction of hypothetical information-processing mechanisms, generally endorsed this interpretation of noise and provided the background framework for many of the subsequent cognitive interpretations of noise and motor control.

The concept of noise has been adopted as a limiting factor in motor control, but this view has never been explored very extensively, beyond a consideration of the standard deviation of a given system variable. The motor control literature has also not distinguished or examined explicitly the different types of noise that have been identified. In white noise, the fluctuations in a given variable are truly random, with there being an equal likelihood of a given value of the variable arising from each measurement. The variability or fluctuations may, however, scale to some other property of the distribution, such as $1/f$, where f is the mean frequency of the distribution. There are many types of noise in system control where the fluctuations scale to different properties of the respective system (Schroeder, 1991), and these are identified by color, such as pink, brown, and black noise. These different types of noise have not been a consideration in the motor control literature, but they hold interesting theoretical implications, as suggested by Turvey, Schmidt, and Beek (chapter 14).

There is also the question of whether there is a trial-to-trial (observation-to-observation) relation in the distribution over the time series of recorded data. Whether there is structure to the trial-to-trial relations has rarely been approached in extant examinations of variability in motor control—in part, we suspect, because of the prevailing theoretical assumptions about the chance fluctuations of noise. However, the successive observations of a given variable over a time series may not be random, which would suggest some structure to the trial-to-trial relations. Spray and Newell (1986) examined this question in the context of the role of knowledge of results in motor learning. They showed that both the knowledge-of-result and the no-knowledge-of-result trials followed a white noise model, with the difference between the groups being the mean of the distribution. The data set available for this test of the trial structure was limited, and much more work is required to understand the trial-to-trial relations in the knowledge-of-result motor learning paradigm and other experimental paradigms that are more central to motor control issues.

Another way to conceptualize noise in motor control is to consider the sensorimotor system as part of a nonlinear dynamical system (Glass & Mackey, 1988). Here noise is viewed as part of the dynamic and not a superimposed component. Thus, noise contributes to the qualitative properties of the dynamical output and not merely the quantitative properties. This formulation holds the interesting property that noise can be beneficial to the system in the sense of facilitating adaptation to task demands. One interesting perspective for motor control is that variation in the output provides information about the state of the system. Little motor control work has been conducted to date within this framework, but this orientation holds considerable promise as a basis to reexamine the concept of noise in motor control. Several of the chapters in this book deal with this issue, particularly chapters 10 through 12.

VARIABILITY AND STABILITY

Variability is often interpreted as an index of the stability of the sensorimotor system. The general assumption is that enhanced variability is a reflection of reduced stability. This relation between stability and variability is often drawn on in studies of posture, where the concept of stability has an intuitive place. The distinction between posture and movement is not absolute, however, and the concept of stability is relevant to movement control, as exemplified in dynamical systems approaches to the problem.

Stability is a slippery concept to define, and it seems that no single variable is sufficient to characterize the stability of a system. In the engineering literature the stability of a system is generally related to that system's facility in accommodating perturbations. Perturbations to the base of support in posture have become one of the main experimental protocols for examining the question of postural stability (Nashner & McCollum, 1985). Variability of the postural response is only one index of the resistance to perturbation and hence the stability of the system.

Postural variability is often indexed by variations in the motion of the center of pressure in a given plane of motion. Thus, like most measures of movement variability, postural variability is measured with respect to some external dimension that is convenient to record, such as the anterior-posterior and lateral-medial planes of motion. Variability of posture, however, needs to be considered in relation to the nature of the dynamical system supporting the posture; otherwise there must be an explicit assumption that each posture has a similar dynamic, so that the respective variability can be interpreted veridically. Newell, van Emmerik, and Sprague show in chapter 17 that the variability of the center of pressure in human bipedal posture is insufficient to characterize the dynamic of the sensorimotor system and hence the stability of posture. The variability of the center of pressure needs to be considered in relation to the attractor dynamics supporting the posture that can be formalized by various techniques of nonlinear dynamics and chaos theory. Without a conceptual link to the underlying dynamical structure of posture, the variability of the center-of-pressure measure can be an equivocal index of the stability of the sensorimotor system.

This equivocal relation between variability and stability in posture extends to the direct study of movement control. The general outcome for motor control is that variability of movement is not a *sufficient* index of the stability of movement. The concept of stability in motor control requires multiple measures for it to be adequately characterized. The chapters in this book that deal directly with a dynamical systems approach to variability suggest new ways to consider the relation between variability and stability in motor control.

VARIABILITY AND SKILL

Variability, or lack thereof, in a given movement parameter is often used as an index of skilled performance. Indeed, in certain situations, variability is the task criterion, as in the activities of rifle shooting and archery. In many other movement tasks, performance variability is only one measure of the reliability with which an individual can reproduce a given task outcome from trial to trial.

The focus on variability in the motor skills literature has traditionally been on the outcome of a given action that is repeated from trial to trial. In many experimental paradigms there has been little interest in the variability of the movement dynamics that produced the accompanying outcome. The typical finding is that outcome variability reduces as a function of practice and increments of skill. The veracity of this finding is, however, task dependent to a certain degree, but it prevails in the spatial and timing accuracy tasks that have dominated the motor skills literature. In tasks where the performer has to maximize the output on a given task variable (e.g., javelin throw, shot put), improved performance leads to an increased potential range of scores for the task variable and, hence, the increased possibility of variability in the resultant performance outcome. As far as we know there is no evidence, for example, as to whether the variability in distance jumped is less in a top-class long jumper as opposed to a mediocre long jumper.

Variability is often used as a marker of individual differences in skilled performance. Variability has also been used to characterize population differences in motor skills, as in the performance of young children and the developmentally disabled and in the various categories of movement disorders. The typical finding is that performance variability is enhanced in these distinct population groups in contrast to some appropriately designated control group. Whether there are clearly distinct populations on the different variability dimensions is an open question.

The examination of performance outcome variability in the skilled performance literature has traditionally taken place without reference to variability in the movement dynamics that have produced the task outcome in the motor skills domain. However, the advent of computer-driven movement analysis systems has increased the propensity to analyze the movement dynamics as well as the movement outcome. This development has helped promote a link between the study of variability in motor skills and the study of movement control. The relation between movement variability and outcome variability has not proved to be straightforward, however, and it is clear that task constraints play a significant role in determining this relation.

One of the major areas to examine the movement-outcome relation is the response variability literature of the movement speed-accuracy trade-off and isometric force production. For example, the impulse variability theory of Schmidt et al. (1979) focuses directly on this link. Many subsequent examinations have been made of these various movement-outcome relations, and several chapters in Parts I and II of this book detail developments in this area. In the main, there appears to be a relatively direct relation between task outcome, movement dynamics, and skill level in these experimental protocols. These response variability task protocols are, however, highly constrained, with often merely a single degree of freedom to be controlled at the biomechanical level.

In tasks where several biomechanical degrees of freedom are to be coordinated in the reduction of variability on a designated task variable, the relation between the movement dynamics and the movement outcome is not so direct. For example, the reduction in the variability of the pistol tip in a shooting task, which accompanies practice and enhanced skill, is accomplished by an increase in the variability of the individual limb joint motions (Arutyunyan, Gurfinkel, & Mirskii, 1968). In this pistol-shooting task and other related tasks, the reduced task-outcome variability is accomplished by increasing the degrees of freedom that are regulated in the coordination mode—or the release of the ban on the degrees of freedom, as Bernstein (1967) described it. The compensatory actions of each single-joint motion lead to reduced outcome variability but also enhanced variability in the individual degrees of freedom. Considerably more work is required to understand the movement-outcome variability relation in tasks requiring the coordination of many biomechanical degrees of freedom. The general implication of the Arutyunyan et al. (1968) finding is that enhanced variability in the individual degrees of freedom is not the negative factor that it is usually portrayed as being in the skilled performance literature.

OUTLINE OF THE BOOK

We now provide a few more details on the organization of the content of this book. There are four main parts, which provide convenient categorizations of the role of variability in the control of movement. These parts are force variability, variability and the movement speed-accuracy trade-off, dynamical systems approaches to movement variability, and variability and movement disorders. The parts tend to reflect the loci of current research activity on variability and motor control.

The opening chapter of Part I, "Force Variability and Characteristics of Force Production," by Carlton and Newell (chapter 2), provides a summary of the different mathematical relationships that have been proposed between the force that is produced when making a muscular contraction and the associated force variability. They conclude by presenting an equation that relates force variability to both peak force and the time it takes to reach peak force. Several lines of evidence suggest that the types of isometric contractions discussed by Carlton and Newell can be generated by either height modulation or width modulation of a hypothetical control signal, depending on the demands of the task. In chapter 3, Ulrich and Wing develop such a model, in which either impulse parameter can be changed, and show simulations of time-series data and the associated variability.

The approach taken in chapters 2 and 3 describes the variability associated with performing a given task in terms of characteristics of the task. The premise is that there are lawful relationships among variables that apply to populations. Worringham (chapter 4) further pursues this idea by showing that the amplitude of the initial phase of an aiming movement is related to the spatial variability of this phase. That is to say, the more variable the movement, the further the distance from the target of the initial phase of the movement. Worringham then turns the question around and asks whether measures of variability can be used to explain individual differences. To do this, he shows that individuals who are ranked the fastest in average movement times tend to rank lowest in spatial variability. He proceeds to discuss the paradox between this finding and the general movement speed-accuracy relation that has variability increasing with speed.

Chapters 2 and 3 present findings that have been derived from mechanical single-degree-of-freedom contractions. The findings of Worringham in chapter 4 are derived from movements that have multiple degrees of freedom but in which the only measure of interest is the movement trajectory. Darling and Stephenson (chapter 5) tackle the considerably more complex problem of variability in multi-degree-of-freedom movements. They provide evidence in support of the idea that the relationship between force and force variability that is observed at one joint and at the movement endpoint is *not* seen at all joints taking part in the movement.

In Part II, ''Variability and the Movement Speed-Accuracy Trade-Off,'' one of the most robust findings in the motor control literature is examined, the movement speed-accuracy trade-off. The four chapters of this section address the speed-accuracy trade-off from four different perspectives. Zelaznik (chapter 6) first reviews the literature on the linear speed-accuracy trade-off. He suggests that neither the empirical data nor the kinematic data in the motor control literature are in accord with the models of Schmidt et al. (1979) and Meyer, Abrams, Kornblum, Wright, and Smith (1988). He then proceeds to develop a model that requires independent control of both the acceleration and the deceleration phases of the movement trajectory. In chapter 7, Agarwal, Logsdon, Corcos, and Gottlieb provide an alternate explanation for Fitts' law, based on the dynamic constraints on the limb movement and optimal control theory. In chapter 8, Latash and Gutman explore the idea that errors of movement can arise from different levels of a control system, and they develop two levels of such a multilayered control system. The first is a kinematic model that simulates the intended movement. This model can produce different kinds of speed-accuracy trade-offs. In addition, they consider motor variability in the framework of the equilibrium-point hypothesis. Here, they show that the final position of a movement (and hence the dispersion of movement endpoints) is influenced by the types of instructions a subject is given. The approach of Bullock and Contreras-Vidal, in chapter 9, is to model many of the aspects of the mammalian neural circuitry that are involved in the control of movement. The model works in cooperation with another model (the Vector Integration to Endpoint, VITE) of a central neural network. The interesting observation that arises from all four chapters in this section is that the description of the speed-accuracy trade-off is preserved, but new computational accounts are provided to explain it.

Parts I and II are dominated by the idea that measured variability is lawfully related to experimental manipulations such that increased variability is generally associated with movements or contractions that are faster or more forceful. The implication of the findings is that variability is a hindrance to performance. This theoretical perspective emanates from the assumption that variability is a reflection of noise in some mechnaism(s) of the sensorimotor system.

Part III, "Dynamical Systems Approaches to Movement Variability," contains five chapters that are diverse in their approaches to variability and motor control. However, the general message that emerges is that variability is essential for many aspects of biological performance. Chapter 10 is as much a treatise on the scientific method in general as it is on variability. In this chapter, Mpitsos and Soinila summarize a diverse set of findings on the sea slug *Pleurobranchaea* and compare these to findings on other animals, including humans. They make the point that information is shared and distributed via diverging and converging networks. These networks are enormously complicated, and their control is error-prone and emergent as opposed to explicitly regulated. In chapter 11, Kelso and Ding continue in the same theme. They argue that the process of self-organization underlies the process of coordination in biological systems. They also show that chaos can be controlled and discuss the advantages and disadvantages of such a control scheme for motor control.

In chapter 12, Riccio emphasizes the role of perception in the control of movement, making the point that variability may serve the very important role of constituting "a pattern of stimulation that provides task-relevant information about the dynamical interaction between an animal and its environment." Riccio reemphasizes the position that understanding perception is necessary for understanding motor control.

As presented in Parts I and II, models have been developed that account for the relationship between response variability and selected dependent measures. Most of these studies have addressed variability in motor control in tasks that directly involve only a single limb. Bimanual tasks allow for the investigation of an additional source of variability, namely, that caused by the influence of one limb upon another. Much of the research on bimanual coordination has involved tasks that require oscillatory movements and has been interpreted within a dynamical systems perspective. In chapter 13, Walter, Swinnen, and Franz summarize the research on continuous bimanual tasks and then proceed to discuss research on discrete bimanual actions. They make the point that there is a wealth of research that shows that bimanual tasks are drawn to a common trajectory and that this may be mediated by the kinetics of the movements.

In chapter 14, Turvey, Schmidt, and Beek focus on interlimb coordination. They argue for a three-tiered structure for understanding rhythmic pendular movements, each level of which can be characterized dynamically and is governed by one or more attractors. The lower level consists of the two limbs involved in the pendular movements. The upper level consists of intention, and the middle level consists of the interlimb coordination. They also discuss the fluctuations observed in interlimb coordination in the framework of "clock" and "motor" variances. They show that both clock and motor variances are influenced by the dynamics of the task.

The three chapters in Part IV, "Variability and Movement Disorders," reveal how describing and understanding variability can further our understanding of movement impairment. Ivry and Corcos (chapter 15) also use the framework of clock and motor variances and synthesize a large number of studies that have shown that variability in performing a tapping task can be divided into two components, one central and one peripheral. When selected groups of patients perform the tapping task, patients with motor impairments of central origin (i.e., cerebellar lesions) display performance deficits different from those of patients whose locus of impairment is peripheral (i.e., peripheral neuropathies). From this perspective, the partitioning of response variability is useful for understanding the nature of motor performance deficits. Roy, Brown, and Hardie (chapter 16) discuss the role of variability in understanding apraxia—a movement disorder in which individuals cannot imitate gestures upon request. They argue that the study of variability in apraxia serves two useful functions. The first is to determine the extent to which a performance falls within a normal profile. The second is to determine what insight can be obtained from trial-to-trial variability in terms of the mechanisms that underlie the generation of the movement. Finally, Newell, van Emmerik, and Sprague (chapter 17) raise the issue of how movement variability and movement stereotypy are related. They investigate the movements of individuals with tardive dyskinesia and show that movements that are stereotypic can also have high degrees of variability, and they suggest that movement adaptability will provide greater insight into understanding tardive dyskinesia than will understanding movement variability.

In summary, an alternative way to conceptualize the four parts of the book is as follows. Parts I and II present deterministic models either of the relationship between kinematic and kinetic measurements and their associated measures of variability or of models of movement that can capture these relations. In Part III, the underlying assumption is that the study of movement is more fruitfully approached by a dynamical systems perspective in which the underlying premise is that coordination will not be understood as a deterministic process. Part IV has chapters that adopt both perspectives in the study of movement impairment.

REFERENCES

Arutyunyan, G.H., Gurfinkel, V.S., & Mirskii, M.L. (1968). Investigation of aiming at a target. *Biophysics*, **13**, 536-538.

Bernstein, N. (1967). *The co-ordination and regulation of movements.* Oxford: Pergamon.

Glass, L., & Mackey, M.C. (1988). *From clocks to chaos: The rhythms of life.* Princeton: Princeton University Press.

Hebb, D.O. (1949). *The organization of behavior.* New York: Wiley.

Hoyle, G. (1983). *Muscles and their neural control.* New York: Wiley.

Kugler, P.N., & Turvey, M.T. (1987). *Information, natural law, and the self-assembly of rhythmic movement.* Hillsdale, NJ: Erlbaum.

Lashley, K.S. (1929). *Brain mechanisms and intelligence.* Chicago: University of Chicago Press.

Meyer, D.E., Abrams, R.A., Kornblum, S., Wright, C.E., & Smith, J.E.K. (1988). Optimality in human motor performance: Ideal control of rapid aimed movements. *Psychological Review*, **95**, 340-370.

Nashner, L.M., & McCollum, G. (1985). The organization of human postural movements: A formal basis and experimental synthesis. *Behavioral and Brain Sciences*, **8**, 135-172.

Newell, K.M., & Hancock, P.A. (1984). Forgotten moments: A note on skewness and kurtosis as influential factors in inferences extrapolated from response distributions. *Journal of Motor Behavior*, **16**, 320-335.

Schmidt, R.A., Zelaznik, H.N., Hawkins, B., Frank, J.S., & Quinn, J.E., Jr. (1979). Motor-output variability: A theory for the accuracy of rapid motor acts. *Psychological Review*, **86**, 415-451.

Schroeder, M. (1991). *Fractals, chaos, power laws: Minutes from an infinite paradise*. New York: Freeman.

Sherrington, C.S. (1906). *The integrative action of the nervous system*. New Haven: Yale University Press.

Spray, J.A., & Newell, K.M. (1986). Time series analysis of motor learning: KR versus no KR. *Human Movement Science*, **5**, 59-74.

Turvey, M.T., Shaw, R.E., & Mace, W. (1978). Issues in the theory of action: Degrees of freedom, coordinative structures, and coalitions. In J. Requin (Ed.), *Attention and performance VII* (pp. 557-598). Hillsdale, NJ: Erlbaum.

Wing, A.M., & Kristofferson, A.B. (1973). Response delays and the timing of discrete motor responses. *Perception and Psychophysics*, **14**, 5-12.

PART I

FORCE VARIABILITY

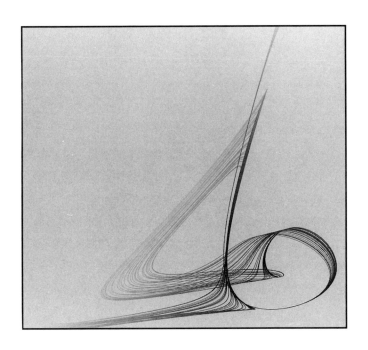

Chapter 2

Force Variability and Characteristics of Force Production

Les G. Carlton, Karl M. Newell

University of Illinois at Urbana-Champaign

Performers that are skilled at a particular motor activity are able to consistently produce the desired motor pattern and accomplish the task goal. In tasks where the quality of performance is assessed from the deviation between an established criterion and a performance outcome, consistency is crucial. Examples include sport activities such as target archery, where skill is based on the deviation of outcomes from a target, and school figures in ice skating, where consistency in the figures drawn on the ice by the blade of the performer's skate is evaluated. In fact, in many activities that have an accuracy demand, the performer is asked to perform a number of repetitions or trials. In golf, high-level competition is usually played over 72 holes and involves over 200 individual strokes. A poor outcome on even one or two strokes will often alter the outcome of the competition.

The importance of consistency for skilled performance has led to its inclusion in definitions of motor skill dating back at least to Guthrie (1935). Of course, sport activities are only one type of activity where consistency, or alternatively, variability, is important for successful motor performance. In handwriting, variations in the heights of the letters can make reading difficult or impossible, and in locomotion, stride-to-stride variations in stride length, stride rate, or height of the swing leg as it passes over the ground can lead to inefficient locomotion or even falls.

In order to understand the factors that influence movement accuracy, performance measures, including measures of variability, have been related to physical dimensions of the task being performed. In isometric force production tasks, variability in producing consistent forces has been related to force characteristics, such as the magnitude (peak force) or aggregate (impulse) of force produced, and temporal variables, such as the time to peak force and impulse duration. In movement tasks much of the research has focused on speed-accuracy trade-offs in limb movements. An example of this approach is Fitts' law (Fitts, 1954), where spatial accuracy, movement distance, and the time of movement are related by a specific mathematical description.

Relating variability to characteristics of response production can lead to descriptions that are in different measurement categories, such as movement tasks in kinematics and isometric tasks in kinetics. Schmidt, Zelaznik, Hawkins, Frank, and Quinn (1979) argued that, given that movement about a joint results from force generated by muscular action, variability in movement tasks could be directly related to variability in force production. Spatial and temporal variability in a variety of rapid actions, including discrete aimed movements, reciprocal movements, and timing tasks, were linked to variability in force generated by the neuromuscular system. This was, to our knowledge, the first attempt to directly associate variability in movement, or movement kinematics, to variability in force production. We, like Schmidt et al. (1979), also assume that a complete description of variability at the kinetic level can be used to explain variability in movement space-time tasks. Models of movement variability should, therefore, be able to account for variability in both the kinetics and the kinematics of response output. A diagrammatic depiction of this rationale is provided in Figure 2.1.

Over the years a variety of theories and models have been developed to account for variability in the motor system. For the most part, these accounts have followed general trends in psychology and psychological approaches to movement control. Initial experimentation on motor variability (Fullerton & Cattell, 1892) was designed to test the validity of Weber's law in the movement domain. As such, it was an initial measure of the psychophysics of the motor system, and it is interesting to note that motor psychophysics has seen a resurgence of interest of late (e.g., Gandevia & Kilbreath, 1990; Soechting & Terzuolo, 1990). Subsequent models of motor variability have used information processing (Fitts, 1954), error detection processes (Crossman & Goodeve, 1963/1983; Keele, 1968), movement dynamics (Schmidt et al., 1979), and cognition (Meyer, Abrams, Kornblum, Wright, & Smith, 1988) as the basis for movement outcome predictions. As the boundaries between the psychology and the physiology of motor control have diminished, some recent approaches have used physiologically based models to account for variability in behavior (e.g., Darling & Cook, 1987; Ulrich & Wing, 1991).

It is obvious from the above discussion that variability in the motor system can be examined at several levels. Although observations of variability are usually in terms of performance outcome or the pattern of movement, these variations are related to variations in force production, which, in turn, are influenced by variations in the state of the muscle at the time of activation, excitability of motor neurons or spinal interneurons, and command signals from higher nervous centers. It is not clear what the primary source of motor output variability is, or what is the most appropriate level for study, although it is clear that the regulation of force is a critical function of the motor system (e.g., Evarts, 1968).

The focus of this paper is on the description of variability in response kinetics. We will restrict ourselves, for the most part, to the study of variability in isometric force production. A complete description of the relation between characteristics of force production and force variability is not available, just as there is not a complete

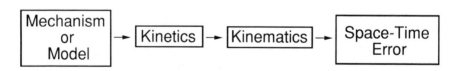

Figure 2.1   Errors in space and time are related to variability in force generation. Ultimately, models of variability should address variability at each level.

description of the speed-accuracy relation in movement tasks (cf. Hancock & Newell, 1985). We will start by presenting a brief overview of various force variability descriptions that have emerged over the last 100 years. A more detailed account of the early force variability studies is provided in Newell, Carlton, and Hancock (1984). Recent findings indicating that the temporal characteristics of force production have a significant impact on variability will be presented. Based on a synthesis of a number of experiments, a force variability description is presented that accounts for variability in peak force, impulse, and rate of force production across a range of force production conditions. This description provides boundary conditions for models of variability in kinetic and kinematic characteristics of motor output. Limitations of this description as well as its possible generalization to a variety of force production situations will be discussed.

FORCE VARIABILITY AND LEVEL OF FORCE

Studies of force variability have been primarily concerned with how the amount of force produced influences variability. This has been examined by having subjects attempt to produce a given level of force over a repeated number of trials. Typically the criterion "force level" is specified, and compared to the *peak* force (highest force level in the force-time profile) produced by the muscular contraction on a given trial. The criterion forces used in experiments have ranged from less than 1 N to the maximum force the subject is capable of generating. A typical force-time curve produced by subjects is graphically displayed in Figure 2.2 and is labeled to help define terms used in this paper. The force curve is modeled as a Gaussian curve that approximates the shape of the force curve when subjects are asked to make short isometric contractions against a force transducer. Peak force (*PF*) is the

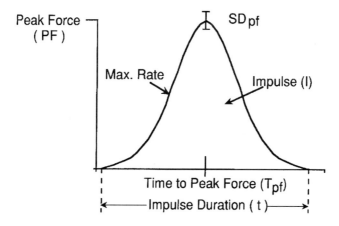

Figure 2.2   A prototypic force-time curve produced by subjects attempting to match a criterion peak force. *Max. rate* is the maximum rate of force produced between the initiation of force and peak force, *impulse* is the area under the force-time curve, and SD_{pf} represents the standard deviation in the peak force over a series of trials.

highest force produced on a trial, and time to peak force (T_{pf}) is the time from the initiation of force to the time at which peak force is reached. Other force-time characteristics are also described. Impulse (I) is the integral of force and time, and impulse duration (t) is the time between the initiation of force and the time of force termination. The maximum rate of force production and other higher order derivatives can also be computed. Each of these dependent measures can be obtained from an individual trial. The standard deviation of peak force (SD_{pf}) is calculated from the distribution of peak forces produced over a series of trials.

The study of variability in force production dates back to Fullerton and Cattell (1892), who examined the within-subject variability in producing several levels of force against a spring dynamometer. The criterion force level ranged from 2 to 16 kg, which subjects attempted to match over a series of trials by producing elbow flexion against the external resistance provided by the dynamometer. Fullerton and Cattell examined whether variability in the level of force produced over repeated trials was proportional to the level of force produced (Figure 2.3a). They were interested in whether Weber's law, which states that the least noticeable difference (or error of observation) is proportional to the magnitude of the stimulus, was valid for the motor domain. They found that variability in the force generated increased as the level of force increased, but at a less than proportional rate (Figure 2.3b). They substituted for Weber's law the statement *The error of observation tends to increase as the square root of the magnitude, the increase being subject to variations whose amount and cause must be determined for each special case* (pp. 153-154).

Cattell favored a stochastic model to account for these results, although, apparently, Fullerton was not entirely convinced (Fullerton & Cattell, 1892, n. 1). Cattell argued that the estimation of one unit of force produced an error of observation.

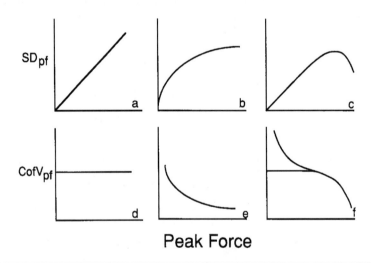

Figure 2.3   (a, b, c) The proposed relation between the standard deviation of peak force and the level of peak force. (d, e, f) The respective relation between the coefficient of variation of peak force and peak force. Proportional (a, d), square-root (b, e), and inverted-U (c, f) functions are displayed.

When two units of force are estimated, each unit will be subject to its own error, that is, the errors for each unit are independent, and, as a result, the total error will be less than the sum of the error for each unit. Instead, the sum of the errors would be the average error for each unit multiplied by the square root of the number of units (or stimulus magnitude). This is essentially a statement of the statistical additivity of variance with the assumption of independence (see Newell et al., 1984). Fullerton and Cattell argued that this description would hold for the estimation of any physical magnitude that is composed of equal units. This has not proven to be the case, with errors of observation generally increasing as a power of the stimulus magnitude (Stevens, 1983).

Fullerton and Cattell (1892) provided a basic psychophysical description for force production, and later Woodworth (1899) produced the same for movement amplitude and movement speed. Interest in the variability in force production waned until practical demands associated with vehicle control in World War II increased interest in the study of human-machine systems. A number of studies again demonstrated that force variability increased as the level of force produced increased, but at less than a proportional rate (Jenkins, 1947; Noble & Bahrick, 1956; Provins, 1957).

The descriptions provided so far have been based on the variability in producing a peak force level or static level of force with no task-specified time constraint. Schmidt et al. (1979) proposed a theory of variability in rapid motor acts based on the assumption that variability in forces produced over a series of trials was proportional to the level of force produced (Figure 2.3a). Thus, one of the bases of the theory was that Weber's law holds for rapid motor action. Although there was some evidence against this viewpoint (e.g., Fullerton & Cattell, 1892; Jenkins, 1947), there was little data corresponding to the production of rapid forces. Initial tests of the force variability assumption upheld a proportional relation between force and force variability. Force variability scaled linearly with force level over small ranges of force (Schmidt et al., 1979, Experiments 1 & 2; Sherwood & Schmidt, 1980), but the slope of the regression line was steeper in experiments using low criterion force levels, which is consistent with a negatively accelerating relation between force level and force variability.

Subsequent experiments (Sherwood & Schmidt 1980, Experiment 2) demonstrated that variability in peak force decreased at high force levels. This decrease in variability occurred at about 65% of the maximum peak force achievable for a subject and led to the prediction of an inverted-U function (Figure 2.3c). Thus, force variability increased proportionately with increases of force at low force levels, leveled off at about 65% of maximal peak force achievable for the measured action, and decreased at higher force levels.

The decrease in variability at high force levels has not been a consistent finding. Holding time to peak force constant and measuring peak force variability with peak forces ranging between 2% and 90% of maximum (Newell & Carlton, 1985) results in a relation between peak force and peak force variability that is consistent with the findings of Fullerton and Cattell (1892; see our Figure 2.3b). It was proposed (Newell et al., 1984), based on an extensive review of the existing literature, that the form of the relation between peak force and peak force variability was exponential (Figure 2.3b). That is,

$$SD_{pf} \propto PF^{1/n}, \tag{2.1}$$

where SD is the standard deviation, and PF is the peak force. As we will see, a number of force production characteristics other than the absolute level of peak force could influence the value for the exponent in Equation 2.1 and could even change the nature of the function (e.g., the inverted-U function; Schmidt & Sherwood, 1982; Sherwood & Schmidt, 1980).

Figure 2.3, a-f, provides the three basic descriptions discussed above, along with the predictions for the resulting coefficient of variation functions. The coefficient of variation is the standard deviation of a distribution of scores divided by the mean for the distribution. The coefficient of variation is a measure of relative variability and is sensitive to deviations from proportionality of the standard deviation to the mean. A linear and proportional relation between force level and force variability (Figure 2.3a) would result in a constant coefficient of variation (Figure 2.3d). A negatively accelerating increase in peak force variability (Figure 2.3b) leads to a negatively accelerating decrease in the coefficient of variation of peak force (Figure 2.3e). The inverted-U function (Figure 2.3c) might lead to either of the two coefficients of variation functions displayed (Figure 2.3f), depending on whether the initial increase in variability at low force levels was proportional to the level of force produced.

TEMPORAL PARAMETERS AND FORCE VARIABILITY

Why are there different descriptions for the relation between force level and force variability? One possibility is that there are factors, other than the level of force (e.g., peak force), that influence peak force variability. One important factor may be the time taken to produce the criterion force level. This explanation has been used to distinguish between linear and curvilinear force variability functions (Newell & Carlton, 1985; Newell et al., 1984; Schmidt, Sherwood, Zelaznik, & Leikind, 1985). Early experiments demonstrating a square-root function (e.g., Fullerton & Cattell, 1892) were interested in sensation and allowed the time to peak force to be 500 ms or longer. In contrast, Schmidt and colleagues (e.g., Schmidt et al., 1979, 1985) were interested in variability associated with motor programming and therefore used short duration pulses. The actual time taken to reach peak force in these experiments was not controlled, and, as we will see, this may have contributed to the reduced variability observed at high force levels (Sherwood & Schmidt, 1980). The range of forces used in these experiments could also be a factor. The inverted-U function, for example, was seen only when very high force levels were used, and the force levels in previous work may have been too low to elicit a decrease in force variability. A limited range of forces in any one experiment could also mask the curvilinear characteristic of the variability function.

Varying Time to Peak Force

As part of a systematic attempt to determine the contribution of various force-time parameters to force variability, Newell and Carlton (1988) varied the time to peak force with subjects attempting to match a criterion peak force of 54 N. The response studied was elbow flexion, and time to peak force varied between 100 and 600 ms. The mean peak forces did not significantly differ across time to peak force conditions, but peak force variability decreased systematically as time to peak force increased. The coefficients of variation (standard deviation of peak force divided by the mean peak force) were .145, .127, .103, and .099 for times to peak force of 100, 200, 400, and 600 ms, respectively.

How might increasing the time taken to reach the criterion peak force influence force variability? There are at least two rationales, both leading to the prediction that a longer time to peak force, at any given force level, results in lower peak force variability. One possibility is that peak force variability is related to the percentage of maximum force the performer is working at. Increasing the time to peak force, at a fixed force level, reduces the relative percentage value of the force criterion compared to the maximum force that could be produced at the criterion time. A second possibility is that force variability is influenced by temporal variability. If temporal variability increases with time but at less than a proportional rate, force variability could decrease at longer times to peak force.

Percentage of Maximum Force

The amount of force that can be produced by a particular muscular action varies with the time allowed to produce force, the greatest possible force taking from 2 to 4 s to generate (Kamen, 1983). The maximum peak force that can be generated with a time to peak force constraint of 400 ms may be more than twice that with a time to peak force of 100 ms (Newell & Carlton, 1985). Lengthening the time to peak force at a constant criterion peak force, therefore, reduces the percentage of maximum being worked at, possibly resulting in lower variability. If subjects increase time to peak force in order to reach high force levels, variability at these high force levels may decrease. This is apparently what happened in the experiment of Sherwood and Schmidt (1980; see Sherwood, Schmidt, & Walter, 1988).

Although this hypothesis seemed intuitive to us, there are two sets of data providing evidence that variability does *not* scale directly to the percentage of peak force. Carlton and Newell (1985) measured peak force variability with times to peak force of 100, 200, and 400 ms and force levels ranging between 5% and 90% of individual subjects' maximums for each of the respective times to peak force. We suspected that plotting the variability of peak force as a function of the percentage of maximum force for the corresponding time to peak force would result in a single variability curve. As can be seen in Figure 2.4, three separate curves emerged, one for each movement time. Variability was greatest at the longest time to peak force (400 ms), but these combinations of peak force and time to peak force also had the greatest criterion force levels.

A second set of data comes from an experiment where subjects were separated based on their ability to produce large forces (Carlton, Kim, Golden, & Newell, in preparation). The criterion peak force level was equal for all subjects. For the strongest subjects this force level was about 35% of their maximum force at the designated time to peak force, and for the weaker subjects the force level was about 60% of maximum. Even though the strongest subjects could produce twice the peak

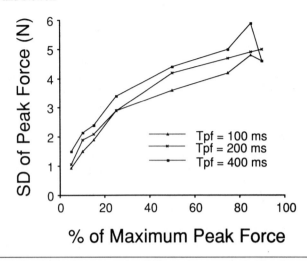

Figure 2.4   Variability in peak force for different times to peak. The data are plotted as a function of the percentage of maximum peak force achievable by the subject at the specified time to peak force.

Note. From "Force Variability in Isometric Tasks" by L.G. Carlton and K.M. Newell. In *Biomechanics IX-A* (p. 130) by D.A. Winter, R.W. Norman, R.P. Wells, K.C. Hayes, and A.E. Patla (Eds.), 1985, Champaign, IL: Human Kinetics. Copyright 1985 by Human Kinetics Publishers, Inc. Adapted by permission.

force of the weakest subjects, the stronger and weaker subjects were not significantly different in peak force variability.

Temporal Variability

The second rationale starts with the assumption that peak force is a function of the rate of force produced and the time over which it is produced. Variability in time estimation increases with time (e.g., Michon, 1967), and this is also the case for variability in the time to peak force in isometric contractions. If the increase in variability of time to peak force increases with time but at less than a proportional rate, it can lead to decreases in peak force variability at longer times to peak force.

To demonstrate this, we will start with a rather simplistic model, where force level increases linearly until the peak force is attained, and it is assumed that there is no variability in the rate of force production (Figure 2.5, a-c). Under these conditions the variability in peak force is solely a function of variability in the time to reach peak force. This hypothetical relation is depicted in Figure 2.5a. If the time to peak force is shorter than the criterion time, the peak force produced is too small and, similarly, is too great when the time to peak force is too long. Thus, variability in the time to peak force results in variability in peak force. Figure 2.5b demonstrates what happens if the standard deviation of time to peak force is proportional to time to peak force. The standard deviation of time to peak force at T_3 is three times as great as the standard deviation at T_1. The standard deviation of peak force in this example is identical for the two time to peak force conditions represented. Therefore, if the standard deviation of time to peak force is proportional to time to peak force, the standard deviation of peak force is constant at a particular force level. For conditions with different force levels and/or different times to peak force, the standard deviation of peak force would be proportional to the level of peak force ($SD_{pf} \propto PF$).

A different picture emerges if the variability in time to peak force increases at a less than proportional rate. Figure 2.5c shows two conditions where the standard deviation of time to peak force is proportional to time to peak force to the one-half power ($SD_{tpf} \propto T_{pf}^{1/2}$). The variability in peak force decreases with longer times to peak force for a particular force level. In this case, the standard deviation of peak force would be proportional to the level of force produced, or peak force, divided

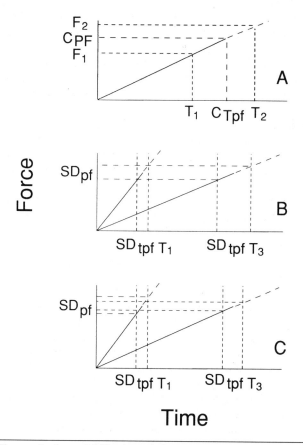

Figure 2.5   Changes in the variability in peak force with variations in time to peak force. (A) Two trials with times to peak force corresponding to T_1 and T_2. C_{Tpf} is the criterion time to peak force, C_{PF} is the criterion peak force, and F_1 and F_2 are the peak forces produced on trials T_1 and T_2, respectively. (B) The standard deviation of the time to peak force (SD_{tpf}) increases linearly with increases in time to peak force. (C) The standard deviation of time to peak force increases as the square root of time to peak force.

by time to peak force to the one-half power ($SD_{pf} \propto PF/T_{pf}^{1/2}$). As we will see, this prediction is the same as that put forth by Newell and Carlton (1988).

Variability in Force Without Time Constraints

In initial experimentation on force variability, Fullerton and Cattell (1892) were as interested in the perception associated with force production as in the execution of force production, and, as a result, the rapidity of force production was not an important concern. Times to peak force were about 1 s for high force levels and were somewhat shorter for low force levels. Thus, time to peak force changed, depending on the force level, with subjects normally extending time to peak force when greater forces were required. Allowing time to peak force to vary makes interpretation of the force variability findings difficult, because it is not clear whether the observed changes in variability are related more to errors of perception, as might be expected with long time intervals, or to movement production.

What is the relation between force variability and force level when the influence of time and other force production factors such as rate of force are minimized? A body of research related to this question has been conducted under the general topic of weight estimation. Recent experiments have examined the sensation of heaviness and voluntary motor commands associated with force production (Gandevia & Kilbreath, 1990; Kilbreath & Gandevia, 1989). The experimental paradigm calls for subjects to match a criterion weight lifted by the action of one muscle or group of muscles in the left arm, with the corresponding muscle(s) of the right arm. This was accomplished by having weights added to, or subtracted from, the weight lifted by the right arm until a match in weights for the two arms was perceived. Subjects were allowed to lift the weights in any manner they wished, and as many times as they wished, until they perceived that the weights matched.

The accuracy of weight estimation was assessed based on the coefficient of variation for repeated trials with the same reference weight. Muscle action of the first dorsal interosseous in lifting the index finger, the flexor pollicis longus in lifting the terminal phalanx of the thumb, the elbow flexors, and a composite task involving much of the musculature of the arm, was examined. Two criterion weights equaling 3% and 15% of the maximal voluntary contraction for each muscle action were used. The results showed that the standard deviation of the estimates increased with heavier reference weights, and the coefficients of variation decreased at the higher percentage of maximum for each of the muscle actions studied. The reduction in coefficient of variation was not as large as is typically seen in rapid force production (e.g., Carlton & Newell, 1985), but the range of force levels used was small, due to practical operational restrictions. In particular, subjects were not able to hold and repeatedly lift high levels of force without inducing significant muscular fatigue.

In summary, the results from a number of experiments indicate that variability increases at a decreasing rate with increases in force level. This holds for force levels produced under a variety of task conditions and experimental protocols. The data come from two psychophysical techniques: average error, measuring the accuracy of repeated trials at the same condition; and stimulus matching, measuring the accuracy in matching two sensations. Although it is apparent that time to peak force and rate of force production have a significant impact on variability, the general form of the variability function remains remarkably constant.

TIME AS A PARAMETER
IN DESCRIPTIONS OF FORCE VARIABILITY

Varying time to peak force has a significant impact on peak force variability (Newell & Carlton, 1988), and we reasoned, as did Schmidt et al. (1979), that force variability could be due to variations in temporal characteristics of force production. Schmidt et al. (1979) were more concerned with how variations in the impulse duration influenced variations in response kinematics, but it is evident that temporal variations could directly influence peak force variability as well. As outlined earlier, establishing a peak force and time to peak force criterion requires subjects to produce a specified rate of force for a time fixed by the temporal criterion (see Figure 2.5). Keeping the rate of force constant and varying the time of force production results in variations in peak force. The assumption that rate of force remains constant for a particular peak force and time to peak force combination does not hold. However, this line of reasoning leads us to believe that manipulation of temporal components of force production could account for decreases in peak force variability with longer times to peak force (Newell & Carlton, 1988, Experiment 4).

Based in part on this rationale and the empirical finding that peak force variability decreases as time to peak force increases, Newell and Carlton (1988, Equation 1) proposed that

$$SD_{pf} \propto PF/T_{pf}^{1/2}, \tag{2.2}$$

where T_{pf} represents time to peak force. This equation was extended to account for conditions where the initial level of force was not zero, but a discussion of preload is beyond the scope of this paper (see Kim, Carlton, & Newell, 1990). The inclusion of time to peak force in the denominator indicates that peak force variability decreases as time to peak force increases. The proposal that the appropriate exponent for time to peak force is .5 was largely speculation. Previous accounts of timing variability had implicated an exponential function (Michon, 1967; Wing & Kristofferson, 1973), and Fullerton and Cattell had proposed a square-root function for time prediction. Equation 2.2 provided a good fit to results from experiments manipulating time to peak force at a constant force level (Newell & Carlton, 1988, Experiment 4) and manipulating time to peak force with a constant rate of force production (Newell & Carlton, 1988, Experiment 5).

There are two aspects of Equation 2.2 that were oversimplified. The first is that the variability of peak force is not proportional to peak force as suggested by Equation 2.2. That is, when time to peak force is constant, Equation 2.2 holds that peak force variability is proportional to peak force. As was reviewed earlier in this paper and elsewhere (Newell et al., 1984), peak force variability increases at a negatively accelerating rate with increases in peak force (Equation 2.1). The second aspect of Equation 2.2 that may have been oversimplified was the time to peak force exponent. As was mentioned above, the exponent of .5 was largely speculative. Given the minimal amount of evidence available at that time, it might have been better, in retrospect, to have assigned a nonspecific constant to the exponent. Making the same argument for the level of peak force gives

$$SD_{pf} \propto PF^{1/m}/T_{pf}^{1/n}, \qquad (2.3)$$

where m and n are > 1.

IMPULSE VARIABILITY

Peak force represents only one point on the force-time curve. Even so, peak force has been the measured variable in nearly all studies on force variability. There are a number of reasons for this, the most important probably being that peak force is relatively easy to measure. Over 100 years ago Fullerton and Cattell were able to make a simple dynamometer using a spring mechanism that allowed peak force to be read off after each pull. More recent work has been influenced by the impulse variability model (Schmidt et al., 1979), which made specific predictions concerning peak force variability. The initial experiments by Schmidt et al. (1979) tested the prediction that variability of force is proportional to force by examining peak force variability as a function of peak force level.

Also important for the variability model was the assumption that variability in the aggregate of force production, or impulse, is proportional to the total impulse produced. This prediction was used to explain variability in response kinematics and to account for the relation between movement speed and accuracy. Most of the evidence concerning this prediction has come from movement tasks where impulse is considered to be proportional to the area under the acceleration-time curve (Carlton & Newell, 1988; Newell, Carlton, & Carlton, 1982; Sherwood, 1986). This measure was termed the "impulse for acceleration" (Schmidt et al., 1979), and it follows from Newton's laws of motion that impulse size is proportional to the impulse for acceleration.

In a recent set of experiments, we set out to examine impulse variability directly in isometric force production (Carlton, Kim, Liu, & Newell, in press). Our earlier experiments had suggested that the rate of force production has an important influence on variability, and we sought to determine if rate of force production also influenced impulse variability. Previous experiments examining rate of force production allowed other force production parameters such as impulse and impulse duration to also vary. Therefore, it was unclear whether rate of force production would influence impulse variability if other important characteristics of force production were held constant.

The experimental strategy was to generate force production conditions that had equal peak force, impulse duration, and impulse size requirements but different initial rates of force production. This was achieved by having subjects attempt to match a criterion force-time curve that varied in shape. The curve was parabolic and was skewed in order to generate different initial rates of force. In all, 10 conditions were used across two experiments. Figure 2.6 provides a visual description

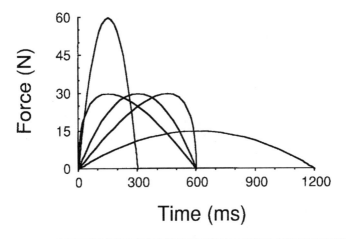

Figure 2.6   Force-time templates.
Note. From ''Impulse Variability in Isometric Tasks'' by L.G. Carlton, K.-H. Kim, Y.-T. Liu, and K.M. Newell, in press, *Journal of Motor Behavior*. Reprinted with permission of the Helen Dwight Reid Educational Foundation. Published by Heldref Publications, 1319 18th Street N.W., Washington, DC 20036-1802. Copyright 1992.

of the conditions used in Experiment 1. In Experiment 2 the peak forces were doubled, but the temporal characteristics were not changed. As can be seen in Figure 2.6, there were three conditions with equal peak force, impulse, and impulse duration, and conditions with either half or double the peak force. Impulse size was constant for all conditions.

The results from these experiments revealed that rate of force production influenced variability in peak force and impulse size, although this was statistically significant only at the higher force levels (Experiment 2). Coefficients of variation for peak force and impulse were largest for the positively skewed force-time condition. The experimental manipulations also resulted in variability differences for timing characteristics of the movement, such as impulse duration and time to peak force, and differences in the variability of rate of force production. Thus, impulse variability was not proportional to impulse size over the force-time conditions. Perusal of the data also indicated that the coefficient of variation for impulse followed the same trend as the coefficient of variation for peak force. This suggested to us that there might be a single description of force variability encompassing both the level of force (peak force) and the aggregate of force (impulse).

A SINGLE FORCE VARIABILITY DESCRIPTION

We were interested in how well the data from the impulse variability experiments (Carlton et al., in press) fit the peak force variability description developed earlier (Newell & Carlton, 1988; our Equation 2.2). The data from all 10 conditions (5 conditions for each experiment) were combined. The resulting correlation between peak force variability and Equation 2.2 was .87. As a more stringent test of the proportionality, we examined whether

$$CofV_{pf} \propto 1/T_{pf}^{1/2}. \tag{2.4}$$

The coefficient of variation (CofV) prediction was obtained by dividing both sides of Equation 2.2 by peak force. The correlation between Equation 2.4 and the coefficient of variation for peak force was only .14. This provided further evidence that Equation 2.2 could not account for peak force variability in a wide range of force production conditions.

The biggest reason for the poor fit between Equation 2.4 and the data from Carlton et al. (in press) was that variability in peak force was clearly not proportional to peak force. We attempted to improve the fit by manipulating the exponents in Equation 2.3. Existing data from a number of sources suggested that the exponent for peak force could be approximated by .5, and .5 was also used for the exponent for time to peak force. This exponent for time was a first approximation and followed from our previous prediction (Equation 2.2). The resulting fit with the data from Carlton et al. (in press) was better than that for Equation 2.2, but it was still not satisfactory. It appeared that the influence of time to peak force was overestimated, and we adjusted for this by changing the exponent for time to peak force to .25, leading to the predictions:

$$SD_{pf} \propto PF^{1/2}/T_{pf}^{1/4}), \tag{2.5}$$

and

$$CofV_{pf} \propto 1/(PF^{1/2} \times T_{pf}^{1/4}). \tag{2.6}$$

This improved the data fit significantly. The correlations for Equations 2.5 and 2.6 were .92 and .93, respectively. This fit seems remarkably good considering the range of force and time constraints, including some force-time profiles with asymmetric shapes.

Equations 2.5 and 2.6 were compared to previous sets of data from our laboratory, where the standard deviation and coefficient of variation for peak force were available along with the time to peak force. These included Experiments 4 and 5 from Newell and Carlton (1988) and data from Carlton and Newell (1985). The correlations were high in each case, ranging from .94 to .95 (Table 2.1). This is a significantly better fit than the prediction of a proportional increase in variability with increasing force (Figure 2.4) or the prediction from Newell and Carlton (1988; our Equation 2.2). With additional post hoc curve fitting—adjusting the exponents for force and for time—the fit may be improved, but the fit was sufficiently good to demonstrate how the level of force and time of force production contribute to peak force variability.

Variability in the rate of force production and impulse is also influenced by variations in peak force. Rate of force, for example, is a function of the change in force divided by the time over which the force occurs. For an individual isometric contraction, the average rate of force can be calculated as the change in force from the initial or resting force level to peak force, divided by the time to peak force. If Equations 2.5 and 2.6 can account for peak force variability, dividing both sides of

Table 2.1 Correlations Between Various Predictive Equations for the Variability of Peak Force and Data From Three Sets of Experiments

Predicted function	Carlton, Kim, Liu, & Newell (in press)	Newell & Carlton (1988)	Carlton & Newell (1985)
$SD_{pf} \propto PF/T_{pf}^{1/2}$	.87	.91	.92
$CofV_{pf} \propto 1/T_{pf}^{1/2}$	.19	.89	.14
$SD_{pf} \propto PF^{1/2}/T_{pf}^{1/4}$	.92	.95	.95
$CofV_{pf} \propto 1/PF^{1/2}T_{pf}^{1/4}$	.92	.94	.95
$SD_{pf} \propto PF$	.81	.78	.94
$CofV_{pf} = k*$	−.83	−.92	−.92

*The variable k is used here as a constant. The correlation coefficient is predicted to be zero.

Equation 2.5 by time to peak force would provide a prediction for variability of rate of force (rf) production:

$$SD_{rf} \propto PF^{1/2}/T_{pf}^{5/4}. \tag{2.7}$$

Further, dividing both sides of Equation 2.7 by rate (PF/T_{pf}) gives a prediction for the coefficient of variation.

$$CofV_{rf} \propto 1/(PF^{1/2} \times T_{pf}^{1/4}). \tag{2.8}$$

Equation 2.5 can be further extended to predict impulse variability. Impulse is the integral of force over time. As the height or the duration of force increases, impulse increases. If the force curve is modeled as a triangle, the impulse (I) can be described as

$$I = \frac{1}{2}PF \times t, \tag{2.9}$$

where PF represents the peak force, corresponding to the height of the triangle, and where t is impulse duration, corresponding to the base of the triangle. From Equation 2.5 we know how the variability in the height (SD_{pf}) of the force curve (triangle) varies with different combinations of peak force and time to peak force. If there is no variability in t,

$$SD_I = \frac{1}{2}SD_{pf} \times t \propto \frac{\frac{1}{2}PF^{1/2}}{T^{1/4}} \times t. \tag{2.10}$$

Dividing both sides of Equation 2.10 by impulse (I) gives

$$CofV_I = \frac{SD_I}{I} = \frac{SD_I}{\frac{1}{2}PF \times t} \propto \frac{\frac{\frac{1}{2}PF^{1/2}}{T^{1/4}} \times t}{\frac{1}{2}PF \times t} = \frac{1}{PF^{1/2}T^{1/4}}. \qquad (2.11)$$

As has been noted, an assumption is made that there is no variability in the temporal characteristics of the produced force-time curves. For any specific criterion force-time condition, variability in impulse duration is assumed to be zero. This is clearly not the case, but this procedure does allow for a direct prediction about the variability in impulse (we will return to the issue of temporal variability in a later section). The shape used to model the force-time curve is arbitrary. The rationale used above is generalizable to other conditions where the impulse can be defined from the values of peak force and impulse duration (Carlton et al., in press). This includes rectangles (pulse steps), parabolas, and Gaussian curves. The standard deviation prediction changes slightly, depending on the shape used, but the coefficient of variation remains unchanged, as specified in Equation 2.11. This holds even under changes in shape as in the experiments by Carlton et al. (in press).

It should be pointed out that the prediction for the coefficient of variation is identical across peak force, impulse, and rate of force. For each of these force parameters, correlating the coefficient of variation with the general predictive equation

$$CofV \propto 1/(PF^{1/2} \times T_{pf}^{1/4}) \qquad (2.12)$$

produces a good fit. Using the data from Carlton et al. (in press), the correlations ranged between .85 and .92. These correlations are high, considering that the coefficient of variation is extremely sensitive to small changes in variability and deviations from proportionality. The points that lie significantly above the regression fit for each measure tended to be associated with asymmetric force production conditions. It appears that producing asymmetric force-time curves adds a small amount of additional variability beyond that expected from peak force and time to peak force conditions. The relative variability (coefficient of variation) was lowest for peak force and was significantly higher for rate of force. In these experiments, subjects appear to put most of their effort into matching the most obvious force-time characteristics, focusing on peak force, time to peak force, and impulse duration. Requiring subjects to pay more direct attention to rate of force production might also reduce its variability.

GENERALIZATION AND LIMITATIONS OF THE VARIABILITY DESCRIPTION

The relation between force production and force variability provided in Equation 2.12 provides an accurate fit to data from a number of experiments. Although these data cover a wide range of force levels and times to peak force, the available data are still rather limited in scope. Most of the data have come from single-joint actions, requiring the isolation of a single muscle or set of muscles. In fact, most of the data have come from two single one-degree of biomechanical freedom actions, namely elbow flexion and extension! There is limited evidence as to whether the force variability description provided in Equation 2.12 extends to other force production tasks.

There are two aspects of the generalization question. The first is whether descriptions of force variability hold *within* different motor actions. More specifically, does Equation 2.12 hold reasonably well for different joints or for multiarticular actions? In the movement domain, speed-accuracy relations appear to hold across different joint actions. For example, the relation between movement speed and movement accuracy described by Fitts' law (Fitts, 1954) holds for different effectors of the arm (fingers, wrist, arm; Langolf, Chaffin, & Foulke, 1976), for leg movements (Drury, 1975), and for head movements (Jagacinski & Monk, 1985). The second aspect of the generalization question is whether force variability descriptions hold *across* different actions or motor effectors. Although a description of variability may hold for individual effectors, ''pooling'' data from different motor actions may cause the predictive quality of the description to decrease. This appears to be the case for Fitts' law, with the slope of the regression changing systematically for different effectors (Langolf et al., 1976).

Evidence for a General Variability Function

There is not a great deal of data available to address the generalization issue, but there are some indications that a single description may hold across a variety of force production situations. For example, it appears that strength of the subject does not influence variability as long as the criterion force level is submaximal (Carlton et al., 1992). In addition, the joint angle used in the production of force also does not influence force variability. We (Carlton et al., 1992) manipulated elbow angle and had subjects produce a submaximal peak force of 60 N with a time to peak force of 200 ms. There were six elbow angles ranging between 175° and 45° of rotation (180° = full elbow extension). The force transducer was positioned so that it recorded the force generated in attempting to rotate the lower arm, and, because of changes in mechanical advantage, subjects were able to produce significantly more force at intermediate joint angles. Peak force variability, however, was not significantly influenced by the joint angle used to produce the force.

One study that has examined variability across different effectors is that conducted by Gandevia and Kilbreath (1990). As was discussed in a previous section, the methodology used was different from that of most studies examining force variability, with the focus being the accuracy of weight estimation. In some of their experiments, variability in estimating weights was compared for different muscle groups and joint actions. For each muscle group examined, the coefficient of variation decreased with larger reference weights and, therefore, greater force. This is consistent with the proposed force variability description (Equation 2.12) and invites the suggestion that this description may hold for different motor effectors. Figure 2.7 shows the coefficient of variation for the different effectors expressed as a function of the reference weight used for weight estimation. As the reference weight increased, the coefficient of variation decreased systematically. A line of best fit based on a logarithmic regression indicates that there is a single variability function that scales to the level of force and is independent of the effector used. This suggests that a single variability description could hold *across* effectors. Gandevia and Kilbreath (1990) also examined variability in weight estimation for an action requiring the use of a combination of the upper and lower arm, hand, and fingers. Again the coefficient of variation decreased as the reference weight increased. Unfortunately, information about the actual reference forces used in this experiment is not available, and, as a result, the coefficient of variation cannot be added to Figure 2.7.

Limitations of the Force Variability Description

The focus of the variability description provided by Equation 2.12 is variability in force-related parameters, including peak force, rate of force, and impulse. Although these are three important force production characteristics, they do not provide the entire force-time variability function (e.g., see Ulrich & Wing, this volume, chap. 3). One critical aspect of this, which is often overlooked, is the variability in the temporal components of force production. The most obvious of these are time to peak force and impulse duration, and there has been little work addressing the variability of these temporal characteristics as force and time requirements change. There are some data (Carlton et al., in press) that suggest that the variability of

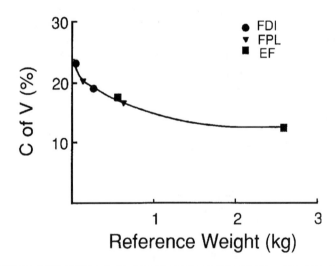

Figure 2.7   The coefficient of variation for weight estimation as a function of the heaviness of the reference weight. Coefficients of variation for three muscle groups are presented: first dorsal interosseous (FDI, circles); flexor pollicis longus (FPL, triangles); and elbow flexors (EF, squares). The line of best fit is based on a logarithmic regression.
Note. From "Accuracy of Weight Estimation for Weights Lifted by Proximal and Distal Muscles of the Human Upper Limb" by S.C. Gandevia and S.L. Kilbreath, 1990, *Journal of Physiology*, **423**, p. 307. Copyright 1990 by The Physiological Society. Adapted by permission.

time to peak force is strongly influenced by the level of force produced. In order to extend descriptions of force variability to space-time phenomena such as the speed-accuracy trade-off for aimed movements, temporal variability associated with force production must also be considered.

We also have little information about the boundary conditions for the variability description. There is some evidence that the description may be accurate up to 90% of maximum force production (Carlton & Newell, 1985; Newell & Carlton, 1985) and down to nearly zero force (Sherwood & Schmidt, 1980), but there are few data available at the upper and lower extremes of force production. Although the range of forces that a subject can produce is fixed at both the lower and the upper boundary, the time of force production is not. There is a limitation in how short in duration an impulse can be, with subjects having difficulty in producing times to peak force under 50 ms and impulse durations less than 100 ms under isometric conditions. The upper time limit of force production can, on the other hand, be very long. Most studies have focused on rather short-duration impulses, partly on theoretical grounds (Schmidt et al., 1979) and partially based on experimental pragmatics. As the time of muscular contraction increases, fatigue builds up, and it is difficult to obtain a reliable estimate of variability independent of muscular fatigue. It is possible that after 1 or 2 s, lengthening the time to peak force does not reduce variability but causes it to increase. Further experimentation is needed to establish more definitively the influences of an upper time boundary on force variability.

Theories or Models of Variability and the Proposed Variability Function

We have not proposed in this paper a specific model to account for variability in force production as described by Equation 2.12. However, the apparent generalizability of the proposed variability functions reveals limitations to extant models of force variability, in addition to providing constraints to future modeling efforts. The basic findings are that variability in peak force increases nonlinearly as force increases, and manipulation of time to peak force, and, hence, rate of force production, also influences peak force variability. These findings are counter to theories or models that predict a proportional relation between force and force variability (e.g., Schmidt et al., 1979) or assume this relation (e.g., Meyer et al., 1982) as part of a model of movement accuracy.

Ulrich and Wing (1991; see also this volume, chap. 3) have recently proposed a parallel force unit model that predicts both of the above force variability findings and also accounts for a number of other force production phenomena. In this model, force is controlled by increasing the number of force units activated, or by increasing the time of activation for each unit. Variability in the time of activation of the individual force units leads to force scaling and force variability predictions similar to the empirical observations reviewed here. It is not clear whether the parallel force unit model can account for the specific interaction between time to peak force and peak force proposed in Equation 2.12, or can account for variability in impulse size or rate of force production. An examination of this relation might be easier for conditions with nearly symmetrical force-time functions.

To date, most models or theories of both kinetic and kinematic variability have been developed on the basis of only two degrees of freedom. In the force variability situation these variables are space (e.g., amplitude) and time. Our work in these variability domains (Hancock & Newell, 1985; Newell et al., 1984), and also as reviewed here, has clearly demonstrated the important role of rate of force production in determining both kinetic and kinematic variability. However, there remains the possibility that we are coming close to the boundaries of the force variability variance (and also movement outcome variance) that can be accommodated with only two degrees of freedom.

Obviously, more of the outcome variability variance can be accommodated with the addition of more degrees of freedom in curve-fitting techniques to the variability functions. However, it may be useful to consider, in addition, dynamic variables such as damping and stiffness. These dynamic variables are essential ingredients of most linear and nonlinear modeling of limb movements. The addition of these variables also overcomes one of the major limitations to the extant modeling techniques, namely, that of being restricted to the assumption of similar impulse shape across experimental conditions. Indeed, damping and stiffness manipulations can produce systematic changes in the form of the force-time profile over a range of force-time conditions in a fashion similar to the actual shifts in the canonical force-time form that are produced by subjects over a range of task demands. These shifts are small but systematic and consistent in the extant force production literature and are usually ignored in most data description and modeling efforts. A move to consider these variables in force variability would also link the force variability literature to the muscle contraction literature.

REFERENCES

Carlton, L.G., Kim, K.-H., Liu, Y.-T., & Newell, K.M. (in press). *Impulse variability in isometric tasks*. Manuscript in press.

Carlton, L.G., Kim, Y.-S., Golden, N., & Newell, K.M. (1992). *Joint angle influences on motor output variability*. Manuscript in preparation.

Carlton, L.G., & Newell, K.M. (1985). Force variability in isometric tasks. In D.A. Winter, R.W. Norman, K.C. Hayes, R.P. Wells, & A. Patla (Eds.), *Biomechanics IX* (pp. 128-132). Champaign, IL: Human Kinetics.

Carlton, L.G., & Newell, K.M. (1988). Force variability and movement accuracy in space-time. *Journal of Experimental Psychology: Human Perception and Performance*, **14**, 24-36.

Crossman, E.R.F.W., & Goodeve, P.J. (1983). Feedback control of hand-movement and Fitts' law. *Quarterly Journal of Experimental Psychology*, **35A**, 251-278. (Original work presented at the meeting of the Experimental Psychology Society, Oxford, England, July 1963)

Darling, W.G., & Cooke, J.D. (1987). A linked muscular activation model for movement generation and control. *Journal of Motor Behavior*, **19**, 333-354.

Drury, C.G. (1975). Application of Fitts' law to foot-pedal design. *Human Factors*, **17**, 368-373.

Evarts, E.V. (1968). Relation of pyramidal tract activity to force exerted during voluntary movement. *Journal of Neurophysiology*, **31**, 14-27.

Fitts, P.M. (1954). The information capacity of the human motor system in controlling the amplitude of movement. *Journal of Experimental Psychology*, **47**, 381-391.

Fullerton, G.S., & Cattell, J. McK. (1892). *On the perception of small differences.* (Philosophical Monograph Series No. 2). Philadelphia, PA: University of Pennsylvania Press.

Gandevia, S.C., & Kilbreath, S.L. (1990). Accuracy of weight estimation for weights lifted by proximal and distal muscles of the human upper limb. *Journal of Physiology*, **423**, 299-310.

Guthrie, E.R. (1935). *The psychology of learning.* New York: Harper.

Hancock, P.A., & Newell, K.M. (1985). The movement speed-accuracy relationship in space-time. In H. Heuer, U. Kleinbeck, & K.H. Schmidt (Eds.), *Motor behavior: Programming, control, and acquisition* (pp. 153-188). Berlin: Springer-Verlag.

Jagacinski, R.J., & Monk, D.L. (1985). Fitts' law in two dimensions with hand and head movements. *Journal of Motor Behavior*, **17**, 77-95.

Jenkins, W.O. (1947). The discrimination and reproduction of motor adjustments with various types of aircraft controls. *American Journal of Psychology*, **60**, 397-406.

Kamen, G. (1983). The acquisition of maximal isometric plantar flexor strength. A force-time curve analysis. *Journal of Motor Behavior*, **15**, 63-73.

Keele, S.W. (1968). Movement control in skilled motor performance. *Psychological Bulletin*, **70**, 387-403.

Kilbreath, S.L., & Gandevia, S.C. (1989). Accuracy of weight estimation for proximal and distal muscles of the upper limb. *Neuroscience Letters*, **34**(suppl.), S104.

Kim, S., Carlton, L.G., & Newell, K.M. (1990). Preload and isometric force variability. *Journal of Motor Behavior*, **22**, 177-190.

Langolf, G.D., Chaffin, D.B., & Foulke, J.A. (1976). An investigation of Fitts' law using a wide range of movement amplitudes. *Journal of Motor Behavior*, **8**, 113-128.

Meyer, D.E., Abrams, R.A., Kornblum, S., Wright, C.E., & Smith, J.E.K. (1988). Optimality in human motor performance: Ideal control of rapid aimed movements. *Psychological Review*, **95**, 340-370.

Meyer, D.E., Smith, J.E.K., & Wright, C.E. (1982). Models for the speed and accuracy of aimed movements. *Psychological Review*, **89**, 449-482.

Michon, J.A. (1967). *Timing in temporal tracking.* Soesterberg, The Netherlands: Institute for Perception RVO-TNO.

Newell, K.M., & Carlton, L.G. (1985). On the relationship between peak force and peak force variability in isometric tasks. *Journal of Motor Behavior*, **17**, 230-241.

Newell, K.M., & Carlton, L.G. (1988). Force variability in isometric responses. *Journal of Experimental Psychology: Human Perception and Performance*, **14**, 37-44.

Newell, K.M., Carlton, L.G., & Carlton, M.J. (1982). The relationship of impulse to response timing error. *Journal of Motor Behavior*, **14**, 24-45.

Newell, K.M., Carlton, L.G., & Hancock, P.A. (1984). A kinetic analysis of response variability. *Psychological Bulletin*, **96**, 133-151.

Noble, M.E., & Bahrick, H.P. (1956). Response generalization as a function of intertask response similarity. *Journal of Experimental Psychology*, **51**, 405-412.

Provins, K.A. (1957). Sensory factors in the voluntary application of pressure. *Quarterly Journal of Experimental Psychology*, **9**, 28-41.

Schmidt, R.A., & Sherwood, D.E. (1982). An inverted-U relation between spatial error and force requirements in rapid limb movements: Further evidence for the impulse-variability model. *Journal of Experimental Psychology: Human Perception and Performance*, **8**, 158-170.

Schmidt, R.A., Sherwood, D.E., Zelaznik, H.N., & Leikind, B.J. (1985). Impulse-variability theory: Recent developments and implications for the control of rapid motor responses. In H. Heuer, U. Kleinbeck, & K.H. Schmidt (Eds.), *Motor behavior: Programming, control, and acquisition* (pp. 79-123). Berlin: Springer-Verlag.

Schmidt, R.A., Zelaznik, H.N., Hawkins, B., Frank, J.S., & Quinn, J.T. (1979). Motor-output variability: A theory for the accuracy of rapid motor acts. *Psychological Review*, **86**, 415-441.

Sherwood, D.E. (1986). Impulse characteristics in rapid movement: Implications for impulse-variability models. *Journal of Motor Behavior*, **18**, 188-214.

Sherwood, D.E., & Schmidt, R.A. (1980). The relationship between force and force variability in minimal and near-maximal static and dynamic contractions. *Journal of Motor Behavior*, **12**, 75-89.

Sherwood, D.E., Schmidt, R.A., & Walter, C.B. (1988). The force/force variability relationship under controlled temporal conditions. *Journal of Motor Behavior*, **20**, 106-116.

Soechting, J.F., & Terzuolo, C.A. (1990). Sensorimotor transformation and the kinematics of arm movements in three-dimensional space. In M. Jeannerod (Ed.), *Attention and performance XIII* (pp. 479-494). Hillsdale, NJ: Erlbaum.

Stevens, S.S. (1983). *Psychophysics: Introduction to its perceptual, neural, and social prospects*. New York: Wiley.

Ulrich, R., & Wing, A.M. (1991). A recruitment theory of force-time relations in the production of brief force pulses: The parallel force unit model. *Psychological Review*, **98**, 268-294.

Wing, A.M., & Kristofferson, A.B. (1973). Response delays and the timing of discrete motor responses. *Perception and Psychophysics*, **14**, 5-12.

Woodworth, R.S. (1899). The accuracy of voluntary movement. *Psychological Review Monographs*, **3**(13).

Acknowledgments

We would like to thank Yeou-Teh Liu and Howard Zelaznik for helpful comments on an earlier draft of this manuscript.

Chapter 3

Variability of Brief Force Impulses

Rolf Ulrich
Fachgruppe Psychologie an der Universität Konstanz
Konstanz, Germany

Alan M. Wing
MRC Applied Psychology Unit, Cambridge, UK

In modeling complex behavioral phenomena, there is often a tension between choosing a theory in which the constituent elements have a restricted set of properties and choosing one in which there is a variety of elements with a range of contrasting properties. In the first case the range of behaviors that can be generated by the model would be much more limited than in the second case. However, if those behaviors map on to what is actually observed in real life, the explanatory power of the model with fewer assumptions is generally recognized as greater. Another point in favor of the simple approach is that, if the model's predictions and behavior diverge, the choice of what to change in the model is more constrained and, so, potentially easier.

In this chapter we review a theoretical model—the parallel force unit model—that we have recently proposed as an account of force control (Ulrich & Wing, 1991). The model follows the route of choosing very simple elements; they are all identical, and no interactions are assumed. Even though this may sound unpromising as a structural basis for accounting for behavior, the model does capture a number of previously documented aspects of the mean force-time function of brief, isometric force impulses. However, the model also makes a prediction about the variability of the force-time function, and this has not previously been investigated. In the second part of this paper we describe a preliminary study of variability in isometric force impulses. Broadly speaking, the results support the theoretical predictions. However, there is an important discrepancy, and we discuss the implication of this for future development of the model.

Our original focus in developing the parallel force unit model was the way a brief voluntary force impulse rises and falls over time. There are a number of reasons for choosing to study force control. One reason is that the regulation and modulation of force is fundamental to our ability to grasp, and so to manipulate, objects in the world around us (e.g., Johansson & Westling, 1990). Another reason for looking at the control of brief force impulses is that this may contribute to our understanding of overt movement. It has, for example, been suggested that an important element in the speed-accuracy trade-off in rapid movements of the hand aimed at a target is variability in the amplitude and duration of acceleration (Schmidt, Zelaznik,

Hawkins, Frank, & Quinn, 1979). Underlying the accelerative burst there is necessarily a force (or torque) impulse. It is possible that the form of the force impulse may best be understood by first looking at the simpler case of force impulses generated under conditions where no movement can take place. In these conditions a number of complicating factors that introduce nonlinearities in these functions for different magnitudes of force impulse are less likely to cloud the picture.

PEAK TIME INVARIANCE
IN BRIEF FORCE IMPULSES

As an example of the kind of behavior in which we are interested, consider the study reported by Freund and Büdingen (1978). Subjects produced brief isometric force pulses of 1 to 2 kg with the muscles of the index finger. Two conditions were run. In the so-called target condition, subjects attempted to produce peak forces within 10% of a target value. In the other condition, designated nontarget, subjects simply produced pulses of as short a duration as possible, with the instruction to produce a range of peak values over trials. The illustrative functions in Figure 3.1, a and b, show that the average force as a function of time was generally a smooth, single-peaked function of time, more time being taken by the decay of force than by the buildup of force to the peak value. An observation made by Freund and Büdingen was that, to a first approximation, the form of the function did not depend on peak force. In particular, the time to peak force did not depend on peak force.

The peak time invariance of force impulses prompted Freund and Büdingen (1978) to suggest that subjects exert control over the rate of force rise. But there is another interpretation. From neurophysiology we know that muscle tension arises as the sum of a large number of muscle fiber contractions. Muscle fibers are grouped into functional groupings termed motor units, in which all the component fibers are

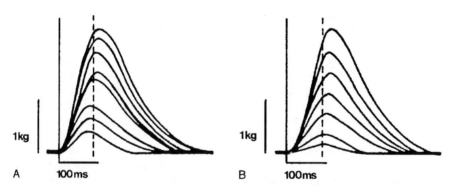

Figure 3.1   Isometric force functions produced by index finger extension for targets selected by the subject (A) and by the experimenter (B).
Note. From "The Relationship Between Speed and Amplitude of the Fastest Voluntary Contractions of Human Arm Muscles" by H.J. Freund and H.J. Büdingen, 1978, *Experimental Brain Research*, **31**, p. 4. Copyright 1978 by Springer-Verlag. Reprinted by permission.

driven by a single motoneuron. The production of a brief increase in tension may therefore be viewed as a problem of simultaneously activating a large set of motor units (Ulrich & Wing, 1991). Across a large set of units there will be variability in the delay between the central command to initiate action and the time at which each unit starts to contribute tension. The variability of such delays and the number of active units are then factors determining the rate of tension rise. Another factor that can also influence the level of muscle tension is the duration of the activity in each unit; if each unit contributes tension for only a very short time, summation will be reduced, because there is less overlap between units. In this view of muscle tension development, the number of units and the duration of their activity have the status of control variables, whereas the rate of change of tension is an emergent property.

An analogy may help the reader better appreciate the significance of this important point concerning the form of the overall force-time function. Imagine that, in moving a piano from one side of a room to another, one must lift it over an obstruction such as the edge of a carpet. The weight of a piano is too much for any one person. However, if several people are brought in, their action is sufficient to cope with the piano's weight, provided they attempt to lift it together. To achieve the necessary simultaneity of action, somebody will give a verbal command to lift. Each person's response to this command will be subject to a reaction time delay, and that delay is likely to be variable. But provided the variability is small relative to the duration of each individual's lift, the combined force will move the piano. In the next section we summarize the parallel force unit model, which provides an explicit formulation for the expected value and the variability of force-time functions generated by the summation of a number of independent units, each contributing force subject to variable delays. (For a formal presentation of the model, the reader is referred to Ulrich & Wing, 1991).

THE PARALLEL FORCE UNIT MODEL

The parallel force unit model is intended as an account of the rise and decay of force with time in tasks where subjects are required to produce brief impulsive changes in force. On a given trial, it is assumed that a central command recruits a subset of b force units from a pool of such units (see Figure 3.2). The force provided by each unit at time t is considered to be a random variable $\mathbf{F}_i(t)$. It is assumed that there is a random delay, $\mathbf{L}_i$, following the central command, during which $\mathbf{F}_i(t)$ is zero; thereafter the unit produces force according to a deterministic, nonnegative force-time function, $u_i(.)$ that is displaced by $\mathbf{L}_i$ time units to the right along the time axis, and thus $\mathbf{F}_i(t) = u_i(t - \mathbf{L}_i)$.

If all force units in the pool are assumed to be identical (i.e., the force-time functions are the same, and the latencies have identical probability density functions), and the overall observed force at time t is taken to be a random variable $\mathbf{F}_o(t)$ equal to the sum (over b) of the force contributions of the $\mathbf{F}_i(t)$, it may be shown that the expected (mean) force-time function is given by

$$E[\mathbf{F}_O(t)] = b.A.h(t), \qquad (3.1)$$

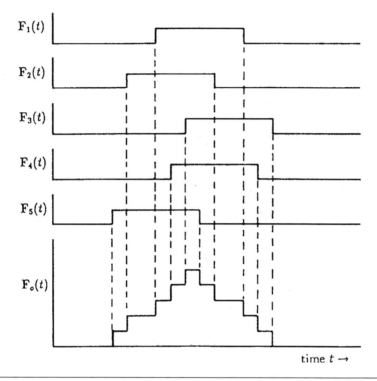

Figure 3.2   The summation principle in the parallel force unit model. On a given trial a central command at time t = 0 results in force production in a number of force units after random delays. At each point in time the overall force developed in a muscle equals the force summed over all recruited units.

where A is the impulse of the force unit, and $h(t)$ is the convolution of the probability density function of the latency, $\mathbf{L}$, with the normalized unit force-time function, $z(t) = u(t)/A$.

Equation 3.1 may be seen as separating out size and shape aspects of the mean force-time function. The force at any given time, t, increases with the number, b, and/or the impulse size, A, of the active force units. The shape of the mean force-time function is determined both by the probability density function, $f(.)$, of the unit onset latency, $\mathbf{L}$, and by the normalized unit force-time function, $z(.)$. If the variance of $\mathbf{L}$ is small relative to the range of the normalized force-time function, then the shape of the mean force-time function is largely determined by the shape of the normalized unit force-time function. However, the larger the variance of $\mathbf{L}$, the less the expected force-time function resembles the shape of the individual unit force-time function.

In order to better appreciate this formal statement about the size and shape of the mean force-time function in terms of the form of the individual force-time function $u(.)$ and the probability density function $f(.)$ of the latencies of force unit onset times, consider the results of the simulation depicted in Figure 3.3, a-d. Each simulation run produced an observed force-time function based on the activation of 400 force units with onset latencies sampled at random from a special Erlangian distribution with fixed mean. In Figure 3.3, a and b, the form of the underlying unit force-time function was symmetric triangular, whereas in Figure 3.3, c and d, the underlying unit force-time function was rectangular. In both cases the duration of the unit force-time function was 80 ms, and the area (the impulse) was set at 5 Newton milliseconds (Nms). The standard deviations of the force unit latencies were less in Figure 3.3, a and c, than in Figure 3.3, b and d: 20 ms and 40 ms, respectively.

A number of points will be noted. First, the temporal dispersion of the activation times "smears" the shape of the underlying form of the unit force-time functions. There is little indication of the underlying force unit force-time functions in any of

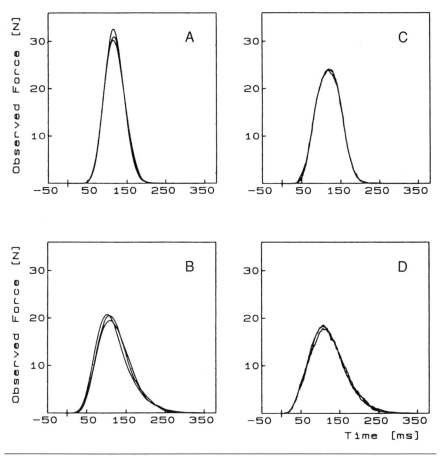

Figure 3.3   Simulation of force development. Each panel depicts the results of three simulations as described in the text. The form of the underlying force unit force-time function was symmetric triangular (A, B) and rectangular (C, D). The standard deviation of the force-unit onset latencies was 20 ms (A, C) and 40 ms (B, D).

the three single-trial force summations depicted in each panel. However, it will be observed that the summed force functions derived from the triangular and rectangular underlying functions are less similar in the upper panel, where there is less variability in onset latency. Thus, with many force units in the summation, the shape of the overall force-time function is heavily dependent on force unit asynchronicity.

SCALABILITY OF FORCE-TIME FUNCTIONS

From the equation for the mean force-time function there follows an important consequence. If peak force is controlled by recruiting varying numbers of force units, and if the onset latency distribution is fixed, the mean force-time functions for different levels of peak force should have the same basic form. Suppose we define a rescaled mean force-time function as the expected force-time function divided by the area. If the corresponding force levels for force-time functions at various amplitudes are achieved by varying the number of units recruited, then all rescaled mean force-time functions should superimpose. However, if the rescaled mean force-time functions do not coincide, then we might suppose that there has been a change in the normalized function $z(.)$.

Clear evidence of scaling of the force-time function is to be found in Gordon and Ghez (1987). In their experiment, subjects produced elbow flexion force impulses to targets at three different levels, with the highest force being 40% to 50% of the maximum voluntary force. Instructions emphasized production of a single smooth impulse of force and that, once initiated, responses should not be amended. Data from single trials aligned at force onset are shown in Figure 3.4, a and b. The similarity of the traces, normalized by peak force, shown in Figure 3.4b led Gordon and Ghez to state that "trajectories of responses to different targets were scalar multiples of a common waveform."

MEAN FORCE-TIME FUNCTION

In the simulation described earlier (Figure 3.3) we showed that, with appreciable variability in onset latencies, the form of the unit force-time function has relatively little effect on the observed force-time functions. However, to take the model further, it is helpful to assume a specific form for the single unit force-time function. We assume a rectangular shape; thus, at time, L_i, following the central command to recruit force unit i, we suppose that constant force a_i acts for duration d_i. With this additional assumption the mean force-time function may then be written as

$$E[F_0(t)] = b.a.[F(t) - F(t - d)], \qquad (3.2)$$

where $F(.)$ is the cumulative distribution function of latency L.

The model stated in the form of Equation 3.2 indicates that there are three modes for increasing peak force. First, more force units may be recruited; that is, the number, b, may be increased. Second, the duration, d, of each unit that contributes constant force, a, may be lengthened. Third, the value of a may be increased.

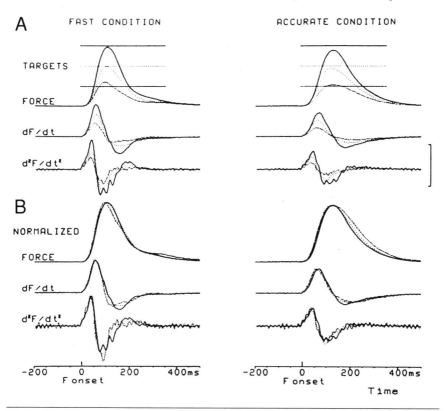

Figure 3.4 (A) Mean observed force-time functions for responses to three target levels in fast (left) and accurate (right) conditions. (B) The same functions as in (A) but normalized by peak force. Vertical calibration 100 N, 2,920 N/s, 116,800 N/s^2.

Note. From "Trajectory Control in Targeted Force Impulses: II. Pulse Height Control" by J. Gordon and C. Ghez, 1987, *Experimental Brain Research*, **67**, p. 245. Copyright 1987 by Springer-Verlag. Reprinted by permission.

Figure 3.5, a-c, illustrates the first two of these contrasting modes of peak force adjustment. As we did earlier, we assumed an Erlangian-distributed random variable **L** for the onset latencies with parameters yielding the mean onset latency of 80 ms and a standard deviation of 40 ms. The left side of Figure 3.5 shows the impact on the mean force function when *d* is fixed. The greater *b* is, the larger the area under the mean force-time function. Observe how the shape of the function stays the same. Thus, for example, zero crossings of the first and second derivatives (corresponding to peak force and peak rate of change of force) coincide in time, as may be seen in the lower rows (b and c) in the left column.

The right side of Figure 3.5 shows the consequence of varying *d* while *b* is held constant. Two effects of *d* on the mean force-time function may be noted. First, an increase of *d* raises the overall force levels. Second, beyond the initial force rise, an increase in *d* raises the mean force-time function at every point in time; that is, the shape of the function varies with *d*. Thus for example the zero crossing of the

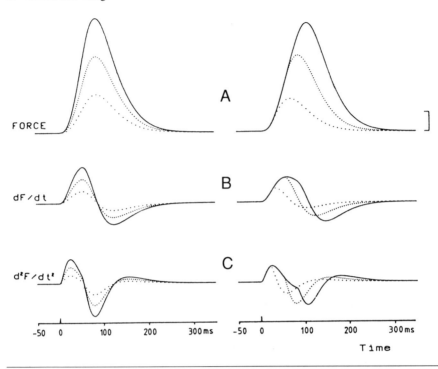

Figure 3.5   Predicted mean force-time functions (A), with their first (B) and second (C) derivatives, based on Equation 3.2. Force is changed on the left by varying the number of force units; $b = 600$ (solid line), 400 (closely spaced dots), and 200 (widely spaced dots). On the right are shown the effects of varying the duration of the rectangular unit force-time function, $d = 80$ ms (solid line), 50 ms (closely spaced dots), and 20 ms (widely spaced dots). Vertical calibration is 2.4 N, 163 N/s, 7,163 N/s^2, for A, B, and C, respectively.

first and second derivatives shown in the lower rows (b and c) in the right column do not exhibit time invariance. The different functions produced by changes in d are not rescalable.

The data from Gordon and Ghez supported the invariance shown on the left of Figure 3.5. However, there are data that accord with the functions on the right of the figure. Thus, for example, in Freund and Büdingen's data (see Figure 3.1), small increases in the time to peak force at larger peak forces may be seen, particularly in the nontarget condition. Such increases were also reported by Gordon and Ghez in a condition where speed rather than accuracy was emphasized. Hefter, Homberg, Lange, and Freund (1987) have reported that when patients with Huntington's disease are asked to produce brief isometric finger flexions, some exhibit force-time functions in which rises in peak force are accompanied by increases in force duration (see Figure 3.6, a-c). However, this is not true in all cases; although the force-time functions in Figure 3.6b are very variable, they do seem to exhibit peak time invariance. Finally, in a rather different paradigm, Carlton, Carlton, and

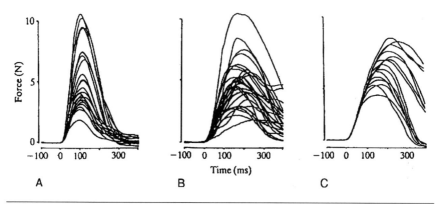

Figure 3.6 Isometric force functions produced by (A) a normal subject and (B and C) two patients with Huntington's disease.
Note. From ''Impairment of Rapid Movement in Huntington's Disease'' by H. Hefter, V. Homberg, H.W. Lange, and H.J. Freund, 1987, *Brain*, **110**, p. 591. Copyright 1987 by Oxford University Press. Reprinted by permission of Oxford University Press.

Newell (1987) have reported a relation between time to peak force and peak force. Subjects were required to generate brief force impulses of a prespecified duration, which ranged between 160 and 600 ms, while being allowed to freely vary other factors such as peak force. It was found that time to peak force increased with impulse duration.

VARIABILITY OF THE FORCE-TIME FUNCTION

We now turn from the expected force-time function of brief force impulses to their variability. A formal relation between the variability of the force-time function and the properties of the elementary force units is simply stated if (as before) we assume that the underlying force units can be approximated by the rectangular function of duration, d:

$$\text{Var}[\mathbf{F}_0(t)] = b.a^2.[F(t) - F(t - d)].[1 - F(t) + F(t - d)]. \tag{3.3}$$

This relationship is particularly interesting because it predicts a local minimum in the variability function that becomes more apparent as force unit duration increases. This may be observed in the simulation results in Figure 3.7, a-c. As before we assumed the special Erlangian distribution of unit latencies ($E[L] = 80$ ms, $\text{Var}[L] = 40$ ms^2, $a = 5$ cN), and the number of units, b, was set at 900. As the duration, d, of the rectangular force unit waveform is systematically changed from 50 to 150 ms, we see a local minimum developing in the variability function. This coincides with the maximum in the force-time function.

There is a simple intuitive explanation of the local minimum in the variability of force-time function. Suppose the durations of the underlying unit force-time functions were infinitely long. In such a case it is clear that variability in the

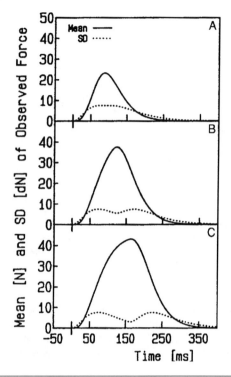

Figure 3.7 Predicted mean $E[\mathbf{F}_0(t)]$ and standard deviation $SD[\mathbf{F}_0(t)]$ as a function of time t. Note that the unit of $SD[\mathbf{F}_0(t)]$ is dN = N/10. In all three panels the parameters are $E[\mathbf{L}]$ = 80 ms, $Var[\mathbf{L}]$ = 40 ms², a = 5cN, and b = 900. The panels differ in the durations of the rectangular force-time functions of the force units. (A) d = 50 ms, (B) d = 100 ms, (C) d = 150 ms.

Note. From "A Recruitment Theory of Force-Time Relations in the Production of Brief Force Impulses: The Parallel Force Unit Model" by R. Ulrich and A.M. Wing, 1991, *Psychological Review*, **98**, p. 281. Copyright 1991 by the American Psychological Association. Reprinted by permission.

summation over many units during the force rise phase would necessarily drop to zero as the summation asymptotes at a steady level corresponding to all units providing continuous force. Indeed, there is an asymptote-like ending to the rise phase visible in the lowest panel of the figure. However, a full asymptote is not obtained, because, as this point is approached, the individual force units begin to turn off and the force function drops down again.

The peak in variability during force rise occurs at the first point at which 50% of the force units become active. This time point corresponds to the median latency L, which is located at 73 ms, in Figure 3.7. The second variability peak occurs later, at the point where the number of active units first drops back below 50%. On the average, the delay will correspond to the duration of the force unit force-time function. Elsewhere (Ulrich & Wing, 1991) we review a number of studies that have documented various aspects of the variability of brief force impulses in relationship to the mean peak force. However, with the prediction so clearly set out in the previous figure for a local minimum in the variability function, we spend the rest of this chapter outlining a new experiment (Ulrich & Wing, 1992) that provides direct supportive evidence.

EXPERIMENT

Three subjects were tested individually and asked to produce brief finger-flexion force impulses to match target durations of time to peak force of 100, 150, or 200 ms. These durations were combined factorially with values for target peak force of 20%, 40%, and 60% of maximum voluntary force (MVF). The time course of a single trial was as follows: 2 s after the subject initiated a trial with a key depression using the left hand, 3 clicks separated by 100 ms provided a warning at the beginning of the trial and were followed 1 s later by a loud warning click, and 800 ms later by a second click. The latter served as an imperative signal, and the subject was instructed to respond to its onset. Force output data were sampled at 1 khz over a period extending from 100 ms before to 1,200 ms after response onset. After each trial, feedback (knowledge of results, KR) was given as shown in Figure 3.8. Subjects were instructed to give equal weight to peak force and time to peak force in the KR display.

Each subject participated in a total of nine sessions; three target forces were blocked within a single session, and the order was counterbalanced according to a

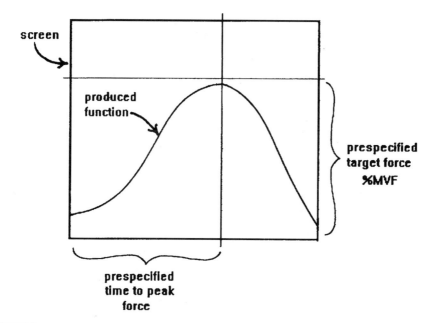

Figure 3.8   Feedback display given to the subject, indicating temporal course of a single trial.

latin square across subjects and sessions. Each block of trials started with 30 practice trials followed by 40 experimental trials. There were a total of 120 observations per subject for each combination of target time to peak force and target peak force.

Figure 3.9 illustrates the results for the mean force-time functions for each of the three subjects in the condition with the middle target time to peak force of 150 ms. The separate panels for each subject show three traces representing 20%, 40%, and 60% MVF. The individual traces superimpose remarkably well. For a given target time we therefore see further confirmation of the rescalability of peak force impulses as predicted by the model, with peak force adjusted by changing the number of active units, *b*.

We now turn to the mean and variability functions from the 40% MVF condition for individual subjects shown in Figure 3.10. Consider first the mean force-time functions shown by the continuous lines. These vary in shape as target time to peak force increases from 100 to 200 ms. In the model this change may be interpreted as due to changes in the duration, *d*, of the underlying force units. To maintain the same level of force across the middle durations, fewer units are required as the duration increases from left to right across the figure. When these results are taken with the results depicted in Figure 3.9, we thus see, according to the model, evidence of subjects using both modes of force-time control within one experiment.

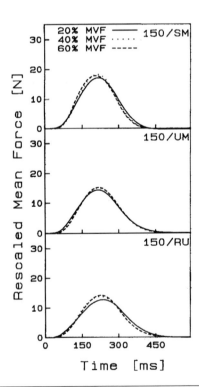

Figure 3.9   Rescaled mean force-time functions as a function of % MVF for a pre-specified time to peak force of 150 ms for each of three subjects, SM, UM, and RU.

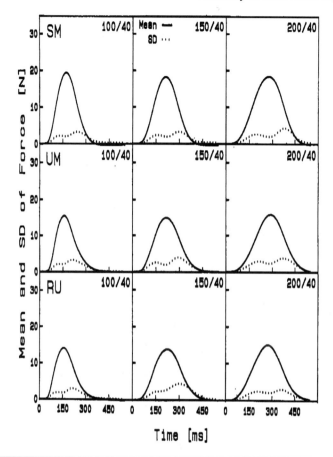

Figure 3.10   Mean and standard deviation of $F_0(t)$ as a function of % MVF and pre-specified time to peak force at 40% MVF for the three subjects, SM, UM, and RU.

Figure 3.10 also shows, with the dotted line, standard deviation of force as a function of time. A clear local minimum in standard deviation is evident in the data of all three subjects, coinciding with the maximum of the mean force-time function. This is also in agreement with the predictions of the parallel force unit model. Of course, there are other possible interpretations of the local minimum, and to select between these will require further experimentation. For example, it is possible the particular form of KR provided to the subjects resulted in their exercising tight control only over the maximum of the force-time function. Another, less likely, possibility is that the obtained force-time functions result from a mixture of two underlying fixed forms, one with an early peak, the other with a late peak, but the variability of either is directly related to level of force.

Finally we should point to one unexpected feature of the data in Figure 3.10. There is a consistent difference in the maxima in standard deviation before and after the local minimum, whereas the parallel force unit model as stated would predict that they should be equal. This asymmetry has a number of possible interpretations. One interesting possibility is that, at the end of a force impulse, subjects introduce a second set of force units whose sign is negative. This would have the benefit of bringing the force-time function down more rapidly than would otherwise be the case, but there would be a cost, in that these "antagonist" force units would add extra variability. Another interpretation is that there is variability in the duration of the force-time units. This also would lead to greater variability in the offset of force than in the onset.

SUMMARY AND CONCLUSIONS

In summary, the parallel force unit model has displayed a fair degree of success in accounting for the results of this new experiment as well as the earlier published results of other researchers. We have been able to relate the form of the expected force-time function and the variability of force as a function of time to variability at the lower level of the force units.

At the outset of the chapter we spoke of a tension between models whose elements are simple and those whose elements have more complex behaviors. Our reductionist approach might appear to conform with the former, and certainly we have presented our model as simple. However, the observant reader will no doubt have noticed that a moderately lengthy list of assumptions was needed to make quantitative predictions and run the simulations. Some of these assumptions are not critical; thus, for example, it would make little difference if we had assumed that the underlying unit force-time functions were triangular rather than rectangular. Nonetheless the question might reasonably be asked, To what extent is the model we propose simple?

One perspective from which our model might be viewed as simple is that of the neurophysiologist. For researchers working at lower levels of analysis of motor control, there is known to exist a variety of elements whose behavioral properties we have not considered in our model. We believe that our assumptions are not entirely contradictory with respect to the physiology but have captured the essential flavor (much as connectionist models attempt to do with their assumptions about neural units). However, some might be tempted to add to the complexity of our model to take account of the different classes of motor units operating at different force levels, for example. Elsewhere (Ulrich & Wing, 1991) we have pointed to a number of reasons why we feel that some of the physiological data may not apply to the rapid force impulses that are the domain of our model (e.g., physiological data usually stem from situations where subjects make slow changes in force level or maintain steady force levels).

We see various possible directions for future research in this area. One would be to explore further phenomena at a behavioral level, encompassing perhaps the same task with different measures. It would be interesting, for example, to know to what extent there is dependence between various parameters of observed force-time functions and how such dependence might be understood within the context of the model. For example, if the controlling parameters of the number of force units, b, and the duration of the force unit waveform, d, are constant, then over a number of trials for a given condition, the model predicts that the area of the overall waveform should be constant. This leads to the prediction of a negative correlation between peak force and duration of impulse.

Another area ripe for exploration at the behavioral level is to consider what might be called the dynamics of behavior: What are the consequences of change in a force production task? The model may form a very useful basis for understanding the strategic control and learning that we imagine allows subjects either to operate in open loop mode, perhaps to adjust their movement in advance according to experience gained over a number of trials, or to make use of feedback during the force change to achieve the target level with a number of corrections.

Finally, we would like to remind the reader that it was our hope that the model might serve as a useful basis for understanding force impulses that in turn are the building blocks of overt movement. The transition from an analysis of isometric force development to tasks with other movement based on isotonic contractions is tricky. There are many nonlinearities that arise, depending on, for example, the speed with which movement is carried out and the initial conditions from which the movement starts. Many of these require detailed mechanical models such as those described by Winters and Stark (1987). Perhaps our model will provide a useful forum by which the different disciplines—physiology, biomechanics, and psychology—can better communicate, in order to understand a fundamental problem of interest to us all: What is the nature of variability in motor control?

REFERENCES

Carlton, L.G., Carlton, M.J., & Newell, K.M. (1987). Reaction time and response dynamics. *Quarterly Journal of Experimental Psychology*, **39**, 337-360.

Freund, H.-J., & Büdingen, H.J. (1978). The relationship between speed and amplitude of the fastest voluntary contractions of human arm muscles. *Experimental Brain Research*, **31**, 1-12.

Gordon, J., & Ghez, C. (1987). Trajectory control in targeted force impulses. II. Pulse height control. *Experimental Brain Research*, **67**, 241-252.

Hefter, H., Homberg, V., Lange, H.W., & Freund, H.-J. (1987). Impairment of rapid movement in Huntington's disease. *Brain*, **110**, 585-612.

Johansson, R.S., & Westling, G. (1990). Tactile afferent signals in the control of precision grip. In M. Jeannerod (Ed.), *Attention and performance XIII: Motor representation and control* (pp. 677-713). Hillsdale, NJ: Erlbaum.

Schmidt, R.A., Zelaznik, H.N., Hawkins, B., Frank, J.S., & Quinn, J.T. (1979). Motor output variability: A theory for the accuracy of rapid motor acts. *Psychological Review*, **86**, 415-451.

Ulrich, R., & Wing, A.M. (1991). A recruitment theory of force-time relations in the production of brief force pulses: The parallel force unit model. *Psychological Review*, **98**, 268-294.

Ulrich, R., & Wing, A.M. (1992). *The scalability of brief force impulses*. Manuscript in preparation.

Winters, J.M., & Stark, L. (1987). Muscle models: What is gained and what is lost by varying model complexity. *Biological Cybernetics*, **55**, 403-420.

Chapter 4

Predicting Motor Performance From Variability Measures

Charles J. Worringham
University of Michigan, Ann Arbor, Michigan

A primary task confronting students of motor behavior is to reduce the bewildering complexity and variety of observations about movement to a set of lawful and general principles. It is ironic that randomness, or variability, a feature of movement that would appear to thwart this goal, is instead proving to be a key to finding ordered phenomena (see, for several examples, Wing, in press). This paper is concerned with three aspects of spatial variability in trajectory formation and their potential for "predicting" aspects of motor performance. First, it will be shown that the initial phase of aiming movements, variously given the epithets *ballistic*, *initial impulse*, or *transport*, have an average amplitude that is related to—and may well be selected on the basis of—the spatial variability of that phase. Secondly, it will be argued that a long-held (but oft-postponed) goal of motor behavior researchers—understanding individual differences—can be aided by paying attention to measures of variability. Finally, it will be shown that trajectory variability influences not only the widely studied task of discrete aiming (whether with spatial, force, or temporal goals), but the related task of object avoidance or "clearance," that is, the planning of a trajectory that takes the limb (or body) into the vicinity of an object that must be negotiated without contact occurring, as in the attempt to pass rapidly around the corner of a hallway, duck under a branch in the woods, or move a wineglass over a flower vase on the dining-room table. Such tasks are experimentally less tractable than discrete aiming but are significant in being common forms of motor behavior sharing a phenomenon—clearance—for which neither lawful description nor theoretical explanation exists.

VARIABILITY AND INITIAL SUBMOVEMENT AMPLITUDE

In this section an "organizing principle" for one important aspect of aiming movements is outlined, namely, how the amplitude of the initial submovement is selected in cases where two or more submovements are likely to be needed (i.e., when moderate or high accuracy constraints apply). This idea is the backdrop for the ideas outlined in the following two sections; it will be presented only briefly, however, because it is more fully described elsewhere (Worringham, 1991).

The proposed principle is that the average amplitude of the initial submovement in many aiming tasks is chosen to reflect the degree of spatial variability associated with that submovement. Specifically, the distance to the target from the end of the initial submovement is proportional to its variability. This is a type of optimization whose purpose is to bring the limb as close to the target as possible without prematurely (and inaccurately) hitting it. The reader will notice in this idea an echo of "stochastic submovement optimization," a recent theoretical proposal (Meyer, Abrams, Kornblum, Wright, & Smith, 1988; Meyer, Smith, Abrams, Kornblum, & Wright, 1990) holding that many discrete aiming movements—again, those that require two or more submovements—are temporally structured so as to optimize the durations of these submovements. Given that each is subject to some variability, an optimal initial submovement should be neither so fast that its variability hinders the conduct of (or necessitates more) subsequent submovements, nor so slow that its increased duration erases any benefits of low variability. Meyer and colleagues present some evidence for this mechanism (Meyer et al., 1988). It should be noted, however, that the Meyer et al. model gives no prediction about the *distance* traveled in the initial movement to the target, assuming a more or less random scatter around the target. Thus their model concentrates on temporal, not spatial, features of trajectories, and the idea discussed here explicitly considers the latter.

Evidence for this organizing principle comes from two aiming tasks. In the first, seated subjects moved a hand-held stylus from one of four starting positions toward the body, before reversing direction to hit a target (oriented vertically in the transverse plane) moving away from the body. These movements therefore produced "hook-shaped" trajectories. The position at which the reversal of motion in the sagittal plane occurred (the "reversal point") was used as a reference position at which to estimate the spatial variability associated with the initial movement toward the target (mean, across subjects, of the standard deviations of the reversal point coordinates calculated over trials within condition). Necessarily, each movement started beyond the plane of the target with respect to the subject and passed this plane before reversal. Four different distances were used (6.48, 10.8, 18, and 30 cm behind the target plane, and offset 7.5 cm to its right). Trajectories were recorded with an ultrasonic tracking system (Berners, 1986).

The relevant findings here concern the positions at which reversals occurred for more or less spatially variable initial movements. The longer, faster movements were more spatially variable in all three dimensions, but primarily in the sagittal, a finding that extends to three dimensions those of Gordon and Ghez (1989) for planar movements. Significantly, the average location of the reversal was also "biased" farther from the target with increasing variability; that is, the reversal occurred closer to the body and farther from the target. The average straight-line distance from the reversal point to the center of the target increased significantly, from 2.42 cm (6.48-cm condition) to 3.86 cm (30-cm condition). It should be noted that this task carries with it no explicit instructions or requirements about the required position of the reversal point, nor how or even whether it should vary with target amplitude. Reversal could take place anywhere within a substantial volume between the target and the subject's torso. It would be very costly, however, to select too short an initial movement—it may fail to pass the target plane on its approach to the body, and thereby invoke the need for an additional submovement. Similarly, too long an initial movement would position the hand and pointer too far from the target, adding time both by having the limb cover an unnecessarily long distance and by making the target more "difficult" in Fitts's (1954) index of difficulty terms. It is reasonable to suppose that the position actually—and spontaneously—selected reflects the outcome of an optimization process aimed at minimizing overall duration by the avoidance of either extreme. In this task, movement duration had to be kept to a minimum, given the requirement to miss the target only infrequently. To accomplish this it is useful for the subject to "predict" the limits (short and long) likely to bound some high proportion of initial movements, for this allows its average (goal or intended) amplitude to be selected optimally.

Such behavior, if shown consistently across tasks, could constitute an organizing principle in motor behavior. To investigate its generality, a second task was examined: rotating a pointer to a target with a wrist rotation. Subjects aimed to move a pointer, attached to a rotating handle, into a target zone without striking a physical stop that demarcated the far boundary of that zone. The starting position was varied, producing five amplitudes (30° to 90°). The initial submovements for longer conditions were both significantly faster (both average and peak velocity) and more spatially variable. As before, they were also scaled so as to end short of the target, by an amount very nearly proportional to their level of variability ($r = .99$, intercept = 1.002°, slope = 2.245 × variability measure). This task differs from the first in several respects: one rather than three degrees of freedom; no requirement to produce a reversal; no physical impact required on the target; and manipulation of a pointer with the hand unseen. The predicted relationship between initial submovement amplitude and variability still applied, however, thus supporting, if not establishing, the universality of the phenomenon.

These results show that subjects systematically scale the amplitude of initial submovements in a fashion that appears to take into account their spatial variability. This has been cast here as an organizing principle. Of course, further work is needed to establish this, especially tests that use quite different manipulations. One possibility is the use of conditions in which "noise"—expressed in variability—is directly varied rather than being manipulated indirectly through amplitude and velocity. This can be accomplished with tasks using controls and displays, in which the system's signal-to-noise ratio is open to manipulation. Nothing in the foregoing suggests that subjects would behave differently if the source of spatial variability were in a control system rather than being of biological origin.

The remaining sections consider two extensions of this principle—to the understanding of individual differences and the study of "clearance" tasks.

VARIABILITY AND INDIVIDUAL DIFFERENCES

Often characterized as belonging to two entirely different psychological traditions, the search for lawful behavior in whole populations and the attempts to characterize and explain individual differences have rarely been undertaken jointly. This has been as true in motor behavior as in other human performance domains. The pioneering efforts of Franklin Henry (1968) and Edwin Fleishman (1964) are now decades old, and rather little sustained effort has been seen among motor behaviorists to take account of—let alone account for—individual differences in movements. Some recent exceptions must be acknowledged. Keele and colleagues (Keele, Pokorny, Corcos, & Ivry, 1985), for example, have demonstrated correlations between timing variance in different limb structures and have also noted a relationship between timing in motor tasks and timing in perceptual tasks. These observations fuel the notion of a modular timing process (attributed to the lateral cerebellum on the basis of movement disorder studies [Ivry, 1989]). In general, however, it is fair to characterize most research in motor behavior as being in the mold of traditional experimental psychology. Some differences in a measure are hypothesized for a set of experimental manipulations and are tested for using inferential statistical procedures. The use of more than a single subject is motivated primarily by the need for statistical power. These are entirely legitimate procedures, but they fail to capitalize on subject differences in performance by treating these as a mere error term. Ultimately, we would wish to know not only, for example, that greater accuracy requirements in aiming cost additional time, but how and why they cost disproportionately more time for some people than for others.

In each of the two experiments mentioned in the previous section, mean movement times (MTs) of different subjects differed widely. In the last, the slowest subject had an average MT (across conditions) of about 600 ms, whereas the value for the fastest of these 12 people was approximately 400 ms. Are there measurable features of these movements that allow prediction of who is fast and who is slow? A further implication of the ideas already presented is that an individual whose trajectories are more variable than another's will (typically) pay the penalty of additional time. Consider the effect of an individual with high levels of variability, who produces wrist rotation trajectories inconsistently from trial to trial. The task requirements place a limit on the proportion of times the subject may hit the block at the end of the target zone. To meet this requirement our more variable performer must plan an "intended" trajectory in which the initial submovement ends at a distance more short of the target than would otherwise be necessary. The remaining portion of the movement—the second, and any subsequent, submovements—will therefore occupy more time, on the average.

This theoretical analysis leads to a prediction about individual differences. Individuals with higher levels of variability should have longer mean movement times and should, in addition, have greater undershoots of the target with their initial submovements. In a previously published data set involving 8 subjects, the first part of this prediction was examined using a rank correlation procedure (Spearman's rho): Those who ranked fastest in average movement times tended to rank lowest in spatial variability (rho = .833, $p < .05$) (Worringham, 1991). A second, independent, data set involved 12 subjects. The experiment differed only in using 30 trials per amplitude condition and being repeated for each of the two arms. Here, too, a relationship between these variables may be observed. Using a simple Pearson product-moment correlation on the mean MT, variability, and undershoot values for each subject, a relationship between average MT and average spatial variability at the end of the submovement of .867 is obtained. Moreover, the three-way relationship of variability, undershoot, and MT emerges, as depicted in Figure 4.1, with the value for the MT-undershoot correlation being .762.

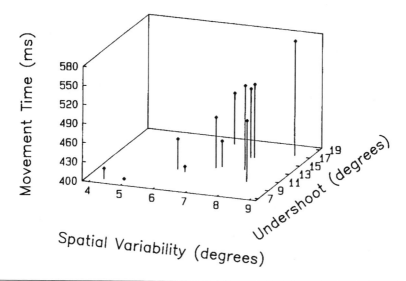

Figure 4.1   Mean movement times, spatial variability measures, and undershoot of target with initial submovement. Each point represents data for a single subject.

One interesting feature of these results is that they are opposite to those expected within a single individual. Working from the premise that, over a large range, spatial variability increases monotonically with speed (whether or not it be linear), a person making slower movements would normally accrue less spatial variability. Between individuals, however, slowness and variability are positively related. This contradiction may be more apparent than real, however; the relationship presented here and in Figure 4.1 involves not the speed of the initial first submovement, but the duration of the entire movement.

A second, more noteworthy finding is that spatial variability emerged as the best predictor of overall movement time. One might expect measures of peak acceleration or velocity to be negatively correlated with MT—as indeed they were ($r = -.767$ and $-.828$, respectively). This relationship could be thought of as an index of how different individuals "preprogram" speed, especially peak acceleration, given that this is normally attained by 70 or 80 ms from movement onset in this task. As Ghez and colleagues have shown for isometric aiming, an equivalent measure (peak dF/dt) is a good predictor of peak force, and they have also argued that it can be seen as an index of preprogramming (see Ghez, Hening, & Favilla, 1990, for a review). Although it is not unexpected that different subjects' mean peak acceleration and velocity are associated with their mean movement times, it is rather striking that spatial variability at the end of the submovement fared better as an MT predictor than either ($r = .867$) and better than did three components of the time being predicted! Correlations between average MT, on the one hand, and average time to peak acceleration, average time to peak velocity, and average time to the end of the first submovement, on the other, were .675, .711, and .788, respectively. Only average time from peak velocity to the end of the movement was a better predictor of whose times would be fast or slow ($r = .905$), a fact that is scarcely surprising, because this component accounts for up to half of the overall MT (the predicted measure) and is the one showing by far the greatest variance. As a check on the possibility that this was a chance finding, an alternative index of spatial variability was also used to examine these individual differences. The position at which peak velocity was reached was obtained on every trial, its standard deviation calculated for each condition, and the average of each individual's values were obtained. Again, this variability index was well correlated with average movement time ($r = .832$), and, again, this was (slightly) more closely related with MT than peak acceleration or peak velocity or three of the four components of MT (time from peak velocity to the end of the movement being, as before, the single exception). That spatial variability at the end of the first submovement is a better MT predictor than time to peak acceleration or peak velocity might not be seen as too surprising, because this estimate occurs closer in time to the end of the movement than either of the latter two. This account cannot explain the superiority of spatial variability at peak velocity over time to peak velocity as an MT predictor, however, thus leaving spatial variability as a strong candidate in the search for indices of individual differences.

This analysis of a single data set on a relatively small number of subjects remains well short of the ultimate goal: accounting for lawful behavior at both individual and group levels. It does, nevertheless, illustrate (generally) another facet of variability in performance that can be used as a tool for understanding motor behavior rather than being dismissed as bothersome statistical "noise" and (specifically) a potential avenue by which the relatively dormant "individual differences" domain might be reawakened. It would be especially informative if the act of discrete aiming—very thoroughly studied at the group level—could be scrutinized with a view to understanding the sometimes rather large individual differences in performance.

VARIABILITY AND CLEARANCE

This discussion of spatial variability has so far dealt with its manifestations in discrete aiming tasks. Motor behaviorists have had something of a preoccupation with discrete aiming, which epitomizes the reductionist approach, being, after all, one of the simplest forms of goal-directed movement behavior. It is clear that many movements have more complex and subtle goals, however. In this section, data on spatial variability in a somewhat different context are presented, and the case will be made that similar processes occur here as in discrete aiming—namely, trajectories are planned in a manner that takes account of spatial variability.

The general class of movements under discussion here can be labeled "clearance tasks," defined as cases in which a trajectory (be it of a limb, the whole body, or even a projectile) must negotiate an obstacle interposed somewhere between its initial position and its desired final position. At one extreme, the obstacle may be given a wide berth so as to eliminate any possibility of contact. In terms of efficiency, considered either as energy expenditure or as time, however, this may be unduly profligate. Conversely, too close an approach will necessarily risk contact with the obstacle. The specific nature of the object and task will determine how detrimental such contact may be. Ideally, a trajectory should be planned that avoids the undue costs that accrue to these opposing constraints. The nearest approach (lowest clearance) should be some function of the variability in that clearance over "trials." The most extreme examples of this optimization come from sport: Slalom skiing and hurdling require the closest possible approach to the gate or barrier. In both, slight contact does not pose a serious problem (though hurdling an immovable steeplechase barrier is quite different from brushing a freestanding hurdle), but a fall, injury, or major time loss is risked. Indeed, similar behavior may be in evidence in other highly skilled activities. A soccer penalty kick should be placed as far away from the goalie as possible—just barely inside the post. The proximity to the post of the position that can be chosen as the target for the kick is an inverse function of the kicker's variability. Thus an "optimal" penalty kicker would have—in addition to other attributes—not only relatively consistent kicks, but a well-calibrated sense of his or her *inconsistency*. Other, more routine actions (rounding a doorway, passing a milk jug over a flower vase on the breakfast table) may be considered in the same light. Given that increased levels of neural noise have been put forward as a possible cause of age-related slowing (Gregory, 1957; Salthouse, 1985; Welford, 1984), and that the elderly have many accidents in tasks involving clearance (e.g., steering a car, stepping over obstacles, climbing stairs), this phenomenon may be of special interest in the study of aging.

The laboratory version of this task required each of eight young adult subjects to move his or her right index finger back and forth over the top of a 10-cm-high foam rubber barrier, to contact a padded surface approximately 12 cm either side of the barrier. (These reciprocal motions were carried out for 20 s and were preceded by a practice trial.) The barrier was 10 cm high, 5.5 cm deep (sagittal dimension), and 1.3 cm wide. One infrared light-emitting diode was attached to the fingernail, and a second to the barrier, allowing three-dimensional movement recording at 200 Hz with the Optotrak system (Northern Digital). Subjects performed under 3 different instruction sets, and at 4 different speeds, for a total of 12 combinations. The instruction sets were labeled "*spontaneous*," "*low*," and "*high*," respectively. All subjects performed in the *spontaneous* condition first, in which the requirement was to make "efficient, comfortable" movements without hitting the barrier, but with no instruction as to the size of the clearance. The *low* and *high* conditions required the subject to pass the finger over the barrier with as low or as high a clearance as possible, respectively, without hitting it. Three of the four speeds were paced, using an electronically produced auditory beat. The *slow*, *medium*, and *fast* conditions had half-periods of 450, 370, and 290 ms, respectively. (The subject made one leftward or rightward pass in that time). The other speed was unpaced and simply required the subject to move as rapidly as possible under each set of clearance instructions. Data on this final speed are not presented here.

The use of different speeds was intended to induce different levels of variability in the trajectories (*medium* and *fast* being approximately 27.5% and 55% faster than the *slow* condition). This manipulation is different from those already presented, in that variability was manipulated through directly varying speed rather than amplitude. One reason this is important is that any change in clearance cannot be the result of the use of different amplitudes.

It was hypothesized that the effect of speed on the average clearance would vary with the instruction set. (Clearance is defined here as the straight-line distance between the fingertip and the point on the barrier vertically below the finger when the latter is above the barrier. This was obtained separately for each direction—left-to-right and right-to-left—because of an asymmetry in clearances, examples being presented in Figure 4.2. All subjects approached the barrier more closely when moving the right limb rightward than when moving it leftward across the barrier, by an average of 0.4, 0.4, and 1.1 cm in *low, spontaneous,* and *high* conditions).

Under the *high* condition, clearance should decrease with higher speed, presumably because of limitations in force production. By contrast, under the *low* condition, clearance should *increase* with higher speed. Slower, less variable trajectories could be planned with a lower clearance than more variable ones. No explicit hypothesis was formulated for the *spontaneous* condition, other than that the average clearance should be in between the other conditions for all speeds.

Data are presented here for both mean clearance and a measure of spatial variability—the standard deviation of the vertical coordinate. Data are averaged over the two directions.

The effects of speed on clearance were essentially as expected, as shown in Figures 4.3 and 4.4. Large decreases in clearance were observed with increasing speed under the *high* condition: averaging nearly 8 cm for each of the two speed increments. This effect is analogous to a "frequency response" characteristic of a mechanical system. As the period decreases (frequency increases), the maximal (vertical) amplitude diminishes, presumably a reflection of a "ceiling" effect in

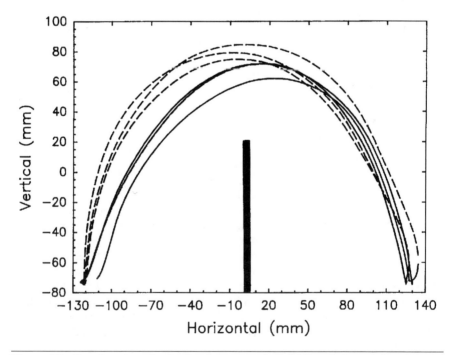

Figure 4.2 Sample trajectories for the clearance task. Solid lines indicate left-to-right movements. Dashed lines indicate right-to-left movements.

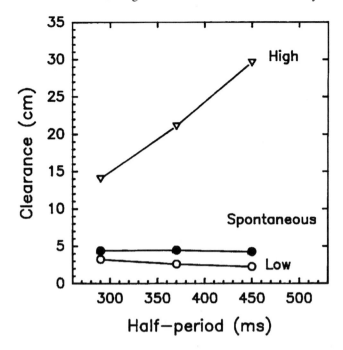

Figure 4.3   Mean clearance of the barrier is shown as a function of half-periods for the three instruction sets.

force production: Within some range the limb cannot be accelerated as far when the time available for that acceleration is reduced. In this case, the highest point is reached at approximately 200, 160, and 120 ms into each motion. The opposite behavior in the *low* condition is seen more clearly in Figure 4.4. Here the clearance is plotted against the variability measure.

For the averaged data, a very linear (though not proportional) relationship emerges ($r = .999$, intercept = −2.18 cm, slope = 5.6 × variability measure), in which more spatially variable conditions tend to have higher mean clearances: The *fast* condition has a mean clearance approximately 1 cm greater than the *slow* condition. Though much more subtle than the nearly 16-cm drop seen in the *high* condition, this was a statistically significant effect ($F_{2,14} = 12.08, p < .001$). When the subject was given the instructions *spontaneous*, average clearance was independent of speed, although variability increased slightly. Note from Figure 4.2 that the average clearance was slightly higher in the *spontaneous* than in the *low* condition, by an average of 1.7 cm. It is possible that a simplifying strategy is used in the *spontaneous* case. The modest energy cost of a moderately higher clearance effectively excludes any possible contact, and thus one no longer sees the tight coupling of average clearance and variability that is evident in the *low* case. As this phenomenon is examined more fully, it will be important to quantify the optimization involved. In principle, wherever the costs of a large clearance are small, variability should exert no, or only a weak, influence on the average trajectory, and vice versa.

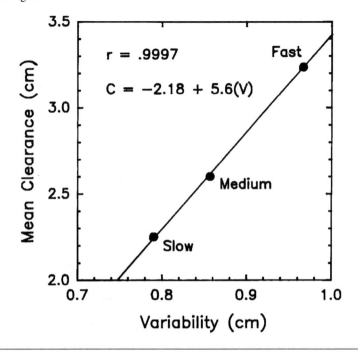

Figure 4.4 Mean clearance of the barrier as a function of spatial variability for the *low* condition.

In any event, these data suggest that the organizing principle for initial sub-movement amplitude selection in discrete aiming might be extended to "variability-related clearance" in the kinds of tasks described here—tasks currently lacking any significant attention in the motor behavior literature.

SUMMARY

The central idea presented here is that spatial variability measures can be used to predict certain apparently lawful facets of movement behavior and should not be regarded only as statistical noise. First, it can provide clues about the ways in which individuals structure the trajectories of aiming movements—specifically, that they tend to engage in a kind of "spatial optimization" that can be seen as the counterpart of the "temporal optimization" theoretically derived and demonstrated by Meyer and colleagues (1988). Second, this spatial optimization can be applied at the individual as well as the group levels, allowing some insights to be gained as to why some individuals tend to be poorer (slower) performers in aiming tasks than others. Finally, these ideas can be extended to another class of movement tasks—those involving the clearance of objects—so as to allow some quantitative predictions about how much clearance will, on the average, be allowed.

REFERENCES

Berners, A.C. (1986). *An ultrasonic time-of-flight system for hand movement detection.* Unpublished master's thesis, University of Wisconsin–Madison.

Fitts, P.M. (1954). The information capacity of the human motor system in controlling the amplitude of movement. *Journal of Experimental Psychology,* **47,** 381-391.

Fleishman, E.A. (1964). *The structure and measurement of physical fitness.* Englewood Cliffs, NJ: Prentice Hall.

Ghez, C., Hening, W., & Favilla, D. (1990). Parallel interacting channels in the initiation and specification of motor response features. In M. Jeannerod (Ed.), *Attention and performance XIII.* Hillsdale, NJ: Erlbaum.

Gordon, J., & Ghez, C. (1989). Independence of direction and amplitude errors in planar arm movements. *Society for Neuroscience Abstracts,* **15,** 50.

Gregory, R.L. (1957). Increase in "neurological noise" as a factor in ageing. *Fourth Congress of the International Association of Gerontology,* **1,** 314-324.

Henry, F.M. (1968). Specificity vs. generality in learning motor skill. In R.C. Brown & G.S. Kenyon (Eds.), *Classical studies on physical activity.* Englewood Cliffs, NJ: Prentice Hall.

Ivry, R. (1989). Timing functions of the cerebellum. *Journal of Cognitive Neuroscience,* **1,** 136-152.

Keele, S.W., Pokorny, R.A., Corcos, D.M., & Ivry, R. (1985). Do perception and motor production share common timing mechanisms? *Acta Psychologica,* **60,** 173-191.

Meyer, D.E., Abrams, R.A., Kornblum, S., Wright, C.E., & Smith, J.E.K. (1988). Optimality in human motor performance: Ideal control of rapid aimed movements. *Psychological Review,* **95,** 340-370.

Meyer, D.E., Smith, J.E.K., Kornblum, S., Abrams, R.A., & Wright, C.E. (1990). Speed-accuracy trade-offs in aimed movements: Toward a theory of rapid voluntary action. In M. Jeannerod (Ed.), *Attention and performance XIII,* Hillsdale, NJ: Erlbaum.

Salthouse, T.A., & Lichty, W. (1985). Tests of the neural noise hypothesis of age-related cognitive change. *Journal of Gerontology,* **40,** 443-450.

Welford, A.T. (1984). Between bodily changes and performance: Some possible reasons for slowing with age. *Experimental Ageing Research,* **10,** 73-88.

Wing, A.M. (in press). The uncertain motor system: Perspectives on the variability of movement. In D.E. Meyer & S. Kornblum (Eds.), *Attention and performance XIV.* Cambridge: MIT Press.

Worringham, C.J. (1991). Variability effects on the internal structure of rapid aiming movements. *Journal of Motor Behavior,* **23,** 75-85.

Acknowledgment

Thanks are expressed to Bob Dennis for his help in the conduct and analysis of the "clearance" experiment reported here.

Chapter 5

Directional Effects on Variability of Upper Limb Movements

Warren G. Darling, M. Stephenson
University of Iowa, Iowa City, Iowa

Previous studies of movement variability of either simple or compound motions have focused primarily on variability at the endpoint of movements in one or two directions. There has been relatively little consideration given to two important issues: (a) directional variability and (b) how variability develops during motions. With regard to the first issue, the results of many recent studies suggest that movement direction is encoded by the activity of neuronal populations in several different regions of the brain, including the motor cortex, cerebellum, and parietal cortex (Fortier, Kalaska, & Smith, 1989; Georgopolous, Kalaska, & Massey, 1981; Kalaska, Caminiti, & Georgopolous, 1983). The observation that this neuronal coding predicts movement direction even for parallel movements that vary in distance and in start and end points suggests that measures of variability in movement direction are related to variations in neuronal activity involved in movement production. Thus, variability associated with two- or three-dimensional movements of different speeds and directions may develop as a function of variations in the specification of movement direction both at the beginning of motion and throughout the movement. Whether directional variability is related to movement speed in the same way as positional variability has not been studied.

In terms of the second issue, previous research of single-joint elbow movements and of multijoint digit movements has shown that kinematic variability increases rapidly during acceleration (Darling, Cole, & Abbs, 1988; Darling & Cooke, 1987a, 1987b). However, variability decreases during deceleration, probably due to compensatory changes in the timing and amplitude of antagonist muscle actions in single-joint movements (Darling & Cooke, 1987b, 1987c). Such compensatory action is limited in single-joint movements to the actions of agonist-antagonist muscle pairs about a joint. However, in multijoint movements, compensatory actions may occur at different joints, as well as at individual joints, to accurately control the direction and speed of movement of the terminal segment (hand or finger, for arm movements). For example, if movement is initiated with velocity too high at the shoulder, this may be compensated by increased antagonist muscle activity (and stronger deceleration of the shoulder motion to maintain movement amplitude at the shoulder nearly constant), or motion at other joints (elbow, wrist) may be slower and smaller in amplitude to compensate for the large motion of the shoulder.

Both of these mechanisms for reducing variability have been observed in multijoint pinch/grasp movements of the index finger and thumb. Study of point-to-point variability of individual finger and thumb joint motions showed a rapid increase in angular position variability during acceleration that was compensated by a rapid decrease in positional variability during deceleration of the motion at each joint (Darling et al., 1988). Also, variability of the grasp "target" on the distal pulps of the index finger and thumb is maintained very low, regardless of movement speed, by way of a mechanism that involves correlated motions of the finger and thumb joints, as demonstrated by Cole and Abbs (1986). It should be noted, however, that such grasp movements differ from pointing movements in that there is no fixed spatial target and the movements involve the touching of two body parts rather than being toward an external target. Such self-directed movements are probably controlled quite differently from pointing movements to targets in extrapersonal space, because an intrinsic frame of reference can be used rather than an extrinsic frame of reference. Indeed, these digit movements showed no evidence of a speed-accuracy trade-off in terms of individual joint motions or in terms of endpoint variability of the location at which the fingertip strikes the thumbtip.

Relatively few studies have examined the relationships among the kinematics of joints of the upper limb and movement variability during pointing movements to external targets. Coordination of the shoulder and elbow for vertical plane movements has been described in terms of a close relationship between angular velocities at these joints during deceleration (Soechting & Lacquaniti, 1981). However, the relation between shoulder and elbow final joint angles was not described, nor was any relationship between endpoint variability at the individual joints and accuracy of the pointing movement. Using a different task, Arutyunyan, Gurfinkel, and Mirskii (1969) showed that experienced markspersons exhibit reciprocal covariation of motions produced by shoulder and wrist joints, which produced a low endpoint variability. Inexperienced markspersons, in contrast, show a weaker covariation of shoulder and wrist motions and higher endpoint variability. Thus, appropriate coordination of the participating joints can permit low endpoint (task) variability while allowing higher variability in the final angles at the individual joints. It should be noted, however, that these motions were measured only in a vertical plane (frontal) and that the reciprocal correlations were attributed only to movements in the medial-lateral direction. Correlations of motions in the vertical directions were low in magnitude for both the experienced and the inexperienced markspersons. Also, the movement amplitudes were very small in this aiming task.

The results of a recent study of dart throwing, in contrast, suggest that high correlations among joint kinematic patterns can be a property of an unpracticed performer (McDonald, van Emmerik, & Newell, 1989). Comparison of the effects of practice of the nondominant and dominant limbs by competent (dominant limb only) dart throwers showed that improvements in performance after practice were associated with reductions in correlations among the kinematic (joint angle, angular velocity) patterns. Practice of the nondominant limb produced no such reductions in correlation values, although performance improved somewhat. This suggests that motions of the elbow and shoulder were tightly linked in the nondominant limb and would require considerable practice (more than 10 to 14 days, 250 throws per day) for uncoupling their motions to allow large improvements in performance. It should be recognized, however, that in dart throwing the point of release of the dart in space is only one aspect related to performance and, indeed, need not be a fixed point for good performance. Velocity and angle of release are the most important factors in determining performance in dart throwing. Thus, coordination of joint motions is likely to differ from tasks with a fixed spatial target as used by Soechting and Lacquaniti (1981) and in many previous studies of movement variability (Meyer, Smith, & Wright, 1982; Schmidt, Zelaznik, Hawkins, Frank, & Quinn, 1979). It is important to note that the issues of point-to-point variability throughout motion, relationships of joint motion variability to endpoint variability, and directional variability have not been studied in these investigations.

DIRECTIONAL VARIABILITY STUDY

The aim of this investigation was to study quantitatively the variability of upper limb pointing movements in different directions in three-dimensional space. Variability at movement endpoint and midpoint were compared to determine whether the processes identified in single-joint movement studies were also observed in multijoint movements. Specifically, we hypothesized that positional variability at the midpoint of movements would be greater than at the endpoint, showing that corrective processes act to reduce variability during deceleration of the motion. We also hypothesized that movements in all directions would not demonstrate a simple relationship between endpoint variability and movement speed as has been shown for movements in one or two (opposite) directions. The final hypothesis was that movement direction variability would be high shortly after movement onset, because of the apparent high rate of increase in trajectory variability observed in single-joint motions. This directional variability should be reduced substantially during the middle portion of the movement but may increase toward the end of the movement due to corrective actions to ensure target acquisition and thereby reduce variability at movement endpoint. These hypotheses were tested in two experiments, one involving targeted pointing movements limited to the horizontal plane and the second involving free movements to targets in three-dimensional space.

METHODS

A total of 13 normal, healthy individuals participated in the two experiments, 7 male and 6 female University of Iowa students. Five subjects (3 males, 2 females) participated in the first experiment, in which movements were limited to the horizontal plane. Eight subjects (4 males, 4 females) participated in the second experiment, in which movements were not constrained to a plane.

Both experiments involved upper limb pointing movements to visual targets under two instructions: (a) ''Make movements smoothly and accurately'' (emphasis on accuracy) and (b) ''Make movements fast and accurately'' (emphasis on speed). In the first experiment, three sets of targets were used to cause movements in three different directions with the following properties: (a) large-amplitude shoulder motion and small-amplitude elbow motion, (b) small-amplitude shoulder motion and large-amplitude elbow motion, and (c) about-equal shoulder and elbow motion. Movement start and finish locations and distance were different for each type of movement. Also, subjects were instructed to simply place a hand-held stylus over the target, without actually touching it, so that voluntary deceleration of the motion was required. Subjects performed 20 consecutive movements to each target under each instruction for a total of 120 movements. Subjects were allowed to rest whenever they desired. The order in which the targets were presented was varied for each subject.

In the second experiment, nine targets arranged about a central starting position were used. This apparatus was constructed using a wooden base and six wooden dowels upon which the targets were painted. The descriptions of the distance and direction of each target from the starting position are shown in Table 5.1. Target direction from the start position was defined using two angles: (a) azimuth—in a horizontal plane, the angle between a vector joining the target and start position relative to the anterior direction (positive azimuth angles are for the right, ipsilateral targets)—and (b) elevation in a vertical plane, the angle between a vector joining the target and start position and the projection of that vector into the horizontal plane. Subjects terminated the movements with the pointing finger lightly touching the target.

Movements of the three upper limb segments (arm, forearm, hand) and the pointing finger in three-dimensional space were recorded using a two-camera optoelectronic (WATSMART) system (Northern Digital). Infrared light-emitting diodes (IREDs) were placed on the arm, forearm, and hand segments and on the pointing device (index finger or stylus) (Figure 5.1, a and b).

The two-dimensional coordinates of these IREDs for each camera were first transformed into three-dimensional coordinates. Because of the possibility that reflections of the infrared light emitted by the IREDs would cause inaccurate recording of IRED position, segment angles were calculated only from data in which the distances between IREDs on the body segments were within ± 10% of the actual

Table 5.1   Characteristics of Target Locations in Three-Dimensional Space

Target	Distance (cm)	Direction (rad)	
		Yaw	Elevation
1	27.0	−1.50	0.55
2	51.4	0.64	0.99
3	29.4	0.64	−0.66
4	30.0	0.00	0.52
5	51.1	−0.66	1.04
6	31.0	−0.66	−0.46
7	28.0	−1.64	0.55
8	54.0	−2.16	0.97
9	35.5	−2.16	−0.37

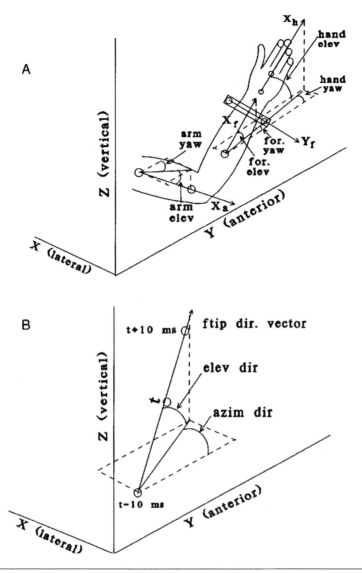

Figure 5.1 Calculation of segment angles and fingertip direction during movement. (A) The upper limb is shown with open circles indicating the positions of the IREDs. The angles of the longitudinal axes of each segment (X_a, arm; X_f, forearm; X_h, hand) projected into the horizontal plane relative to the anterior axis defines the segment yaw angle. The angles of the longitudinal axis relative to the projections of the longitudinal axes into the horizontal plane are used to define elevation angles. (B) This graph shows a vector drawn between fingertip positions at the times t − 10 ms, t + 10 ms. Fingertip position at time t, for which direction is being calculated, is also shown. The angles of this vector in a horizontal plane relative to the anterior direction, and in a vertical plane relative to the projection of the direction vector in the horizontal plane, are used to describe direction of fingertip motion.

distances (maximum allowance of ± 1 cm of the actual distance). The distance of the IRED positioned on the fingertip from the IRED placed over the head of the second metacarpal was used to test for the accuracy of recording of the IRED position on the fingertip; however, due to the possibility of finger flexion, a 20% reduction from the actual distance was allowed. The data were smoothed by digital low-pass filtering, usually with a 20-Hz cutoff, although in some cases a 10-Hz cutoff was used for the slower movements.

The orientations of the upper limb segments in three-dimensional space were calculated from the recorded IRED positions that were used to define longitudinal axes of each segment (Figure 5.1a). Specifically, segment yaw and elevation angles (for three-dimensional movements) were calculated as follows: (a) yaw—the angle of the segment longitudinal axis projected into the horizontal plane relative to the anterior axis (positive yaw angles to the right), and (b) elevation—in a vertical plane, the angle of the segment longitudinal axis relative to the projection of that axis into the horizontal plane (positive elevation angles are upward). In the experiment in which movements were constrained to the horizontal plane, shoulder joint angle was defined simply as the arm yaw angle; joint angles at the elbow and wrist were calculated as the difference between yaw angles of the adjacent segments at each joint. The directions of fingertip motion in the experiment allowing three-dimensional motion were calculated from the time derivative of the fingertip (XYZ) motion to allow calculation of instantaneous fingertip azimuth and elevation angles. Figure 5.1b shows how these angles were calculated.

Movement onset and termination were determined from the fingertip or stylus speed records (obtained from digital differentiation of the fingertip distance records for individual movements—see Figure 5.2) using thresholds of 10 to 15 cm/s. The midpoint of the movement was defined as the time at which one half of the movement distance toward the target had been traveled. Segment angles and fingertip position were measured at movement midpoint and endpoint for calculation of kinematic variability of the movements.

Fingertip positional variability was calculated as the area of ellipses with radii equal to one standard deviation (s.d.) in the X and Y directions (horizontal planar motions) or as the volume of ellipsoids with radii equal to 1 s.d. in the X, Y, and Z directions of the final fingertip positions. Angular position variability was measured as the standard deviation of joint angle (horizontal planar motions) or as the area of ellipses with radii equal to 1 s.d. of the segment elevation and yaw angles (3-D movements). Fingertip direction variability (calculated as the area of an ellipse with radii equal to 1 s.d. in the elevation and azimuth directions) was evaluated 30 ms after movement onset, at movement midpoint, and 30 ms prior to movement termination. This directional variability measure was sensitive to the time interval relative to movement onset or termination at which it was taken. This sensitivity results from the fact that early in the movement, the direction changes rapidly from zero (prior to movement onset) toward the desired direction toward the target, and at the end of the movement directional changes are also relatively rapid. Variability measures were calculated only if there was a minimum of seven trials that were free from infrared reflection errors during the movements. The relationships among variability measures and movement speed and target direction were studied using simple and multiple correlation and regression techniques.

RESULTS

Figure 5.2 shows averaged records of stylus motion and shoulder, elbow, and wrist joint angular motions during horizontal planar movements from one subject performing movements involving large-amplitude shoulder motion and small-amplitude elbow motion. Wrist joint angular motion generally contributed little to these movements, because only wrist abduction/adduction was permitted. Variability throughout the movements is indicated by the dots overlying the averaged records. Clearly, the records of stylus position indicate a pattern of increasing variability from the time of movement onset until about the time of peak speed (end of acceleration). During deceleration of the stylus motion, variability decreases as the target is approached. Records of individual joint motions show a similar pattern of increasing variability during acceleration, but during deceleration variability does not always decrease.

Variability of joint angles at the midpoint of movement distance and at movement termination is shown in Figure 5.3, a-c. Endpoint joint-angle variability was not sensitive to average stylus speed, as indicated by the observation that there were no significant differences in joint angle variability for the six different average movement speeds studied (ranging from 49 to 275 cm/s). Note also that joint angle variability was quite low at movement endpoint, usually much less than 0.1 rad. Joint angle variability at the movement midpoint or at the end of acceleration typically did not differ significantly from that at movement endpoint. This contrasts with the findings of simple elbow joint and digit multijoint movement studies in which joint angle variability at the end of acceleration was always greater than that at movement endpoint (Darling & Cooke, 1987a; Darling et al., 1988). Shoulder and elbow joint angle variability at movement endpoint were also not sensitive to average joint angular velocity (Figure 5.4, a and b). Thus, joint angle variability does not show a speed-accuracy trade-off in these planar multijoint motions.

The relationship between variability of final stylus position (measured in cm^2 as the area of an ellipse with radii of 1 s.d. in the anterior and lateral directions) and average movement speed for all subjects is shown in Figure 5.5. Regression analysis indicated a linear relation between stylus endpoint variability and average speed (Figure 5.5; $r = .96$, $p < .01$). Stylus position variability at the movement distance midpoint also increased with average movement speed, but with a different slope (Figure 5.5; $r = .76$, $p < .1$).

These data suggest that, similar to the observations in studies of single-joint movements, variability of position of the pointing segment (in this case the stylus held by the hand) rises rapidly during acceleration and then decreases during deceleration of the movement. We hypothesized that the mechanism underlying this reduction in variability during deceleration would be appropriate coordination of the joint angular motions involved in producing the movement. Because most of the stylus motion was produced by shoulder and elbow motion, one would expect negative correlations between the final joint angles at these joints, because such a relationship would help to preserve stylus position at movement termination relatively constant in spite of the variability in joint angles. That is, if flexion at the shoulder were too large, decreased flexion at the elbow would partially compensate so that the stylus

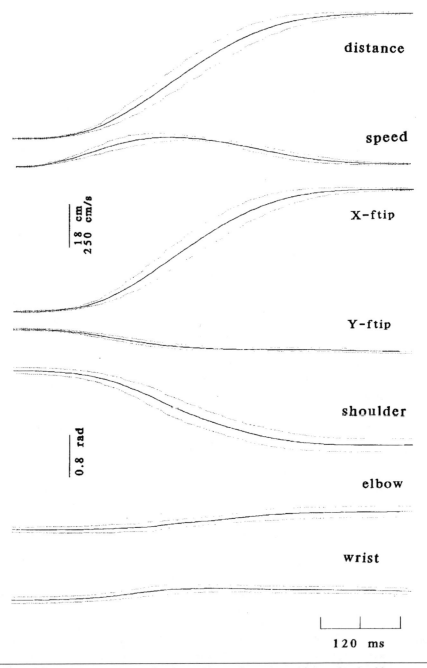

distance

speed

18 cm
250 cm/s

X-ftip

Y-ftip

shoulder

0.8 rad

elbow

wrist

120 ms

Figure 5.2 Averaged records of stylus motion and angular motion at the shoulder, elbow, and wrist joints for 20 horizontal-plane pointing movements by one subject. Movements were aligned at movement onset prior to averaging. Superimposed on each record are dots positioned 1 s.d. above and 1 s.d. below the movement record at 5-ms intervals.

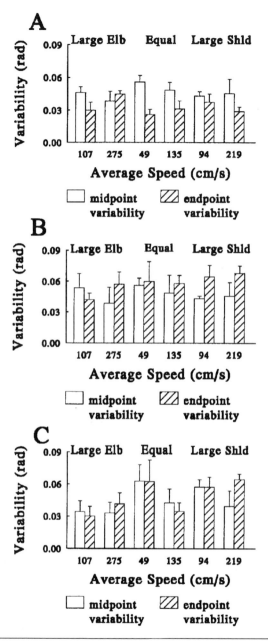

Figure 5.3 Variability of shoulder (A), elbow (B), and wrist (C) joint angles at movement midpoint and at movement endpoint, for horizontal-plane pointing movements. Each bar graph shows the mean from 5 subjects of joint angle variability for the three different types of motions under two speed conditions. The average speed is shown on the abscissa for smooth and fast movements within each type of movement. Error bars are 1 standard error (SE).

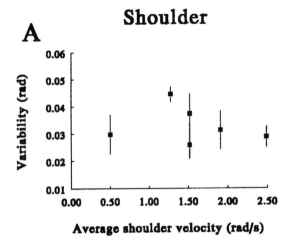

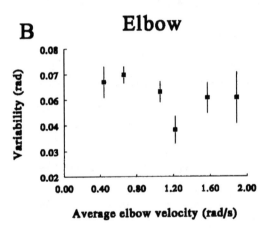

Figure 5.4   The relationships between shoulder (A) and elbow (B) joint angle variability at movement endpoint and average shoulder and elbow angular velocities. Data are from five subjects for horizontal-plane pointing movements. Error bars are 1 SE.

could be positioned close to the target. This was indeed observed, as can be seen in Figure 5.6, which contains scatter plots of final shoulder, elbow, and wrist joint angles. Final shoulder and elbow joint angles were usually highly negatively correlated (often exceeding −.9). In contrast, wrist joint angles were generally poorly correlated with final elbow and shoulder angles, and even the direction (positive/negative) of these relations varied across subjects (Figure 5.6, a-f).

The preceding results suggest that the relationship between endpoint variability of the pointing segment and average speed is independent of movement direction. Because only two-dimensional motions of three different directions were studied, a stronger test of this hypothesis would be to examine unconstrained motions in three-dimensional space with many more directions, as in the second experiment.

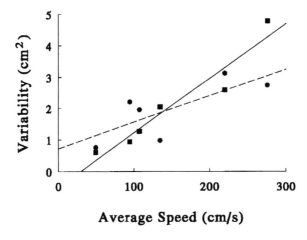

Figure 5.5   The relationship between variability of stylus position and average movement speed for horizontal-plane pointing movements. Each plotted point represents the mean from five subjects. The closed circles represent variability at the midpoint of movement distance, and the closed squares represent the variability at movement endpoint. The regression lines are for the relationships of midpoint variability (dashed line) and endpoint variability (solid line) to average movement speed.

This experiment also differed from the study of planar motions in that subjects pointed with the fingertip rather than a stylus and were permitted to lightly touch the movement target.

Averaged records of fingertip motion of fast movements toward Target 4 by one subject are shown in Figure 5.7. Superimposed dots show variability throughout the averaged motions in these records. It is clear that variability rises during acceleration and decreases during deceleration, as was observed in the planar motions. Figure 5.8, a and b, shows midpoint and endpoint variability for the fingertip for smooth (Figure 5.8a) and fast (Figure 5.8b) movements to all of the targets. Variability at movement midpoint was usually, but not always, greater than at movement endpoint (for the group averages, variability was always greater at movement midpoint, but for individual subjects this was not always true). Note that for all targets except Target 9 (contralateral low target), endpoint variability was greater for faster movements than for slower movements.

Variability of fingertip direction was measured 30 ms after movement onset, at movement midpoint, and 30 ms prior to movement endpoint (Figure 5.9, a and b). This variability measure was calculated as the area of an ellipse with radii equal to 1 s.d. in the azimuth and elevation directions at these points in the movements. Start and endpoint variability were relatively high (group averages range from 0.05 to 0.2 rad^2), but variability at movement midpoint was always substantially lower. Fingertip azimuth and elevation directions during motion were nearly constant during the middle parts of the movements but could vary over a large range early and late in the movements (Figure 5.7). Such large variations in direction in the early phases of the movements presumably reflect variability in the initial specification

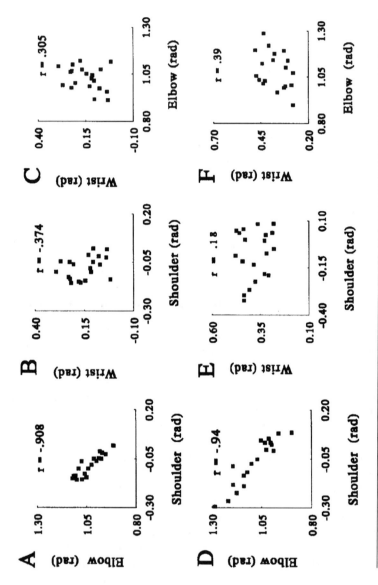

Figure 5.6 The relationship between final joint angles of the shoulder, elbow, and wrist joints for smooth (A, B, C) and fast (D, E, F) movements involving a large range of shoulder joint motion and a small range of elbow joint motion. Data are from one subject.

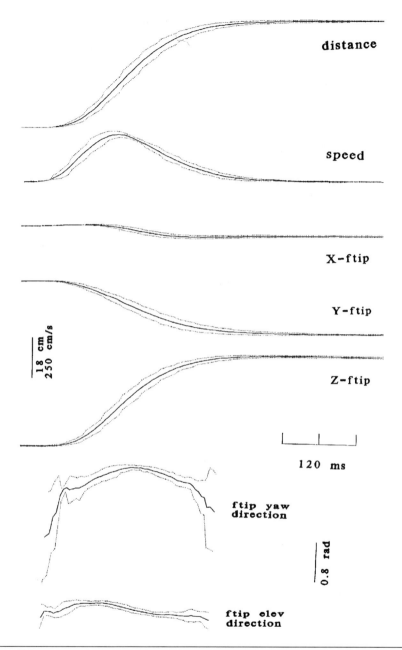

Figure 5.7 Averaged records of fingertip motion for movements toward Target 3 (ipsilateral, anterior, downward motion) by one subject. Movements were aligned to time of onset prior to averaging. Superimposed on each record are dots positioned 1 s.d. above and 1 s.d. below the movement record at 10-ms intervals. Fingertip direction records are shown only during the movement, because prior to and after movements the fingertip is stationary and has no direction.

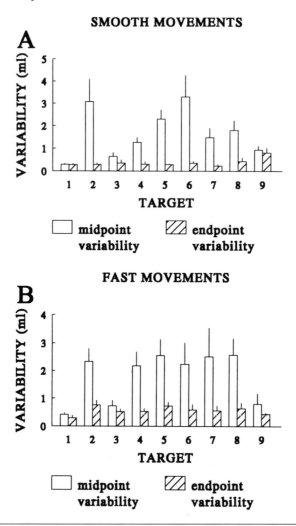

Figure 5.8   Midpoint and endpoint positional variability of the fingertip for smooth (A) and fast (B) movements to all targets, in three-dimensional space. Each bar represents the mean positional variability from seven or eight subjects (technical problems led to a loss of data from one subject). Positional variability was calculated as the volume of an ellipsoid with radii equal to 1 s.d. in the lateral, anterior, and vertical directions. Error bars are 1 SE.

of movement direction. Large variations in direction near the endpoint suggest corrective motions to ensure that the fingertip strikes the target. Such corrective motions, however, appear to be carried out rather smoothly (i.e., without submovements), because inspection of individual trials of fast movements showed little evidence of submovements. Slower movements (''smooth and accurate'' instruction) showed some evidence of submovements as multipeaked velocity records.

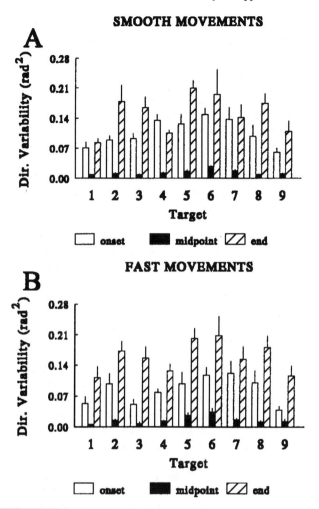

Figure 5.9   Fingertip directional variability during acceleration, at movement midpoint, and during deceleration, for smooth (A) and fast (B) movements to each target. Each bar represents the mean direction variability from seven or eight subjects (technical problems led to a loss of data for one subject for some targets). Error bars are 1 SE.

Directional variability during acceleration, at movement midpoint, and during deceleration was not well correlated with movement speed (Figure 5.9, a and b—compare variability of smooth and fast movements for each target). Correlation analyses also indicated no strong relationship between directional variability and movement speed.

Trial-to-trial variability of individual upper limb segment motions was also assessed at movement midpoint and endpoint; however, due to technical problems caused by loss of view of IREDs from the cameras, only hand angle variability could be measured in all subjects for most targets. The general pattern was that midpoint and endpoint variabilities were about equal for all three segment angles.

Comparing smooth and fast movements, midpoint and endpoint segment angle variability (measured as the area of ellipses with radii of 1 s.d. in elevation and yaw angles for the segments) did not appear to increase with movement speed, for most targets. Indeed, the endpoint variability was very low and was generally less than 0.01 rad^2 for smooth movements to all targets and for fast movements to most targets. Correlation of segment angles at movement endpoint for each target was carried out in the limited number of subjects for whom such data were available. These analyses showed that there were usually negative correlations (sometimes exceeding $-.8$) among arm and forearm angles (i.e., arm yaw vs. forearm yaw, arm elevation vs. forearm elevation) but that correlations with final hand angles were usually low and varied in direction. However, correlation of contributions of individual joints to segment yaw and elevation final angles produced consistently high negative correlations (always exceeding $.8$ in magnitude) between shoulder and elbow joint contributions (Figure 5.10, a and d). Lower correlations between wrist and shoulder and wrist and elbow motions were observed (Figure 5.10, b, c, e, and f). These contributions of the elbow and wrist to the amplitude of segment yaw and elevation motions during movement were computed as follows:

1. Elbow: (forearm termination angles − arm termination angles) − (forearm onset angles − arm onset angles)
2. Wrist: (hand termination angle − forearm termination angle) − (hand onset angle − forearm onset angle).

Note that the contributions assigned to the elbow result partially from internal/external rotation at the shoulder, and contributions assigned to the wrist result partially from forearm pronation/supination.

The relationship between fingertip endpoint variability and average movement speed for averaged data from all subjects is shown in Figure 5.11. Endpoint variability of movements to all targets was only weakly related to movement speed (Table 5.2; $r = .61$, $p < .05$). There appeared to be a strong dependence of endpoint variability on target location (compare circles and squares in Figure 5.11), thus correlations involving targets in ipsilateral body space were computed separately from those for targets in contralateral body space. This analysis showed that endpoint variability of movements to ipsilateral targets was strongly related to movement speed ($r = .8$, $p < .05$), but for movements to contralateral targets the relation was rather weak ($r = .48$, $p > .1$). Furthermore, multiple regression analysis showed that including target directions (yaw and elevation directions) as predictors of endpoint variability improved these correlations (Table 5.2).

DISCUSSION

The results of the two experiments show clearly that variability of motion at individual joints during a multijoint pointing movement does not show a speed-accuracy trade-off. Indeed, variability in angular position of joint/segment motions at movement endpoint is relatively insensitive to either movement speed or movement direction. Similarly, fingertip direction variability was relatively insensitive to either

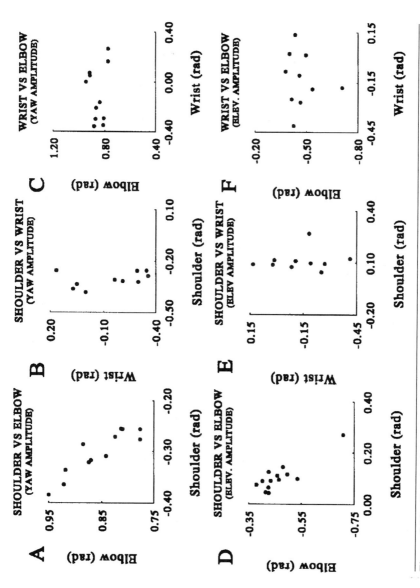

Figure 5.10 Relationships among shoulder, elbow, and wrist positions to yaw (A-C) and elevation (D-F) motions of the segments of the upper limb during three-dimensional pointing movements. Data are from one subject performing fast movements to Target 2.

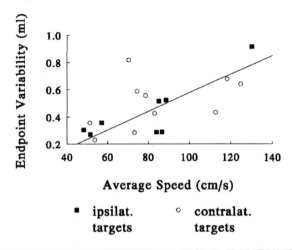

Figure 5.11   The relationship between fingertip endpoint variability and movement speed for three-dimensional movements to nine different targets. Each plotted point represents the mean from seven or eight subjects.

Table 5.2 R^2 Correlation and Multiple Regression Coefficients for Prediction of Endpoint Variability From Average Speed and Target Direction

	Targets		
	Ipsilateral	Contralateral	All
R_a^2	0.64*	0.23	0.37*
R_b^2	0.78*	0.32	0.50*
Yaw coefficient	−0.95	−0.01	−0.03
Elevation coefficient	−0.08	−0.09	−0.09

R_a^2: Average speed is predictor. R_b^2: Average speed, target directions are predictors.

*Significant at $p < .05$.

movement speed or target direction. The observation that motions of the arm and forearm segments (or shoulder and elbow joints) are correlated in such a way as to decrease variability of the pointing segments (fingertip/hand) suggests a different mechanism from that used to reduce variability at movement endpoint in single-joint movements. The finding that variability of motion of the pointing segment increased during acceleration and decreased during deceleration, in a manner similar to that observed in single-joint forearm motions, indicates that factors affecting control of the motion of the pointing segment are similar in single- and multijoint movements. However, endpoint variability of the pointing segment clearly depends on direction and target location, as shown by the second experiment, in which nine different movement directions were studied. These findings will be discussed in relation to the models of motor output variability and issues of control of multijoint movements.

It appears from correlation analyses that motion at the shoulder and elbow joints is coupled in such a way as to maintain variability of the pointing segment at movement endpoint relatively low. Such coupling of shoulder and elbow motions in multijoint planar movements has been described previously and was attributed to inertial coupling of the two joints (Soechting & Lacquaniti, 1981). That is, because the motions at one joint in a multijoint limb affect the movement of the other joint, due to inertial interactions (acceleration-dependent torques), such motions are necessarily related, especially in two-joint motions. Also, the fact that several muscles controlling elbow joint motion also span the shoulder would contribute to a mechanical coupling of the motion of the arm and forearm.

Wrist joint angular motion did not appear to be coupled to either the elbow or the shoulder joint motion. Possible reasons why wrist joint motion would be uncorrelated to motions at the elbow and shoulder include these: (a) Hand inertia is relatively small in comparison to the arm and forearm, thus inertial coupling would be less likely; (b) muscles causing wrist flexion/extension and abduction/adduction have small moment arms at the elbow (An, Hui, Morrey, Linscheid, & Chao, 1981); and (c) because the wrist is a more distal joint and is closer to the pointing segment, it may be controlled in such a way as to modulate the final positioning of the pointing segment on the basis of sensory information (visual and proprioceptive) regarding the ongoing movement. Certainly wrist joint motion allows more flexible positioning of the pointing segment than when motion is limited to the shoulder and elbow, because two-joint motions in a plane result in any stylus position requiring a unique combination of shoulder and elbow angles. With the addition of the wrist joint, such a unique combination of shoulder and elbow joint angles to attain the target is no longer required. Thus, the shoulder and elbow may be used primarily for transport of the pointing segment, whereas the wrist may be used in final positioning of the pointing segment (cf. Jeannerod, 1984; Lacquaniti & Soechting, 1982).

The finding that variability of joint or limb segment angles at movement termination does not depend on the average speed of pointing segment motion was surprising. Impulse-variability models would presumably predict a dependence of variability of final joint angles on average movement speed. In impulse-variability models of single-joint motions (Darling & Cooke, 1987c; Meyer et al., 1982), it is assumed that the magnitude of accelerative and decelerative torque impulses produced by agonist and antagonist muscles varies in proportion to their mean magnitude. The variations in these muscular torque patterns produce variations in acceleration patterns that lead to spatial variability at movement endpoint. Trial-to-trial variations in peak muscular torques and duration of torque production presumably arise from variations in the controlling neural signals. If one assumes that the controlling neural signals retain such variability in multijoint motions, one might predict a dependence of variability of final joint angles on average movement speed. Indeed, the observation that variability rises during acceleration for each of the involved joint motions suggests that the controlling neural signals do retain considerable variability in multijoint movements. It is also clear that variability of the shoulder and elbow angles at movement endpoint do not depend on the average speed of motion at the individual joints (Figure 5.5). Thus, it seems that variability of individual joint motions in a multijoint pointing task is unrelated to movement speed. How can such a result be interpreted on the basis of impulse-variability models?

It should be remembered that motion at individual joints in multijoint movements is due not only to the muscular torques exerted at those joints but also to the motions occurring at other joints, as discussed earlier. It is possible that the torques arising from motions at other joints act in some way to compensate for variability in muscular action. Torques arising from angular acceleration at one joint tend to produce acceleration in the opposite direction at neighboring joints because of the oppositely directed torque produced at the other joint (Darling & Cole, 1990). Such interaction torques may contribute to the negative correlation observed between final shoulder and elbow joint angles that allows for low variability of the final pointing segment spatial location.

Direction of the pointing segment movement during multijoint motions appears to be a kinematic variable specified in a number of brain regions, as discussed at the start of this paper. Indeed, it has even been shown that changes in firing rate of populations of cortical neurons throughout movement precede changes in direction by about 20 ms (Georgopolous, Schwartz, & Kettner, 1986). Thus, one might suggest that variations in movement directions represent a measure of the variability of the firing of populations of cortical neurons involved in controlling the movement.

The observation that directional variability is high early in the movements indicates that initial specification of direction is a highly variable process. This is consistent with the findings of previous reports of high variability in the amplitude of the agonist EMG burst that initiates single-joint movements (Darling & Cooke, 1987b). Such high variability likely results from variability in the initial movement commands and variations in membrane excitability of brain neurons and spinal motor neurons on which these commands are superimposed. It is interesting, however, that the initial variability in fingertip direction appears uncorrelated to speed of movement. This suggests that the variability of commands specifying direction of motion of the distal (pointing) segment is not dependent on movement speed, although the variation of commands to muscles that produce the motion is presumably speed dependent (Darling & Cooke, 1987b). However, once the desired movement direction is attained, it remains nearly constant for a period of time and is nearly constant from trial to trial (Figure 5.7). Thus, specification of the desired movement direction appears to be associated with little variability and, presumably, results from a low degree of trial-to-trial variability in neuronal firing during this phase of the movement.

During deceleration as the target is approached, directional variability again rises substantially, perhaps due to corrective motions that ensure target acquisition. Because these changes in fingertip direction are apparently associated with smooth changes in the movement trajectories, rather than submovement-type corrections (cf. Meyer et al., 1982; Meyer, Abrams, Kornblum, Wright, & Smith, 1988), it is possible that such actions depend on predictive use of information regarding the locations of the target and pointing fingertip. Such information may arise from proprioceptive and/or visual information regarding the initial motion of the pointing segment or may be derived from corollary discharge information regarding the actual movement commands (cf. Cooke & Diggles, 1984). In any case, it appears that such corrective action in the late stages of the movement can occur without stopping, or even extreme slowing, of the movement from its normal path or trajectory.

Data from the experiment in which motions were constrained to the horizontal plane clearly showed a linear speed-accuracy trade-off for the group average data. In contrast, when unconstrained movements in three-dimensional space were studied, a much weaker relationship between pointing segment endpoint variability and average speed was found. There are a number of possible reasons underlying these differing results.

The most likely reason is the larger number of movement directions used in the three-dimensional motion experiment. Evidence for this interpretation comes from the finding that endpoint variability of movements to ipsilateral targets was closely related to movement speed but for movements to contralateral targets the relationship was weak. Inclusion of target direction as a predictor of endpoint variability improved correlations between endpoint variability and speed for all targets and for targets within ipsilateral and contralateral body-space (Table 5.2). In addition, other studies have suggested that pointing movements into contralateral body-space are less accurate than pointing to targets in ipsilateral body-space (e.g., Fisk & Goodale, 1985). Thus, contralateral motions may be associated with a different slope to the speed-accuracy trade-off than ipsilateral motions. Indeed, the coefficient for yaw direction in the multiple regression equation was always negative, which, because contralateral motions were in the negative yaw direction, suggests that variability was generally higher for contralateral motions of the same speed as ipsilateral motions.

Another issue in terms of directionality is the vertical direction; that is, downward movements may be more variable than upward movements, or vice versa. The coefficient for elevation direction in the multiple regression equations for contralateral and ipsilateral targets was also negative. Because downward motions were in the negative elevation direction, this indicates that downward motions were, on the average, more variable than upward motions for a given movement speed. Both of these suggestions regarding directional effects on endpoint variability must be viewed with caution, however, because although target directions as predictors significantly increased the percentage of unexplained variance in the prediction of endpoint variability (Table 5.2), the actual regression coefficients for target yaw and elevation did not differ significantly from zero. Thus, further research is needed to demonstrate whether motions directed upward and toward targets located in ipsilateral body-space are indeed less variable than motions directed downward or toward targets in contralateral body-space.

A second possible reason for the stronger relationship between endpoint variability and average speed in the horizontal-plane motions is that the range of average movement speeds was greater in that experiment (50 to 275 cms^{-1} vs. 50 to 125 cms^{-1}). This probably occurred because the movements involving large elbow and large shoulder-joint motions in the horizontal-planar movements were larger in amplitude than in the three-dimensional movements study. Because average speed tends to increase with distance, this would lead to a greater range of movement speeds. Thus, the higher correlation between endpoint variability and average speed may be a range effect. However, inspection of Figure 5.11 indicates a large range of endpoint variabilities for movements with average speeds in the range of 75 to

90 cms^{-1} in the three-dimensional motion study. This suggests that the reason for the poorer relation in three-dimensional movements can be attributed to a direction effect on endpoint variability, because it is doubtful that a high correlation would have been attained for all targets even if a greater range of movement speeds had been studied.

Finally, it should be noted that the spatial variability of the pointing segment measured at movement midpoint was not always greater than that at movement endpoint. Thus, large reductions in spatial variability during deceleration did not always occur. This was especially the case for the larger amplitude horizontal plane motions. However, one should be careful not to confuse movement midpoint with the end of acceleration, because that would be the case only if movements are perfectly symmetrical in acceleration and deceleration. Most pointing movements show an asymmetry such that more of the movement distance is traveled in the deceleratory phase of the movements. Thus, the midpoint of movement distance often occurs during deceleration. The abilities of subjects to reduce spatial variability during deceleration may be related to factors such as movement distance and direction.

CONCLUSION

In conclusion, the findings of the present investigation provide strong support for the concept of motor equivalence (Abbs & Cole, 1987). Specifically, it is apparent that final angles of individual joints or limb segments of the upper limb are not controlled in the same manner as is the terminal pointing segment. If this were so, one would expect to observe similar speed-accuracy trade-offs for all of the joints participating in a particular movement. Instead, the final position of the pointing segment is closely regulated to ensure target acquisition, whereas the other segments are positioned more flexibly. Such a mechanism is similar to that observed in speech and multijoint digit motions where targets are acquired with variable positioning of each of the articulators contributing to the overall motion (Abbs & Cole, 1987; Darling et al., 1988). The advantage of such a control system is that this flexibility allows accurate target acquisition in spite of variability in the motions at the several joints that contribute to motion of the pointing segment. However, incorporation of such flexibility in individual joint motions of a multijointed limb is a difficult challenge for those attempting to develop models of movement control to explain motor output variability.

REFERENCES

Abbs, J.H., & Cole, K.J. (1987). Neural mechanisms of motor equivalence and goal achievement. In S.P. Wise (Ed.), *Higher brain functions: Recent exploration of the brain's emergent properties* (pp. 15-43). New York: Wiley.

An, K.N., Hui, F.C., Morrey, B.F., Linscheid, R.L., & Chao, E.Y. (1981). Muscles across the elbow joint: A biomechanical analysis. *Journal of Biomechanics*, **14**, 659-669.

Arutyunyan, G.H., Gurfinkel, V.S., & Mirskii, M.L. (1969). Organization of movements on execution by man of an exact postural task. *Biophysics*, **14**, 1162-1167.

Cole, K.J., & Abbs, J.H. (1986). Coordination of three-joint digit movement for rapid finger-thumb grasp. *Journal of Neurophysiology*, **55**, 1407-1423.

Cooke, J.D., & Diggles, V.A. (1984). Rapid error correction during human arm movements: Evidence for central monitoring. *Journal of Motor Behavior*, **16**, 348-363.

Darling, W.G., & Cole, K.J. (1990). Muscle activation patterns and kinetics of human index finger movements. *Journal of Neurophysiology*, **63**, 1098-1108.

Darling, W.G., Cole, K.J., & Abbs, J.H. (1988). Kinetic variability of grasp movements as a function of practice movement and speed. *Experimental Brain Research*, **73**, 225-235.

Darling, W.G., & Cooke, J.D. (1987a). Movement related EMGs become more variable during learning of fast accurate movements. *Journal of Motor Behavior*, **19**, 311-331.

Darling, W.G., & Cooke, J.D. (1987b). A linked muscular activation model for movement generation and control. *Journal of Motor Behavior*, **19**, 333-354.

Darling, W.G., & Cooke, J.D. (1987c). Changes in the variability of movement trajectories with practice. *Journal of Motor Behavior*, **19**, 291-301.

Darling, W.G., & Gilchrist, L. (1991). Is there a preferred coordinate system for perception of hand orientation in three-dimensional space? *Experimental Brain Research*, **85**, 405-416.

Fisk, J.D., & Goodale, M.A. (1985). The organization of eye and limb movements during unrestricted reaching of targets in contralateral and ipsilateral visual space. *Experimental Brain Research*, **60**, 159-178.

Fortier, P.A., Kalaska, J.F., & Smith, A.M. (1989). Cerebellar neuronal activity related to whole-arm reaching movements in the monkey. *Journal of Neurophysiology*, **62**, 198-211.

Georgopolous, A.P., Kalaska, J.F., & Massey, J.T. (1981). Spatial trajectories and reaction times of aimed movements: Effects of practice, uncertainty, and change in target location. *Journal of Neurophysiology*, **349**, 725-743.

Georgopolous, A.P., Schwartz, A.B., & Kettner, R.E. (1986). Neuronal population coding of movement direction. *Science*, **233**, 1416-1419.

Jeannerod, M. (1984). The timing of natural prehension movements. *Journal of Motor Behavior*, **16**, 235-254.

Kalaska, J.F., Caminiti, R., & Georgopolous, A.P. (1983). Cortical mechanisms related to the direction of two-dimensional arm movements: Relations in parietal area 5 and comparison with motor cortex. *Experimental Brain Research*, **51**, 247-260.

Lacquaniti, F., & Soechting, J.F. (1982). Coordination of arm and wrist motion during a reaching task. *Journal of Neuroscience*, **2**, 399-408.

McDonald, P.V., van Emmerik, R.E.A., & Newell, K.M. (1989). The effects of practice on limb kinematics in a throwing task. *Journal of Motor Behavior*, **21**, 245-264.

Meyer, D.E., Abrams, R.A., Kornblum, S., Wright, C.E., & Smith, J.E.K. (1988). Optimality in human motor performance: Ideal control of rapid aimed movements. *Psychological Reviews*, **95**, 340-370.

Meyer, D.E., Smith, J.E.K., & Wright, C.E. (1982). Models for the speed and accuracy of aimed movements. *Psychological Reviews*, **89**, 449-482.

Schmidt, R.A., Zelaznik, H., Hawkins, B., Frank, J.S., & Quinn, J.T. (1979). Motor output variability: A theory for the accuracy of rapid motor acts. *Psychological Review*, **86**, 415-451.

Soechting, J.F., & Lacquaniti, F. (1981). Invariant characteristics of a pointing movement in man. *Journal of Neuroscience*, **1**, 710-720.

PART II

VARIABILITY
AND THE MOVEMENT
SPEED-ACCURACY TRADE-OFF

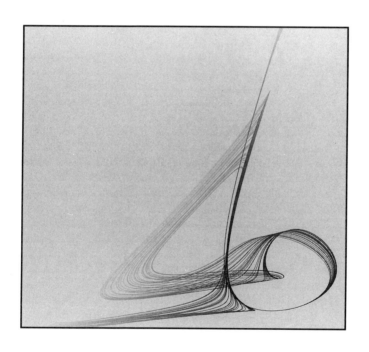

Chapter 6

Necessary and Sufficient Conditions for the Production of Linear Speed-Accuracy Trade-Offs in Aimed Hand Movements

Howard N. Zelaznik
Purdue University, West Lafayette, Indiana

One goal of any research domain is the establishment of lawful relations between independent and dependent variables. In motor behavior, the investigation of movement distance, time, and accuracy has produced such a lawful relation: As movement speed increases, there is an accompanying increase in the variability, or error, in movement distance. Woodworth (1899) examined this speed-accuracy trade-off and attributed it to the limitations in the current control phase of motor control, the phase during which feedback is used to terminate the movement at or near the target. According to Woodworth, the current control phase follows an initial impulse, a ballistic open-loop control process that drives the limb toward the vicinity of the end location.

Fitts (1954) attributed the speed-accuracy relation to noise in the neural signal for motor output. He proposed that there is a logarithmic relation between minimum movement time and the ratio of movement distance to target width. This relationship, now known as Fitts' law, states that $T = a + b [\log_2(2D/W)]$, where T denotes movement time, D and W are the movement distance (or amplitude) and target width, respectively, and a and b are empirical constants. The Fitts relation between T, D, and W has been called the *logarithmic speed-accuracy trade-off* (Meyer, Abrams, Kornblum, Wright, & Smith, 1988).

Research has indicated the robustness of the relationship Fitts proposed (see Rosenbaum, 1991; Schmidt, Zelaznik, Hawkins, Frank, & Quinn, 1979). For example, Fitts' law holds for tapping movements viewed with a microscope (Langolf, Chaffin, & Foulke, 1976), movements made in an underwater environment (Kerr, 1973), and movements of a joystick controlling a cursor on a computer screen (Kantowitz & Elvers, 1988). Furthermore, one of the leading explanations of Fitts' law has been derived from closed-loop theory (Crossman & Goodeve, 1963/1983; Keele, 1981, 1986), making explanations of Fitts' law consistent with a major line of theorizing concerned with motor control (Adams, 1971).

LINEAR SPEED-ACCURACY TRADE-OFF

Although it is true that explanations of Fitts' law are consistent with the closed-loop perspective of control, there appears to be a class of actions that are not as easily explained by closed-loop processes. Thus, Keele (1968) developed the idea of a motor program. In motor program control, a central representation of a skill produces the necessary instructions to control a movement from initiation until completion. Within the motor program framework, error can result from the motor program possessing a set of incorrect instructions, or from a set of instructions being executed incorrectly. Within the past dozen years Schmidt, Zelaznik, and Frank (1978; Schmidt et al., 1979) and Meyer, Smith, and Wright (1982)[1] theorized about speed-accuracy trade-offs when motor program control is present. These models predict that the relation between D, T, and W is *linear*, not logarithmic as found by Fitts. Before the models are described in detail, it will be necessary to describe the task that was utilized by Schmidt et al. (1979; Schmidt, Zelaznik, & Frank, 1978) that produced what is known as the *linear speed-accuracy trade-off*, as well as the task utilized by Fitts that produced the *logarithmic speed-accuracy trade-off*.

Distance-Constrained and Velocity-Constrained Tasks

In the Fitts task a subject moves a stylus (or just the hand) between two targets of width W and distance D apart, as rapidly as possible without making too many spatial errors. In the discrete version of this task, a subject moves from a home position to a target at a distance D with a target width of W, attempting to make each movement as fast as possible without missing the target. Meyer, Smith, Kornblum, Abrams, and Wright (1990) have called these tasks *distance-constrained* because the subject must complete each movement within the experimenter-defined target width, trying as well to minimize T. This distance-constrained paradigm produces Fitts' law—a logarithmic relation between minimum time, T, and the ratio of distance to target width.

Schmidt, Zelaznik, and colleagues (Schmidt et al., 1978, 1979; Zelaznik, Mone, McCabe, & Thaman, 1988; Zelaznik, Shapiro & McColsky, 1981) utilized a *velocity-constrained* task initially developed by Woodworth (1899). In this task the subject attempts to execute a movement on time, so that the measured movement time meets an experimenter-defined goal. In the velocity-constrained situation, the dependent variable is the dispersion of endpoints, called W_e, defined as the standard deviation in distance. W_e is related linearly to the ratio of D to T—the average velocity of the movement. This relation has been termed the linear speed-accuracy trade-off. Because the present chapter is concerned with the speed-accuracy trade-off in velocity-constrained movements, and in particular the aimed-hand movement, a

---

[1] I am concentrating on the model of D.E. Meyer et al. (1982) because it explicitly attempts to understand and explain the linear speed-accuracy trade-off. Their subsequent theorizing (Meyer et al., 1988; Meyer et al., 1990) takes as their starting point the linear relation between speed and accuracy to explain the logarithmic speed-accuracy trade-off in Fitts tasks. This subsequent work is not under debate in the present chapter.

brief description of the aimed-hand movement task and experimental procedure is now presented.

Aimed-Hand Movement Paradigm

In an aimed-hand movement a subject moves a penlike instrument, called a stylus, from a home position to a target a fixed distance, D, away. The target is typically a small dot, a cross, or a line. In the movements studied by Schmidt et al. (1979) and Zelaznik et al. (1981, 1988) the movement unfolds in all three dimensions. In the vertical dimension (z) the subject lifts the stylus off the plane of the tabletop to initiate the movement, while at the same time moving the stylus horizontally (x) toward the target. Movements in the lateral dimension (y) are extremely small (less than 1 cm) and have yet to be modeled. The movement ends when the stylus returns to the plane of the tabletop. Movement time, T, is defined as the interval between lift-off and touchdown of the stylus. The movement distance on a trial is defined as the x-displacement from the home position.

A subject performs a series of trials with a fixed D and T goal. During the first 25 trials or so, the subject is provided with his or her T score (called knowledge of results, KR). During the remaining 75 to 100 trials, KR is provided if and only if the subject's absolute T error is greater than 10% of the goal T. For example, if the goal T were 200 ms, KR would be provided if the absolute error were greater than 20 ms (T scores greater than 220 ms or less than 180 ms). After one set of trials, the subject is switched to a different D-T condition. Usually about four D-T conditions are performed in a 50-min session. Subjects are tested over several days until all D-T conditions are performed. The dependent variables are average T, the within-subject standard deviation in T, average D, and W_e (effective target width), defined as the within-subject standard deviation in movement endpoint.

This velocity-constrained task produces a linear relation between W_e and D/T. The two current models for linear speed-accuracy relations utilize the notion of stochastic behavior of motor program control. In the following sections I describe these two models, known as impulse-variability models.

IMPULSE-VARIABILITY THEORIES

In the velocity-constrained task, both the Schmidt et al. (1979) and the Meyer et al. (1982) theories assume that movements are not comprised of a set of discrete submovements. Under these conditions, a linear relation between speed and accuracy results, not the logarithmic one that is produced in the distance-constrained Fitts task.

Schmidt et al. (1978, 1979) and Meyer et al. (1982) have proposed that Fitts' law is not applicable to all movement control situations. Movements that do not utilize feedback-based corrections will not exhibit a logarithmic relation between speed and accuracy (Meyer et al., 1988, 1990).

According to Meyer et al. (1982, 1988) and Schmidt et al. (1979), those movements not governed by current control exhibit a linear relation between speed and accuracy.

Movements that exhibit a linear speed-accuracy relation are theorized to be governed by a motor program (Keele, 1968; Pew, 1974; Schmidt, 1975, 1988). The stochastic behavior of motor program execution, called motor output variability, or noise by Schmidt et al. (1979), produces variability in a set of movement trajectories. It is assumed that the result of these stochastic processes grows in magnitude as movement speed increases, thus producing a speed-accuracy trade-off. Assuming that the force-time functions in motor program control have a characteristic shape, these models predict a linear relation between movement speed and spatial variability.

Schmidt et al. (1979)

Impulse-variability theory is based upon the following assumptions. First, movements that are velocity constrained are produced with a sinusoidal-like force-time function (called an *impulse*). Second, this force-time function is controlled as a unit. In other words, when a subject alters the force to the muscles, all portions of the forcing function are scaled proportionally upward or downward. Third, the nature of the force-time function can be "captured" by two parameters, the duration of the impulse and its amplitude or intensity. Fourth, that a well-practiced subject attempts to produce a particular movement distance-time constraint by executing an identical force-time function over trials. Fifth, as distance and time goals change over trials, a subject scales the intensity and duration of his or her impulse in a manner similar to scaling the amplitude and period of a sine function. The amplitude and time scaling assumptions have been called force rescalability and time rescalability, respectively (Meyer et al., 1982). Sixth, variability in the spatial trajectory of these movements is the result of the variability in the force-time function (i.e., variability in the impulse).

Because the fifth assumption of force and time scaling properties is crucial to both Meyer et al. (1982) and Schmidt et al. (1979; Schmidt, Sherwood, Zelaznik, & Leikind, 1985), it will be examined next. Furthermore, because all of the reported work measures acceleration, and because acceleration is proportional to force, acceleration will be the main dependent variable to be discussed.

According to Schmidt et al. (1979), when movement time and or distance demands change, a subject modifies the entire acceleration-time profile as a unit. Thus, the acceleration profile produced by the subject should maintain its characteristic shape across changes in time or distance demands. Given this assumption of a prototypical shape, we can "capture" the qualities of the acceleration profile by discussing its "height" (peak amplitude, both positive and negative) and its "length" (the duration of either positive or negative acceleration). Schmidt et al. (1979) analyzed impulse functions this way.

The Schmidt et al. model predicts that the duration of the impulse should be proportional to movement time, and that the height of the impulse should be proportional to D and inversely proportional to T^2. Support for this model, termed impulse timing (Wallace, 1981), has been provided by Wallace and Wright (1982). They showed that EMG durations were scaled with goal T. Meyer et al. (1982) incorporated the impulse-timing model in their force-time decomposition assumption. This assumption assumes that the subject has independent control over the intensity, f, and the durations, t, of the force-time function driving the stylus.

Meyer et al. (1982)

Meyer et al. (1982) presented an impulse-variability model that is very much in the spirit of the model of Schmidt et al. (1979). They provided some more stringent assumptions about force and time scaling, in addition to postulating a weighting function that is utilized to scale the intensity of the force-time function when movement time is changed. Meyer et al. assumed that acceleration-time profiles are symmetric. The shape of the negative acceleration portion was assumed to be an inverted version of the positive acceleration portion. Furthermore, it was assumed that the acceleration profile appears to be sinusoidal. The movement starts with zero acceleration, and a period of positive acceleration commences. Halfway through the movement (in terms of distance *and* time), acceleration returns to zero (peak velocity), and negative acceleration begins, with an impulse equal and opposite to positive acceleration. The movement is over when acceleration returns to zero. Furthermore, Meyer et al. (1982) assume that a prototypical acceleration-time profile exists and can be scaled in intensity (amplitude) and time to produce a family of profiles for different distance and time constraints. Their model is tied to the shape and properties of these profiles: ''As Appendix B shows, a linear speed-accuracy trade-off will occur under our model if and only if the prototype has a certain precise shape'' (p. 462). With these assumptions it was proven mathematically that a force-time function obeying a particular equation will produce a linear relation between speed and accuracy.

Force Rescalability

The force-rescalability assumption allows the subject to move a new distance in the same T. The assumption is that the amplitude of the acceleration-time profile will be proportional to D. To produce a movement with twice the distance in the same T requires the limb to have twice as much velocity at all points in time. Thus the size of the impulse must be doubled, and this is accomplished, theoretically, by doubling the height of the force-time impulse at all values of t (see Schmidt et al., 1979; Meyer et al., 1982). The force-rescalability assumption has received empirical support (Ghez, 1979).

Time Rescalability

Time rescalability allows the impulse to retain its shape when the overall T changes. Figure 6.1 shows two idealized acceleration-time curves, each with a different T goal. As can be observed in the figure, these two kinematic profiles have the same form, although their absolute T is different.

According to the time-rescalability assumption, when the duration of a movement changes, the subject alters the acceleration-time profile proportionally, maintaining the same relative time of events in the acceleration profile. The assumption is that movements controlled by the same generalized motor program possess a spatial-temporal invariance that allows for movements to be produced at different speeds and sizes without changing the overall spatial-temporal form (Pew, 1974; Schmidt, 1975, 1988).

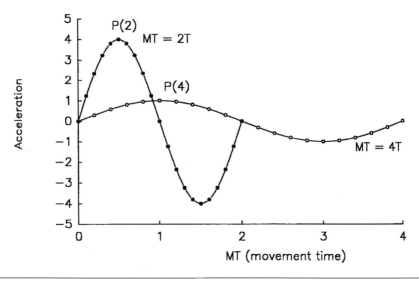

Figure 6.1   Depiction of two idealized acceleration-time profiles that exhibit the time-rescalability property. P(2) is the peak acceleration for the movement that has a movement time of 2T units, and P(4) is the peak acceleration for the 4T movement. Notice that when time rescalability is observed, the relative times of P(2) and P(4) are unchanged.

Weighting Function

The impulse, which is the area under the force-time function, must be proportional to average velocity. When movement time is halved but distance remains the same, a subject must generate twice the average velocity but has only half the time to do it. In other words, the subject must increase the amplitude of the force-time function by a factor of four, in order to double the area of the force-time function when the duration of the impulse has been reduced by a factor of two. In general, the amplitude of the force-time function is proportional to $1/T^2$ (for a more detailed discussion, see Meyer et al., 1982; Schmidt et al., 1979, 1985). This scaling of the amplitude of the force-time function when movement time goals are changed is accomplished by a weighting function, ω (Meyer et al., 1982). The weighting function thus serves to scale the amplitude to match the change in the period of motion. Without the weighting function the movement distance would not be accomplished with changes in T. Furthermore, the Meyer et al. (1982) model of the linear speed-accuracy trade-off requires a weighting function, and that it be of a particular form. Without the weighting function, there would be a distance-accuracy trade-off, not a speed-accuracy trade-off.

Variability Relationships in Both Models

Both the Schmidt et al. (1979) and the Meyer et al. (1982) models assume that the relationship between the standard deviation in the amplitude of the impulse and the amplitude of the impulse is proportional and, furthermore, that a proportional relation exists between the standard deviation in the duration of the impulse and the impulse duration. In other words, the within-subject standard deviation in each parameter is proportional to the mean. This is a very strong assumption, especially because Meyer et al. utilized this assumption to demonstrate that the Schmidt et al. model produces only a linear distance-accuracy trade-off, not a linear speed-accuracy trade-off.

The evidence in support of these assumptions is at best circumstantial. Schmidt et al. (1979) verified the force–force standard deviation relationship in isometric conditions. However, at maximum levels of force the relationship significantly deviates from a linear one (Sherwood & Schmidt, 1980). Furthermore, Carlton and Newell (this volume, chap. 2) provide evidence that time to peak force (impulse height) is a major contributor to impulse-height variability.

The relationship between the variability in timing and the duration of the impulse also is not without problems. The only data from Schmidt et al. (1979) that bear on this issue are from the reciprocal movement experiment. Subjects produced reciprocal movements between 16° and 48° in half-period times of 200, 300, 400, and 500 ms. In this experiment there was a proportional relationship between the standard deviation in timing and the duration of the movement. On the other hand, Newell, Hoshizaki, Carlton, and Halbert (1979) suggest that the timing–variability in timing relation is not proportional. Instead, small-amplitude movements (and of course, small impulse area) have relatively large timing variability. Furthermore, Michon (1967) provides evidence that the relationship between standard deviation in timing and the duration of the timed interval is a square-root one. Thus, it is not clear that the timing–variability in timing relationship is as posited by Schmidt et al. and later by Meyer et al.

Behavioral Support for the Linear Speed-Accuracy Trade-Off

Because the impulse variability models posit a relationship between speed and accuracy, I'll now present behavioral data that relate W_e to D/T. After reviewing evidence that shows a linear relationship between W_e and D/T, the kinematic data will be reviewed.

Schmidt et al. (1979)

Schmidt et al. (1979) presented 10 experiments that examined their model in great detail. The two that concern us most are those of short movement times (p. 426) and long movement times (p. 425). In the short-movement-time experiment, aimed-hand movements 10, 20, and 30 cm in distance and with T goals of 140, 170, 200 ms were performed by four subjects. Each subject showed a strong linear speed-accuracy trade-off, with the individual correlation coefficients all being higher than .85. Averaged over the four subjects, the correlation between W_e and D/T was .97. There was a slight tendency for the slope of the W_e-D/T function to decrease as T increased.

Hancock and Newell (1985) claimed that these data are better fit by a curvilinear function. Although it might be true that there is some increased variance accounted for by adding an extra parameter over the linear trade-off, the fact that the linear speed-accuracy trade-off explained 94% of the variance and that the extra variance accounted for is small and, as of yet, without theoretical explanation, causes us to accept the adequacy of the linearity of the trade-off.

In the long-movement-time experiment the values of T ranged from 200 to 500 ms. In this experiment there were clear departures from a linear speed-accuracy trade-off, in that the overall correlation between W_e and D/T was only .91. Inspection of Figure 7 in Schmidt et al. (1979) shows that one of the reasons for the smaller correlation was that as T increased, the slope of the W_e-D/T relation became smaller. Thus, the overall correlation was depressed due to the shifts in slopes of the speed-accuracy relation with T.

Wright and Meyer (1983) and Zelaznik et al. (1981, 1988)

These three papers all were concerned with understanding the conditions under which the linear speed-accuracy trade-off is observed. Wright and Meyer discussed two competing viewpoints for the linear relation. The first, *movement brevity*, posits that only when movement durations are short and peripheral feedback is not utilized to produce within-movement error corrections will the linear speed-accuracy trade-off apply. The second, *temporal precision*, posits that the linear speed-accuracy trade-off will be observed when the subject has to be ''on time.'' In other words, when there is a T goal, and the subject is encouraged to meet this goal by producing movements in which the average T is isomorphic with the goal T and the variability in T is small, a linear speed-accuracy tradeoff will be obtained. Wright and Meyer (1983) posited that only when temporal-precision demands are high will the subject choose to control the movement with a single accelerative and decelerative impulse. This strategy is a requirement of the impulse-variability models in order to predict a linear speed-accuracy trade-off. The behavioral evidence supports the temporal-precision hypothesis. Wright and Meyer (1983) observed a linear speed-accuracy trade-off for movements with durations up to 420 ms with temporal precision stressed. In fact, the strength of the linear speed-accuracy trade-off was unaffected by T.

In Zelaznik et al. (1981), subjects produced aimed-hand movements of either 200 or 500 ms in duration, at various values of D. In addition, for half of the trial blocks, subjects performed a reaction-time secondary task concurrently with the aimed-hand movement. Theoretically, the secondary task should occupy limited capacity attention. Zelaznik et al. (1981) assumed that longer duration movements (500 ms) were more attention demanding than shorter duration movements (200 ms), because the longer duration movements utilize feedback for the closed-loop control processes near the target. If attention were occupied with the secondary task, subjects would not be capable of producing within-movement error corrections. Thus, the longer duration movements should exhibit a steeper slope between W_e and D/T when subjects are performing the secondary task. However, it is assumed that limited-duration movements are controlled by more automatic, motor-programming processes and thus should not be affected by the secondary task. The steepness of the

W_e-D/T slope for 500-ms aimed-hand movements increased when the secondary task was performed concurrently with the aimed-hand movement. The slopes of the 200-ms aimed-hand movements were unaffected by the introduction of the secondary task. Thus, brief movement duration is not a necessary condition for the linear speed-accuracy trade-off.

Direct support for the temporal-precision hypothesis comes from Zelaznik et al. (1988). In these experiments, subjects produced aimed-hand movements with movement durations ranging from 150 to 450 ms. The independent variable was temporal precision, manipulated by the relative size of the temporal window. In the 10% conditions, the acceptable T values were ± 10% of the goal T. Thus, if T was supposed to be 250 ms, the acceptable temporal window was 225 to 275 ms. In the 40% conditions, the same T goal had an acceptable window of 150 to 350 ms. The idea behind these experiments was that in the 40% window conditions, the very relaxed temporal constraints would diminish the strength of the linear relation between W_e and D/T. The 40% temporal window conditions exhibited a stronger logarithmic relation between speed and accuracy than the 10% temporal window condition. Conversely, the 10% conditions exhibited a stronger linear relation between speed and accuracy than the 40% temporal window conditions. This experiment, therefore, provides direct support for the temporal-precision hypothesis.

Patla et al. (1985) and Abrams et al. (1989)

Saccadic eye movements exhibit a linear relation between movement speed and accuracy. In saccadic eye movements, T cannot be an independent variable, because saccadic velocity is not voluntarily controlled. When D increases, T also increases (Robinson, 1964). Patla, Frank, Allard, and Thomas (1985) and Abrams, Meyer, and Kornblum (1989) measured the W_e in saccadic movements at different distances. In both cases there was a linear relation between D and W_e. Furthermore, Abrams et al. observed that the trajectories (acceleration profiles) looked like those predicted by the Meyer et al. (1982) symmetric impulse-variability theory. Thus, the ocular-motor system also exhibits a linear relation between speed and accuracy.

In summary, experiments designed to test the efficacy of the symmetric impulse-variability model at the behavioral level (W_e-D/T relation) have been very successful in confirming the predictions of the Meyer et al. (1982) model. However, it is more important that the kinematic assumptions of the model be examined to determine if the trajectory patterns of these movements follow the tenets of the Meyer et al. (1982) theory.

Kinematic Evidence for Impulse-Variability Theory

The behavioral evidence supports the predictions about the speed-accuracy relationship, but the kinetic and kinematic evidence should support the assumptions of this theory as well. A theory that correctly predicts behavior but whose assumptions are not correct does not deserve exploration.

General Issues

Figure 6.2, a-c presents some typical kinematic profiles, depicting the displacement, velocity, and acceleration of the stylus in the major axis of motion. As you can see, the displacement path is ogival and smooth. The velocity profile is bell shaped with only one peak. The acceleration profile has two components—positive acceleration, which operates until the stylus reaches peak velocity, and negative acceleration, which operates until the stylus reaches zero velocity and the movement is completed. In a discussion of the relevant kinematic data and my description of a new theory, the acceleration profiles, because they are directly related to the forces driving the stylus toward the target, will be analyzed.

The most important assumption of impulse-variability theories resides in the time-rescalability assumption. Recall that this assumption has its origins in the relative-time invariance hypothesis of generalized motor program theory (see Schmidt, 1988). Aside from the aimed-hand-movement literature to be discussed later, it is clear that there are small, albeit serious, violations of the relative-time invariance hypothesis. Gentner (1987) operationalized this assumption by searching for a multiplicative rate parameter in movements of the same class, but produced at different rates.

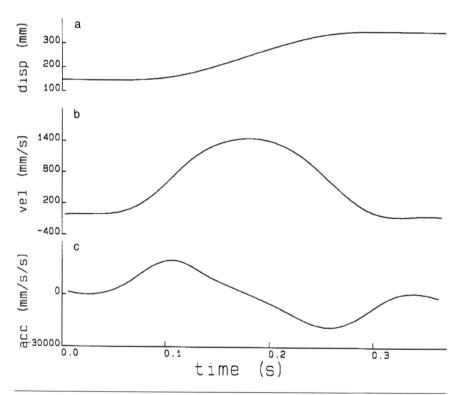

Figure 6.2   Typical kinematic profile of an aimed-hand movement. This trace is for a 25-cm movement with an instructed goal movement time of 200 ms. This movement is a discrete line drawn movement showing displacement (a), velocity (b), and acceleration (c).

Gentner showed that in almost all cases there are violations of the multiplicative rate parameter model.

The multiplicative rate parameter model also has difficulty in describing transfer effects in motor learning. Langley and Zelaznik (1984) had subjects knock down three barriers in a sequential order. Some subjects were trained to produce an overall movement duration; other subjects were trained to produce a temporal pattern of interbarrier durations. A transfer-of-training paradigm was utilized to ascertain the difficulty that subjects have in performing a movement with a different relative-timing pattern than was practiced during the acquisition trials. Subjects were able to adopt a new relative-timing pattern when given novel interbarrier durations during transfer—suggesting that time invariance is not obligatory.

Kinematic Properties of Aimed-Hand Movements

Although the deviations of relative-time invariance are at odds with the time-rescalability assumption, they are not necessarily damaging to the theory, because the above-studied movements were not examined for their linear speed-accuracy trade-off properties. Two series of studies, conducted by Gielen and Zelaznik (Gielen, Oosten, & Pull ter Gunne, 1985; Zelaznik, Schmidt, & Gielen, 1986), are more problematic for current impulse-variability theories.

Gielen et al. (1985) observed that rapid aimed-hand movements do not produce time-rescalable acceleration-time functions. They found that the time of peak positive acceleration was fixed in absolute time, and that the duration of electromyographic (EMG) activity was not scaled with the goal T. Because Gielen et al. were concerned with generalized motor program issues and not the linear speed-accuracy trade-off, they did not perform a detailed analysis of the movement trajectories as they related to impulse-variability theory.

On the other hand, Zelaznik et al. (1986) were concerned with examining the details of kinematic profiles of aimed-hand movements in order to examine impulse variability for the linear speed-accuracy trade-off. They conducted two experiments, one in which an infrared on-line kinematic recording device (SELSPOT) was utilized to gather kinematic data and one in which a pulsed-infrared light-emitting diode was photographed with a 35-mm camera to gather trajectory information. A strong linear relation was observed between W_e and D/T. However, in both experiments the time to peak positive acceleration was not related to T but instead was fixed across different values of T. In other words, the data from Zelaznik et al. did not support the time-rescalability assumption. Time rescalability is not a necessary condition for the production of a linear speed-accuracy trade-off.

In addition to the lack of time rescalability, Zelaznik et al. (1986) noted two important violations of the assumptions of previous linear speed-accuracy trade-off theories. First, the final velocity at impact was linearly related to the movement's average velocity. The data from Zelaznik et al. (1986, Experiment 2) are plotted in Figure 6.3. As can be seen in the figure, as D/T increased, the final velocity at impact increased. If the trajectory exhibited symmetrical but inverted mirror-image acceleration profiles for positive and negative acceleration, the integral of positive acceleration would equal the integral of negative acceleration. Thus, these equal

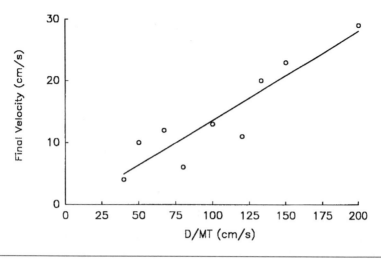

Figure 6.3   Final horizontal velocity at impact as a function of average velocity (D/MT).
Note. Data from "Kinematic Properties of Aimed Hand Movements" by H.N. Zelaznik, R.A. Schmidt, and C.C.A.M. Gielen, 1986, *Journal of Motor Behavior,* **18**, pp. 353-372.

and opposite integrals would have produced zero velocity at the end of the movement. The fact that there was positive horizontal velocity at the completion of the movement and that this velocity was linearly related to D/T suggests that the asymmetry between positive and negative acccleration grows as average velocity increases.

Another important aspect of the symmetric impulse-variability model is that the total effect on W_e is the sum of the effect on W_e at the end of positive acceleration and negative acceleration. In fact, Meyer et al. (1982) assume that both positive and negative acceleration make equal contributions to W_e. However, there is evidence that such an assumption is not correct. Schmidt et al. (1979) examined W_e at the end of a rapid-timing task in which the subject passed a target line. This type of task involves only positive acceleration (pp. 435-436) and can thus provide a good estimate of the effects of D and T on W_e at the end of positive acceleration. Only two T-goals were tested, 125 and 200 ms, and two levels of D, 16° and 32°. W_e at the end of T was independent of T but linearly related to distance. These data appear to argue against the symmetric impulse-variability model.

Zelaznik et al. (1986) computed the time of peak velocity (i.e., when negative acceleration began), determined the displacement of the stylus at that time, and computed the average displacement and its standard deviation. The standard deviation in distance at the end of positive acceleration is the effective target width at the end of positive acceleration. Table 6.1 reproduces the relevant data from both experiments. As seen in the table, W_e at the end of positive acceleration was independent of T. In other words, there was not an inverse relation between T and W_e at the end of positive acceleration.

Table 6.1 Effective Target Width at the End of Positive Acceleration (in mm) for Experiments 1 and 2 From Zelaznik et al. (1986)

Goal movement time	125	150	175	200	225	250
Experiment 1	6.7	4.7	4.5	7.8	6.5	
Experiment 2		6.3		5.6		5.9

Note. Experiment 1 utilized five movement time goals, between 125 and 225 ms; Experiment 2 utilized only three movement time goals, between 150 and 250 ms.

Data from "Kinematic Properties of Aimed Hand Movements" by H.N. Zelaznik, R.A. Schmidt, and C.C.A.M. Gielen, 1986, *Journal of Motor Behavior*, **18**, pp. 353-372.

A TRIGGERED DECELERATION TIMING (SLIDE) THEORY FOR THE LINEAR SPEED-ACCURACY TRADE-OFF

In the previous section on kinematic data related to the linear speed-accuracy trade-off, evidence was presented that the kinematic profiles of these aimed-hand movements do not exhibit time rescalability. Given that aimed-hand movements do not exhibit this property, it is important to develop a new theory to explain linear speed-accuracy trade-off behavior that is not constrained by a particular kinematic description. In this section, the theory for the aimed-hand-movement case is described. In the following section some general conditions for the production of a linear speed-accuracy trade-off are presented.

The Triggering of Deceleration—A Kinematic Description

In the limited-duration aimed-hand movements, what strategy might a subject employ to move at very high average velocities? Suppose that when average velocity requirements are high, subjects delay the onset of negative accelerative forces that slow a limb down. The subject would be able to achieve increases in average velocity with a smaller expenditure of agonist muscular force, compared to when the negative accelerative forces were equal and opposite to the positive accelerative forces and, of course, were initiated halfway through the movement. In other words, if the impulses for positive acceleration and negative acceleration, during the measured T, were not equal and opposite, but instead the positive acceleration impulse were larger, and this asymmetry grew as average velocity increased, the integral of the impulse for positive acceleration would not have to be proportional to D/T.

In general, the position presented here is that in the control of an aimed-hand movement, subjects have independent control of positive and negative accelerative forces to achieve the D and T demands of a task. Specifically, as the T goal decreases, the subject increases the relative delay in the onset of negative acceleration forces. In other words, there is a timing process that determines the onset of the negative acceleration forces, that is, a trigger. When T decreases, the relative time of the onset of the negative acceleration forces increases. The control of this trigger process for the onset of negative acceleration we term "slide," because the subject can, in effect, temporally slide the onset of negative acceleration relative to the offset of positive acceleration forces.

According to this model, an aimed-hand movement is controlled by two sets of pulses, one for positive and one for negative acceleration. The pulses comprise the summation of all muscular forces that act upon the device being moved. Each acceleration pulse is characterized by an exponential function. For positive acceleration the function has the following form: $a_p(t) = \alpha_p t \exp(-t/\tau_p)$. The parameter α_p is an amplitude-scaling parameter. The parameter τ_p determines the time of the peak positive acceleration. For negative acceleration the kinematic description has the same general form, although the acceleration equation becomes $a_n(t) = \alpha_n (\omega - t) \exp(-(\omega - t)/\tau_n)$. When ω equals t, acceleration is zero. The parameter, ω, is called the virtual movement time, because in the aimed-hand-movement situation, the measured movement time, T, occurs before the subject has returned to zero horizontal velocity. Therefore, a portion of the negative accelerative force generated by the muscles occurs after target impact. The virtual movement time is defined as the time at which the movement would have reached zero velocity given that the movement was stopped solely by passive and active muscular forces (i.e., without the tabletop imparting a stopping force). The quantity $\omega - T$ is called slide because it represents the delay of the antagonist decelerative forces. In situations in which slide is equal to zero, the movement "ends" coincident with the muscular forces' returning the limb to zero velocity and zero acceleration. The Meyer et al. (1982) model posits that slide always equals zero.

The overall acceleration pattern is determined by the summation of the positive and negative acceleration functions, so that

$$a(t) = \alpha_p t \exp(-t/\tau_p) - \alpha_n(\omega - t) \exp(-(\omega - t)/\tau_n)$$

Figure 6.4 presents some simulated acceleration-time profiles for movements of one unit of distance performed in 150 to 450 ms. These profiles look similar to those observed by Zelaznik et al. (1986) as well as those posited by Meyer et al. (1982).

The model, at this stage in its development, is grounded in a kinematic description of the aimed-hand movement. No formal dynamical models are utilized. However, because of the kinematic description utilized, the current model makes some assumptions radically different from generalized motor program theory.

In generalized motor program theory the duration of the impulse for positive and negative acceleration is scaled with T, because the central nervous system "times" the duration of the agonist and the antagonist burst of activity (Wallace, 1981). This timing is scaled proportional to T. In the present model, the duration of positive acceleration is determined by the values of α_p and τ_p. The duration of negative acceleration is determined primarily by the values of α_n, ω, and τ_n. Furthermore, the duration of positive acceleration also is determined by when the negative accelerative forces are initiated via the antagonist electromyographic activity.

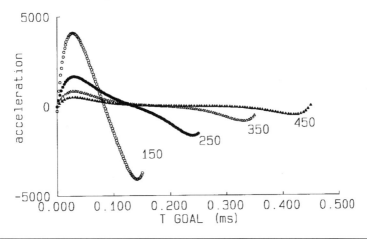

Figure 6.4   Simulated acceleration profiles for the trajectories generated by the slide model (acceleration is in arbitrary units).

From a dynamical perspective, the current model is akin to a mass-spring system with a decelerative trigger that determines movement time. The exponential functions used in the model are of the proper form for the solution of the differential equations of a mass-spring system with damping and stiffness. Thus, although the model is not based on a dynamical model, the model is consistent with one type—a mass-spring system.

Predictions of the Linear Speed-Accuracy Trade-Off

From a stochastic perspective the model is much like its predecessors. There can be variability in acceleration-time profiles that will produce variabilities in the end location of the movement. Such variability can derive from variability in the times of peak positive and negative acceleration τ_p and τ_n, respectively, and variability in setting the magnitude of peak positive and negative acceleration, α_p and α_n, respectively. There also is variability in ω. Variability in the duration of positive acceleration is the result of variability in all of the model's parameters. In the symmetric impulse-variability model of Meyer et al. (1982), variability in the time domain was the result of variability in t (duration of the impulse), whereas variability in the intensity is the result of the f (height of the impulse) parameter. The present model does not predict independence between f and t. The lack of independence between impulse height and length has been confirmed by Sherwood (1986).

How does the slide model produce linear speed-accuracy trade-off behavior? First, we must distinguish between the effects of distance versus the effects of time. The previous models of Schmidt et al. (1979) and Meyer et al. (1982) both view the distance-accuracy relation to be the result of variability in the intensity of the force-time function. The slide model does as well. However, there is a crucial difference between the former models and the currently proposed one for the inverse relation between T and W_e. This inverse relationship between T and W_e is viewed in the slide model as the result of variability in the onset of the braking force. In contrast, impulse-variability models view the effect of T on W_e as the result of the overall variability in impulse timing. The Meyer et al. (1982) model attributes the effects of $1/T$ on W_e to occur in both positive and negative acceleration processes. In the current slide model the effect of $1/T$ on W_e is the result of the variability in the process of triggering the onset of decelerative forces, and thus for positive acceleration there need not be an inverse relation between W_e and T.

Simulations of Speed-Accuracy Trade-Off Behavior

Before presenting some empirical evidence concerning the slide model, we will divert our attention to whether time-rescalable force-time functions, in general, are sufficient to produce speed-accuracy trade-off behavior. Furthermore, we examine whether the relation between T and variability in T is critical in determining the speed-accuracy relation. These simulations are important because Meyer et al. (1982) claimed that only a special function could produce linear speed-accuracy trade-off behavior, a quasi-sinusoidal one. In all of these simulations, it is assumed that the acceleration-time function can be captured by the two parameters of the original Schmidt et al. (1979) model, namely, impulse height and impulse duration. In these simulations the relationship between T and the variability in T is manipulated.

The acceleration-time function was modeled as either a square wave (as in Schmidt et al., 1979) or as a sine function ($a[t] = K \sin[\omega t]$), with ω being equal to $2\pi/T$. The square wave was at $a(t) = K$ when $0 \leq t \leq 0.5T$, and at $a(t) = -K$ when $0.5T < t \leq T$. Three levels of distance, 1, 2 and 3, and five levels of T, 0.5 to 0.1 s, in 0.1-s decrements, were used. There were 1,000 trials within each D-T condition. The expected values of K and t were chosen, and then K and t were given variability. In all cases the variability in K was proportional to K. In one simulation the standard deviation in t was proportional to t (as assumed by Schmidt et al., 1979.) In another simulation the standard deviation in t was proportional to the square root of t (Michon, 1967). The distributions of K and t were normal, and values of K and t were randomly selected from these normally distributed values.

Figure 6.5 presents the relevant results. As is clear from the figure, when distance increased there was an increase in W_e. However, the effect of T on W_e depended on whether the standard deviation of T was proportional or increased as a square root function of T. When there was a proportional relationship between T and the standard deviation of T, there was no effect of T on W_e. On the other hand, with a square-root relation, a linear speed-accuracy trade-off was produced.

These results help us understand some of the underlying causes of the linear speed-accuracy trade-off. Meyer et al. (1982) and Schmidt et al. (1979) argued that the nature of the speed-accuracy relation is a function of the type of force-time function driving the limb from one position to the other. In these simulations we demonstrated that it is not necessarily the nature of the underlying kinematics, but it also is the nature of the relation between timing and timing variability, that determines the speed-accuracy relation. In one case (the sine wave representation) the kinematic acceleration profile produced strict time rescalability, but in the other situation (the square wave simulation) there was no time rescalability. Therefore,

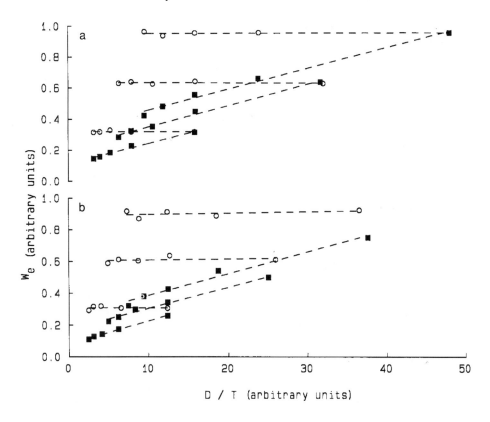

Figure 6.5   Simulated relationship for W_e and average velocity for sinusoidal (a) and square wave (b) acceleration functions. In both panels the open circle symbols represent the case in which the variability in timing is proportional to the goal T. The filled squares represent the case when the variability in timing is proportional to the square root of goal T. The dashed lines represent the best fitting linear relationship between D/T and W_e for a given movement distance. Thus, it shows the effect of 1/T on the W_e relation. The solid line represents the overall regression. As can be seen in both panels, only when the relationship between the variability in time and the goal T is a square-root one, is there an inverse relation between T and W_e.

we offer the hypothesis that for movement trajectories that can be characterized by a single function, it is the nature of the relation between T and the variability in T that determines the nature of the speed-accuracy relation.

In another series of simulations, the slope of the relation between the amplitude and the variability in amplitude of the acceleration function was manipulated. The effect of an increase in variability in acceleration amplitude was an increase in spatial variability W_e. However, there was no effect of this increased variability on the effect of T on W_e. In other words, the effect of increased variability in impulse amplitude contributed to the distance effect on the speed-accuracy relation, but not to the movement-time effect. In the two previous impulse-variability models, variability in the impulse produced W_e, regardless of whether the variability was produced by impulse height or impulse length. The present simulations suggest that the effect of distance and the effect of time on the speed-accuracy trade-off are the results of different processes. This inference runs counter to the intuitions of Schmidt, Zelaznik, and colleagues (Schmidt et al., 1979). In that paper we argued that when T changes, the amplitude and duration of the impulse changes in a particular fashion. These changes then produce changes in variability of the impulse that mediate the speed-accuracy trade-off. The fact that increased variability in impulse amplitude does not produce increased W_e at smaller values of T (in the simulated experiments) suggests that our intuitions were not correct. The new theory, presented in this chapter, suggests that the effect of T on W_e resides in some other variability process—namely, in the triggering of deceleration forces. Let's now turn our attention to how the slide model produces linear speed-accuracy trade-off relations.

Slide Model and the Linear Speed-Accuracy Trade-Off

There are two ways in which the slide model can produce a linear speed-accuracy trade-off. The first is to assume that the relation between the standard deviation of the virtual movement time and T is a square-root function of T. This "solution" can be applied to any of the previous models, such as Meyer et al. (1982) and Schmidt et al. (1979). The second way in which the slide model can produce a linear speed-accuracy relation concerns the nature of the relation between the virtual movement time and T. If the virtual movement time was not a constant proportion of the goal T, but instead increased proportionally as T decreased, then the relation between the standard deviation in t and T would approximate a square-root function. Thus, the speed-accuracy trade-off would be linear. We now turn our attention to some experimental evidence from the discrete aimed-hand-movement paradigm to examine the model.

EXPERIMENTAL EVIDENCE

Because current kinematic data concerning the aimed-hand-movement trajectories have only examined landmarks in the kinematic record, and the number of trials within any D-T condition has been small (less than 20), it was not possible to evaluate all the predictions of the slide model against the data that have been reported in the extant literature. Therefore, an experiment was conducted so that the efficacy of the slide model could be evaluated.

Six subjects performed aimed hand movements of either 5, 20, or 30 cm in distance with movement time goals of 150, 175, 200, 225, and 250 ms. The target was a 0.4-cm-wide pair of 5-cm-long parallel lines. The stylus was an art burnisher with a metal ball tip. This stylus was spring loaded. The start position was a small metal disk. When the subject broke contact with the disk, a clock started. Movement time was measured as the interval of time between this signal and the closing of a switch within the stylus upon contact with the tabletop.

Subjects performed a set of 96 trials within each D-T condition. All five T goals were performed at one D on each day. These blocks of trials were performed over two sets of three 1-hour sessions. For the second set of D-T conditions, a WATSMART system recorded the trajectory of the stylus at a sampling rate of 333 Hz. Forty trials, selected at random, were recorded. The displacement data were filtered at 10 Hz (Butterworth, low-pass digital filter in a forward and then backward direction to remove phase lags). Only trials 33 through 96 on Set 2 are reported. Furthermore, as all current models for linear speed-accuracy trade-off data are one-dimensional, we report only data for horizontal motion.

Figure 6.6 presents the speed-accuracy trade-off relation between W_e and D/T, averaged across subjects. There was a strong relation between these variables; the overall correlation coefficient was .98. Within a distance the correlations were .74, .99, and .96 for the 5-, 20-, and 30-cm movements, respectively. Of great interest is the fact that the slopes of the W_e-D/T relations were equal for the three movement distances. These slopes were .0044, .0041, and .0040 for the 5-, 20-, and 30-cm movements, respectively. Therefore a single speed-accuracy relation is capturing the description of the data.

There was no effect of T upon the time to peak positive acceleration. The average times to peak acceleration, averaged across the three distances, were 28, 28, 29, 30, and 28 ms, for the 150-, 175-, 200-, 225-, and 250-ms conditions, respectively. Distance did not produce any changes in the time of peak positive acceleration. These results are very similar to those of Zelaznik et al. (1986) in similar aimed-hand-movement experiments.

The slide model was fit to the kinematic profiles of these movements. The model fit was done on a trial-by-trial basis, not across the averaged trajectories. The program STEP-IT (developed in the late 1960s by the Department of Chemistry,

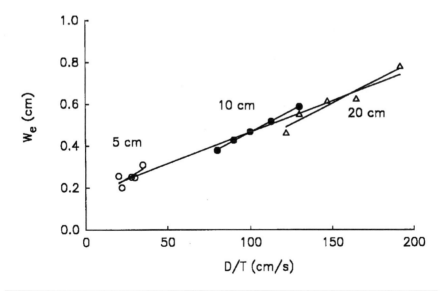

Figure 6.6 W_e as a function of average velocity for discrete aimed-hand movements.

University of Illinois at Urbana-Champaign) was modified to fit the model equation. The only constraint in the fitting procedure was that the values of τ_p or τ_n could not be negative. The STEP-IT program determined the value for each of the five parameters α_p, α_n, τ_p, τ_n, and ω, for each trial, by determining the values of the five parameters that produced a least squares fit to the acceleration profiles. The number of data points fit depended upon the movement time. For example, a trial with a movement time of 200 ms had 66 data points in the kinematic record. The fitted parameters were analyzed to determine how they behaved when movement distance and movement time goals changed. Because STEP-IT became locked into local minima of fitted parameters for the 5-cm movements, we report only the fit for the 10- and 20-cm movements.

On an individual trial basis the r^2 values were never less than .95 and typically were .98 and above. In other words, the fit of the slide model to the data was strong on an individual trial basis. The fit of the model appeared to be unaffected by D or T goals. There was a slight decrease in r^2 as T increased. Thus, we are confident that the slide model is capturing the kinematic profiles.

For the model parameters, the time to peak positive acceleration, τ_p, was about 42 ms and did not change as T changed. The time of peak negative acceleration, relative to the end of the movement, captured by τ_n, was about 40 ms. However, there was a significant increase in τ_n as T increased from 150 to 250 ms, being 32 ms for the 150-ms movements, and 50 ms for the 250-ms movements.

Figure 6.7 depicts the model's values for peak positive and negative acceleration, which are captured by the values of α_p and α_n. As can be seen in the figure, the model captured the general form of these parameters but consistently underestimated their magnitude.

The STEP-IT program also calculated the virtual movement time, ω, for the 10 conditions. The values of ω were 180, 204, 227, 254, and 275 ms for the 150- through 250-ms conditions, respectively. This effect was significant. There was no effect of D on ω. Figure 6.8 presents these values in a different fashion—as a ratio of ω to the actual time T. What is clear in the figure is that the relative size of the virtual movement time increased as the actual time, T, decreased. Note that, according to the time-rescalability assumption of the impulse-variability models, ω should be proportional to T. Of greater importance is that the relative ω increases as T decreases—this condition is necessary for the slide model to produce linear speed-accuracy trade-off behavior. I take this as strong preliminary support for the slide model.

SUMMARY AND FUTURE DIRECTIONS

The present chapter has reviewed theory and data concerning the linear speed-accuracy trade-off. It has been shown that the kinematic profiles of aimed-hand movements do not conform to the tenets of symmetric impulse-variability theory. A new model has been proposed based on the idea of triggering negative acceleration forces. This triggering process was instantiated in a model that makes distinct predictions about kinematics that are largely borne out in the available data. The model also makes predictions about electromyographic activity underlying speeded

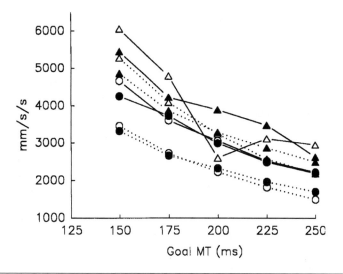

Figure 6.7 Observed and predicted values of peak positive and negative acceleration. Solid symbols are the positive acceleration values, and open symbols are negative accleration values. Circles represent 20-cm movements, and triangles represent 30-cm movements. The solid lines are the model values, and the dotted lines are the observed values.

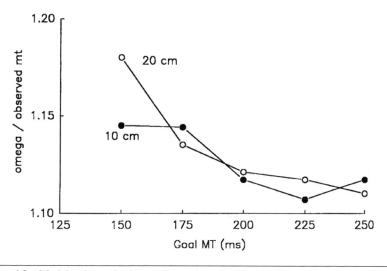

Figure 6.8 Model values of relative slide values for 10- and 20-cm discrete movements.

movements, namely: (a) Agonist muscular activity should be less affected by manipulations of distance, time, and strategy than antagonist muscular activity; and (b) the relative onset of antagonistic EMG activity should become larger as movement time becomes smaller, where relative onset of antagonistic activity is defined as the time between offset of agonist activity and onset of antagonist activity, relative to overall movement duration. To my knowledge, the latter variable has not been investigated in the studies of motor control of the arm. There is evidence that antagonist activity is modulated according to speed, inertial conditions, and instructions (Brown & Cooke, 1981; Ghez & Martin, 1982; Lestienne, 1979). Unfortunately most models of motor control based on EMG concentrate on the agonist burst of activity (see Gottlieb, Corcos, & Agarwal, 1989; Wadman, Denier Van Der Gon, Geuze, & Mol, 1979). Therefore we know very little about the timing of the onset of negative acceleration.

The lack of confirmatory data from the EMG level of analysis should not cloud the achievement of the present work. It has been shown that the linear speed-accuracy trade-off can be explained by a process of the triggering of the onset of negative acceleration forces. Future work will examine the validity of this idea.

REFERENCES

Abrams, R.A., Meyer, D.E., & Kornblum, S. (1989). Speed and accuracy of saccadic eye movements: Characteristics of impulse variability in the oculomotor system. *Journal of Experimental Psychology: Human Perception and Performance*, **15**, 529-543.

Adams, J.A. (1971). A closed-loop theory of motor learning. *Journal of Motor Behavior*, **3**, 111-150.

Brown, S.H.C., & Cooke, J.D. (1981). Amplitude- and instruction-dependent modulation of movement-related electromyogram activity in humans. *Journal of Physiology*, **316**, 97-107.

Crossman, E.R.F.W., & Goodeve, C. (1983). Feedback control of hand-movement and Fitts' law. *Quarterly Journal of Experimental Psychology*, **35A**, 251-278. (Original work presented at the meeting of the Experimental Psychology Society, Oxford, England, July 1963)

Fitts, P.M. (1954). The information capacity of the human motor system in controlling the amplitude of movement. *Journal of Experimental Psychology*, **47**, 381-391.

Gentner, D.R. (1987). Timing of skilled motor performance: Tests of the proportional duration model. *Psychological Review*, **94**, 255-276.

Ghez, C. (1979). Contributions of central progams to rapid limb movement in the cat. In H. Asanuma & V.J. Wilson (Eds.) *Integration in the nervous system* (pp. 305-320). Tokyo: Igaku-Shoin.

Ghez, C., & Martin, J.H. (1982). The control of rapid limb movement in the cat. III. Agonist-antagonist coupling. *Experimental Brain Research*, **45**, 115-125.

Gielen, C.C.A.M., Oosten, K., & van Pull ter Gunne, F. (1985). Relation between EMG activation patterns and kinematic properties of aimed arm movements. *Journal of Motor Behavior*, **17**, 421-442.

Gottlieb, G.L., Corcos, D.M., & Agarwal, G.C. (1989). Strategies for the control of voluntary movements with one mechanical degree of freedom. *Behavioral and Brain Sciences*, **12**, 189-250.

Hancock, P.A., & Newell, K.M. (1985). The movement speed-accuracy relationship in space-time. In H. Heuer, U. Kleinbeck, & H.H. Schmidt (Eds.), *Motor behavior: Programming, control, and acquisition* (pp. 153-188). Berlin: Springer-Verlag.

Kantowitz, B.H., & Elvers, G.C. (1988). Fitts' law with an isometric controller: Effects of order of control and control-display gain. *Journal of Motor Behavior*, **20**, 53-66.

Keele, S.W. (1968). Movement control in skilled motor performance. *Psychological Bulletin*, **70**, 387-403.

Keele, S.W. (1981). Behavioral analysis of movement. In V. Brooks (Ed.), *Handbook of physiology. Section I. The nervous system. Vol. 2. Motor control* (Part 2). Baltimore: American Physiological Society.

Keele, S.W. (1986). Motor control. In J.K. Boff, L. Kaufman, & J.P. Thomas (Eds.), *Handbook of human perception and performance* (Vol. 2). New York: Wiley.

Kerr, R. (1973). Movement time in an underwater environment. *Journal of Motor Behavior*, **5**, 175-178.

Langley, D.J., & Zelaznik, H.N. (1984). The acquisition of time properties associated with a sequential motor skill. *Journal of Motor Behavior*, **16**, 275-301.

Langolf, G.D., Chaffin, D.B., & Foulke, J.A. (1976). An investigation of Fitts' Law using a wide range of movement amplitudes. *Journal of Motor Behavior*, **8**, 113-128.

Lestienne, F. (1979). Effects of inertial load and velocity on the braking process of voluntary limb movements. *Experimental Brain Research*, **35**, 407-418.

Meyer, D.E., Abrams, R.A., Kornblum, S., Wright, C.E., & Smith, J.E.K. (1988). Optimality in human motor performance: Ideal control of rapid aimed movements. *Psychological Review*, **95**, 340-370.

Meyer, D.E., Smith, J.E.K., Kornblum, S., Abrams, R.A., & Wright, C.E. (1990). Speed-accuracy tradeoffs in aimed movements: Toward a theory of rapid voluntary action. In M. Jeannerod (Ed.), *Attention and performance XII* (pp. 173-226). Hillsdale, NJ: Erlbaum.

Meyer, D.E., Smith, J.E.K., & Wright, C.E. (1982). Models for the speed and accuracy of aimed movements. *Psychological Review*, **89**, 449-482.

Michon, J.A. (1967). *Timing in temporal tracking*. Soesterberg, The Netherlands: Institute for Perception-TNO.

Newell, K.M., Hoshizaki, L.E.F., Carlton, M.J., & Halbert, J.A. (1979). Movement time and velocity as determinants of movement timing accuracy. *Journal of Motor Behavior*, **11**, 49-58.

Patla, A.E., Frank, J.S., Allard, R., & Thomas, E. (1985). Speed-accuracy characteristics of saccadic eye movements. *Journal of Motor Behavior*, **17**, 411-419.

Pew, R.W. (1974). Human-perceptual motor performance. In B.H. Kantowitz (Ed.), *Tutorials in performance and cognition* (pp. 1-39). Hillsdale, NJ: Erlbaum.

Robinson, D.A. (1964). The mechanics of human saccadic eye movement. *Journal of Physiology*, **174**, 245-264.

Rosenbaum, D.A. (1991). *Human motor control.* New York: Academic Press.

Schmidt, R.A. (1975). A schema theory of discrete motor skill learning. *Psychological Review*, **82**, 225-260.

Schmidt, R.A. (1988). *Motor control and learning: A behavioral emphasis* (2nd ed.). Urbana, IL: Human Kinetics.

Schmidt, R.A., Sherwood, D., Zelaznik, H.N., & Leikind, B. (1985). Speed-accuracy tradeoffs in motor behavior: Theories of impulse variability. In H. Heuer, U. Kleinbeck, & H.H. Schmidt (Eds.), *Motor behavior: Programming, control, and acquisition* (pp. 79-123). Berlin: Springer-Verlag.

Schmidt, R.A., Zelaznik, H.N., & Frank, J.S. (1978). Sources of inaccuracy in rapid movement. In G.E. Stelmach (Ed.), *Information processing in motor control and learning* (pp. 183-203). New York: Academic Press.

Schmidt, R.A., Zelaznik, H.N., Hawkins, B., Frank, J.S., & Quinn, J.T., Jr. (1979). Motor-output variability: A theory for the accuracy of rapid motor acts. *Psychological Review*, **86**, 415-451.

Sherwood, D.E. (1986). Impulse characteristics in rapid movement: Implications for impulse-variability models. *Journal of Motor Behavior*, **18**, 188-214.

Sherwood, D.E., & Schmidt, R.A. (1980). The relationship between force and force variability in minimal and near-maximal static and dynamic contractions. *Journal of Motor Behavior*, **12**, 75-89.

Wadman, W.J., Denier Van Der Gon, J.J., Geuze, R.H., & Mol, C.R. (1979). Control of fast goal-directed arm movements. *Journal of Human Movement Studies*, **5**, 3-17.

Wallace, S.A. (1981). An impulse-timing theory for reciprocal control of muscular activity in rapid, discrete movements. *Journal of Motor Behavior*, **13**, 144-160.

Wallace, S.A., & Wright, L. (1982). Distance and movement time effects on the timing of agonist and antagonist muscles: A test of the impulse-timing theory. *Journal of Motor Behavior*, **14**, 341-352.

Woodworth, R.S. (1899). The accuracy of voluntary movement. *Psychological Review*, **3**, 1-114.

Wright, C.E., & Meyer, D.E. (1983). Conditions for a linear speed-accuracy trade-off in aimed movements. *Quarterly Journal of Experimental Psychology*, **35A**, 279-296.

Zelaznik, H.N., Mone, S., McCabe, G.P., & Thaman, C. (1988). Role of temporal and spatial precision in determining the nature of the speed-accuracy trade-off in aimed-hand movements. *Journal of Experimental Psychology: Human Perception and Performance*, **14**, 221-230.

Zelaznik, H.N., Schmidt, R.A., & Gielen, C.C.A.M. (1986). Kinematic properties of aimed hand movements. *Journal of Motor Behavior*, **18**, 353-372.

Zelaznik, H.N., Shapiro, D.C., & McColsky, D. (1981). Effects of a secondary task on the accuracy of single aiming movements. *Journal of Experimental Psychology: Human Perception and Performance*, **7**, 1007-1018.

Acknowledgments

Portions of the described research were supported by NIH grant DC00559 awarded to A. Smith (PI), C. McGilliam, and H.N. Zelaznik. Appreciation is extended to E. Fischbach and C. Talmadge for discussions that produced an earlier model that is the precursor to the model currently being proposed and to D.A. Rosenbaum and J. Slotta for help and support when the model was first devised. Thanks to D. Corcos and K. Newell for comments on an earlier draft of this chapter.

Chapter 7

Speed-Accuracy Trade-Off in Human Movements: An Optimal Control Viewpoint

Gyan C. Agarwal, Joseph B. Logsdon, and Daniel M. Corcos
University of Illinois at Chicago

Gerald L. Gottlieb
Rush Medical College, Chicago, Illinois

The spatial accuracy of rapid, aimed movements is known to be inversely proportional to the speed of movement (Fitts, 1954; Fitts & Peterson, 1964; Woodworth, 1899). Rapid, aimed movements between two points in space are involved in many physical skills and are also typical of robotic movements in pick-and-place activities. The relationship between movement speed and accuracy is a fundamental topic of interest in human motor performance. Several detailed reviews of speed-accuracy trade-offs have been published (Crossman & Goodeve, 1963/1983; Meyer, Abrams, Kornblum, Wright, & Smith, 1988; Meyer, Smith, Kornblum, Abrams, & Wright, 1988; Schmidt, Zelaznik, Hawkins, Frank, & Quinn, 1979; Woodworth, 1899). Our concern in this chapter is with rapid, discrete, aimed movements typical of the Fitts paradigm. The experimental data to be discussed later is taken from papers by Fitts (Fitts, 1954; Fitts & Peterson, 1964) and from our own studies (Corcos, Gottlieb, & Agarwal, 1989; Gottlieb, Corcos, & Agarwal, 1989a, 1989b; Gottlieb, Corcos, Agarwal, & Latash, 1990).

FITTS' LAW AND CHANNEL CAPACITY

The paradigm in Fitts's experiments consisted of two metal target plates, 6 in. long and of four different widths W (0.25, 0.5, 1.0, and 2 in.), placed at four values of center-to-center distances D (2, 4, 8, and 16 in.). There were thus a total of 16 combinations of distance and target sizes. Two metal-tipped styluses (the light stylus weighing 1 oz and the heavy stylus weighing 1 lb) were used (Fitts, 1954). Each subject was given the following instructions: "Strike these two target plates alternately. Score as many hits as you can. If you hit either of the side plates an error will be recorded. You will be given a 2-s warning before a trial. Place your hand here and start tapping as soon as you hear the buzzer. *Emphasize accuracy rather*

than speed. At the end of each trial I shall tell you if you have made any errors''
(Fitts, 1954, p. 384). Fitts defined an index of difficulty ID by Equation 7.1.

$$ID = \log_2 \frac{2D}{W} \tag{7.1}$$

$$MT = a + b \times ID = a + b\left(\log_2 \frac{2D}{W}\right) \tag{7.2}$$

where a and b are constants. The movement time plotted as a function of the index
of difficulty from Fitts's data is shown in Figure 7.1. For the regression fit, a =
3.3 ms, b = 99.74 ms, and r^2 = 0.956. Fitts observed that an index of performance
defined as the ratio of the index of difficulty and the movement time, that is,

$$IP = \frac{ID}{MT} \text{ bits/s,} \tag{7.3}$$

was nearly constant and varied between 10.3 and 11.5 bits/s.

Fitts argued that the speed-accuracy trade-off can be accounted for by assuming
that for the transfer of information the channel capacity of the motor system is
independent of the motor task. When the target width is reduced for the same target
distance, then the subject must either make more errors or slow his or her movements
if the channel capacity cannot be increased.

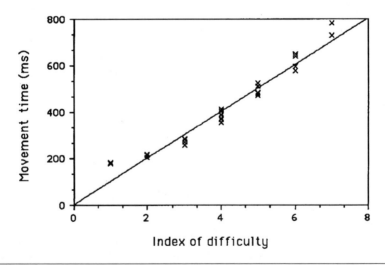

Figure 7.1 Fitts's data showing movement time as a function of the index of difficulty
for both styluses (the light stylus weighing 1 oz and the heavy stylus weighing 1 lb).
Note. Data from ''The Information Capacity of the Human Motor System in Controlling the Amplitude of Movement'' by P.M. Fitts, 1954, *Journal of Experimental Psychology*, **47**(6), p. 385.

Shannon (Shannon, 1948) showed that the capacity, C, of a channel of limited bandwidth, W, and signal power, S, and perturbed by white Gaussian noise of average power, N, is given by

$$C = W \log_2 \frac{S + N}{N} \text{ bits/s.} \qquad (7.4)$$

Fitts considered the movement distance to be equal to average signal plus noise power and the half target width to be equivalent to noise power. Although Fitts's explanation accounts for the speed-accuracy trade-off, it neglects the biomechanical aspects of the limb system and the nature of the controlling signals to the muscle. Equation 7.2 implies that an increase in MT arising from an increase in D results entirely from the consequent increase in information per symbol. Such an implication neglects the fact that even with a perfect channel, MT must increase with D because muscle power output has an upper bound.

Crossman and Goodeve (1963/1983) have shown that the minimum movement time is about 190 ms for index of difficulties less than 2. The minimum value of ID in Fitts's data was equal to 1. In reality, the variables D and W are task parameters, and the central nervous system must translate the task parameters to appropriate muscle activations. The S, N, and W of Shannon's channel capacity theorem (Equation 7.4) correspond to the signals transmitted from the motor control centers to the muscles. These are dependent on the task parameters, but they are not identical to task parameters. *This is a critical distinction in the analysis presented in this paper.*

Fitts and Peterson (1964) used discrete movements rather than the periodic movements used by Fitts (1954). The Fitts-Peterson experimental paradigm was similar to that of Fitts except that two lights were used to indicate whether the subject should move the stylus to the left plate or to the right plate from the central starting position. The degree of uncertainty was varied by varying the probability that each light would be turned on. The widths of the two target plates were allowed to vary independently of one another and could even occur on the same side of the starting point. In the Fitts-Peterson study, the reaction time, RT, increased slightly as movement amplitude increased or the target width decreased. However, the only variable that had an appreciable effect on RT was the relative probability of the two targets, and RT was linearly dependent on the entropy of the source. The MT was reliably predicted by the index of difficulty, and uncertainty about the choice of target had a much smaller effect.

ALTERNATE HYPOTHESES

Several alternative hypotheses have been proposed for the speed-accuracy trade-off. These are briefly discussed here (for a review, see Meyer, Smith, Kornblum, Abrams, & Wright, 1988).

Deterministic Iterative Corrections Model

This model was proposed by Crossman and Goodeve (1963/1983). They argued that movements intended to hit a target region quickly and accurately consisted of several discrete submovements made in rapid succession. These submovements satisfy three assumptions: (a) Each submovement travels a constant proportion (p) of the distance between its starting location and the center of the target; (b) each submovement takes the same constant amount of time (δt) regardless of the distance; and (c) submovements are guided by sensory feedback, and the submovement sequence continues until the target region has been reached.

Let x_k represent the distance to be covered after the kth submovement; then

$$x_{k+1} = (1 - p)x_k,\tag{7.5}$$

where $x_0 = D$. This discrete equation leads to

$$x_k = D(1 - p)^k.\tag{7.6}$$

The measured elapsed time t from the initiation of movement after k submovements is $t = k\ \delta t$. Thus from Equation 7.6, we get

$$x(t) = D(1 - p)^{t/\delta t},\tag{7.7}$$

and, taking the logarithms,

$$t = \delta t\ \log_{(1-p)}\left(\frac{x}{D}\right).\tag{7.8}$$

In Equation 7.8, the logarithm term is to the base $(1 - p)$. Crossman and Goodeve estimated that the submovement time $\delta t = 100$ ms and $p = 0.5$. This is a form of Fitts' law and has an information rate of 10 bits/s for $\delta t = 100$ ms.

A number of recent studies have shown that the deterministic, iterative-corrections model is seriously flawed. The submovements do not travel a constant proportion of the remaining distance, nor are their times constant (Meyer, Smith, Kornblum, Abrams, & Wright, 1988).

Linear Speed-Accuracy Trade-Off Model

Schmidt et al. (1979) had subjects make single, aimed tapping movements whose distances and durations matched specified target values. They showed that the variable errors in movements (measured in terms of the standard deviation of the movement endpoint) was linearly proportional to the average movement velocity; that is,

$$S = A + B\left(\frac{D}{MT}\right),\tag{7.9}$$

where MT is the mean movement time and S is the standard deviation of the final movement position. These observations led them to a new theory, called the *impulse variability model* (Schmidt et al., 1979), based on several assumptions: (a) The rapid movements in the time-matching task are generated by a pulse of force that has a selected amplitude and a specified time parameter; (b) the amplitude and the time parameters are stochastic variables; and (c) the variability, or noise, in the amplitude and time parameters obeys Weber's law. That is, the standard deviations are proportional to their mean values.

Although supporting evidence has been presented in several studies, questions have been raised over the logic of the formal theoretical derivations (Meyer, Smith, & Wright, 1982). Nevertheless, Meyer et al. (1982) demonstrated mathematically that a refined version of the impulse-variability model does yield Equation 7.9.

Stochastic Optimized-Submovement Model

The stochastic optimized-submovement model is an extension by Meyer, Smith, Kornblum, Abrams, and Wright (1988) of the deterministic model of Crossman and Goodeve (1963/1983). We first consider the time-minimization task, which involves either one or two submovements. The first movement is the primary submovement, and if it is unsuccessful, a secondary submovement is made, as shown in Figure 7.2.

The primary submovement is aimed at the center of the target region in a time-minimization task. If the primary submovement lands within the target region, W, then the action terminates. The noise in the motor system affects the primary submovement and may cause it to either overshoot or undershoot the target. Meyer, Abrams, Kornblum, Wright, and Smith (1988) also assumed that the effects of motor noise increase with the average velocity of the submovements. That is, if the

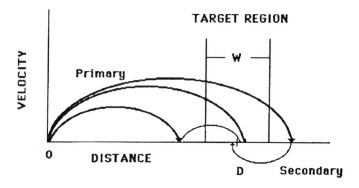

Figure 7.2   The horizontal axis represents movement distance, and the vertical axis represents movement velocity in the primary direction of travel. The heavy lines illustrate three primary submovements. If the primary submovement does not land within the target zone, a secondary submovement (thin lines) is made. The nominal desired movement distance is D, and the target width is W.

Note. From "Optimality in Human Motor Performance: Ideal Control of Rapid Aimed Movements" by D.E. Meyer, R.A. Abrams, S. Kornblum, C.E. Wright, and J.E.K. Smith, 1988, *Psychological Reviews*, **95**(3), pp. 340-370. Copyright 1988 by the American Psychological Association. Adapted by permission.

primary submovement travels a mean distance D_1 in mean time T_1, then the standard deviation S_1 of the endpoint is given by

$$S_1 = KV_1 = K\left(\frac{D_1}{T_1}\right). \tag{7.10}$$

Similarly for the secondary submovement:

$$S_2 = KV_2 = K\left(\frac{D_2}{T_2}\right). \tag{7.11}$$

It is assumed that the average velocities of the primary and secondary submovements are programmed to minimize the average total movement duration $T = T_1 + T_2$. Meyer, Abrams, Kornblum, Wright, and Smith (1988) found that the minimum time T is given by the equation

$$T = A + B \sqrt{\frac{D}{W}}, \tag{7.12}$$

where A and B are nonnegative constants. The primary movement time is given by

$$T_1 = A_1 + B_1 \sqrt{\frac{D}{W}}, \tag{7.13}$$

where $0 < A_1 < A$, $0 < B_1 < B$. That is, the mean duration of the primary (and similarly for the secondary) submovements, not just average total movement times, conforms to a square-root approximation of the ratio D/W. Meyer, Smith, Kornblum, Abrams, and Wright (1988) extended this model to include cases where n submovements are required. They also predicted that the total minimum movement time T will generally be less when n is large than when n is small. Increasing the maximum number of submovements enhances the subject's ability to overcome the effects of the noise (Meyer, Smith, Kornblum, Abrams, & Wright, 1988). Although increasingly complex models have been suggested (Meyer, Abrams, Kornblum, Wright, & Smith, 1988), they are all kinematic. None consider the dynamic aspects of limb mechanics or of muscle force generation.

OPTIMAL CONTROL MODELS

Some recent work on rapid limb movements has taken an optimal control theory approach to minimizing various performance indices (Flash & Hogan, 1985; Hasan, 1986; Hogan, 1984, 1988; Nelson, 1983; Stein, Oguztoreli, & Capaday, 1986). To study the interrelationship between the system dynamics, the physical constraints, and the assumed performance objectives, Nelson (1983) modeled the human limb movement with one degree of freedom. The displacement x of a mass m with instantaneous velocity v is governed by the equations of motion given by

$$\frac{dx}{dt} = v \tag{7.14}$$

$$m\frac{dv}{dt} = f_a(t) - \beta v,$$

where β is the viscous force coefficient, the dissipating force is assumed to be a linear function of the velocity v, and $f_a(t)$ is the net muscle force applied along the direction of movement. To normalize these equations in terms of applied forces per unit mass, define

$$u(t) = \frac{f_a(t)}{m} \tag{7.15}$$

$$b = \beta/m.$$

Thus the equations of motion in normalized form may be written as

$$\frac{dx}{dt} = v \tag{7.16}$$

$$\frac{dv}{dt} = u(t) - bv.$$

The force $u(t)$ is assumed to be bounded by a maximum value U; that is,

$$|u(t)| \leq U. \tag{7.17}$$

The boundary conditions on Equation 7.16 are assumed to be

$$x(0) = 0, \qquad x(T) = D \tag{7.18}$$

$$v(0) = 0, \qquad v(T) = 0$$

Because these equations are time invariant, they may be applied to each segment of the trajectory.

Nelson (1983) considered five performance objectives to derive the controlling input $u(t)$ and the system trajectories. His objective was to find the control $u(t)$ that would minimize one of the cost functions defined by Equations 7.19 to 7.23.

$$\text{Minimum time:} \quad T = \text{movement time} \tag{7.19}$$

$$\text{Minimum force:} \quad A = \max_{t \in (0,T)} |u(t)| \tag{7.20}$$

$$\text{Minimum impulse:} \quad I = \frac{1}{2}\int_0^T |u(t)| \, dt \tag{7.21}$$

$$\text{Minimum energy:} \quad E = \frac{1}{2}\int_0^T u(t)^2 dt \tag{7.22}$$

$$\text{Minimum jerk:} \quad J = \frac{1}{2}\int_0^T \left(\frac{d^3x(t)}{dt^3}\right)^2 dt \qquad (7.23)$$

The minimum jerk cost function of Equation 7.23 was proposed by Hogan (1984). For a double integrator plant with bounded input, the optimization problem for various performance indices has been examined in several studies (Fuller, 1985).

For the elbow joint movements, the moment of inertia of the forearm about the elbow axis has been estimated to be 0.06 kg m² (Amis, Dowson, & Wright, 1980a, 1980b). The maximum isometric torque for an average subject in flexion and extension direction is a function of the elbow angle, as shown in Figure 7.3.

The second-order regression curves plotted in Figure 7.3 are given by

$$T_{flex} = 69.1 - 24.44\theta - 22.5\theta^2 \qquad r^2 = 0.993$$

$$T_{ext} = -50.599 + 3.348\theta + 9.035\theta^2 \qquad r^2 = 0.925$$

where θ is the elbow joint angle (flexion angles are positive). The maximum isometric torques in flexion and extension directions over the range of angles used in our studies ($\pm 45°$) from Figure 7.3 are about 60 Newton meters (N.m.) and -46 N.m.,

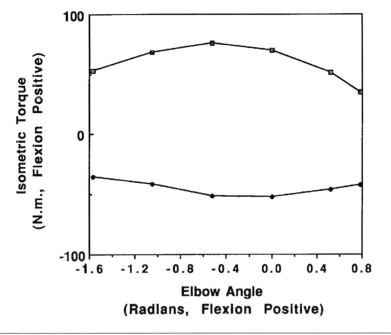

Figure 7.3   Maximum isometric torque in flexion and extension directions for an average subject as a function of the elbow angle. The range of angles is from full extension ($-90°$) to 45° flexion. The angle axis is shown in radians.

Note. From "Analysis of Elbow Forces Due to High Speed Forearm Movements" by A.A. Amis, D. Dowson, and V. Wright, 1980, *Journal of Biomechanics*, **13**, pp. 828-829. Copyright 1980 by Pergamon Press Ltd. Adapted by permission.

respectively. Thus, the normalized torque variable U (torque per unit inertia) for flexion and extension is about 1,000 and 767 rad/s², respectively.

If the damping parameter, b, is neglected, the acceleration corresponds to the input force per unit mass, u(t). The minimization of the movement time, T, for a given D and bounds on U, or maximization of the movement distance, D, for a given T and U, can all be determined geometrically. The minimum-time solution is the classical bang-bang solution with inputs $u(t) = U$ for $0 \le t \le T/2$ and $u(t) = -U$ for $T/2 \le t \le T$. The minimum time for this solution is given by

$$T_{min} = 2\sqrt{\frac{D}{U}}, \tag{7.24}$$

and the maximum distance is given by

$$D_{max} = \frac{UT^2}{4}.$$

For $U = 1,000$ (and assuming symmetrical flexion and extension torques) and $D = 60°$ (1.047 rad), minimum movement time from Equation 7.24 is equal to 65 ms. This number is clearly far from the minimum movement-time values of about 200 ms.

For parameter b nonzero, the relationships for T, D, and U are considerably more complicated (Nelson, 1983). The optimal solution is still bang-bang type with one switching for a second-order system, and the minimum time is given by

$$T_{min} = \frac{bD}{U} + \frac{2}{b}\log_e \left\{ 1 + \left[1 - \exp\left(\frac{-b^2D}{U}\right) \right]^{1/2} \right\}. \tag{7.25}$$

The switching time, s, at which u(t) changes sign is given by

$$s = \frac{1}{2}\left(\frac{bD}{U} + T_{min}\right). \tag{7.26}$$

Note that the input switching does not occur at the midpoint, which would lead to asymmetric velocity and acceleration profiles.

The minimum energy solution without any input magnitude constraint, that is, neglecting Equation 7.17, is obtained using Euler-Lagrange equations. For $b = 0$ and a fixed terminal time T, the position x(t), velocity v(t), and acceleration a(t) for an optimal trajectory are given by

$$x(t) = 3D\left(\frac{t}{T}\right)^2 - 2D\left(\frac{t}{T}\right)^3,$$

$$v(t) = \frac{6Dt}{T^2} - \frac{6Dt^2}{T^3}, \tag{7.27}$$

$$a(t) = \frac{6D}{T^2} - \frac{12Dt}{T^3}.$$

The minimum jerk solution without any input magnitude constraint, that is, neglecting Equation 7.17, is obtained using Euler-Poisson equations (Elsgolc, 1962, pp. 42-45; Hogan, 1984). For a fixed terminal time T, the position x(t), velocity v(t) and acceleration a(t) for an optimal trajectory are given by

$$x(t) = 10D\left(\frac{t}{T}\right)^3 - 15D\left(\frac{t}{T}\right)^4 + 6D\left(\frac{t}{T}\right)^5,$$

$$v(t) = 30\frac{Dt^2}{T^3} - 60\frac{Dt^3}{T^4} + 30\frac{Dt^4}{T^5}, \qquad (7.28)$$

$$a(t) = 60\frac{Dt}{T^3} - 180\frac{Dt^2}{T^4} + 120\frac{Dt^3}{T^5}.$$

Equation 7.28 has a symmetrical velocity profile, and the peak velocity occurs at the midpoint of the trajectory. The acceleration peaks occur at 0.2113 T and at 0.7887 T, one being the maximum acceleration and other being the minimum acceleration or maximum deceleration. In magnitude both are equal, and the acceleration profile is symmetrical with respect to the midpoint.

An alternative model for trajectory optimization has been considered by Hasan (1986). In this model the optimal control solution of the limb movement trajectory is obtained using Feldman's set-point concept (Adamovitch & Feldman, 1984; Feldman, 1966) and regulation of muscle stiffness (Houk, 1979). The problem is considered in two steps. In the first step, the optimal trajectory of movement is determined for a specified value of stiffness parameter. In the second step, the value of the criterion function is minimized with respect to the stiffness parameter. The mechanical properties of the joint are defined by the equations

$$\tau = \sigma(x - s) \qquad (7.29)$$

$$I\frac{d^2x}{dt^2} + \tau = 0,$$

where τ is the net torque about the joint, σ is a stiffness parameter, x is the joint angle, s represents the equilibrium position of the joint, and I is the moment of inertia of the limb segment with respect to the axis of rotation. The damping term has been neglected in this model.

Hasan argued that enhanced stiffness should be associated with greater effort, and therefore the criterion function was considered to be proportional to s. A change in the equilibrium position causes the system to go from one position to another. The criterion function was assumed to be dependent on the rate of change of the set point. The criterion chosen was

$$J = \int_0^T \sigma\left(\frac{ds}{dt}\right)^2 dt, \qquad (7.30)$$

where T is a specified movement duration. From Equation 7.29,

$$s = x + \left(\frac{I}{\sigma}\right)\ddot{x} = x + \frac{\ddot{x}}{k^2}, \qquad (7.31)$$

where $k = \sqrt{\frac{\sigma}{I}}$. Therefore, the optimization problem reduces to

$$J = \int_0^T \sigma(\dot{x} + \ddot{x}/k^2)^2 dt. \tag{7.32}$$

Note that the integrand term is a sum of velocity and jerk. This optimization problem is solved using a Euler-Poisson equation with the boundary conditions that velocity and accelerations be zero at $t = 0, T$, and the displacement be equal to D. Hasan showed that the trajectory $x(t)$ is a function of sine and cosine terms, and J may be written as

$$\frac{J}{(D^2I/T^3)} = \Psi\left(\frac{\sigma}{I/T^2}\right). \tag{7.33}$$

The Ψ function is calculated for different values of the parameter σ to find a minimum solution. Hasan found that the best stiffness value is given by

$$\sigma = 22.54\frac{I}{T^2}, \tag{7.34}$$

and the minimum value for J is

$$J = 28.78\frac{D^2I}{T^3}. \tag{7.35}$$

Thus J depends on the movement magnitude D, its duration T, and the moment of inertia I.

The peak velocity and peak acceleration for various optimal control models for given values of movement distance D and the movement time T compare reasonably well (see Table 7.1).

The ratio of peak to average velocity in elbow movements has been calculated in several experimental studies. Ostry, Cooke, and Munhall (1987) have reported a velocity ratio range (rad/s) of 1.45 to 1.48 in fast continuous flexion movements and a range of 1.85 to 2.12 in fast discrete flexion movements. (For extensions these ratios ranged from 1.49 to 1.55 in continuous, and 1.87 to 1.94 in discrete, movements.) All these numbers are within the theoretical range for various optimal

Table 7.1 Comparison of Peak Velocity and Peak Acceleration for Various Optimal Control Models

Criterion type	Peak velocity/(D/T)	Peak acceleration/(D/T²)
Minimum torque (Nelson)	2	4
Minimum energy (Nelson)	1.5	6
Minimum jerk (Hogan)	1.875	5.774
Minimum stiffness (Hasan)	1.97	6.12

criteria. Karst and Hasan (1987) have reported peak velocities of elbow flexion up to 6.9 rad/s (= 400°/s). Gottlieb et al. (1989a) reported a similar range of peak velocity in flexion movements and acceleration peaks to 157 rad/s^2 (= 9,000°/s^2). From Gottlieb et al. (1989, Figure 1 on p. 343), an estimated jerk value is 3,140 rad/s^3 (= 180,000°/s^3). Because peak-jerk-to-peak-velocity ratio is of the order of 500, Equation 7.32 of the Hasan minimum stiffness criterion is nearly the minimum jerk criterion of Hogan in Equation 7.33.

MODEL PREDICTIONS
AND EXPERIMENTAL OBSERVATIONS

The Nelson model is a classic control problem with a bang-bang solution for minimum-time movements (Bellman, Glicksberg, & Gross, 1959; Sage & White, 1977). This can be interpreted as biphasic muscular activation. That is, a pulse of agonist muscle force accelerates the limb toward the target, and at the midpoint of the trajectory the antagonist muscle force decelerates the movement at the desired target distance. The concept of bang-bang control of muscular activation based on optimal control theory has been suggested in several studies (Bahill, Clark, & Stark, 1975; Fitzhugh, 1977; Hannaford & Stark, 1985; Smith, 1962; Zangemeister, Lehman, & Stark, 1981). Whereas the time-optimal control theory predicts biphasic activation, Hannaford and Stark (1985), based on their simulation study, made the following observations: "In fast (time-optimal) movements about many joint systems, the triphasic EMG pattern has been observed. Although the first agonist burst obviously initiates the movement, the roles of the second and third bursts, appearing in the antagonist and agonist respectively, have been less clear" (p. 619). They defined the first agonist pulse, PA, as the action pulse; the antagonist pulse, PB, as the braking pulse; and the second agonist pulse, PC, as the clamping pulse.

Single-joint, discrete movements of a limb, such as the forearm rotating in the horizontal plane about the elbow, possess a surprisingly rich diversity of behaviors. How are such movements organized and performed? Because the muscular forces cannot be directly measured in human subjects, most studies have compared electromyographic signals under various experimental conditions. The optimal control models assume that u(t) represents force, whereas in most of the papers cited above the input signal is assumed to be a rectangular approximation of the EMG pulses recorded. There is no one-to-one correspondence between the instantaneous EMG and muscle force in dynamic contractions. Although there are no general models currently available to express this relationship (Agarwal & Gottlieb, 1982), it is clear that the muscle force cannot be turned on or off in discrete pulse form. There is at least a first-order dynamics between the muscle EMG and the force produced in rapid contractions (Gottlieb & Agarwal, 1971).

Another difficulty with Nelson's model and Hogan's minimum jerk optimality is the experimental observation that the velocity and acceleration profiles are not symmetrical. This can be seen in Hogan's (1984) original data as well as in several other studies (Gottlieb et al., 1989a; Nagasaki, 1989). Nagasaki (1989) has noted: "Jerk cost consumed by the movements with intermediate speed approximately satisfied minimum-cost criterion predicted by the model but was higher than the criterion for slow and ballistic movements. The results suggested that optimality criteria other than jerk cost also should be considered to predict movement profiles over the entire range of speeds'' (p. 319). Nagasaki suggested using minimum jerk optimization with constraints on jerk magnitudes.

One significant shortcoming of all the optimal control models is their inability to predict the speed-accuracy trade-off, because of their deterministic nature. As mentioned earlier in the review of the Fitts paradigm and related models, the stochastic nature of the input signal must be considered to account for the speed-accuracy trade-off.

NUMBER OF SWITCHINGS
IN THE CONTROL SIGNAL

There are disagreements in the literature concerning whether the EMG signals are biphasic or triphasic (i.e., one or two switchings in the control signal) and also concerning the EMG magnitudes and pulse widths for the agonist and antagonist muscles and their relationship to kinematic data such as movement distance. Some investigators report constant EMG activation duration for different distances; others find that it varies with distance. Some studies show triphasic patterns of EMG; other studies do not (Gottlieb et al., 1989a, 1989b, 1990).

Various studies have provided conflicting results for antagonist scaling. For elbow movements, Brown and Cooke (1981) reported that as distance increases, antagonist EMG duration increases but amplitude remains constant. Mustard and Lee, however, found that for wrist movements, the antagonist muscle EMG increases in amplitude, is of constant duration, but is delayed in time. Hallett and Marsden (1979) found no relationship between antagonist EMG and either distance or peak velocity for thumb movements. However, Marsden, Obeso, and Rothwell (1983) have reported that the antagonist burst is larger and earlier for shorter movements of both elbow and thumb.

Although the literature has numerous conflicting reports, there is little doubt that the optimal control theory as applied to limb movements using the Nelson model and a fair amount of experimental evidence are not in complete agreement. This is neither expected nor particularly surprising. The biomechanics of the limb system are nonlinear, and such a simple model can go only so far. We will reexamine the biphasic versus triphasic EMG problem using a modified Nelson model and the time-optimal control theory problem with different boundary conditions.

A reduced Nelson model (with b = 0 in Equation 7.16) is given by

$$\frac{dx}{dt} = v(t) \tag{7.36}$$

$$\frac{dv}{dt} = u(t).$$

Because u(t) in this equation represents force, and the rate of change of force that a muscle can produce is bounded, there will be no discontinuities in force. We will assume that

$$\left| \frac{du}{dt} \right| = |\dot{u}| \leq U. \tag{7.37}$$

The continuity constraint on u(t) can be incorporated in this model by defining a new state variable f(t) for force and a new control input u(t), which is the neural input to the muscle (in appropriate units, such as N/s), which is bounded. With these modifications, our system model is given by

$$\frac{dx(t)}{dt} = v(t)$$

$$\frac{dv(t)}{dt} = f(t) \tag{7.38}$$

$$\frac{df(t)}{dt} = u(t)$$

$$|u(t)| \leq U.$$

Although the relationship between the input to the muscle u(t) and the force output of the muscle f(t) is unknown and is likely to be nonlinear, this equation represents the simplest possible situation. The minimum-time problem may now be solved using the maximum principle (Sage & White, 1977). The cost function as given in Equation 7.19 may be rewritten as

$$J = T \int_0^T dt. \tag{7.39}$$

The Hamiltonian is given by

$$H = 1 + \lambda_1 v + \lambda_2 f + \lambda_3 u. \tag{7.40}$$

The minimization is carried out with respect to u(t). Thus,

$$u(t) = -U, \quad \text{for } \lambda_3(t) \geq 0 \tag{7.41}$$

$$u(t) = U, \quad \text{for } \lambda_3(t) \leq 0.$$

In Appendix A of this chapter, we show that the number of switchings n_s is given by the equation

$$n_s = n - 1 - n', \tag{7.42}$$

where n is the number of state variables, $n' = 2n - q$, and q is the number of constraints on the initial and final conditions, and the terminal time T is assumed to be unknown for the minimum-time problem.

The initial and final conditions on the variables of Equation 7.38 are typically given by

$$x(0) = 0, \qquad x(T) = D$$
$$v(0) = 0, \qquad v(T) = 0 \qquad\qquad (7.43)$$
$$f(0) = 0, \qquad f(T) = 0$$

The subject starts and stops a movement of distance D with zero velocity and force. This is the paradigm frequently used in experimental studies. Under these conditions, from Equation 7.42, $n = 3$, $q = 6$, $n' = 0$, and $n_s = 2$; that is, there will be two switchings in the control signal. The movement is initiated by the activation of the agonist, switching to antagonist, and finally agonist again, which leads to the triphasic response pattern seen in many studies.

Fuller (1973a, 1973b) has given a general solution of this time-optimal problem with a triple-integrator plant as well as plants with three real distinct eigenvalues (see also Ryan, 1977, for higher order time-optimal solutions). For the model in Equation 7.38, the total minimum time is given by

$$T = \left(\frac{32|D|}{U}\right)^{1/3} = 3.175 \left(\frac{|D|}{U}\right)^{1/3}. \qquad (7.44)$$

The times for the first and the second switchings are given by

$$t_1 = \left(\frac{|D|}{2U}\right)^{1/3} = 0.7936 \left(\frac{|D|}{U}\right)^{1/3} = \frac{T}{4} \qquad (7.45)$$

$$t_2 = 3t_1 = 3 \left(\frac{|D|}{2U}\right)^{1/3} = 2.381 \left(\frac{|D|}{U}\right)^{1/3} = \frac{3T}{4}$$

Although the mathematical relationships for switching surfaces for third- or higher order models can be calculated, physical implementation of switching surfaces is often difficult. Simplified controllers, where switching decision is based on the sign of a simple function of state variables, have been suggested (Billigsley & Coales, 1968; Fuller, 1971). Although such controllers give suboptimal performance, they are easy to implement and nearly optimal. This may very well be true of biological systems where the environmental conditions (internal as well as external) are continually changing. The noise contamination in neural signals is always present in biological systems. The motor control system also has nonlinearities as well conduction time delays. Fuller (1971) defined suboptimal controllers using a modified coordinate system defined by

$$z_1 = |Ux_1|^{1/3} \operatorname{sgn} x_1$$

$$z_2 = |Ux_2|^{1/2} \operatorname{sgn} x_2 \qquad (7.46)$$

$$z_3 = Ux_3,$$

where sgn x is defined as signum function, sgn $x = 1$ if $x > 0$, sgn $x = -1$ if $x < 0$, sgn x not defined for $x = 0$.

In a simplified strategy called constant ratio trajectory, the time interval between two successive switches is a constant, ρ, times the previous switching interval. If T is the duration of a P-trajectory (portion of a trajectory with positive input, $u = U$), then the settling time is given by

$$T_s = T + \rho T + \rho^2 T + \ldots = (1 - \rho)^{-1} T. \tag{7.47}$$

Although the total settling time is finite, an infinite number of switchings are required. This type of control is an alternative to that proposed by Meyer, Smith, Kornblum, Abrams, and Wright (1988), where the distance traveled during each phase is assumed to be a constant ratio of the previous distance rather than the time interval of the previous phase.

An alternative scheme uses a singular sliding motion (Fuller, 1971), where the state space is reduced to a second-order system in polar coordinates, the system remains stationary at a singular point, while the state point in z-space slides along the ray through the singular point. In such a scheme, Fuller has shown that the plant input is constant:

$$u(t) = \frac{dz_3}{dt} = \sigma; \tag{7.48}$$

and in z-space all z_i change at constant rates. With this strategy, Fuller (1973a) has calculated the total movement time to be

$$T = 2\left[\frac{3(1 + \sqrt{2})|D|}{U}\right]^{1/3} = 3.87\left(\frac{|D|}{U}\right)^{1/3}, \tag{7.49}$$

which is 1.219 times the optimal settling time given in Equation 7.44. The first and the second switching times are given by

$$t_1 = 0.8014\left(\frac{|D|}{U}\right)^{1/3} = 0.207 \, T, \tag{7.50}$$

$$t_2 = 2.4324\left(\frac{|D|}{U}\right)^{1/3} = 0.629 \, T.$$

These values compare quite favorably with the optimal values given in Equation 7.45. Other special cases have been analyzed that provide near minimum time optimal solutions for higher order systems.

In the Fitts-Peterson experiments, if the coefficient of friction between the target plate and the stylus is large, such that sliding does not occur, there would be no requirements on values of terminal velocity and force. From Equation 7.42 ($q = 4$, $n' = 2$), the number of switchings will be zero. This would be the case when the movement is stopped against a mechanical barrier that the subject is told to strike in minimum time. If the subject wants to prevent recoil at the target and prevent sliding motion, then the terminal force and the terminal velocity must be brought to zero values.

In Fitts's experiments, the subject did not attempt to minimize the time to reach the target in a single movement. The movements were reciprocal between two target plates. For example, suppose the subject had just reached the right target plate in the minimum time possible, then to return to the left target plate he would have to reverse the direction of the force. He could return to the other target in less time if he allowed the final force to be a little less than it would be if he had minimized the previous movement time. Thus the average time for movement would be minimized for a periodic movement for which

$$x(0) = 0, \qquad x(T) = D$$
$$v(0) = 0, \qquad v(T) = 0 \qquad\qquad (7.51)$$
$$f(0) = -f(T)$$

From Equation 7.42 (q = 5, n' = 1), this set of conditions yields one switching time.

The solution of Equation 7.38 and the analysis of minimum time depends on the boundary conditions and the number of switching times. In our discussion of our proposed model, we will do the initial analysis of Fitts's experiment assuming zero switchings and then consider the other problems with one or two switching times in control.

NATURE OF THE EXCITATION SIGNAL

The variable f(t) in Equation 7.38 (which is same as u(t) in Equation 7.16) is the force applied per unit mass. It is not the descending signal from the higher centers. The exact nature of the controlled variable and the applied inputs to the peripheral apparatus (inputs to alpha and gamma motoneurons) are not well known (Stein, 1982). Gottlieb et al. (1989b) defined an excitation pulse as the net descending presynaptic input, excitatory and inhibitory, that converges and summates in the alpha motoneuron pool. The output of the motoneuron pool is a composite train of action potentials of different frequencies in a variable number of neurons (Adrian & Bronk, 1929). Recruitment and frequency modulation of the motor units are the well-established mechanisms for control of muscle force. Transmission across the neuromuscular junction gives rise to action potentials on muscle fibers and intra-cellular calcium release. These physically distinct and separate processes produce the electromyographic activity and the muscle tension. The exact nature of the relationship between the neural excitation pulse, u(t), and the tension response per unit mass, f(t), is nonlinear with probable hysteresis and is a function of the length and rate of change of length of muscle and prior activation level (history-dependent activation) of the muscle. Hill's equation, an example of the classical lumped models, is a relation between muscle force and velocity of contraction in steady-state conditions. This equation is not directly applicable to a muscle with changing length. Other models based on cross-bridge dynamics, such as models by Wood (1981) or Zahalak (1981), are much too complicated for incorporation into an optimal control system analysis.

In our discussion, we will assume the excitation pulse to be a neural signal applied to motoneuron pools, as proposed by Gottlieb et al. (1989a, 1989b). Following those arguments, we will also assume that there is a minimum of first-order filtering (low-pass filter) between the excitation pulse u(t) and force input f(t) to the model in Equation 7.38. Thus, as an initial approximation, we can model this relationship by a first-order linear differential equation:

$$\frac{du(t)}{dt} + \frac{1}{a}u(t) = \eta(t). \qquad (7.52)$$

Gottlieb et al. (1989b) have proposed that single mechanical degree-of-freedom movements are controlled by one of two strategies: a speed-insensitive strategy or a speed-sensitive strategy. The term *strategy* here implies a set of rules that specify, in terms of task variables and subject instructions, how to compute each excitation parameter. This controlling signal, the excitation pulse, u(t), at the alpha-motoneuron level, is modeled as a rectangular pulse in which modulation occurs in either pulse amplitude or pulse width. This is similar to the concept that Schmidt et al. (1979) had proposed for their impulse-timing model. The controlling signal will be taken to be positive for the agonist and negative for the antagonist. The two strategies differ in that the speed-insensitive strategy is a result of duration modulation of the excitation pulse, whereas the speed-sensitive strategy results from modulation of height. In the next section, we will further reduce Equation 7.52 to be a pure integrator, that is, delete the parameter *a* term, which significantly simplifies the mathematics.

PROPOSED MODEL

Considering second-order dynamics of the limb and first-order dynamics of the muscle process, the system model is given by

$$\frac{dx(t)}{dt} = v(t),$$

$$\frac{dv(t)}{dt} = f(t), \qquad (7.53)$$

$$\frac{df(t)}{dt} = u(t),$$

where u(t) is the excitation pulse. In developing a model of the excitation pulse, we assume that the composite nerve is a set of parallel axons that are being activated from the same source at the same pulse rate, R. To account for differing sizes of the motor units, we assume N active motor units with different force amplitudes A_i, i = 1, 2, . . . , N. We will further assume that the net force produced, f(t), depends on the pulse rate R and the unit amplitudes A_i in such a manner that it can be expressed by a product function. Thus the last expression in Equation 7.53 may be written as

$$\frac{df(t)}{dt} = g(R)h(A_1, A_2, \ldots, A_N) = g(R)h(\underline{A}). \qquad (7.54)$$

It is important to note that $h(\underline{A})$ pertains to all active motor units, which may include both agonist as well as antagonist if they are both active simultaneously. This model is, for mathematical ease, a simplified version of a much more complex physiological system.

We assume that when attempting to move a load in minimum time, the subject adjusts the activation parameters R and N, which define the magnitude of the activation pulse based on the size principle (Henneman, 1979), such that the right-hand side of Equation 7.54 attains a maximum value subject to the condition that the subject strikes the target at least a certain fraction of the time. We will define this maximum value as U_e:

$$U_e = g(R)h(A_1, A_2, \ldots, A_N) = g(R)h(\underline{A}). \qquad (7.55)$$

The subscript e denotes that different error constraints will allow different values of the upper bound on U.

To account for the variability in activation of motor units and recruitment of different motor units, we assume that the amplitudes A_i have a random variation within similar-size motor units. Let this variation be defined as $n_i(t)$, $i = 1, 2, \ldots, N$. We assume that the magnitude of the random component is much smaller than the value of A_i. To account for the fluctuations in motor unit recruitments, the activation Equation 7.54 may be written as

$$\frac{df(t)}{dt} = g(R)h[A_1 + n_1(t), A_2 + n_2(t), \ldots, A_N + n_N(t)]. \qquad (7.56)$$

Expanding the function $h(..)$ in terms of Taylor series and retaining only the first-order terms, we get

$$\frac{df(t)}{dt} = g(R)[h(A_1, A_2, \ldots, A_N) + \underline{n}.\nabla_A h \mid_{\underline{n}=0}]$$

$$= U_e\left(1 + \frac{\underline{n}.\nabla_A h \mid_{\underline{n}=0}}{h(\underline{A})}\right) \qquad (7.57)$$

$$= U_e + U_e\left(\frac{\underline{n}.\nabla_A h \mid_{\underline{n}=0}}{h(\underline{A})}\right).$$

The bar underneath indicates vector notation, and delta is the gradient operator. The variable $U_e = g(R)h(\underline{A})$ represents a nominal value of the neural signal.

The random variations $n_i(t)$ are assumed to be zero mean, independent and identically distributed (iid) random variables with variance σ^2, and the number of active motor units N to be sufficiently large; the second term in Equation 7.57 will be approximately Gaussian with zero mean and variance given by

$$\sigma_N^2 = \frac{U_e^2\sigma^2}{h^2(\underline{A})} \sum_{i=1}^{N} \left(\frac{\partial h}{\partial A_i}\right)^2$$

or

$$\sigma_N = \frac{U_e\sigma}{h(\underline{A})} \left[\sum_{i=1}^{N} \left(\frac{\partial h}{\partial A_i}\right)^2\right]^{1/2}. \tag{7.58}$$

Note that the second term in Equation 7.57 may be rewritten as

$$\frac{n.\nabla_A h|_{n=0}}{h(\underline{A})} = \frac{\sum_{i=1}^{N} n_i(t)\frac{\partial h}{\partial A_i}}{h(\underline{A})} = \frac{n(t)}{h(\underline{A})} = Hn(t), \tag{7.59}$$

where $H = 1/h(\underline{A})$. Thus, Equation 7.57 becomes

$$\frac{df(t)}{dt} = U_e[1 + Hn(t)] \tag{7.60}$$
$$= U_e + U_e Hn(t).$$

In Equation 7.60, the first term represents the mean value of the controlling input generated by the subject to perform a movement task. The second term represents the variability of that input. Note that in this model the variability term is proportional to the mean amplitude of the input. If the subject misses the target too often and does not satisfy the specified error criterion, the solution would be to reduce U_e, thereby reducing the variability of the input signal as well. Thus the speed-accuracy trade-off may be interpreted in terms of adjustment of the input parameter U_e to reach a satisfactory compromise between the conflicting requirements of accuracy and speed.

The first term in Equation 7.60, U_e, represents a composite input neural signal. We will denote this input by the usual notation $u(t)$, which is bounded by some value U, that is, $|u(t)| < U$. The second term represents variability in the input term, which will be termed "noise" and labelled as $v(t)$. This noise is a consequence of fluctuations in the recruitment thresholds of motor units. This noise is zero-mean Gaussian with its standard deviation defined by Equation 7.58. Thus, the proposed system model Equation 7.53 may be rewritten as

$$\frac{dx(t)}{dt} = v(t)$$

$$\frac{dv(t)}{dt} = f(t) \tag{7.61}$$

$$\frac{df(t)}{dt} = u(t) + v(t)$$

$$|u(t)| \leq U$$

SPEED-ACCURACY TRADE-OFF
WITH NO SWITCHINGS

In discrete movements, the boundary conditions are

$$x(0) = 0, \ v(0) = 0, \ f(0) = 0 \tag{7.62}$$

$$E\{x(T)\} = D.$$

We assume that, at the starting time, the subject initiates the movement at zero position with zero initial velocity and zero initial muscle force. The neural input to the model at time $t = 0$ is not required to be zero, the input $u(t)$ will jump from a value of 0 to a finite value at $t = 0$. At the terminal boundary, we require that the expected value of $x(t = T) = D$. The boundary conditions on $v(t = T)$ and $f(t = T)$ are not given, because we are assuming that in the Fitts-Peterson paradigm the motion is terminated by striking the target plate with a stylus. An implicit assumption is that the friction between the target plate and the stylus is sufficiently large that the motion is terminated on striking the target plate.

The minimum-time solution for a triple integrator plant and bounded input was discussed earlier. The bang-bang solution with arbitrary starting condition and origin as the final position in general has two switchings. However, in our problem the input has an associated variable noise component, and two of the final variables are not specified. From Equation 7.42, the Fitts-Peterson paradigm has no switching of the control input. Because of the stochastic nature of the system, only the expected value of the final position must be D, and the movement may be terminated within a certain target window. From Equation 7.61, with $u(t) = U$ (or using Equation 7.60), we get

$$\frac{d^3x(t)}{dt^3} = \frac{d^2v(t)}{dt^2} = \frac{df(t)}{dt} = U + UHn(t). \tag{7.63}$$

The triple integration of Equation 7.63 over the time interval $0 \le t \le T$ is given by the equation

$$x(T) = \frac{1}{6}UT^3 + UH \int_0^T \int_0^{t_3} \int_0^{t_2} n(t_1)dt_1dt_2dt_3. \tag{7.64}$$

Because the noise is assumed to be zero mean, the expected value of $x(T)$, for a given value of T, is given by

$$E\{[x(T)/T]\} = \frac{1}{6}UT^3. \tag{7.65}$$

Setting $E\{[x(T)/T]\} = D$, from Equation 7.65

$$D = \frac{1}{6}UT^3. \tag{7.66}$$

The terminal time T is also a random variable. Equation 7.65 is a conditional expectation for a given T. The expected value of $x(T)$ is given by

$$E\{x(T)\} = E\{E\{x(T)/T\}\} = \frac{1}{6}UE\{T^3\}.$$

In Fitts's original work, as well as in most later studies following this paradigm, the Fitts equation is verified in terms of the average movement time, and this equation does take into account the variability in T within and between subjects. If the variance of T is ignored, Equation 7.65 gives the expected value of the terminal position. In our remaining discussion we will consider T to be the average movement time and ignore the variance of T.

To calculate the variance of x(T) from Equation 7.64, certain covariance properties of the noise must be defined. The additive noise in the control signal will be assumed to be band-limited Gaussian. We will present two models of this noise signal, the first being the simplest possible, and the second being derived from simulation studies based on Fitts's observation of a logarithmic relationship between movement time and index of difficulty.

NOISE MODELS AND VARIANCE
OF THE TERMINAL POSITION

We have investigated two noise models. There is no information available from experimental studies concerning the power spectral density of the noise signal. By physical nature of the transmission on the nerve and muscle fibers, which are generally modeled using resistive and capacitive elements, the power spectral density of the noise signal is band-limited. The high-frequency components are short-circuited by the parallel-capacitive components. The first model is the simplest possible condition where a Gaussian noise is passed through a first-order low-pass filter. The second model is obtained by trial and error and is a fifth-order low-pass filter.

Noise Model I

The first model for the noise part of the input $n(t)$ is a simplest possible description with a power spectral density (PSD) given by

$$S_n(\omega) = \frac{\alpha^2 N}{\alpha^2 + \omega^2}. \tag{7.67}$$

The power density is assumed to be N at low frequencies, and the model represents a first-order low pass with a break frequency of α. The autocorrelation function is given by

$$E\{n(t_1)n(s_1)\} = \frac{N}{2}\alpha e^{-\alpha|t_1-s_1|}. \tag{7.68}$$

Thus the variance of $x(T)$ is given by

$$\text{var}\{x(T)\} = \frac{U^2 H^2 \alpha N}{2} \int_0^T \int_0^{t_3} \int_0^{t_2} \int_0^T \int_0^{s_3} \int_0^{s_2} e^{-\alpha |t_1 - s_1|} \, ds_1 ds_2 ds_3 dt_1 dt_2 dt_3. \tag{7.69}$$

In Appendix B, the value of this integral is calculated and is defined as I, which is given by the equation

$$I = \left(\frac{T^5}{10\alpha} - \frac{T^4}{4\alpha^2} + \frac{T^3}{3\alpha^3} - \frac{2}{\alpha^6} \right) + e^{-\alpha T} \left(\frac{2}{\alpha^6} + \frac{2T}{\alpha^5} + \frac{T^2}{\alpha^4} \right).$$

Thus, Equation 7.69 reduces to

$$\text{var}\{x(T)\} = \frac{U^2 H^2 \alpha N}{2} I. \tag{7.70}$$

The terminal position $x(T)$ is a Gaussian distribution. The value of $x(T)$ is a function of the magnitude of input signal U and the cube of the terminal time T. This value was set equal to D in Equation 7.66. The variance of $x(T)$ is a function of the noise parameters α and N, and the input bound U and terminal time T. Its variance will be denoted in terms of tolerance limits on D, say $\pm \delta D$. Therefore, the target width will be set equal to 2 δD. The variance of $x(T)$ in terms of δD must be an even function. A simplest possible relationship is of the form

$$\text{var}\{x(T)\} = \frac{k_1}{4} (\delta D)^n. \tag{7.71}$$

We expect $\text{var}\{x(T)\}$ to be an increasing function of δD and must go to zero as δD goes to zero. The parameter k_1 must be positive. For simplicity we assume that the subject chooses to move in such a manner that the probability of missing the target (i.e., the probability of an error) is independent of the task condition. In Appendix C we have shown that for $n = 2$ and k_1 independent of D and δD, the probability of error will be independent of D and δD. Thus

$$\frac{[E\{x(T)\}]^2}{\text{var}\{x(T)\}} = \frac{D^2}{k_1(\delta D)^2/4} = \frac{\frac{1}{36} U^2 T^6}{\frac{1}{2} N \alpha U^2 H^2 I}. \tag{7.72}$$

Therefore,

$$\left(\frac{2D}{\delta D} \right)^2 = \frac{k_1 T^6}{18 N \alpha H^2 I} = \frac{k_1 [h(\underline{A})]^2 T^6}{18 N \alpha I}.$$

Taking the logarithm on both sides and rearranging the terms,

$$ln(T) = \frac{1}{6} \, ln \left[\frac{18 N \alpha I}{k_1 [h(\underline{A})]^2} \right] + \frac{1}{3} \, ln \left(\frac{2D}{\delta D} \right). \tag{7.73}$$

Thus for the noise model chosen in Equation 7.67, the logarithm of movement time T is related to the logarithm of the ratio of $2D/\delta D$. However, in the Fitts equation the movement time T is logarithmically related to the ratio of $2D/\delta D$. This simple model of noise does not account for the Fitts relationship.

Noise Model II

The second model for the noise $n(t)$ was obtained by iterative analysis such that the model output will yield a logarithm type of relation similar to the Fitts equation, such that the movement time T is logarithmically related to the ratio of $2D/\delta D$. The power spectral density (PSD) of this noise signal is given by

$$S_n(\omega) = (0.12N\alpha)\frac{60\alpha^7\omega^2 + 116\alpha^5\omega^4 + 20\alpha^3\omega^6 + 28\alpha\omega^8}{(\alpha^2 + \omega^2)^5}. \tag{7.74}$$

The autocorrelation function corresponding to this power density function is given by

$$E\{n(t_1)n(s_1)\} = (0.36N\alpha)\left(1 - \frac{11}{3}\alpha|\tau| + \frac{5}{2}\alpha^2|\tau|^2 - \frac{1}{2}\alpha^3|\tau|^3 + \frac{1}{36}\alpha^4|\tau|^4\right)e^{-\alpha|\tau|}, \tag{7.75}$$

where $\tau = t_1 - s_1$.

The variance of $x(T)$ is given by

$$\text{var}\{x(T)\} = U^2H^2\int_0^T\int_0^{t_3}\int_0^{t_2}\int_0^T\int_0^{s_3}\int_0^{s_2} E\{n(t_1)n(s_1)\}ds_1ds_2ds_3dt_1dt_2dt_3. \tag{7.76}$$

In Appendix B, we also show that

$$I_2 = \int_0^T\int_0^{t_3}\int_0^{t_2}\int_0^T\int_0^{s_3}\int_0^{s_2} E\{n(t_1)n(s_1)\}ds_1ds_2ds_3dt_1dt_2dt_3 \tag{7.77}$$

$$= \frac{1}{36}T^6e^{-\alpha T}(0.36N\alpha).$$

As before, let

$$\text{var}\{x(T)\} = \frac{k_1}{4}(\delta D)^2. \tag{7.78}$$

Thus

$$\frac{[E\{x(T)\}]^2}{\text{var}\{x(T)\}} = \frac{D^2}{k_1(\delta D)^2/4} = \frac{U^2T^6}{(0.36N\alpha)U^2H^2T^6e^{-\alpha T}} = \frac{1}{(0.36\ N\alpha H^2)}e^{\alpha T}. \tag{7.79}$$

Rearranging this equation, we get

$$\left(\frac{2D}{\delta D}\right)^2 = \frac{k_1}{(0.36NH^2\alpha)} e^{\alpha T} = \frac{k_1[h(A)]^2}{(0.36N\alpha)} e^{\alpha T}. \tag{7.80}$$

Rewriting this equation in terms of T, we get

$$T = \frac{1}{\alpha} \ln \left\{ \frac{0.36N\alpha}{k_1[h(A)]^2} \right\} + \frac{2}{\alpha} \ln \left(\frac{2D}{\delta D}\right), \tag{7.81}$$

which has the same form as the Fitts speed-accuracy trade-off equation. The target width W will then be defined as a function of the standard deviation of the final position δD. For the Fitts data shown in Figure 7. 1, the linear regression equation is given by

$$T(ms) = 3.3 + 99.74 \text{ ID}, r^2 = 0.956. \tag{7.82}$$

This gives a value of $\alpha = 13.9$. (Note that Equation 7.81 is given in terms of Naperian logarithms and Equation 7.82 is in terms of base-2 logarithms.)

Interpretation of the Noise Power Spectral Density

From Equation 7.59, we have

$$n(t) = \sum_{i=1}^{N} n_i(t) \frac{\partial h}{\partial A_i}. \tag{7.83}$$

Thus

$$E\{n(t)n(t + \tau)\} = \sum_{i=1}^{N} \sum_{j=1}^{N} E\{n_i(t)n_j(t + \tau)\} \frac{\partial h}{\partial A_i} \frac{\partial h}{\partial A_j}. \tag{7.84}$$

Because the n_j are zero-mean i.i.d., we have

$$E\{n(t)n(t + \tau)\} = \sum_{j=1}^{N} E\{n_j(t)n_j(t + \tau)\} \left(\frac{\partial h}{\partial A_j}\right)^2. \tag{7.85}$$

Taking the Fourier transform of both sides of Equation 7.85 and noting that the PSD of the n_j's are independent of j, we see that the PSD of n(t) is proportional to the PSD of $n_j(t)$ with a proportionality constant that depends on the A_1, A_2, . . . , A_N. Therefore the variation of the pulse amplitude is adequately represented by the PSD of the process n(t). In particular, the peak location of the PSD of n(t) is the same as the peak location of the PSD of $n_j(t)$.

SPEED-ACCURACY TRADE-OFF
WITH TWO SWITCHINGS

Several studies have revealed a triphasic excitation signal. This condition occurs when the model equations and the boundary conditions are given by the following equations:

$$\frac{dx(t)}{dt} = v(t)$$

$$\frac{dv(t)}{dt} = f(t) \qquad (7.86)$$

$$\frac{df(t)}{dt} = u(t)$$

$$\mid u(t) \mid \leq U$$

with the boundary conditions

$$
\begin{aligned}
x(0) &= 0 & E\{x(T)\} &= D \\
v(0) &= 0 & E\{v(T)\} &= 0 \qquad (7.87)\\
f(0) &= 0 & E\{f(T)\} &= 0
\end{aligned}
$$

The minimum-time solution for this system has two switching times, $t = R$ and $t = S$. Let the control values for these intervals be

$$
\begin{aligned}
0 \leq u(t) \leq U_1 & \qquad \text{for } 0 \leq t \leq R, \\
-U_2 \leq u(t) \leq 0 & \qquad \text{for } R \leq t \leq S, \qquad (7.88)\\
0 \leq u(t) \leq U_3 & \qquad \text{for } S \leq t \leq T.
\end{aligned}
$$

For the minimization of the Hamiltonian function, $u(t)$ will take values at the boundary. Thus

$$
\begin{aligned}
u(t) &= U_1 & \qquad \text{for } 0 \leq t \leq R, \\
u(t) &= -U_2 & \qquad \text{for } R \leq t \leq S, \qquad (7.89)\\
u(t) &= U_3 & \qquad \text{for } S \leq t \leq T.
\end{aligned}
$$

For $0 \leq t \leq R$, from Equation 7.63, we get

$$\frac{d^3x(t)}{dt^3} = \frac{d^2v(t)}{dt^2} = \frac{df(t)}{dt} = U_1 + U_1 Hn_1(t). \qquad (7.90)$$

Integration of Equation 7.90 gives the following expresions for the state variables:

$$f(R) = U_1R + U_1H \int_0^R n_1(t_1)dt_1$$

$$v(R) = \frac{1}{2}U_1R^2 + U_1H \int_0^R \int_0^{t_2} n_1(t_1)dt_1dt_2 \tag{7.91}$$

$$x(R) = \frac{1}{6}U_1R^3 + U_1H \int_0^R \int_0^{t_3} \int_0^{t_2} n_1(t_1)dt_1dt_2dt_3 .$$

For the time interval $R \le t \le S$, the system equation is given by:

$$\frac{d^3x(t)}{dt^3} = \frac{d^2v(t)}{dt^2} = \frac{df(t)}{dt} = -U_2 - U_2Hn_2(t). \tag{7.92}$$

Integration of Equation 7.92 gives the following expresions for the state variables:

$$f(S) = f(R) - U_2(S - R) - U_2H \int_R^S n_2(t_1)dt_1$$

$$v(S) = v(R) + (S - R)f(R) - \frac{1}{2}U_2(S - R)^2 - U_2H \int_R^S \int_R^{t_2} n_2(t_1)dt_1dt_2 \tag{7.93}$$

$$x(S) = x(R) + (S - R)v(R) + \frac{1}{2}(S - R)^2f(R) - \frac{1}{6}U_2(S - R)^3$$

$$- U_2H \int_R^S \int_R^{t_3} \int_R^{t_2} n_2(t_1)dt_1dt_2dt_3 .$$

For the final time interval $S \le t \le T$, the system equation is given by

$$\frac{d^3x(t)}{dt^3} = \frac{d^2v(t)}{dt^2} = \frac{df(t)}{dt} = U_3 + U_3Hn_3(t). \tag{7.94}$$

Integration of Equation 7.94 gives the following expresions for the state variables:

$$f(t) = f(S) + U_3(t - S) + U_3H \int_S^t n_3(t_1)dt_1$$

$$v(t) = v(S) + (t - S)f(S) + \frac{1}{2}U_3(t - S)^2 + U_3H \int_S^t \int_S^{t_2} n_3(t_1)dt_1dt_2 \tag{7.95}$$

$$x(t) = x(S) + (t - S)v(S) + \frac{1}{2}(t - S)^2f(S) + \frac{1}{6}U_3 (t - S)^3$$

$$+ U_3H \int_S^t \int_S^{t_3} \int_S^{t_2} n_3(t_1)dt_1dt_2dt_3 .$$

Taking the expected values of Equations 7.91, 7.93, and 7.95, and setting t = T in Equation 7.95, we get

$$E\{f(T)\} = U_1R - U_2(S - R) + U_3(T - S) = 0$$

$$E\{v(T)\} = \frac{1}{2}U_1R^2 + (S - R)U_1R - \frac{1}{2}U_2(S - R)^2$$

$$+ (T - S)[U_1R - U_2(S - R)] + \frac{1}{2}U_3(T - S)^2 = 0 \qquad (7.96)$$

$$E\{x(T)\} = \frac{1}{6}U_1R^3 + (S - R)\frac{1}{2}U_1R^2 + \frac{1}{2}(S - R)^2U_1R$$

$$- \frac{1}{6}U_2(S - R)^3 + (T - S)[\frac{1}{2}U_1R^2 + (S - R)U_1R - \frac{1}{2}U_2(S - R)^2]$$

$$+ \frac{1}{2}(T - S)^2[U_1R - U_2(S - R)] + \frac{1}{6}U_3(T - S)^3 = D.$$

Solving these equations simultaneouly, we obtain the following expressions for R and S:

$$R = \frac{U_2 + U_3}{U_2 + U_1} S - \frac{U_3}{U_2 + U_1} T \qquad (7.97)$$

$$S = \frac{-U_3}{U_1 - U_3} T + T \sqrt{\frac{U_3^2}{(U_1 - U_3)^2} + \frac{U_3(U_1 + U_2 + U_3)}{(U_1 - U_3)(U_2 + U_3)}}.$$

The peak velocity occurs at time t when f(t) = 0. Because f(t) is a random variable, the peak velocity V_p will occur at time T_p where $E\{f(t)\} = 0$. Thus, from Equations 7.95, 7.96, and 7.97,

$$T_p = \left(1 + \frac{U_1}{U_2}\right)R$$

$$V_p = v(R) + \frac{U_1}{U_2}Rf(R) - \frac{1}{2}U_2\left(\frac{U_1}{U_2}R\right)^2 \qquad (7.98)$$

$$= \frac{1}{2}U_1R^2\left(1 + \frac{U_1}{U_2}\right).$$

Different ratios of the control signal U were tried, to give a reasonable fit to the Corcos et al. (1989) data. The chosen values are

$$\frac{U_1}{U_2} = 1.1 \qquad\qquad \frac{U_3}{U_1} = 0.48$$

$$R = 0.21\ T \qquad\qquad S = 0.64\ T \qquad (7.99)$$

$$D = (0.023)U_1T^3 \qquad\qquad V_p = (0.046)U_1T^2$$

Combining V_p and D expressions,

$$V_p = \frac{2D}{T}.$$

(7.100)

Thus the peak velocity is two times the average velocity. As discussed earlier, this value is within the range of observed values as well as theoretical values using other criteria functions.

CONCLUSIONS

An alternate explanation of the Fitts speed-accuracy trade-off equation based on the dynamic constraints on the limb movement and optimal control theory is provided. The input to the model is an excitation pulse rather than a torque input considered in most previous models. Based on this model, the number of switchings in the controlling input may range from zero as in Fitts's original paradigm, where the movement is terminated by a mechanical stop, to two switchings as seen in many studies. The noise model in the excitation signal explains the variability in the terminal movement.

In the process of initial learning of a discrete movement, the subject learns to control the excitation pulse amplitude based on the task requirements. In Fitts-type experiments, where both speed and accuracy are part of the task requirement, the subject must vary the number of active motor units and the frequency of their activation based on the size principle such that a satisfactory compromise between speed and accuracy is achieved. The observed behavior where the accuracy decreases with increase in speed is explained by the model proposed.

The proposed model is a simplistic representation of the actual physical system. Many approximations have been made to simplify the mathematical concepts and to make the model analytically tractable. For example, we deleted the damping term in the mechanical model of the limb. We deleted the time constant in the excitation model, which was proposed by Gottlieb et al. (1989b). We neglected saturation in force production and other nonlinearities of the muscle and the limb. We made simplification in the stochastic activation of the motor units. We assumed the motor units to be independent. In a more detailed publication we will include further analysis of this model to partially account for the dual-strategy concept proposed by Gottlieb et al. (1989b).

Acknowledgment

This work was supported, in part, by National Institutes of Health Grants AR-33189 and NS-23593. The initial part of this study was done as part of the MS thesis by Joseph B. Logsdon at the University of Illinois at Chicago.

APPENDIX A

We will show that the number of switchings n_s for a minimum-time problem is given by the equation

$$n_s = n - 1 - n', \tag{A.1}$$

where n is the number of state variables, $n' = 2n - q$, and q is the number of constraints on the initial and final conditions, and the terminal time T is unknown.
We are limiting our discussion to linear time-invariant systems of the form

$$\dot{x} = Ax + bu, \tag{A.2}$$

where x is an $n \times 1$ state vector, A is the system matrix, b is an $n \times 1$ input parameter vector, and u(t) is a scalar input (extension to vector form is trivial; see Sage & White, 1977). The performance index J is given by

$$J = \int_0^T dt = T. \tag{A.3}$$

The control input u is assumed to be bounded:

$$| u(t) | \le U. \tag{A.4}$$

In most time-optimal control problems, the boundary conditions are assumed to be

$$x(0) = x_0 , \; x(T) = 0. \tag{A.5}$$

The initial condition is arbitrary, and the final state is the origin. The time T is unknown and is to be minimized.

If the eigenvalues of the system matrix A are real, there are at most $n - 1$ switchings or changes of sign of the control. That is, the input will take values of either U or $-U$. This type of solution is known as bang-bang control (Sage & White, 1977). For the general case where all the boundary conditions may not be as given in Equation A.5 and where the control input is piecewise constant, let the number of switchings be at most n_s. Thus, there are $n_s + 1$ time subintervals. For an nth-order system, we will have a total of $n(n_s + 1)$ constants of integration. In addition we need to find n_s switching times and the total time T, which is also unknown. Therefore the total number of unknowns K is given by

$$K = n(n_s + 1) + n_s + 1. \tag{A.6}$$

If all the initial and final conditions on the state are given at the initial time $t = 0$ and the terminal time T, and knowing that the states at the switching times (which are outputs of integrators with bounded inputs) must be continuous, there are a total of $(n_s + 2)$ times at which the states must be matched with the initial conditions or at the switching boundaries. This gives us a total of $n(n_s + 2)$ equations.

To obtain at least one solution, the number of unknowns must be equal to or greater than the number of equations. Thus,

$$K = n(n_s + 1) + n_s + 1 \geq n(n_s + 2) \tag{A.7}$$

$$n_s \geq n - 1.$$

If we choose $n_s = n - 1$, we immediately obtain a unique solution.

General Boundary Condition Problem

For a more general problem, we replace Equation A.5 by the following equation:

$$N[\underline{x}(0), \underline{x}(T)] = \underline{0}, \tag{A.8}$$

where N is a $q \times 1$ vector that determines the relationship between initial states and final states. Therefore, q is the number of equations that can be used to find the constants of integration. To obtain the remaining equations, define the cost function and the Hamiltonian H as

$$J = \underline{v}'N + T \tag{A.9}$$

$$H(\underline{x},u,\underline{\lambda},t) = \underline{\lambda}'(A\underline{x} + \underline{b}u).$$

The maximum principle gives the following equations:

$$H(\underline{x},u,\underline{\lambda},t) \leq H(\underline{x},v,\underline{\lambda},t) \tag{A.10}$$

and

$$u = U \qquad \text{when } \underline{b}'\underline{\lambda}(t) < 0 \tag{A.11}$$

$$u = -U \qquad \text{when } \underline{b}'\underline{\lambda}(t) > 0$$

for all $v(t)$ that satisfy the control variable constraint. The system equations are given by

$$\underline{\dot{x}} = A\underline{x} + \underline{b}u \tag{A.12}$$

$$-\underline{\dot{\lambda}} = A'\underline{\lambda}.$$

The boundary conditions are

$$\underline{\lambda}(0) = \left(\frac{\partial \underline{N}}{\partial \underline{x}(0)}\right)' \underline{v}$$

$$\underline{\lambda}(T) = \left(\frac{\partial \underline{N}}{\partial \underline{x}(T)}\right)' \underline{v} \qquad \text{(A.13)}$$

$$\underline{N} = \underline{0}$$

$$H[\underline{x}(T),u(T),\underline{\lambda}(T),T] + \left(\frac{\partial \underline{N}'}{\partial T}\right)\underline{v} + 1 = 0$$

The set of Equations A.13 provides $2n + q + 1$ equations.

Let n' be the difference between the number of initial and terminal state variables, $2n$, and the number of equations relating the initial and terminal state variables, which is q as given by Equation A.8. Thus:

$$n' = 2n - q. \qquad \text{(A.14)}$$

At any time t, the knowledge of the covariables $\underline{\lambda}(t)$ implies knowledge of the input variable $u(t)$, therefore the covariables must satisfy the remaining boundary conditions. These conditions are $2n - q$ missing state conditions, n_s switching times, and the unknown final time, that is, $2n - q + n_s + 1 = n' + n_s + 1$ conditions. The integration of λ equations in Equation A.12 requires n constants of integration. Thus, for a solution to exist,

$$n \geq n_s + n' + 1, \qquad \text{(A.15)}$$

$$\text{or} \qquad n_s \leq n - 1 - n'.$$

A unique solution exists when $n_s = n - 1 - n'$.

APPENDIX B

In this appendix, we evaluate the integrations given in Equations 7.69 and 7.76. Let

$$I = \int_0^T \int_0^{t_3} \int_0^{t_2} \int_0^T \int_0^{s_3} \int_0^{s_2} e^{-\alpha |t_1 - s_1|} ds_1 ds_2 ds_3 dt_1 dt_2 dt_3. \qquad \text{(B.1)}$$

Consider the first part of this integration,

$$I_1 = \int_0^T \int_0^{s_3} \int_0^{s_2} e^{-\alpha |t_1 - s_1|} ds_1 ds_2 ds_3, \qquad \text{(B.2)}$$

which can be rewritten as

$$
= \int_0^{t_1} \int_0^{s_3} \int_0^{s_2} e^{-\alpha t_1} e^{\alpha s_1} ds_1 ds_2 ds_3
$$

$$
+ \int_{t_1}^{T} \int_0^{t_1} \int_0^{s_2} e^{-\alpha t_1} e^{\alpha s_1} ds_1 ds_2 ds_3
$$

$$
+ \int_{t_1}^{T} \int_{t_1}^{s_3} \int_0^{t_1} e^{-\alpha t_1} e^{\alpha s_1} ds_1 ds_2 ds_3
$$

$$
+ \int_{t_1}^{T} \int_{t_1}^{s_3} \int_{t_1}^{s_2} e^{\alpha t_1} e^{-\alpha s_1} ds_1 ds_2 ds_3 .
$$

(B.3)

Thus,

$$
I_1 = \left(\frac{2}{\alpha^3} + \frac{1}{\alpha} T^2 \right) - e^{-\alpha t_1} \left(\frac{1}{\alpha^3} + \frac{1}{\alpha^2} T + \frac{1}{2\alpha} T^2 \right)
$$

$$
- t_1 \left(\frac{2}{\alpha} T \right) + t_1^2 \left(\frac{1}{\alpha} \right) - \frac{1}{\alpha^3} e^{\alpha t_1} e^{-\alpha T} .
$$

(B.4)

Going back to Equation B.1,

$$
I = \int_0^{T} \int_0^{t_3} \int_0^{t_2} \left[\left(\frac{2}{\alpha^3} + \frac{1}{\alpha} T^2 \right) - e^{-\alpha t_1} \left(\frac{1}{\alpha^3} + \frac{1}{\alpha^2} T + \frac{1}{2\alpha} T^2 \right) \right.
$$

$$
\left. - t_1 \left(\frac{2}{\alpha} T \right) + t_1^2 \left(\frac{1}{\alpha} \right) - \frac{1}{\alpha^3} e^{\alpha t_1} e^{-\alpha T} \right] dt_1 dt_2 dt_3 .
$$

(B.5)

After completing the integration, we obtain

$$
I(\alpha, T) = \left(\frac{T^5}{10\alpha} - \frac{T^4}{4\alpha^2} + \frac{T^3}{3\alpha^3} - \frac{2}{\alpha^6} \right) + e^{-\alpha T} \left(\frac{2}{\alpha^6} + \frac{2T}{\alpha^5} + \frac{T^2}{\alpha^4} \right) .
$$

(B.6)

Equation B.6 is the value of the integration in Equation 7.69.
   Now, for the second noise model and Equation 7.76, let

$$
I_2 = \int_0^{T} \int_0^{t_3} \int_0^{t_2} \int_0^{T} \int_0^{s_3} \int_0^{s_2} E\{n(t_1)n(s_1)\} ds_1 ds_2 ds_3 dt_1 dt_2 dt_3 ,
$$

(B.7)

where the autocorrelation function is given by

$$
E\{n(t_1)n(s_1)\} = (0.36N\alpha) \left(1 - \frac{11}{3} \alpha|\tau| + \frac{5}{2} \alpha^2 |\tau|^2 - \frac{1}{2} \alpha^3 |\tau|^3 + \frac{1}{36} \alpha^4 |\tau|^4 \right) e^{-\alpha|\tau|}
$$

(B.8)

and $\tau = t_1 - s_1$.

From Equations B.1, B7, and B.8, we get

$$I_2 = 0.36N\alpha\left(I(\alpha,T) + \frac{11}{3}\alpha\frac{dI(\alpha,T)}{d\alpha} + \frac{5}{2}\alpha^2\frac{d^2I(\alpha,T)}{d\alpha^2} + \frac{1}{2}\alpha^3\frac{d^3I(\alpha,T)}{d\alpha^3} + \frac{1}{36}\alpha^4\frac{d^4I(\alpha,T)}{d\alpha^4}\right),$$

(B.9)

where $I(\alpha,T)$ is given by Equation B.6. Define

$$I(\alpha,T) = F + G$$

$$F = \left(\frac{T^5}{10\alpha} - \frac{T^4}{4\alpha^2} + \frac{T^3}{3\alpha^3} - \frac{2}{\alpha^6}\right)$$

(B.10)

$$G = e^{-\alpha T}\left(\frac{2}{\alpha^6} + \frac{2T}{\alpha^5} + \frac{T^2}{\alpha^4}\right).$$

From Equations B.9 and B.10,

$$\left(F + \frac{11}{3}\alpha\frac{dF}{d\alpha} + \frac{5}{2}\alpha^2\frac{d^2F}{d\alpha^2} + \frac{1}{2}\alpha^3\frac{d^3F}{d\alpha^3} + \frac{1}{36}\alpha^4\frac{d^4F}{d\alpha^4}\right) = 0$$

(B.11)

and

$$\left(G + \frac{11}{3}\alpha\frac{dG}{d\alpha} + \frac{5}{2}\alpha^2\frac{d^2G}{d\alpha^2} + \frac{1}{2}\alpha^3\frac{d^3G}{d\alpha^3} + \frac{1}{36}\alpha^4\frac{d^4G}{d\alpha^4}\right) = \frac{1}{36}T^6e^{-\alpha T}.$$

(B.12)

From Equation B.9, we get

$$I_2 = 0.36N\alpha\left(\frac{1}{36}T^6e^{-\alpha T}\right).$$

(B.13)

APPENDIX C

In the Fitts equation

$$T = a + b\left(ln\frac{2D}{\delta D}\right),$$

D is the desired movement distance of the mass, and δD is the amount by which the final position of the mass is allowed to vary from trial to trial. We assume the probability of error, P_e, to be independent of the distance D and δD. The specification of δD and the probability distribution of the final position $x(T)$ are closely related. Because $x(T)$ is a Gaussian distribution with zero mean, its variance must be specified in relation to δD as an even power function. A simplest possible relationship is of the form

$$\text{var}\{x(T)\} = \frac{k_1}{4}(\delta D)^n. \tag{C.1}$$

Because we expect $\text{var}\{x(T)\}$ to be an increasing function of δD, it must go to zero as δD goes to zero. The parameter k_1 must be positive. Using a normal distribution for $x(T)$, the probability of error is given by

$$P_e = 1 - \frac{2}{\sqrt{2\pi k_1 (\delta D)^n}} \int_{D-\frac{1}{2}\delta D}^{D+\frac{1}{2}\delta D} e^{\frac{-2(x-D)^2}{k_1(\delta D)^n}} dx. \tag{C.2}$$

Define

$$w = \frac{(x - D)}{(\delta D)^{\frac{n}{2}}} \tag{C.3}$$

and, rearranging Equation C.2, we get

$$1 - P_e = \frac{1}{\sqrt{\frac{\pi k_1}{2}}} \int_{-\frac{1}{2}(\delta D)^{1-n/2}}^{\frac{1}{2}(\delta D)^{1-n/2}} e^{\frac{-2w^2}{k_1}} dw. \tag{C.4}$$

In Equation C.4, if we choose $n = 2$, and if k_1 is independent of D and δD, then the probability of error P_e is independent of D and δD and is given by

$$1 - P_e = \frac{1}{\sqrt{\frac{\pi k_1}{2}}} \int_{-\frac{1}{2}}^{\frac{1}{2}} e^{\frac{-2w^2}{k_1}} dw. \tag{C.5}$$

The integration in Equation C.5 is the error function and can be obtained from mathematics tables by making the following substitutions. Let

$$z = \frac{2w}{\sqrt{k_1}}.$$

Equation C.5 becomes

$$1 - P_e = \frac{1}{\sqrt{2\pi}} \int_{-\frac{1}{\sqrt{k_1}}}^{\frac{1}{\sqrt{k_1}}} e^{\frac{-z^2}{2}} dz = 1 - 2Q\left(\frac{1}{\sqrt{k_1}}\right), \tag{C.6}$$

where

$$Q(\zeta) = \frac{1}{\sqrt{2\pi}} \int_{\zeta}^{\infty} e^{\frac{-z^2}{2}} dz. \tag{C.7}$$

Given the values D and δD and error rates from experimental data, we can easily calculate the values of parameter k_1. Taking $\delta D = W$ and average error rates for various values of 2D/δD equal to P_e, we get the following values for P_e and k_1 (as calculated from Equation C.6):

2D/δD	P_e	k_1
2	0.0000	0.000
4	0.00185	0.103
8	0.01055	0.150
16	0.02244	0.191
32	0.02372	0.196
64	0.02955	0.209
128	0.03865	0.231

Earlier we had suggested that for n = 2 and k_1 independent of D and δD, the probability of error would be independent of D and δD. However, Fitts's original data appear to suggest that the probability of error may be a function of D and δD. The value of P_e increases with the index of difficulty. The error rates as a function of index of difficulty are plotted in Figure 7.4 from Fitts's (1954) data. The first-order regression line is given by

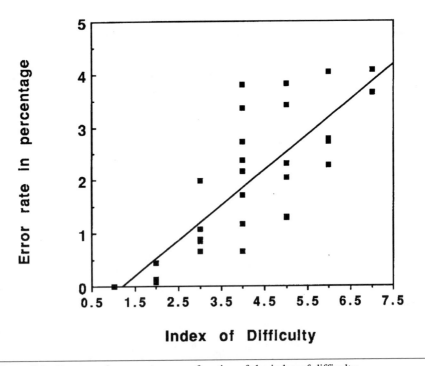

Figure 7.4   Error rate in percentage as a function of the index of difficulty.
Note. Data from ''The Information Capacity of the Human Motor System in Controlling the Amplitude of Movement'' by P.M. Fitts, 1954, *Journal of Experimental Psychology,* **47**(6), p. 385.

Error rate = −0.825 + 0.665 × ID, r^2 = 0.655.

As this data indicates, the error rates are not totally independent of the index of difficulty.

REFERENCES

Adamovitch, S.V., & Feldman, A.G. (1984). Model of the central regulation of the parameters of motor trajectories. *Biofizika*, **29**, 306-309. (English translation, 338-342)

Adrian, E.D., & Bronk, D.W. (1929). The discharge of impulses in motor nerve fibres. Part II. The frequency discharge in reflex and voluntary contractions. *Journal of Physiology* (London), **67**, 119-151.

Agarwal, G.C., & Gottlieb, G.L. (1982). Mathematical modeling and simulation of the postural control loop: Part I. *CRC Critical Reviews in Biomedical Engineering*, **3**, 93-134.

Amis, A.A., Dowson, D., & Wright, V. (1980a). Analysis of elbow forces due to high-speed forearm movements. *Journal of Biomechanics*, **13**, 825-831.

Amis, A.A., Dowson, D., & Wright, V. (1980b). Elbow joint force predictions for some strenuous isometric actions. *Journal of Biomechanics*, **13**, 765-775.

Bahill, A.T., Clark, M.R., & Stark, L. (1975). The main sequence, a tool for studying human eye movements. *Mathematical Bioscience*, **24**, 191-204.

Bellman, R., Glicksberg, I., & Gross, O. (1959). On the "bang-bang" control problem. *Quarterly of Applied Mathematics*, **14**, 1320-1334.

Billigsley, J., & Coales, J.F. (1968). Simple predictive controller for high-order systems. *Proceedings of the Institution of Electrical Engineers* (London), **115**, 1568-1576.

Brown, S.H.C., & Cooke, J.D. (1981). Amplitude- and instruction-dependent modulation of movement-related electromyogram activity in humans. *Journal of Physiology* (London), **316**, 97-107.

Corcos, D.M., Gottlieb, G.L., & Agarwal, G.C. (1989). Organizing principles for single joint movements: I. A speed-sensitive strategy. *Journal of Neurophysiology*, **62**, 358-368.

Crossman, E.R.F.W., & Goodeve, P.J. (1983). Feedback control of hand-movement and Fitts' law. *Quarterly Journal of Experimental Psychology*, **35A**, 251-278. (Originally presented at the meeting of the Experimental Society, Oxford, July 1963)

Elsgolc, L.E. (1962). *Calculus of variations*. Reading, MA: Addison-Wesley.

Feldman, A.G. (1966). Functional tuning of the nervous system during control of movement or maintenance of a steady posture. III. Mechanographic analysis of the execution by man of the simplest motor tasks. *Biofizika*, **11**, 667-675. (English translation 766-775)

Fitts, P.M. (1954). The information capacity of the human motor system in controlling the amplitude of movement. *Journal of Experimental Psychology*, **47**, 381-391.

Fitts, P.M., & Peterson, J.R. (1964). Information capacity of discrete motor responses. *Journal of Experimental Psychology*, **67**, 103-112.

Fitzhugh, R. (1977). A model of optimal voluntary muscular control. *Journal of Mathematical Biology*, **4**, 203-236.

Flash, T., & Hogan, N. (1985). The coordination of arm movements: An experimentally confirmed mathematical model. *Journal of Neuroscience*, **5**, 1688-1703.

Fuller, A.T. (1971). Sub-optimal nonlinear controllers for relay and saturating control systems. *International Journal of Control*, **13**, 401-428.

Fuller, A.T. (1973a). Notes on some predictive control strategies. *International Journal of Control*, **18**, 673-687.

Fuller, A.T. (1973b). Simplified time-optimal switching functions for plants containing lags. *International Journal of Control*, **18**, 1141-1149.

Fuller, A.T. (1985). Minimization of various performance indices for a system with bounded control. *International Journal of Control*, **41**, 1-37.

Gottlieb, G.L., & Agarwal, G.C. (1971). Dynamic relationship between isometric muscle tension and the electromyogram in man. *Journal of Applied Physiology*, **30**, 345-351.

Gottlieb, G.L., Corcos, D.M., & Agarwal, G.C. (1989a). Organizing principles for single joint movements: A speed-insensitive strategy. *Journal of Neurophysiology*, **62**, 342-357.

Gottlieb, G.L., Corcos, D.M., & Agarwal, G.C. (1989b). Strategies for the control of single mechanical degree of freedom voluntary movements. *Behavioral and Brain Sciences*, **12**, 189-210.

Gottlieb, G.L., Corcos, D.M., Agarwal, G.C., & Latash, M.L. (1990). Organizing principles for single joint movements. III. The speed-insensitive strategy as default. *Journal of Neurophysiology*, **63**(3), 625-636.

Hallett, M., & Marsden, C.D. (1979). Ballistic flexion movements of the human thumb. *Journal of Physiology* (London), **294**, 33-50.

Hannaford, B., & Stark, L. (1985). Roles of the elements of the triphasic control signal. *Experimental Neurology*, **90**, 619-634.

Hasan, Z. (1986). Optimized movement trajectories and joint stiffness in unperturbed, inertially loaded movements. *Biological Cybernetics*, **53**, 373-382.

Henneman, E. (1979). Functional organization of motoneuron pools: The size principle. In H. Asanuma & V.J. Wilson (Eds.), *Integration in the nervous system*. Tokyo: Igaku-Shoin.

Hogan, N. (1984). An organizing principle for a class of voluntary movements. *Journal of Neuroscience*, **11**, 2745-2754.

Hogan, N. (1988). Planning and execution of multijoint movements. *Canadian Journal of Physiology and Pharmacology*, **66**, 508-517.

Houk, J.C. (1979). Regulation of stiffness by skeletomotor reflexes. *Annual Reviews Physiology*, **41**, 99-114.

Karst, G.M., & Hasan, Z. (1987). Antagonist muscle activity during human forearm movements under varying kinematic and loading conditions. *Experimental Brain Research*, **67**, 391-491.

Marsden, C.D., Obeso, J.A., & Rothwell, J.C. (1983). The function of the antagonist muscle during fast limb movements in man. *Journal of Physiology* (London), **335**, 1-13.

Meyer, D.E., Abrams, R.A., Kornblum, S., Wright, C.E., & Smith, J.E.K. (1988). Optimality in human motor performance: Ideal control of rapid aimed movements. *Psychological Review*, **95**, 340-370.

Meyer, D.E., Smith, J.E.K., Kornblum, S., Abrams, R.A., & Wright, C.E. (1988). Speed-accuracy tradeoffs in aimed movements: Toward a theory of rapid voluntary action. In M. Jeannerod (Ed.), *Attention and performance XIII*. Hillsdale, NJ: Erlbaum.

Meyer, D.E., Smith, J.E.K., & Wright, C.E. (1982). Models for the speed and accuracy of aimed movement. *Psychological Reviews*, **89**, 449-482.

Mustard, B.E., & Lee, R.G. (1987). Relationship between EMG patterns and kinematic properties for flexion movements at the human wrist. *Experimental Brain Research*, **66**, 247-256.

Nagasaki, H. (1989). Asymmetric velocity and acceleration profiles of human arm movements. *Experimental Brain Research*, **74**, 319-326.

Nelson, W.L. (1983). Physical principles for economies of skilled movements. *Biological Cybernetics*, **46**, 135-147.

Ostry, D.J., Cooke, J.D., & Munhall, K.G. (1987). Velocity curves of human arm and speech movements. *Experimental Brain Research*, **68**, 37-46.

Ryan, E.P. (1977). Time-optimal feedback control of certain fourth-order systems. *International Journal of Control*, **26**, 675-688.

Sage, A.P., & White, C.C. (1977). *Optimal systems control* (2nd ed.). Englewood Cliffs, NJ: Prentice-Hall.

Schmidt, R.A., Zelaznik, H., Hawkins, B., Frank, J.S., & Quinn, J.T. (1979). Motor-output variability: A theory for the accuracy of rapid motor acts. *Psychological Review*, **86**, 415-451.

Shannon, C.E. (1948). A mathematical theory of communication. *Bell Systems Technical Journal*, **27**, 379-423.

Smith, O.J.M. (1962). Nonlinear computations in the human controller. *IRE Transactions on Biomedical Engineering*, **BME-9**, 125-128.

Stein, R.B. (1982). What muscle variable(s) does the nervous system control in limb movements? *Behavioral and Brain Sciences*, **5**, 535-577.

Stein, R.B., Oguztoreli, M.N., & Capaday, C. (1986). What is optimized in muscular movements? In N.L. Jones, N. McMartney, & A.J. McComas (Eds.), *Human muscle power* (pp. 131-150. Champaign, IL: Human Kinetics.

Wood, J.E. (1981). A statistical-mechanical model of the molecular dynamics of striated muscle during mechanical transients. *Lectures in Applied Mathematics*, **19**, 213-259.

Woodworth, R.S. (1899). The accuracy of voluntary movement. *Psychological Review*, **3**(2, Suppl. 13), 1-114.

Zahalak, G.I. (1981). A distribution-moment approximation for kinetic theories of muscular contraction. *Mathematical Biosciences*, **55**, 89-114.

Zangemeister, W.H., Lehman, S., & Stark, L. (1981). Simulation of head movement trajectories: Model and fit to main sequence. *Biological Cybernetics*, **41**, 19-32.

Chapter 8

Variability of Fast Single-Joint Movements and the Equilibrium-Point Hypothesis

Mark L. Latash, Simon R. Gutman

Rush Medical College, Chicago, Illinois

In general, variability is an antonym of reproducibility and is a curse of motor control studies that do not investigate it explicitly. Reducing variability is frequently the goal of training procedures in both "real life" (including training athletes or rehabilitating patients with motor disorders) and the laboratory environment. On the other hand, variability by itself is a fascinating phenomenon that obeys its own laws and demonstrates consistent relations between task and performance variables. Motor variability is a manifestation of the functioning of a neuromuscular system of movement production. As such, it should be considered in a general framework of some model of motor control. We are going, first, to accept a very general scheme of movement production, and, then, to move to a more specific approach based on the equilibrium-point hypothesis (EP hypothesis) (Feldman, 1966a, 1986).

The notion of variability can be introduced in two different ways. The first considers deviations of movement characteristics from those observed during a certain "ideal" or "average" performance. For example, this notion can be applied to any repetitive action, such as cycling or locomotion, for describing patterns of various kinematic, kinetic, or electromyographic (EMG) variables. It can also be used for analysis of a series of discrete motor acts, presuming that conditions of their execution and intentions of the subject (or instruction) are not changing from one attempt to another. Dispersion of a chosen variable (or some related measure) is frequently used as a measure of variability.

The second approach requires the introduction of the notion of a quantifiable target, which imposes explicit restrictions upon changes in certain variables of the performance, such as final position, movement time, or some others. In motor control studies, dimensions of spatial targets are usually manipulated. There is an ambiguity in choosing a measure of variability when an explicit target is presented. Sometimes the percentage of successful trials is used as a measure. In other experiments, the dispersion of a selected variable is used, as is done in the experiments without an explicit target.[1] However, most frequently it is assumed that the target imposes

---

[1] If a nonquantifiable target (e.g., a point) is presented, it leaves the choice of permissible errors to the subject. That is, the subject may have his or her own understanding of how accurate or inaccurate he or she can be. However, in the future, we will consider point targets as absolutely nonrestrictive and will use the term *target* in the sense of an explicitly imposed restriction, that is, only for quantifiable targets.

certain restrictions upon dispersion of the chosen variable, and therefore its physical dimension constrains variability and may be considered its measure. This assumption, which seems to us far from obvious, leads to comparison of the data from the experiments where target size was manipulated (T-experiments) with the data from the experiments where dispersion of the final position was used as a measure of variability (D-experiments). In this paper, we will try to dissociate the effects of target size upon a hypothetical voluntary motor command from the effects of variability in the motor command upon dispersion of the final position. As a result, T-experiments and D-experiments are considered as studying different aspects of movement production.

SPEED-ACCURACY TRADE-OFFS

The most famous general law of motor variability is probably Fitts' law, which deals with a certain aspect of variability, namely, the speed-accuracy trade-off. It was originally introduced for describing the dependence of movement time upon target size during repetitive (Fitts, 1954) and, later, discrete (Fitts & Peterson, 1964) arm movements performed under visual control with explicitly presented targets. *Target size was assumed to restrict the variability and therefore be its measure* (cf. T-experiments, previous section). Fitts suggested a logarithmic relation between movement time (T), movement distance (D), and target width (W):

$$T = a + b \ \log_2 \frac{2D}{W}, \tag{8.1}$$

where a and b are empirically defined constants.

Since the original works by Fitts, logarithmic relations like Equation 8.1 have been used for describing experimental findings in a variety of conditions with movements of different complexity involving different joints (Flowers, 1975; Knight & Dagnall, 1967; Langolf, Chaffin, & Foulke, 1976; MacKenzie, Marteniuk, Dugas, Liske, & Bickmeier, 1987; McGovern, 1974), including, in particular, single-joint movements (Corcos, Gottlieb, & Agarwal, 1988; Crossman & Goodeve, 1983; Meyer, Abrams, Kornblum, Wright, & Smith, 1988).

A different relation among movement time, movement distance, and variability *assessed as the standard deviation* (SD) of final position or force level (cf. D-experiments, previous section) has been reported based on the studies of very fast force pulses or movements (Schmidt, Zelaznik, & Frank, 1978; Schmidt, Zelaznik, Hawkins, Frank, & Quinn, 1979; Sherwood & Schmidt, 1980; Wright & Meyer, 1983):

$$SD = a + b\frac{D}{T}, \tag{8.2}$$

where a and b are constants. In order to compare the results of these experiments with the findings by Fitts, a variable has been introduced characterizing dispersion of the final position, termed "effective target width," W_e. Equation 8.2 has been many times compared to Equation 8.1, although Equation 8.2 relates movement distance and time to a *variable characterizing actual performance* (SD, or W_e), while Equation 8.1 relates the same variables to a *variable characterizing externally imposed requirements* (W).

Relations of the type of Equation 8.2 were observed in past studies in a wide range of muscle forces. However, at forces more than 60% to 70% of the maximal voluntary contraction force, the linear relation breaks in a rather unexpected way: An increase in force leads to a decrease in variability (Schmidt & Sherwood, 1982). We will return to this seemingly paradoxical finding later.

Interpretations of Equation 8.2 have been based mostly on an assumed increase in variability of muscle force with an increase in absolute values of muscle force that is supposed to occur at higher movement speeds. Newell (1980; see also Newell & Carlton, 1988) has demonstrated that force variability does increase with muscle force in isometric conditions, although nonlinearly. However, during isotonic movements, a nonlinear relation between joint torque and its variability (Sherwood, Schmidt, & Walter, 1988) and between variability of movement time with an increase in movement amplitude (Newell, Carlton, & Carlton, 1982) has been reported. A variety of EMG and kinematic parameters have been shown to change nonlinearly with movement speed and amplitude (Carlton, Robertson, Carlton, & Newell, 1985).

Some of the experimental findings failed to obey either the logarithmic or the linear relations (Ferrell, 1965; Kvalseth, 1980) and suggested another form of the trade-off:

$$T = a + b\left(\frac{D}{W}\right)^p, \tag{8.3}$$

where $0 < p \leq 1$ (cf. Equation 8.1).

Recently, Meyer and his colleagues have advanced a dual-submovement model that tries to reconcile different expressions for the speed-accuracy trade-off (Meyer et al., 1988). A common mechanism has been suggested for the experimental data described by Equations 8.1, 8.2, and 8.3. Note that, from our view, such a mechanism cannot exist, in principle, because it is supposed to suggest a universal description of phenomena that are likely to emerge at different levels of a hypothetical system of movement production (see the next section). The model of Meyer et al. is based on a number of assumptions. The most important of them are these: (a) Aimed movement consists of several submovements and each consecutive submovement tries to improve the accuracy of the whole movement; (b) there is "motor noise" that may lead to inaccurate submovements that thus require corrections; and (c) each submovement is organized in such a way that the total movement duration is minimized.

When a subject initiates an aimed movement, he or she preplans the first movement component (the first submovement). In the course of the movement, the subject assesses its possible outcome. If the movement is likely to be successful, no corrections (further submovements) are introduced. If, however, the target is likely to be missed, the subject introduces a correction (a second submovement). Generally speaking, any number of corrections can be used during one movement. Note that the dual-submovement model implies the presence of a continuous feedback (e.g., visual) to the subject during the course of the movement that informs the subject of the possibility of success or failure.

Analysis of variability in the framework of the model has led Meyer and colleagues to a relation similar to Equation 8.3, with p being the inverse of the number of submovements for a particular task. An increase in the number of submovements leads to a decrease in p, so that for movements consisting of only one submovement, p = 1, and a linear relation is observed similar to Equation 8.2. The presence of only one submovement means that visual feedback has not been used for movement corrections, which is likely to happen during very fast movements or force impulses (cf. Schmidt et al., 1978; Schmidt, Zelaznik, Hawkins, Frank, & Quinn, 1979). For movements consisting of two submovements, a square-root relation was predicted and experimentally observed (Meyer et al., 1988):

$$T = a + b\sqrt{\frac{D}{W}}. \tag{8.4}$$

Slow movements performed under continuous visual control are likely to consist of many submovements leading eventually to a nearly logarithmic Fitts relation.

There are many factors that can theoretically affect variability of a performance variable (e.g., final position) by acting at different levels of the formation and realization of the motor command. In order to analyze their potential influence on the variability, one needs at least a hypothetical scheme of motor control. In the next section, we discuss a very general scheme for the production of voluntary movements, emphasizing possible sources of the speed-accuracy relations. Then we analyze the first hypothetical step of the formation of a "descending" motor command using a kinematic model. Later, the next step of processing of the motor command will be analyzed using the framework of the equilibrium-point hypothesis.

WHERE DOES THE VARIABILITY COME FROM?

Let us consider the following, very general scheme of the generation of a voluntary movement (Figure 8.1). A motor task, as defined by an experimenter or by the subject's will, is processed and leads to the formation of a goal of the future movement. Bernstein (1935, 1947, 1967) was the first to suggest that the planning of voluntary movements is performed in terms of kinematics in the external Cartesian space. Since that time, this view has gained considerable support from studies of various movements, although most of this support comes from multijoint movement studies (Berkinblit, Feldman, & Fukson, 1986; Flash, 1987, 1990; Hogan, 1984; Mussa Ivaldi, Morasso, & Zaccaria, 1989). Let us consider that even a simple single-joint movement is also planned at some very high "abstract" level in terms of kinematics in the external space. That is, the goal of the future movement is expressed in terms of its trajectory: The system is assumed to perform an internal simulation of a planned movement, leading to a function directly reflecting the desired trajectory x(t) (and, consequently, amplitude and speed of the future movement). This simulation certainly takes into account the conditions of movement execution, including predictable changes in the external load.

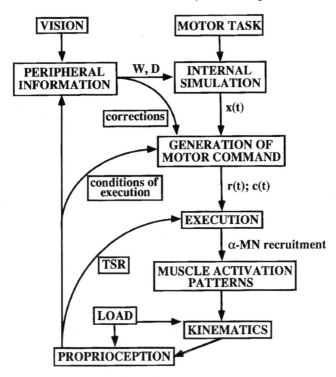

Figure 8.1 A block scheme of generation of a voluntary motor command. Presentation of a motor task leads to internal simulation of the future trajectory in the external Cartesian space. The simulated trajectory is later transformed into a motor command, which, according to the EP hypothesis, can be expressed for single-joint movements in the form of r(t) and c(t) functions. The r(t) and c(t) signals are processed by the peripheral mechanism of the tonic stretch reflex and lead to "actual" kinematics, which are also defined by the external load and its changes.

The next step would be to formulate a motor command in terms interpretable by the "lower" structures. Let us consider motor commands at this level as functions r(t) and c(t), corresponding respectively to reciprocal and coactivation commands originally introduced by Feldman (1980a, b; see details in the next section). Presently, the exact form of these commands does not matter. An important point is that they lead to changes in muscle activation and, consequently, to a movement that would ideally follow the planned trajectory x(t). If an actual trajectory deviates considerably from the planned one, corrections can be introduced at different levels. The internal simulation process can be restarted, leading to a new planned trajectory x'(t) that is processed through the whole chain in Figure 8.1 and leads to a new movement. The r(t) and c(t) functions (or their analogues) can be regenerated based on the same x(t). The feedback also acts at the level of execution of a command specified by the r(t) and c(t) functions (the mechanism of the tonic stretch reflex, TSR, according to the λ-model of Feldman, 1986).

In Figure 8.1, the described steps in generation of voluntary movement are shown as motor task, peripheral information, internal simulation, generation of motor command, and execution. The output of the execution step leads to α-motoneuron (α-MN) recruitment and eventually to limb kinematics. Instruction to the subject is assumed to be one of the components of the motor task. An explicitly presented target defines both the amplitude of the movement (D) and the target dimensions (W). It is mediated by peripheral information and affects the internal simulation step. If a subject has an opportunity to make several practice trials, he or she is likely to create an internal representation of the conditions of movement execution that will be used for transforming the results of the internal simulation into the r(t) and c(t) functions. Let us presently consider only very fast movements, that is, those that are preprogrammed by the subject. In our terms, this means that no corrections are introduced at the levels of internal simulation and generation of motor command.

Where can a speed-accuracy trade-off come from? Let us not consider the possibility of a wrong assessment of the movement amplitude (D) or misunderstanding of the instruction, although both these factors are likely to considerably contribute to the overall variability and consistent inaccuracy, at least in not very experienced subjects. Target width (W) acts only at the level of internal simulation. Therefore, if there is a trade-off induced by W, its mechanism should be described at the level of internal simulation. This mechanism should be able to explain how the prototype of movement speed coded in the x(t) function depends upon W (or rather upon the ratio W/D).

It is conceivable that at each level there is some inherent noise leading to suboptimal processing of the incoming information. That is, x(t) does not exactly correspond to the optimal planned trajectory, r(t) and c(t) do not ideally correspond to x(t), the TSR mechanism is not 100% reproducible, and, finally, the external conditions of movement execution (e.g., load, friction) vary slightly from trial to trial. Each of these factors will contribute to the *observed variability*, in particular to the variability in final position (movement amplitude). Because movement speed can theoretically be reflected at each level (e.g., in x(t), r(t), c(t), and external load), the dependence of variability in movement amplitude on movement speed (the speed-accuracy trade-off) can be defined by many factors related to the mechanisms acting at each of the levels.

So, we have come to two conclusions that are in fact axioms:

1. The dependence of movement speed upon target width (W/V trade-off) emerges at the level of internal simulation, which is the highest level of the planning of a voluntary movement.
2. The dependence of the dispersion of the final position upon movement speed (V/D trade-off) gets contributions from each of the levels shown in Figure 8.1.

The mechanism of the W/V trade-off is just one of the factors influencing the V/D trade-off, and it does so after several steps of neural and mechanical processing. Therefore, the results of the T-experiments (whose data correspond to the W/V trade-off) and D-experiments (whose data correspond to the V/D trade-off) cannot be directly compared even theoretically, because of the difference in the underlying hypothetical mechanisms. Let us note that originally this conclusion emerged as a result of an analysis of a kinematic model for single-joint movements (Gutman & Gottlieb, 1989; Gutman, Gottlieb, & Corcos, 1992). We are going to describe separately the hypothetical mechanisms leading to both kinds of trade-offs. First a kinematic model of Gutman will be used, and later we shall invoke the framework of the EP hypothesis.

THE KINEMATIC APPROACH

One can tentatively classify motor control models into three groups. First, there are physiological models, which try to explain different aspects of voluntary motor behavior basing on a certain hypothetical physiological mechanism of movement generation (λ- and α-versions of the EP hypothesis are examples of this approach; for review, see Feldman, 1986). There are also engineering models, based on some optimization criteria that are likely to be used by a 20th-century engineer asked to design a robot mimicking human motor behavior (for review, see Nelson, 1983; Seif-Naraghi & Winters, 1990). The third group of models is probably the least restrictive, because it does not try to guess how the central nervous system (CNS) controls movements (or rather to suggest it as a solution) but just attempts to describe kinematic aspects of voluntary movements with some "unbiased" mathematical procedure. This approach is founded on the ideas originally expressed by Bernstein (1947) that movements are planned in terms of their kinematics. Having a good mathematical description of movement kinematics can, therefore, be expected to be helpful in making a good guess about the central control mechanisms underlying the system's behavior, which is probably the ultimate goal of motor control studies.

The kinematic approach, by definition, cannot account for the movement dynamics. For example, it cannot describe changes in movement trajectories under different states of loading. Therefore, the results of kinematic modeling can give only an ideal representation of a movement for certain predictable time changes in dynamic variables (e.g., inertia, viscosity, stiffness, and external forces). In this sense, it gives an equivalent of a virtual trajectory at some level of movement planning. However, the results of kinematic modeling are always compared with actual experimentally observed kinematics (Berkinblit, Gelfand, & Feldman, 1986; Gutman et al., 1992), which apparently depend upon the dynamic factors.

This seeming inconsistency can be solved by adding a step of internal simulation to the hypothetical system of motor control (Figure 8.1). It is supposed that, at the level of internal simulation, the kinematic trajectory of the "most important point" (working point) is planned assuming unchanging dynamic conditions of movement execution. This modeled kinematic trajectory is used for the generation of a motor command that further leads to actual patterns of muscle activation and actual kinematics. The actual trajectory should very closely resemble its simulated template if the conditions of movement execution correspond to those used during the simulation. If the external conditions change, the kinematics can be very different from what has been planned centrally. For example, if a limb unexpectedly hits an obstacle, the kinematics will change abruptly, and the simulated virtual trajectory will be translated into a change in force with which the limb presses against the obstacle.

VARIABILITY IN A KINEMATIC MODEL

We are going to discuss the phenomena of variability using a kinematic model suggested recently by Gutman and Gottlieb (1989; Gutman et al., 1992). This model is based on an idea of nonlinear internal time and eventually leads to the following formula for the simulated movement trajectory:

$$x(t) = D \exp\left(-\frac{t^\alpha}{\tau_{mov}}\right), \tag{8.5}$$

where D is movement distance, α is a constant, and τ_{mov} is a movement time constant reflecting the planned movement time (MT). Analysis of simulated movement trajectories for different values of α has demonstrated a good fit to the experimentally observed patterns at $\alpha = 3$ (Gutman et al., 1992).

Figure 8.2 shows examples of trajectories of single-joint aimed movements corresponding to Equation 8.5 at $\alpha = 3$. Note that the simulated trajectories exhibit sigmoid time profile and bell-shaped velocity that are independent of the movement amplitude and duration (cf. Bouisset & Lestienne, 1974; Freund & Budingen, 1978; Hoffmann & Strick, 1986; Milner, 1986). The acceleration time is slightly shorter than the deceleration time (cf. Brown & Cooke, 1990; Nagasaki, 1989). It has been shown that the model can account for the kinematic patterns of the speed-sensitive and speed-insensitive strategies (Gottlieb et al., 1989).

As is known from Fitts's classic experiments (Fitts, 1954; Fitts & Peterson, 1964), MT increases with an increase in the ratio of distance to target width (D/W). Assume that a subject originally plans to perform a movement within a MT corresponding to some value of $\tau_{mov} = \tau^0_{mov}$. Now let us introduce an axiom: *Visual information about the target width and movement amplitude leads to a change in τ_{mov} defined by the value of the D/W ratio.* In the first approximation, this assumption can be modeled as a linear relation between the corrected value of τ_{mov} and D/W:

$$\tau_{mov} = \tau^0_{mov}\left(1 + \frac{D}{W}\right). \tag{8.6}$$

Experimental definition of MT is an ambiguous task mostly because of the difficulties in determining the moment of movement termination. A common although arbitrary approach is based on an assumption that the movement terminates at some moment when the endpoint achieves some location close to the target. Using this definition, MT can be expressed from Equations 8.5 and 8.6, assuming $\alpha = 3$, in the form

$$MT = \sqrt[3]{-\tau^0_{mov}\ln \xi} \ \sqrt[3]{1 + \frac{D}{W}}, \tag{8.7}$$

where ξ is a small positive value.

Equation 8.7 is an expression of the W/V trade-off in our model. It is different from the logarithmic relation of Fitts (see Equation 8.1). Figure 8.3b shows a curve corresponding to Equation 8.7 superimposed on the original Fitts data (Figure 8.3a). Although Equation 8.7 is monoparametric (with only the parameter τ^0_{mov}), it fits the

$$x(t) = D \exp \left(\frac{-t^{\alpha}}{\tau_{mov}} \right)$$

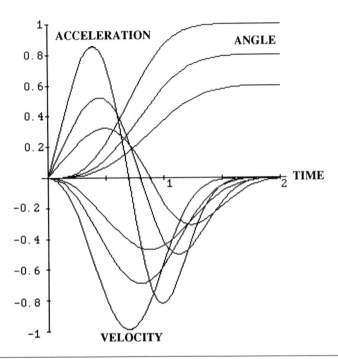

Figure 8.2 Simulated kinematic trajectories for different movement times. Note the sigmoidal shape of the trajectory, the bell-shaped velocities, and the asymmetry of the acceleration and deceleration phases. Scales are in arbitrary units.

data at least as well as the original biparametric Fitts' law (cf. Equation 8.1, dotted line in Figure 8.3). Note that fitting the Fitts data with a power function (Ferrell, 1965; Gan & Hoffman, 1988; Kvalseth, 1980) has led to the best results for the index of the power in the range of 1/2.6 . . . 1/3.2. It is consistent with our value of 1/3.

If a subject is reproducing an aimed movement several times, the standard deviation of the instantaneous position of the endpoint at different times during the movement can be considered as consisting of two components. The first component is due to variable errors in estimating the target location, and the second one results from variable errors in setting the time constant. We do not consider consistent errors in both components that can lead to relatively large errors in kinematic variables with a low variability. The second component reflects a trade-off between movement velocity and positional variability (a V/D trade-off, according to our terminology). Let us denote standard deviation of instantaneous endpoint position

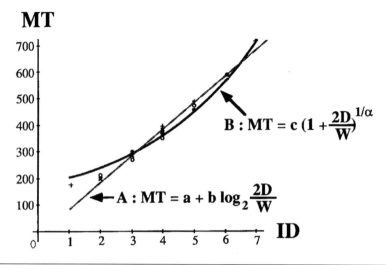

MT

B : MT = c $(1 + \frac{2D}{W})^{1/\alpha}$

A : MT = a + b $\log_2 \frac{2D}{W}$

ID

Figure 8.3   A simulated error curve (B) is superimposed on the original Fitts data. The Fitts curve is shown with the solid line (A). MT = movement time in ms; ID = index of difficulty.

defined by this component as $SD_{tr.-off}$. From Equation 8.5, one gets the following expression for the coefficient of variation of this component:

$$\frac{SD_{tr.-off}}{D} = \frac{SD(\tau_{mov})}{\tau_{mov}} \frac{t^3}{\tau_{mov}} \exp\left(-\frac{t^3}{\tau_{mov}}\right), \tag{8.8}$$

where $SD(\tau_{mov})$ is standard deviation of τ_{mov}.

Figure 8.4a shows a trajectory of a simulated reaching movement x(t) with two curves (dotted lines) corresponding to x(t) + $SD_{tr.-off}$ and x(t) − $SD_{tr.-off}$. An error curve corresponding to Equation 8.8 is shown in Figure 8.4b together with a simulated velocity profile. It has the following major features:

1. The error curve starts with a lag after the velocity curve. This lag reflects the fact that the error is accumulated in the beginning of the movement. Note that a similar conclusion has been drawn by van der Meulen, Gooskens, Denier van der Gon, Gielen, and Wilhelm (1990).
2. The error curve has a maximum in the middle of the movement and then drops down. Van der Meulen et al. (1990) have reported a similar increase in the positional variability in the middle of the movement, which dropped during the deceleration period. They have come to the conclusion that there is some compensatory mechanism correcting the errors accumulated during the first half of the movement by adjusting movement parameters during the deceleration period. Note, however, that in our model, this form of the error curve has emerged as a result of purely kinematic considerations without any compensatory mechanisms. Note also that ellipses of variability reported by

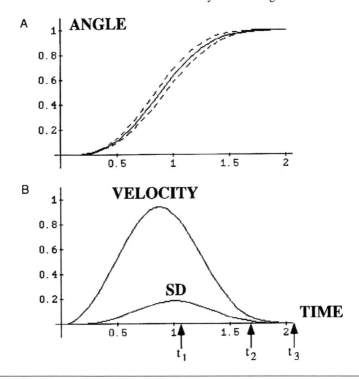

Figure 8.4   (A) A simulated trajectory (x[t], solid line) is shown together with lines corresponding to x(t) + SD(t) and x(t) − SD(t) (broken lines). (B) An error curve (positional error) is shown together with the velocity curve. Only the velocity-dependent component of the error is illustrated, implying that the original position is reproduced and the target location is assessed ideally. Note that the peak of the error curve (t_1) emerges somewhat later than the velocity peak, and the error declines during the later phase of the movement (t_2) and becomes zero at its end (t_3).

Darling and Cooke (1987) are in a good correspondence with the positional error curve in Figure 8.4b.

3. The value of this function at the beginning and end of the movement is zero. This means that the V/D trade-off exists only during the movement.

A positional-error curve shown in Figure 8.4b illustrates that if one measures endpoint location at some distant moment of time (t_3), when the movement has presumably stopped, there will be no velocity-dependent positional errors (i.e., no V/D trade-off). If the measurement is made at some time before the movement has terminated (t_2), a V/D trade-off is likely to occur. It will be maximal at a time (t_1) somewhat after the moment of peak velocity. Let us note that in many experiments studying V/D trade-offs, movements were considered stopped and measurements were made at a moment of touching the target (Schmidt et al., 1978, 1979; Sherwood, & Schmidt, 1980; Sherwood et al., 1988). The velocity of the endpoint at the moment of impact was presumably not zero (cf. with t_2 in Figure 8.4b). According to our model, this "minor" experimental detail could have led to a V/D trade-off that would be otherwise absent.

Until now, we have discussed the possibility of the emergence of two kinds of trade-offs at the level of internal simulation. Other kinds of trade-offs can emerge at later stages of the generation of voluntary movement. The next sections are devoted to the analysis of trade-offs that can emerge at the level of generation of motor command (Figure 8.1). For this purpose, we shall use the framework of the equilibrium-point hypothesis.

THE EQUILIBRIUM-POINT HYPOTHESIS

Let us use a commonly accepted definition of a *complex system* as a system whose behavior cannot be predicted from the properties of its components. For example, a termite is a complex system because it is impossible to predict its behavior based on properties of the cells of its body. A community of termites is a complex system whose behavior cannot be predicted based on characteristics of its components (i.e., individual termites). A jungle as an ecological system is a complex system as compared with groups of animals. However, behavior at each level can be described more or less adequately using an appropriate taxonomy (cf. Turvey, 1990). Such taxonomies are based on certain specific variables that describe general characteristics of behavior at a chosen level. For example, the height of the termite cone or the total amount of food consumed by the colony per day may be good variables for describing a community of termites. Generally, the number of such variables is relatively small, so that using them makes description of a complex system look simple. Defining such variables is a crucial step in the analysis of any complex system.

In a human body, a neuron is a complex system as compared with elementary particles. The system of motor control of one muscle is complex as compared with individual neural and muscular cells. The system of motor control of multijoint movements is likely to be complex as compared with control of individual muscles.

For control of a single muscle, a variable, λ, the threshold of the tonic stretch reflex (TSR), has been suggested by Feldman (1966a, 1986) for describing the muscle behavior. Originally, the EP hypothesis (the λ-model) emerged as a formal language for describing a body of experimental data on muscle force-length characteristic curves in animals (Matthews, 1959) and joint torque-angle characteristic curves in humans (Asatryan & Feldman, 1965; Feldman, 1966a, 1966b). The most important finding was the lack of intersections of the curves recorded with different "fixed" descending commands. These curves were termed "invariant characteristics" (ICs). Although fixing the descending command could not be controlled ideally, and the experimenters relied upon the subject's ability "not to intervene voluntarily," the results themselves, and their reproducibility, corroborated the used approach.

The lack of intersections of the ICs let Feldman introduce a monoparametric description of the process of control of one muscle. The chosen variable represented the muscle length at which recruitment of autogenic α-MNs started during a slow muscle stretch, that is, the threshold (λ) of the TSR. If λ is set by some hypothetical control system, the muscle behaves like a nonlinear spring, and its force and length are dependent upon the external load. λ is a typical complex-system variable in the sense that it describes muscle behavior that is mediated by complex interactions between the receptor cells, interneurons, motoneurons, and muscle cells.

If one considers a joint controlled by an agonist and an antagonist muscle, its central control can be described with a pair of variables corresponding to values of λ for the two muscles, λ_{ag} and λ_{ant}. Torque-angle characteristics of a pair of muscles are illustrated in Figure 8.5, a and b. Behavior of the joint will be defined by the algebraic sum of the two ICs and the external load. In the length range where both muscles are activated, the resultant characteristic of the net joint compliance can be moved along the length axis by unidirectional shifts of λ_{ag} and λ_{ant}, or its slope can be changed by contradirectional shifts of λ_{ag} and λ_{ant} due to the nonlinear form of the ICs.

Another pair of variables can be used for describing central regulation of a joint (Feldman, 1980a, 1980b; Feldman & Latash, 1982). These variables represent half a sum and half a difference between λ_{ag} and λ_{ant} and were introduced in order to draw close parallels with physiological notions of reciprocal activation and coactivation:

$$r = \frac{1}{2} (\lambda_{ag} + \lambda_{ant}) \qquad (8.9)$$

$$c = \frac{1}{2} (\lambda_{ag} - \lambda_{ant}) \qquad (8.10)$$

Simultaneous shift of both λ_{ag} and λ_{ant} in one direction increases the level of activation (as reflected by muscle torque) for one of the muscles and decreases it for the other muscle, which can be associated with reciprocal excitation r of one of the muscles (Figure 8.5a). Shifts of both λs in opposite directions increase or decrease activation of both muscles simultaneously, which can be associated with changes in a coactivation command c (Figure 8.5b). The pairs $(\lambda_{ag}, \lambda_{ant})$ and (r, c) are equivalent, and either one can be used for analysis of single-joint control.

There are three basic variables characterizing a single-joint movement: distance (D), speed (e.g., average speed V_a), and time (t). They are obviously interrelated so that only two of them can be chosen independently:

$$D = V_a t. \qquad (8.11)$$

Planning a movement at the internal simulation level (Figure 8.1) leads, according to our assumptions, to specifying the kinematic trajectory in the external space. Therefore, all three variables (D, V_a, and t) are represented at this level as well. Further processing of a simulated trajectory x(t) leads to generation of motor commands that, according to the λ-model, can be expressed in the form of r(t) and c(t) functions.

Two patterns of r(t) and c(t) have been suggested for control of fast single-joint movements (Figure 8.6, a-c). One of them (Figure 8.6a; Abdusamatov & Feldman, 1986; Feldman, Adamovitch, Ostry, & Flanagan, 1990) includes a shift of r at a constant rate. Another pattern implies a more complex nonmonotonic shift in r (an N-curve; see Figure 8.6b), which has been supported by both experimental

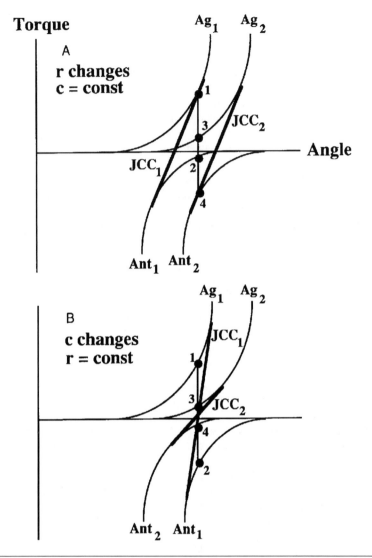

Figure 8.5 (A) Unidirectional shift of the agonist and antagonist torque-angle characteristics (thin lines) leads to a change in the reciprocal variable r and an unchanged coactivation variable c. The position of the joint compliant characteristic (JCC) changes, whereas its slope remains constant. In the illustrated example, for a given joint position (thin vertical line), the level of agonist activation decreased (compare torque levels corresponding to Points 1 and 3), whereas the antagonist activation level increased (compare torque levels corresponding to Points 2 and 4). (B) Contradirectional shifts of the agonist and antagonist torque-angle characteristics (thin lines) lead to a change in the coactivation variable c and an unchanged reciprocal variable r. The slope of the joint compliant characteristic (JCC) changes, whereas its position remains relatively constant. In the illustrated example, for a given joint position (thin vertical line), the level of both agonist and antagonist activation decreased (compare torque levels corresponding to Points 1, 2 and 3, 4).

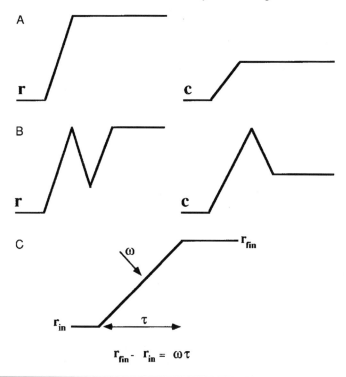

Figure 8.6   Two hypothetical patterns of r(t) and c(t) controlling a fast single-joint movement. (A) The Feldman version implies a rapid shift in r to the desired final value and a simultaneous increase in c. (B) The Latash and Gottlieb version implies a delayed shift in λ_{ant}, leading to a nonmonotonic N-shaped r and a transient increase in c. (C) The simplest version of control of a single-joint movement with a shift in r from an initial value r_{in} to a final value r_{fin} at a rate ω over a time τ.

observations (Latash & Gottlieb, 1991) and modeling based on certain optimization techniques (Hasan, 1986; Hogan, 1984). However, even in this case, one may consider the first arm of the N-curve as being characteristic for movement speed and amplitude. Let us presently simplify the analysis by ignoring the possible contribution of variations of the coactivation component c(t) and of the latter parts of the N-shaped function r(t).

Let us consider, for simplicity, that a simulated trajectory x(t) for an isotonic movement leads to a shift in r at a rate ω over a time τ from an initial value of r_i to a final value r_f (Figure 8.6c). Two of the three interrelated variables (r_f, τ, and ω; $r_f - r_i = \omega\tau$) can rather directly be associated with parameters of the trajectory; r_f with movement amplitude, and τ with movement time. One may associate ω with a loosely defined notion of movement speed. Although the speed of any movement is a time function, we suggest that ω can in fact be considered a central variable. That is, when subjects want to control movement speed, they act in terms of controlling ω. Naive subjects have no problems in understanding an instruction to perform movements of different amplitudes "with the same speed," although, strictly speaking, this instruction is senseless, because movement velocity is a time function. These subjects show different peak and average velocities while being confident that they obey the instruction. It has been demonstrated (Gottlieb, Corcos, Agarwal, & Latash, 1990) that the instruction to move "with the same speed" leads to predictable kinematic and EMG patterns in a wide range of "speeds" (e.g., characterized by peak speeds).

THREE STRATEGIES

In this section, we are going to assume that the internal simulation process has been completed and move to the next level, generation of motor command (Figure 8.7; cf. Figure 8.1). Therefore, the hypothesized influence of target width (W) on simulated movement kinematics is supposed to be already taken into account. As a result, we are going to deal only with the V/D trade-offs, that is, changes in the variability of the final position with movement speed, and with other kinds of trade-offs that are likely to emerge *at the level of execution of a voluntary movement* rather than at the level of its planning.

Assume that at the level of generation of motor command, two of the three control variables can be specified with some level of variability that is inherent to the control process and cannot be reduced (cf. "motor noise" in Meyer et al., 1988). We assume this inherent variability to be independent of the absolute values of the variables. Variability of the third, dependent variable will be defined not only by the variability of the first two, but also by their magnitude, which can theoretically lead to different kinds of trade-offs. This approach suggests three basic strategies corresponding to three pairs of independently chosen controlled variables. Note that a new assumption has emerged: The motor task can exert its influence not only at

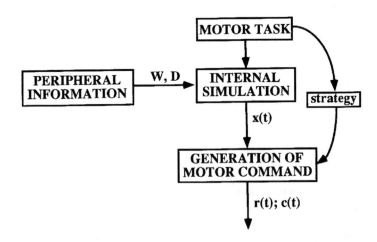

Figure 8.7   The upper part of the scheme in Figure 8.1 is supplemented with a possibility of choosing a strategy at the level of generation of motor command.

the level of internal simulation but also at the level of generation of motor command (the curved arrow in Figure 8.7).

It is not evident, a priori, whether a subject is able to choose any of the three variables (r_f, ω, and τ) with a lower inherent variability than the others. Everyday experience suggests that "going to the same point" is easier than "going with the same speed," which, in turn, is easier than "going within the same time." This assumption can be formally expressed as

$$\frac{d\tau}{\tau} > \frac{d\omega}{\omega} > \frac{dr_f}{r_f}, \tag{8.12}$$

where $d\tau$, $d\omega$, and dr_f are hypothesized variable errors in specifying τ, ω, and r_f characterizing inherent variability of the nervous system. We will assume that these errors are mutually independent and also independent of the absolute values of τ, ω, and r_f.

A series of experiments was performed trying to validate, at least partially, Inequality 8.12 (Latash, 1989; Latash & Gottlieb, 1990). During nearly isotonic single-joint elbow flexion movements, the subjects were required in different series to reproduce one of the following three parameters: (a) movement amplitude starting from different initial positions and moving "as fast as possible," (b) final position starting from different initial positions and moving "as fast as possible," and (c) movement time for movements of different amplitudes. During a short training period (10 practice trials before the first series in each of the three conditions), the subjects knew in advance which parameter they would be asked to reproduce and tried to "remember" it. The results have demonstrated a marked difference in the coefficients of variation of the instructed variable for the three experimental conditions (Table 8.1). The lowest values of the coefficients of variation were observed during reproduction of final position, whereas the highest values corresponded to reproduction of movement time.

Better reproduction of final position rather than movement distance has been demonstrated in a number of studies (Keele & Ells, 1972; Marteniuk, 1973; Marteniuk, Shields, & Campbell, 1972; Wrisberg & Winter, 1985), although the opposite results have also been reported (Bock & Eckmiller, 1986). Better reproduction of final position is a natural consequence of the EP hypothesis, because in isotonic movements, final position is assumed to be directly related to the value of one of the main control variables, r_f. So, in order to reproduce a final position, the subjects could "remember" r_f and shift r at the highest available rate ω (as required by the instruction).

On the other hand, reproduction of movement amplitude while moving at the highest speed requires control of time of r changes (τ). The subjects supposedly choose the highest available value of ω and reproduced τ. Higher variability in these experiments suggests that a hypothesized irreducible error in τ is higher than in r ($d\tau/\tau > dr_f/r_f$; cf. Inequality 8.12).

Movement time, in the way it was measured in the experiments (from the first deflection of the acceleration to the crossing of the zero line by a linear extrapolation of the descending deceleration curve), also depends upon two control variables, τ and r_f or τ and ω. However, changing movement distance required the subjects not

Table 8.1   Averaged Values of the Coefficients of Variation (CV) and Standard Deviations (SD) for the Experiments With Movement Time (MT), Amplitude (AMP), and Final Position (FP) Reproduction

	MT-1	AMP-1	FP-1	MT-all	AMP-all	FP-all
Mean CV	9.4	6.1	3.5	10.8	7.0	4.1
SD	3.3	2.6	1.4	4.3	3.4	1.5

t-Values for the Student's t-Test

	First series		All series	
	t	p	t	p
MT vs. AMP	1.9	.12	3.5	0.003
MT vs. FP	3.6	0.015	6.9	0.0001
AMP vs. FP	2.4	0.06	4.2	0.007

Note. The coefficients of variation (in %) and their standard deviations are shown in the upper part of the table separately for the three instructions requiring reproduction of movement time (MT), amplitude (AMP), and final position (FP). Data for the first sets of trials (MT-1, AMP-1, and FP-1) and for all the trials in all three series (-all) are shown separately. t-values are based on the Student's two-tailed paired t-tests.

to reproduce, but rather to adjust, both variables. So it is not surprising that the variability in these experiments was higher than in the first two series.

Let us introduce the following assumption:

> In different situations, humans have a choice of specifying any two of the three variables (r_f, ω, and τ). This choice depends upon requirements particular to the movement task and will influence the interrelation between movement speed (or movement time), distance, and accuracy.

ω/τ-Strategy (Open-Eyes Strategy)

If a task is to move quickly to a visually presented stationary target, an "ω/τ" strategy will be used. A visual target implies a possibility of using visual information for planning and terminating the movement at an appropriate time; that is, by fixing ω in advance and adjusting τ based on the visual feedback (cf. iterative correction models, Keele & Posner, 1968). This strategy will also be used for movement amplitude control (Bock & Eckmiller, 1986). The error in the final position (Δr_f) will depend upon the errors in the regulated variables ($d\omega$ and $d\tau$). We will use "d" for inherent errors (see above) in the centrally regulated variables, and "Δ" for the calculated error in the dependent variable.

Let us assume that a subject specifies τ with an error $d\tau$ and ω with an error $d\omega$. Then, the error in specifying r_f will be

$$\Delta r_f = \omega d\tau + \tau d\omega. \qquad (8.13)$$

Equation 8.13 implies a close-to-linear relation between spatial variability (Δr_f) and movement time if movement amplitude is increased without changing the speed variable ω. Note that we consider *smooth* movements that, in the framework of the dual-submovement model (Meyer et al., 1988), correspond to a single submovement, which, in turn, leads to the Schmidt relation (Equation 8.3). An increase in movement speed from ω to ω_1 without changing movement distance ($r_1 = r_f$) is associated with a decrease in the time variable from τ to τ_1, because $\tau_1 = r_1/\omega_1$. However, because it is assumed that $\omega d\tau > \tau d\omega$ (cf. Inequality 8.12), the first term on the right side of Equation 8.13 will dominate, and an increase in movement speed will lead to an increase in the variability of final position. This might be the basis of an aspect of the speed-accuracy trade-off that predicts increased variability for movements at higher speeds.

r/ω-Strategy (Closed-Eyes Strategy)

Another strategy will be used in tasks requiring preplanning a movement to a point (cf. Meyer, Smith, & Wright, 1982; Schmidt et al., 1979; Schmidt & McGown, 1980) if, for example, visual feedback is not available. Let us suppose that a subject chooses values of r_f and ω with accuracy dr_f and $d\omega$. According to an assumption above, these errors are independent of the values of r_f and ω. Then timing errors can be calculated as

$$\Delta\tau = \frac{\omega dr_f - r_f d\omega}{\omega^2}. \qquad (8.14)$$

The original assumption (Inequality 8.12) leads to $\omega dr_f < r_f d\omega$ and $\Delta\tau < 0$. Then, Equation 8.14 implies that an increase in movement speed, preserving distance, will lead to a decrease in absolute timing variability. Similarly, an increase in movement distance, preserving the speed variable ω, will lead to an increase in absolute value of timing variability. Because dr_f and $d\omega$ are postulated to be independent of both r_f and ω, there is *no* reason to expect any kind of speed-accuracy trade-off when the subject uses this strategy.

We have induced this strategy in a pilot series of experiments in which the subjects were required to make movements of different amplitudes to a small target with their eyes closed (Latash, 1989; Latash & Gottlieb, 1990). The data were compared to performance of the same subjects in a more classical paradigm when the eyes were open. Standard deviation of the final position was used as an index of variability; peak speed was used as an index of movement speed. Two findings are of interest. First, movements with closed eyes were characterized by considerably higher peak speeds (Figure 8.8) *without an increase in variability of the final position*. Second, movements over the largest amplitude performed with open eyes had a higher index of variability than movements over smaller amplitudes, whereas movements with closed eyes did not show an apparent increase in variability with movement amplitude (note that an increase in variability with movement amplitude was reported by Chapanis, Garner, & Morgan, 1949, for "blinded" movements although in a different experimental situation).

A similar lack of effect of movement speed upon the variability of the location of the "working point" (finger and thumb joint end-positions during grasping) has been reported by Darling, Cole, and Abbs (1988) for both fast and relatively slow movements. The slower movements in these experiments demonstrated a number of submovements (two to four) that, according to the dual-submovement hypothesis, should lead to different kinds of trade-offs. However, no detectable trade-off was observed. This result can be expected from our model if the subjects were using the closed-eyes strategy although their eyes were open (i.e., not using visual feedback for corrections during the grasping task).

The classical Fitts paradigm implies that the subjects are trying their best to perform at the highest possible speed when some level of accuracy is imposed. However, our results suggest that changing the instruction can "force" the subjects to perform better, that is, to increase the speed without deteriorating the accuracy. For example, the first blocks of movements in our experiments were performed by the subjects in both conditions in virtually identical situations after the same number of practice trials. The only difference was in open or closed eyes and somewhat different instructions. The instruction in the experiments with open eyes was typical

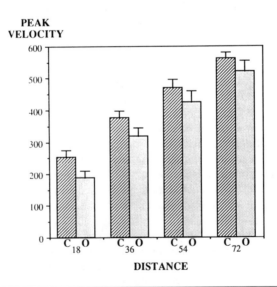

Figure 8.8 Averaged across nine subjects, peak velocity as a function of movement distance in the experiments with open eyes (O) and closed eyes (C). Standard errors are shown with vertical bars. Each value is based on 45 trials.

for the Fitts-type experiments (i.e., "be fast and accurate"; Corcos et al., 1988; Fitts, 1954). In the experiments with closed eyes, the instruction was to "go as fast as possible to the final position; consider the final position as a point." Therefore, it is not surprising that the subjects in the latter series performed at higher peak velocities (and correspondingly shorter movement times). However, they did it without an increase in the final position variability. In fact, even the subjects with plenty of experience in similar experiments were quite surprised by their very high accuracy and consistency in the experiments with closed eyes. After participating a couple of times in such experiments, the subjects became "spoiled," realizing that it was possible to use the closed-eyes strategy even with the eyes open, that is, not to pay any attention to the target and to go to a final position (the middle of the target bar) as fast as possible, thus improving the performance in seeming violation of Fitts' law.

In fact, all these experiments originated from the experience of one of the authors (MLL) as a "bad subject" in a modification of the classical Fitts-type experiments that required reproduction of the same movement hundreds of times and, therefore, were extremely boring. Somewhere in the middle of the experiment, the subject started to invent different ways of "doing the same," in particular tried to move without looking at the screen. To his surprise (and, later, to the justified dismay of the experimenter!), the performance improved dramatically.

One may suggest another explanation for these findings. We have already mentioned reports by Schmidt and Sherwood (1982; see also Sherwood & Schmidt, 1980) that an increase in force amplitude during force pulses over 60% to 70% has led to a paradoxical drop in force variability. An increase in peak speed in our experiments suggests a considerable increase in inertial joint torques (and, correspondingly, muscle forces). Inertial torques during very fast and large movements have been shown to be close to those observed during attempts at isometric maximal voluntary contraction (Gottlieb et al., 1989). Therefore, the lack of increase in variability in our experiments and observations by Schmidt and Sherwood can share the same nature. Newell and his colleagues (Carlton et al., 1985; Newell et al., 1982) have suggested that the cause of the drop in variability of kinetic and kinematic variables is the reduction of variable timing error with an increase in movement velocity. However, this explanation seems to substitute one mystic phenomenon with another, at a "higher" level of the system of movement production.

r/τ-Strategy (Badminton Strategy)

The third option is to fix r_f (distance) and τ. The error in ω is

$$\Delta\omega = \frac{\tau dr_f - r_f d\tau}{\tau^2}.$$

(8.15)

Using this assumption (Inequality 8.12), $\tau dr_f < r_f d\tau$ and $\Delta\omega < 0$. Taking Equation 8.15 into account, variability in ω (or related speed parameters, e.g., peak speed) will increase with an increase in movement distance and will decrease with an increase in movement time.

This strategy is likely to be used in tasks that require getting to a point at a certain moment of time, as when playing badminton. The prediction above implies that hitting a shuttlecock with a desired force (relative speed at the impact) is harder for longer and faster swing movements. We have not consulted professional badminton players, but our own amateur experience suggests that this is likely to be true. This strategy can probably be modeled in experiments with single-joint movements when a subject is asked to hit a *moving* target at a certain point on the screen, being both fast and accurate.

Note that, for all three strategies, the primary task was the same: "Move to a target, being both fast and accurate." However, the details of the experimental situation and of the instruction have been shown to exert significant effects upon the subject's performance. It seems amazing that, even during maximally constrained smooth single-joint movements with only one degree of freedom, there is still room for choosing a strategy of how to perform "as fast as possible." It seems that the word *strategy* is appropriate for describing this situation, because there is apparently a choice in meeting the primary task requirements.

CONCLUDING COMMENTS

Analysis of any aspect of behavior of a complex system, including motor behavior of humans, requires an absolutely necessary first step of introduction of a general scheme and an appropriate taxonomy. We have attempted to make this step for the analysis of motor variability. Consideration of a very general scheme of production of voluntary movement has led to the possibility of emergence of two aspects of motor variability. One of them, the dependence of movement velocity upon target width (the W/V trade-off, or Fitts' law) is determined only at the level of internal simulation of the planned movement. A simple kinematic model of the processes at this level has demonstrated the experimentally observed features of the W/V trade-off. The other aspect, the dependence of final position variability on movement velocity (the V/D trade-off), can theoretically be affected at all the hypothetical steps of the generation of voluntary movement. We have come to a conclusion that the V/D trade-off is influenced at the level of internal simulation only if positional variability is measured during the movement but not after the movement termination. Analysis of the next step of processing of a hypothetical motor command has been performed using the framework of the λ-model, although any kind of control of equilibrium states would lead to similar conclusions. This analysis has suggested that, even during very simple single-joint movements, there is room for control strategies that can be used by the subjects based on certain specific features of the motor task and can manifest themselves at the level of motor variability.

REFERENCES

Abdusamatov, R.M., & Feldman, A.G. (1986). Description of the electromyograms with the aid of a mathematical model for single joint movements. *Biophysics*, **31**, 549-552.

Asatryan, D.G., & Feldman, A.G. (1965). Functional tuning of the nervous system with control of movements or maintenance of a steady posture. I. Mechanographic analysis of the work of the limb on execution of a postural task. *Biophysics*, **10**, 925-935.

Berkinblit, M.B., Feldman, A.G., & Fukson, O.I. (1986). Adaptability of innate motor patterns and motor control mechanisms. *Behavioral and Brain Sciences*, **9**, 585-638.

Berkinblit, M.B., Gelfand, I.M., & Feldman, A.G. (1986). A model for the control of multijoint movements. *Biofizika, **31**, 128-138.

Bernstein, N.A. (1935). The problem of interrelation between coordination and localization. *Archives of Biological Sciences*, **38**, 1-35. (In Russian)

Bernstein, N.A. (1947). *On the construction of movements*. Moscow: Medgiz. (In Russian)

Bernstein, N.A. (1967). *The co-ordination and regulation of movements*. Oxford: Pergamon Press.

Bock, O., & Eckmiller, R. (1986). Goal-directed arm movements in absence of visual guidance: Evidence for amplitude rather than position control. *Experimental Brain Research*, **62**, 451-458.

Bouisset, S., & Lestienne, F. (1974). The organization of simple voluntary movement as analyzed from its kinematic properties. *Brain Research*, **71**, 451-457.

Brown, S.H., & Cooke, J.D. (1990). Movement-related phasic muscle activation. I. Relations with temporal profile of movement. *Journal of Neurophysiology*, **63**, 455-464.

Carlton, M.J., Robertson, R.N., Carlton, L.G., & Newell, K.M. (1985). Response timing variability: Coherence of kinematic and EMG parameters. *Journal of Motor Behavior*, **17**, 301-319.

Chapanis, A., Garner, W.R., & Morgan, C.T. (1949). *Applied experimental psychology: Human factors in engineering design*. New York: Wiley.

Corcos, D.M., Gottlieb, G.L., & Agarwal, G.C. (1988). Accuracy constraints upon rapid elbow movements. *Journal of Motor Behavior*, **20**, 255-272.

Crossman, E.R.F.W., & Goodeve, P.J. (1983). Feedback control of hand movement and Fitts' law. *Quarterly Journal of Experimental Psychology*, **35A**, 251-278.

Darling, W.G., Cole, K.J., & Abbs, J.H. (1988). Kinematic variability of grasp movements as a function of practice and movement speed. *Experimental Brain Research*, **73**, 225-235.

Darling, W.G., & Cooke, J.D. (1987). A linked muscular activation model for movement generation and control. *Journal of Motor Behavior*, **19**, 333-354.

Feldman, A.G. (1966a). Functional tuning of the nervous system with control of movement or maintenance of a steady posture. II. Controllable parameters of the muscle. *Biophysics*, **11**, 565-578.

Feldman, A.G. (1966b). Functional tuning of the nervous system with control of movement or maintenance of a steady posture. III. Mechanographic analysis of execution by man of the simplest motor task. *Biophysics*, **11**, 667-675.

Feldman, A.G. (1980a). Superposition of motor programs. I. Rhythmic forearm movements in man. *Neuroscience*, **5**, 81-90.

Feldman, A.G. (1980b). Superposition of motor programs. II. Rapid flexion of forearm in man. *Neuroscience*, **5**, 91-95.

Feldman, A.G. (1986). Once more on the equilibrium-point hypothesis (λ model) for motor control. *Journal of Motor Behavior*, **18**, 17-54.

Feldman, A.G., Adamovitch, S.V., Ostry, D.J., & Flanagan, J.R. (1990). The origin of electromyograms—explanations based on the equilibrium point hypothesis. In J.M. Winters & S.L.-Y. Woo (Eds.), *Multiple muscle systems. Biomechanics and movement organization* (pp. 195-213). New York: Springer-Verlag.

Feldman, A.G., & Latash, M.L. (1982). Afferent and efferent components of joint position sense: Interpretation of kinaesthetic illusions. *Biological Cybernetics*, **42**, 205-214.

Ferrell, W.R. (1965). Remote manipulation with transmission delay. *IEEE Transactions on Human Factor in Electronics*, **6**, 24-32.

Fitts, P.M. (1954). The information capacity of the human motor system in controlling the amplitude of movements. *Journal of Experimental Psychology*, **47**, 381-391.

Fitts, P.M., & Peterson, J.R. (1964). Information capacity of discrete motor responses. *Journal of Experimental Psychology*, **67**, 103-112.

Flash, T. (1987). The control of hand equilibrium trajectories in multi-joint arm movements. *Biological Cybernetics*, **57**, 257-274.

Flash, T. (1990). The organization of human arm trajectory control. In J.M. Winters & S.L.-Y. Woo (Eds.), *Multiple muscle systems. Biomechanics and movement organization* (pp. 282-301). New York: Springer-Verlag.

Flowers, K.A. (1975). Ballistic and corrective movements on an aiming task. *Neurology* (Minneapolis), **25**, 413-421.

Freund, H.J., & Büdingen, H.J. (1978). The relationship between speed and amplitude of the fastest voluntary contractions of human arm muscles. *Experimental Brain Research*, **55**, 167-171.

Gan, K., & Hoffman, E.R. (1988). Geometrical conditions for ballistic and visually controlled movements. *Ergonomics*, **31**, 829-839.

Gottlieb, G.L., Corcos, D.M., & Agarwal, G.C. (1989). Strategies for the control of voluntary movements with one mechanical degree of freedom. *Behavioral and Brain Sciences*, **12**, 189-250.

Gottlieb, G.L., Corcos, D.M., Agarwal, G.C., & Latash, M.L. (1990). Organizing principles for single joint movements. III: Speed-insensitive strategy as a default. *Journal of Neurophysiology*, **63**, 625-636.

Gutman, S.R., & Gottlieb, G.L. (1989). A solution for the problem of multi-joint redundancy by minimizing joint angle increments. *Abstracts of the Society for Neuroscience*, **15**, 606.

Gutman, S.R., Gottlieb, G.L., & Corcos, D.M. (1992). Exponential model of a reaching movement trajectory with non-linear time. *Comments on Theoretical Biology*, **2**, 357-384.

Hasan, Z. (1986). Optimized movement trajectories and joint stiffness in unperturbed, inertially loaded movements. *Biological Cybernetics*, **53**, 373-382.

Hoffmann, D.S., & Strick, P.L. (1986). Steptracking movements of the wrist in humans: Kinematic analysis. *Journal of Neuroscience*, **6**, 3309-3318.

Hogan, N. (1984). An organizational principle for a class of voluntary movements. *Journal of Neuroscience*, **4**, 2745-2754.

Keele, S.W., & Ells, J.G. (1972). Memory characteristics of kinesthetic information. *Journal of Motor Behavior*, **4**, 127-134.

Keele, S.W., & Posner, M.I. (1968). Processing of visual feedback in rapid movements. *Journal of Experimental Psychology*, **77**, 155-158.

Knight, A.A., & Dagnall, P.R. (1967). Precision in movements. *Ergonomics*, **10**, 321-330.

Kvalseth, T.O. (1980). An alternative to Fitts' law. *Bulletin of the Psychonomical Society*, **16**, 371-373.

Langolf, G.D., Chaffin, D.B., & Foulke, J.A. (1976). An investigation of Fitts' law using a wide range of movement amplitudes. *Journal of Motor Behavior*, **8**, 113-128.

Latash, M.L. (1989). Implications of the equilibrium-point hypothesis for the variability of aimed hand movements. In C.J. Worringham (Ed.), *Spatial, temporal, and electromyographical variability in human motor control* (pp. 16-17). Ann Arbor, MI: University of Michigan.

Latash, M.L., & Gottlieb, G.L. (1990). Equilibrium-point hypothesis and variability of the amplitude, speed and time of single-joint movements. *Biofizika*, **35**, 870-874.

Latash, M.L., & Gottlieb, G.L. (1991). Reconstruction of joint compliant characteristics during fast and slow movements. *Neuroscience*, **43**, 697-712.

MacKenzie, C.L., Marteniuk, R.G., Dugas, C., Liske, D., & Bickmeier, B. (1987). Three-dimensional movement trajectories in Fitts' task: Implications for control. *Quarterly Journal of Experimental Psychology*, **39A**, 629-647.

Marteniuk, R.G. (1973). Retention characteristics of motor short-term memory cues. *Journal of Motor Behavior*, **5**, 249-259.

Marteniuk, R.G., Shields, K.W., & Campbell, S. (1972). Amplitude, position, timing, and velocity as cues in reproduction of movement. *Perception and Motor Skills*, **35**, 51-58.

Matthews, P.B.C. (1959). The dependence of tension upon extension in the stretch reflex of the soleus of the decerebrate cat. *Journal of Physiology*, **47**, 521-546.

McGovern, D.E. (1974). *Factors affecting control allocation for augmented remote manipulation*. Unpublished doctoral dissertation, Stanford University.

Meyer, D.E., Abrams, R.A., Kornblum, S., Wright, C.E., & Smith, J.E. (1988). Optimality in human motor performance. Ideal control of rapid aimed movements. *Psychological Review*, **95**, 340-370.

Meyer, D.E., Smith, J.E.K., & Wright, C.E. (1982). Models for the speed and accuracy of aimed movements. *Psychological Review*, **89**, 449-482.

Milner, T.E. (1986). Controlling velocity in rapid movements. *Journal of Motor Behavior*, **18**, 147-161.

Mussa Ivaldi, F.A., Morasso, P., & Zaccaria, R. (1989). Kinematic networks. A distributed model for representing and regularizing motor redundancy. *Biological Cybernetics*, **60**, 1-16.

Nagasaki, H. (1989). Asymmetric velocity and acceleration profiles of human arm movements. *Experimental Brain Research*, **74**, 319-326.

Nelson, W. (1983). Physical principles for economies of skilled movements. *Biological Cybernetics*, **46**, 135-147.

Newell, K.M. (1980). The speed-accuracy paradox in movement control: Errors of time and space. In G.E. Stelmach & G. Requin (Eds.), *Tutorials in motor behavior* (pp. 501-510). Amsterdam: North-Holland.

Newell, K.M., & Carlton, L.G. (1988). Force variability in isometric responses. *Journal of Experimental Psychology: Human Perception and Performance*, **14**, 24-36.

Newell, K.M., Carlton, L.G., & Carlton, M.J. (1982). The relationship of impulse to timing error. *Journal of Motor Behavior*, **12**, 47-56.

Schmidt, R.A., & McGown, C. (1980). Terminal accuracy of unexpected loaded rapid movements: Evidence for a mass-spring mechanism in programming. *Journal of Motor Behavior*, **12**, 149-161.

Schmidt, R.A., & Sherwood, D.E. (1982). An inverted-U relation between spatial error and force requirements in rapid limb movements: Further evidence for the impulse-variability model. *Journal of Experimental Psychology: Human Perception and Performance*, **8**, 158-170.

Schmidt, R.A., Zelaznik, H.N., & Frank, J.S. (1978). Sources of inaccuracy in rapid movement. In G.E. Stelmach (Ed.), *Information processing in motor control and learning* (pp. 183-203). New York: Academic Press.

Schmidt, R.A., Zelaznik, H., Hawkins, B., Frank, J.S., & Quinn, J.T. (1979). Motor output variability: A theory for the accuracy of rapid motor acts. *Psychological Review*, **86**, 415-451.

Seif-Naraghi, A.H., & Winters, J.M. (1990). Optimized strategies for scaling goal-directed dynamic limb movements. In J.M. Winters & S.L.-Y. Woo (Eds.), *Multiple muscle systems. Biomechanics and movement organization* (pp. 312-334). New York: Springer-Verlag.

Sherwood, D.E., & Schmidt, R.A. (1980). The relationship between force and force variability in minimal and near-maximal static and dynamic contractions. *Journal of Motor Behavior*, **12**, 75-89.

Sherwood, D.E., Schmidt, R.A., & Walter, C.B. (1988). Rapid movements with reversals in direction. II. Control of movement amplitude and inertial load. *Experimental Brain Research*, **69**, 355-367.

Turvey, M.T. (1990). The challenge of a physical account of action: A personal view. In H.T.A. Whiting, O.G. Meijer, & P.C.W. van Wieringen (Eds.), *The natural-physical approach to movement control* (pp. 57-92). Amsterdam: Vrije Universitat Press.

Van der Meulen, J.H.P., Gooskens, R.H.J.M., Denier van der Gon, J.J., Gielen, C.C.A.M., & Wilhelm, K. (1990). Mechanisms underlying accuracy in fast goal-directed arm movements in man. *Journal of Motor Behavior*, **22**, 67-84.

Wright, C. E. & Meyer, D.E. (1983). Conditions for a linear speed-accuracy trade-off in aimed movements. *Quarterly Journal of Experimental Psychology*, **35A**, 279-296.

Wrisberg, C.A., & Winter, T.P. (1985). Reproducing the end location of a positioning movement: The long and short of it. *Journal of Motor Behavior*, **17**, 242-254.

Acknowledgments

The authors are grateful to Drs. Daniel Corcos, Gerald Gottlieb, Karl Newell, and Chuck Walter for many helpful suggestions and to Mr. Om Paul for the miraculous programming support. The study was partially supported by NIH grant AR 33189.

Chapter 9

How Spinal Neural Networks Reduce Discrepancies Between Motor Intention and Motor Realization

Daniel Bullock, José L. Contreras-Vidal

Boston University, Boston, Massachusetts

This chapter attempts a rational, step-by-step reconstruction of many aspects of the mammalian neural circuitry known to be involved in the spinal cord's regulation of opposing muscles acting on skeletal segments. Mathematical analyses and local circuit simulations based on neural membrane equations are used to clarify the behavioral function of five fundamental cell types, their complex connectivities, and their physiological actions. These cell types are α-*MNs*, γ-*MNs*, *IaINs*, *IbINs*, and Renshaw cells. It is shown that many of the complexities of spinal circuitry are necessary to ensure near-invariant realization of motor intentions when descending signals of two basic types independently vary over large ranges of magnitude and rate of change. Because these two types of signal afford independent control, or factorization (F), of muscle length (LE) and muscle tension (TE), our construction was named the FLETE model (Bullock & Grossberg, 1988b, 1989). The present chapter significantly extends the range of experimental data encompassed by this evolving model.

INTRINSIC THREATS TO INVARIANT REALIZATION OF MOVEMENT PLANS

When motor variability measured with respect to some criterion of accuracy is of significant magnitude and cannot be attributed to across-trial variability in the performer's representation of the criterion, then it indicates an inability of the movement control system to perfectly realize a movement plan. Most physically realizable plans that go seriously awry do so because the environment changes unpredictably across and within occasions of action. Correspondingly, much of what we call skill is constituted by outright avoidance of, or anticipatory compensations for, environmental contingencies that would otherwise prevent accurate realization of plans.

When we speak of environmental contingencies in discussions of motor variability, we usually think of inertial loads, gravity fields, support-surface properties, and the like. However, as Fitts (1954) realized when he discussed variability in terms of limitations on information transmission within the motor control system, all within-organism processes interposed between internal representation of the criterion of accuracy and the measured effector are in a sense part of the environment that affects realization of the plan.

If so, part of any comprehensive theory of motor variability will be a specification of what may go awry along the series of neural, neuromuscular, and musculoskeletal transductions interposed between movement plan specification and movement realization. Such a theory should ultimately encompass a full array of sources of "normal" variability as well as the kind of variability exhibited in motor system diseases such as Parkinson's syndrome. In this chapter, we summarize recent work on a mathematical model of neural networks we believe to be involved in animal movement planning and realization. Our general thesis is that much of the structure of the neural networks used to realize movement plans can be understood only as adaptations that compensate for intrinsic sources of motor variability.

COMPETENCIES OF THE SKELETOMOTOR CONTROL SYSTEM

To begin, it is necessary to clarify the range of motor control tasks the biological system needs to perform. We then show how to build the system's competence incrementally from biological materials. The tasks we address can be clarified by a series of questions: How can a limb be rotated to, and stabilized at, a desired angle? How can joint stiffness be varied independently of joint angle? How can the speed of movement from an initial to a desired final angle be controlled under conditions of low joint stiffness? Simultaneous achievement of these abilities requires a rather complex neuromuscular system, with several identifiable subsystems. However, all of these tasks require that each muscle be able to generate a wide range of tensions at any of the lengths it may assume as the limb (into which it inserts) rotates. More stringently, all of these tasks require factorization (F), or independent control, of muscle length (LE) and muscle tension (TE). This overarching theme led us to choose *FLETE* as the acronym for our original mathematical model of the neuromuscular system (Bullock & Grossberg, 1988b, 1989). An enhanced version of the FLETE model, shown in Figure 9.10, is assembled in steps in the next several sections.

HOW A LIMB CAN BE ROTATED TO, AND STABILIZED AT, A DESIRED JOINT ANGLE

Figure 9.1 schematizes a system in which two opposing muscles insert into a distal limb segment connected by a rotary or hinge joint to a more proximal limb segment, in a manner reminiscent of the human forearm's connection to the upper arm. Suppose such a forearm segment is initially at rest and that $F_1 = F_2$, where the F_i, $i = 1, 2$, denote the pulling forces exerted by the opponent muscles. Then the limb can be set in motion by making the forces F_1 and F_2 unequal. The limb can be halted and stabilized at a new joint angle if the forces reequilibrate as it approaches that angle and if the system is capable of automatically generating whatever new muscle force imbalance may be needed to return it to the desired angle after any deviation (e.g., after the rotating limb initially overshoots the desired angle).

As many observers have noted (Cooke, 1980; Feldman, 1986; Polit & Bizzi, 1979), muscle itself seems to have evolved to help provide this basic functionality. Essentially, muscle is springy tissue with a neurally controllable contractile component, which gives it a neurally modifiable threshold length for force development (Rack & Westbury, 1969). To highlight this essence, at risk of oversimplification, we can assume that the force F_i developed by a muscle is a threshold-linear function of its length L_i, its fixed resting length Γ_i, its stiffness, k, and its neurally modifiable contractile state, C_i:

$$F_i = k[L_i - (\Gamma_i - C_i)]^+, \qquad (9.1)$$

where notation $[\omega_i]^+$ means max $(0, \omega_i)$. So if $\omega_i = L_i - (\Gamma_i - C_i) > 0$, $F_i = k \cdot \omega_i$; if $\omega_i \leq 0$, $F_i = k \cdot 0 = 0$.

Equation 9.1 shows that a muscle is springlike in that it develops a force only when stretched to a length L_i greater than the effective threshold length, $\Gamma_i - C_i$. However, it also shows that muscle is more versatile than an ordinary spring, because this threshold can be neurally adjusted by varying the muscle's state of contraction, C_i.

To see the implications for movement production and postural stabilization, suppose that at time $t = 0$, the limb is at rest and not subject to any nonmuscular forces. Then $F_1(0) = F_2(0)$, so by Equation 9.1,

$$k[L_1(0) - (\Gamma_1 - C_1(0))]^+ = k[L_2(0) - (\Gamma_2 - C_2(0))]^+, \qquad (9.2)$$

or equivalently

$$k[L_1(0) - \Gamma_1 + C_1(0)]^+ = k[L_2(0) - \Gamma_2 + C_2(0)]^+. \qquad (9.3)$$

Now suppose that a neural process increments C_1 from $C_1(0)$ to $C_1(0) + \Delta C$ and decrements C_2 from $C_2(0)$ to $C_2(0) - \Delta C$. These changes create the inequality

$$k[L_1(0) - \Gamma_1 + C_1(0) + \Delta C]^+ > k[L_2(0) - \Gamma_2 + C_2(0) - \Delta C]^+, \qquad (9.4)$$

which by Equation 9.1 implies that $F_1 > F_2$. This force imbalance will set the limb in motion. Because the force imbalance favors Muscle 1, and because the lengths L_1 and L_2 of the two muscles are rigidly linked by their common insertion, the resultant rotation will shorten Muscle 1 and lengthen Muscle 2 by the same amount, ΔL.

When ΔL becomes equal to ΔC,

$$k[L_1(0) - \Delta L - \Gamma_1 + C_1(0) + \Delta C]^+ = k[L_2(0) + \Delta L - \Gamma_2 + C_2(0) - \Delta C]^+. \quad (9.5)$$

Thus the rotation itself annihilates the force imbalance, and a force balance reappears. Moreover, it is easy to see that if the inertia of the limb causes the rotation to continue further, then Muscle 1 shortens to $L_1 < L_1(0) - \Delta L$, and Muscle 2 lengthens to $L_2 > L_2(0) + \Delta L$, for $\Delta L = \Delta C$. Thus $F_2 > F_1$, and the limb will slow, then stop, and ultimately reverse the overshoot. Assuming some damping, the limb settles down at the joint angle corresponding to the set of muscle lengths $(L_1(0) - \Delta L, L_2(0) + \Delta L)$. Generally, for fixed Γ_i and k, a full range of stable points (L_1, L_2) between (L_1^{max}, L_2^{min}) and (L_1^{min}, L_2^{max}) can be created by changing the sign and magnitude of the difference between C_1 and C_2.

To gain control over contractile states C_1 and C_2, there must exist (see Figure 9.1) opposing alpha-motoneuron pools $\alpha\text{-}MN_1$ and $\alpha\text{-}MN_2$ whose axons project to, and allow differential activation of, the opposing muscles. Let the activation levels of the opposing motoneuron pools be designated by M_1 and M_2. Then, as shown in Figure 9.1, a motor intention—a neural state corresponding to specification of a desired joint angle—can take the form of a pattern of signals (A_1, A_2) suitable for

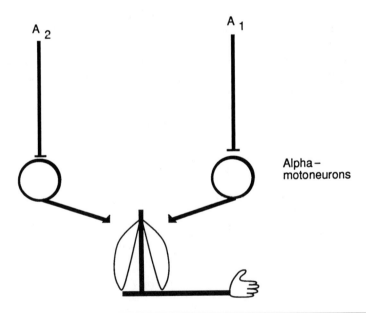

Figure 9.1   Opponent alpha-motoneuron pools provide neural control over muscle contractile states and thereby the balance of forces acting across the joint. A motor intention can take the form of a pattern of descending signals (A_1, A_2) to the $\alpha\text{-}MN$ pools.

inducing a differential pattern of activation (M_1, M_2) across the motoneuron pools, which in turn creates a pattern (C_1, C_2) of contractile states, thereby creating a new stable point (L_1, L_2) for the limb. So if nothing goes wrong along the way, motor intention (A_1, A_2) will invariably lead to desired joint angle $\theta(L_1, L_2)$. But we now show that many things can go wrong along the way, and all the circuitry that distinguishes Figure 9.10 from Figure 9.1 will be motivated by the animal's need to reduce errors of motor realization to a minimum.

HOW CAN JOINT STIFFNESS BE VARIED INDEPENDENTLY OF JOINT ANGLE?

Historically, analyses of what can go wrong in motor realization have focused on how nonmuscular forces imposed by the external world can complicate the story we were able to keep simple by assuming that only muscular forces were acting on the limb. We now supplement such analyses by turning our attention inward, to neural, neuromuscular, and musculoskeletal sources of error variance.

Suppose that we want to improve upon the Figure 9.1 system by adding the ability to stiffen a joint in varying degrees while holding joint angle constant. Such joint stiffening is known (e.g., Humphrey & Reed, 1983) to involve simultaneous increments to the contractile states of the joint's opponent muscles, which results in cocontraction.

The simplest way for the higher nervous system to effect a cocontraction is to add a signal, whose magnitude we will denote by P, to both components of the signal pattern (A_1, A_2). Then the net input to the opponent α-MNs would be $(A_1 + P, A_2 + P)$. This modification is shown in Figure 9.2. If variations in P always have the same effect on muscle force production in both opponent channels, then a limb initially at equilibrium at a desired angle θ will remain there as P varies: Though F_1 and F_2 will both increase or decrease, their difference will remain unchanged. Such an invariant relationship between (A_1, A_2) and θ under variations of cocontraction signal P can be summarized by

$$\theta(A_1, A_2) = \theta(A_1 + P, A_2 + P). \tag{9.6}$$

Threats to this desirable invariance property can arise in each channel at both transduction steps interposed between the convergence of signal A_i with signal P and the generation of muscular force F_i. First consider whether the pattern of α-MN activities (M_1, M_2) remains sensitive to the difference $(A_1 - A_2)$ as P increases. Maintaining such sensitivity is a necessary (but not sufficient) condition for invariance property (Equation 9.6). In fact, it is easy to show that without compensatory neuronal circuitry the opponent α-MN pattern (M_1, M_2) becomes insensitive to $A_1 - A_2$ as P increases.

To demonstrate this, we need to write a differential equation to describe fluctuations of motoneuron activation M_i through time. The simplest biologically plausible form is

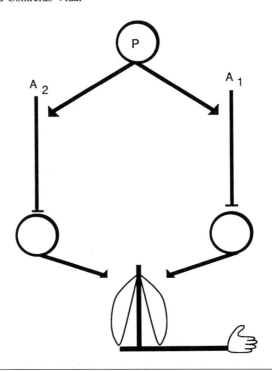

Figure 9.2   Joint stiffness can be controlled by adding descending signal P to both signals A_1 and A_2. The signal P is capable of producing high levels of cocontraction of the opponent muscles.

$$\frac{d}{dt}M_i = -\delta M_i + (B - M_i)(A_i + P).$$ (9.7)

If $(A_i + P) = 0$, that is, if there is no input, this equation reduces to

$$\frac{d}{dt}M_i = -\delta M_i,$$ (9.8)

which for $\delta > 0$ implies that activities M_i spontaneously decay toward zero. For a constant positive input $A_i + P$, M_i will approach an equilibrium value found by setting $(d/dt)M_i = 0$ in Equation 9.7 and solving for M_i, which yields

$$M_i = B\frac{A_i + P}{A_i + P + \delta}.$$ (9.9)

When $A_i + P \gg \delta$, we have

$$M_i \approx B.$$ (9.10)

This result means that when the total excitatory input $A_i + P$ to an α-MN grows large enough, both opponent α-MN activities M_1 and M_2 will approximate the same maximal activity level B even if $A_1 \neq A_2$. Thus in the network of Figure 9.2, invariance (Equation 9.6) cannot be preserved, because any difference $M_1 - M_2$ corresponding to the difference $A_1 - A_2$ is progressively eroded as increments in signal P make both the $A_i + P$ progressively larger.

This loss of sensitivity to input differences near the upper bound of neuronal activity is often called *saturation* in the literature on neural networks (e.g., Grossberg, 1973). Grossberg noted decades ago that saturative loss of sensitivity to differences existing across pattern-processing channels can be prevented by allowing the channels to interact laterally via inhibitory signals. These inhibitory signals ensure that M_i remains less than B, as can be seen by examining the equilibrium states implied by a differential equation for the motor pool activation M_i when that pool is subject to an inhibitory input I_j in addition to the excitatory input $A_i + P$. One model for how inhibition acts is captured by

$$\frac{d}{dt}M_i = -\delta M_i + (B - M_i)(A_i + P) - M_i I_j , \qquad (9.11)$$

where $i, j = \{1, 2\}$. Here the new term, $-M_i I_j$, says that the effect of the inhibitory input on M_i is proportional to M_i (a situation that arises naturally when inhibitory signals act by changing membrane conductance; cf. Grossberg, 1973, 1982; Hodgkin & Huxley, 1952). When $(d/dt)M_i = 0$ in Equation 9.11, we find

$$M_i = B\frac{A_i + P}{A_i + P + \delta + I_j} . \qquad (9.12)$$

Now, even if $A_i + P \gg \delta$, M_i remains less than B as long as I_j is an increasing function of P.

Figure 9.3, a-c, illustrates three designs by which lateral inhibitory interactions might be added. In 9.3a, an inhibitory copy of the excitatory descending signal $A_i + P$ is sent to the opponent α-MN pool. Unfortunately, in this design the inhibitory inputs to the α-MNs would grow as fast with increasing P as the excitatory inputs. As a result, the equilibrium pattern of M_i values (M_1, M_2) would tend to add to a constant, with M_1/M_2 linearly related to the ratio $(A_1 + P)/(A_2 + P)$, as in Grossberg (1973). This design would prevent loss of sensitivity due to saturation, but would not allow independent control of $M_1 + M_2$ versus $M_1 - M_2$, as shown in Figure 9.4, a and d, respectively.

In Figure 9.3b, inhibitory interneurons are introduced to mediate the between-channel interactions. If these interneurons obey an equation like

$$\frac{d}{dt}I_i = -\delta I_i + (B - I_i)(A_i + P), \qquad (9.13)$$

then their activity grows as a slower-than-linear function of $A_i + P$. As a result, inhibitory inputs to α-MNs do not grow as quickly with increasing P as do excitatory inputs to α-MNs, and the sum $M_1 + M_2$ remains sensitive to P while $M_1 - M_2$ remains sensitive to $A_1 - A_2$, as shown in Figure 9.4, b and e, respectively.

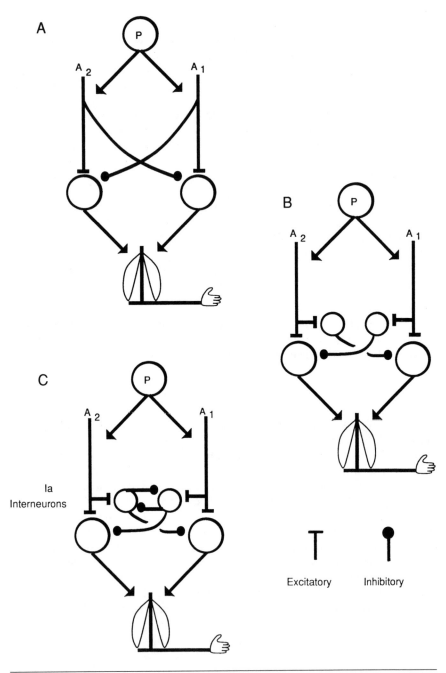

Figure 9.3 (A-C) Three designs for alleviating saturative loss of sensitivity by α-*MN* pools to the difference $A_1 - A_2$ when signal P becomes large. Current experimental evidence accords with the design in (C); the added model interneurons have the same connectivity as *Ia* interneurons known to exist in vivo.

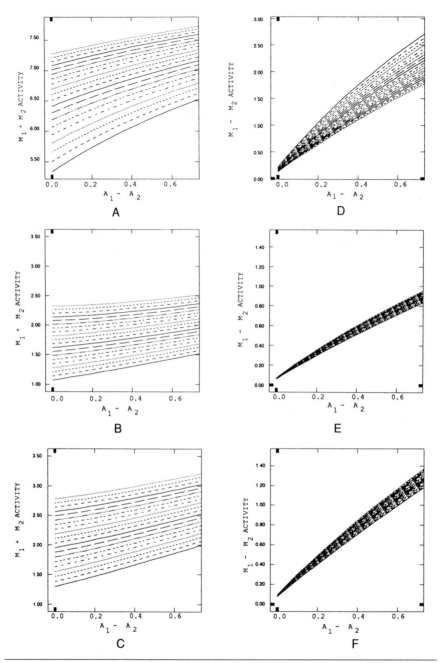

Figure 9.4 (A-C) Total motoneuron activity $M_1 + M_2$ as a function of positioning command $A_1 - A_2$ and stiffness command P for the designs shown in Figure 9.3, A-C, respectively. (D-F) The difference in motoneuron activities $M_1 - M_2$ as a function of positioning command $A_1 - A_2$ and stiffness command P for the networks shown in Figure 9.3, A-C, respectively. (See legend in Figure 9.12.)

Figure 9.3c shows an even better design, in which each inhibitory interneuron pool inhibits the other as well as the opponent α-MN pool. This mutual-inhibitory design prevents the possibility that very large values of P will lead to saturation of the interneurons themselves, which if allowed would reintroduce the problem of $M_1 - M_2$ insensitivity to $A_1 - A_2$ due to α-MN saturation. Consistent with these expectations, Figure 9.4, c and f, shows that, relative to network 9.3b, network 9.3c exhibits a 25% greater dynamic range in the response of $M_1 + M_2$ to variation in P and a 50% greater dynamic range in the response of $M_1 - M_2$ to $A_1 - A_2$.

In vivo, inhibitory interneurons called IaINs are known to exist with the signed connectivity, vis-à-vis α-MNs and each other, shown in Figure 9.3c. This close correspondence motivates the label given these model interneurons in Figure 9.3c. The need for a pathway to mediate reciprocal inhibition between opponent muscle channels was demonstrated by Sherrington in his experiments on the stretch reflex. The IaINs of Figure 9.3c are known also to receive feedback from stretch receptors and from Renshaw cells (both of which are introduced into the model in later sections). Our remarks on the computational necessity for IaINs are compatible with, but also extend, prior proposals regarding their function. In particular, we agree that for rapid movements to be energetically efficient, it is important to prevent an antagonist muscle from retarding the action of an agonist muscle. This would be difficult if the only process for lowering α-MN activation levels were the passive decay process of Equation 9.8, especially if, as appears to be true in vivo, δ is in the range $0 < \delta \ll 1$ (see Kiehn, 1991). Reciprocal inhibition via IaINs allows rapid decrementing of activity in antagonist alpha-motoneuronal pools. Both the rapid antagonist resetting and the saturation-prevention functions of IaINs are implicit in Equation 9.11.

We next need to ask whether the pattern (M_1, M_2) induced by motor intention $(A_1 + P, A_2 + P)$ is faithfully registered in the pattern (C_1, C_2) of contractile states induced by activities M_1, M_2. To see why it would not be, in the absence of further structure, consider first a simple differential equation describing changes in contractile state through time:

$$\frac{d}{dt}C_i = (B_i - C_i)M_i - \delta C_i. \qquad (9.14)$$

This says that a sufficiently large neural input M_i can push contractile state C_i up to the limit B_i and that contractile state relaxes at rate δ. In vivo, B_i corresponds to the maximal number of muscle fibers that can be simultaneously activated.

The presence of an upper bound B_i means that the ability of the C_i to remain sensitive to differences across the M_i can saturate if the range of M_i is too large relative to B_i. This problem can be avoided, given the neural provisions that avoid M_i saturation in Figure 9.3c, if B_i is itself a function of M_i. In fact, this is assured in vivo by a motor unit design principle together with a progressive recruitment rule. Motor units are composed of distinct alpha-motoneurons that project to distinct sets of contractile fibers. Moreover, within the motoneuron pools, there exist distributions of activation thresholds such that larger net excitatory inputs to the pool recruit larger numbers of motor units. Because smaller α-MNs are recruited earlier, and larger later, this rule has been called the size principle of motoneuron recruitment

(Henneman, 1957, 1985). Figure 9.5 schematizes the addition of a size principle to our model by showing a stacked series of α-MN cells with increasing diameter.

Another aspect of the size principle becomes comprehensible if we revise Equation 9.14 to better describe the real-time behavior of muscle whose shortening is being opposed. When muscle is stretched by a contraction-opposing force, it exhibits an erosion of its contractile state and a consequent reduction of produced force called "yielding" (e.g., Nichols, 1984). To accommodate this effect we modify Equation 9.14 to

$$\frac{d}{dt}C_i = \beta_i[(B_i - C_i)M_i - \delta C_i] - [F_e - \Gamma_F]^+, \tag{9.15}$$

where F_e is an external force that erodes contractile state C_i if it exceeds a threshold force Γ_F. Now observe that under conditions of cocontraction, F_e can be a large force produced by the antagonist muscle!

Given Equation 9.15, we can find the equilibrium contractile state associated with a given value of M_i under conditions that cause yielding. Setting $(d/dt)C_i = 0$ and solving for C_i, we have

$$C_i = \frac{M_i B_i - \left(\dfrac{[F_e - \Gamma_F]^+}{\beta_i}\right)}{M_i + \delta}. \tag{9.16}$$

A consideration of the negative term in the numerator of this expression shows that the erosive effect of the opposing force is smaller for larger values of β_i. Because parameter β_i in Equations 9.15 and 9.16 corresponds to the rate at which fiber twitches change contractile state, Equation 9.16 helps us understand why twitch contraction rate is another facet of the size principle of motor unit recruitment: Unless β_i also covaries with M_i and B_i, yielding would often cause a premature saturation in the development of contractile state, and thereby in force production. With the covariation, an appropriately innervated muscle is able to develop and maintain a very wide range of force-production levels at any fixed muscle length, even in the presence of a highly activated antagonist muscle or a contraction-opposing external force.

Earlier we noted that reciprocal inhibition via IaINs ensured rapid decrementing of activity in antagonist α-MN pools. Though this helps prevent an antagonist muscle force from retarding the action of an agonist muscle, it would be insufficient if high antagonist contractile states had long relaxation times, corresponding to a small passive decay term δ in Equation 9.15. The passive decay rate is critical, because there is no process (other than agonist-induced antagonist yielding!) for actively supressing antagonist muscle force. However, it can be seen in Equation 9.15 that β_i multiplies both the contraction and the relaxation rates. Therefore, the covariation of M_i, B_i, and β_i also assures that higher antagonist contractile states decay more quickly than smaller contractile states when antagonist α-MNs are inhibited. IaIN reciprocal inhibition and the parametric covariation that constitutes the size principle therefore work together to enable rapid movement reversals that can be energetically efficient because of minimal opposition by lingering antagonist forces.

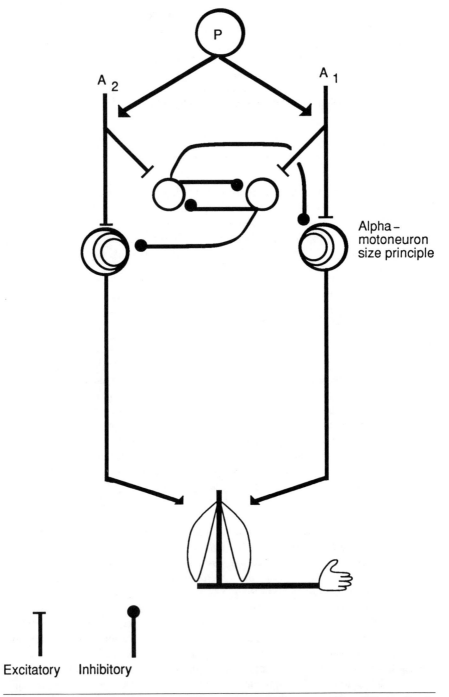

Figure 9:5   Alpha-motoneurons have different sizes, which correspond to different thresholds for recruitment.

INDEPENDENCE LOST—AND REGAINED

Unfortunately, introduction of the size principle by itself causes a loss of independent control of joint angle by (A_1, A_2) and of joint stiffness by signal P. Figure 9.6, a and b, illustrates the problem. In the scenario of Figure 9.6a, we show that signals A_1 and A_2 have been set to values sufficient to activate only small α-MNs, and we suppose that the limb has rotated to that angle θ at which the force imbalance created by the difference $A_1 - A_2$ would be annihilated by the length changes in L_1 and L_2 associated with the rotation as in Equation 9.5. Now suppose that the animal anticipates a need to hold the assumed posture more rigidly, and that it attempts to do so by stepping up cocontractive signal P, as shown in Figure 9.6b.

Under all initial choices of A_1, A_2, other than $A_1 = A_2$, signal P will cause deeper recruitment in one muscle channel than the other. Because of the size principle, part of the signal P is subjected to greater amplification in that channel where recruitment is deeper, and a resultant force imbalance develops in that channel's favor. In consequence, the animal who had hoped to further stabilize its limb at its initial posture by stiffening the joint would instead experience a large, unwanted, limb rotation!

Figure 9.7, a and b, shows illustrative plots of $M_1 - M_2$ and $M_1 + M_2$ as functions of $A_1 - A_2$ for a wide range of choices of cocontraction signal P. Figure 9.7a illustrates the desirable property that $M_1 + M_2$ increases with P for any fixed value of $A_1 - A_2$. However, Figure 9.7b shows that the function relating $M_1 - M_2$ to

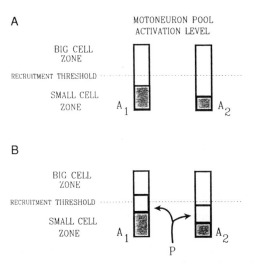

Figure 9.6 When opponent motoneuron populations obey the size principle, a cocontractive signal P sent to both can disrupt the joint position code: (A) Signals A_1 and A_2 activate only small motoneurons, and thereby determine a balance of muscular forces and a corresponding equilibrium joint position. (B) With $A_1 > A_2$, cocontractive signal P is subjected to greater amplification in Channel 1 than in Channel 2. Unless compensated, this would cause an undesired joint rotation.

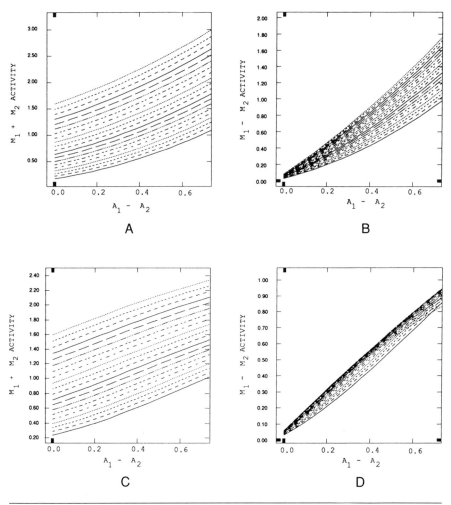

Figure 9.7   (A, B) Total motoneuron activity $M_1 + M_2$ and the difference between moto-neuron activities $M_1 - M_2$ as a function of $A_1 - A_2$ and P for the network in Figure 9.5. (C, D) The same variables for the network in Figure 9.8.   (See legend in Figure 9.12.)

$A_1 - A_2$ is far from invariant across different choices of P: The unequal amplification wrought by the size principle makes the slope of this function five times as sensitive to variation in P as the analogous function in Figure 9.4, e and f.

Such an unequal amplification could be neurally compensated if it could be measured. Because the α-MNs, which are directly linked to muscle, are usually looked upon as the last stage of the nervous system, it might be supposed that the unequal amplification could be measured only by its effect on muscle, that is, by way of stretch receptors embedded in the opponent muscles. However, because muscle contraction is slow relative to the unequal neural amplification, a significant rotation error could develop before it could be halted by feedback from stretch receptors.

In fact, the α-*MNs* project both directly to muscle and directly to a class of cells called Renshaw cells, whose function has not been well understood (Shepherd, 1990). In earlier papers (Bullock & Grossberg, 1988b, 1989), we proposed that these Renshaw cells were perfectly situated to measure and compensate for unequal amplifications of a cocontractive signal *P* sent to both opponent muscle channels. As shown in Figure 9.8, each muscle control channel has its own Renshaw cell pool, which receives excitatory inputs from its channel's α-*MN* pool. The Renshaw pool in turn sends inhibitory signals to its own channel's α-*MN* and *IaIN* pools, as well as to the opponent channel's Renshaw pool.

Consider the consequences of this signed connectivity under the conditions depicted in Figure 9.6b. When *P* causes deeper recruitment in α-*MN* pool 1, the Renshaw population in channel 1 becomes much more active than in channel 2. This causes α-MN_1 to be subjected to significantly greater Renshaw inhibition than α-MN_2, thus partially correcting channel 1's expected force advantage. Simultaneously, α-MN_2 is disinhibited by two pathways:

$$R_1 \rightarrow R_2 \rightarrow \alpha\text{-}MN_2$$

and

$$R_1 \rightarrow IaIN_1 \rightarrow \alpha\text{-}MN_2 .$$

This further compensates for channel 1's expected force advantage by increasing the force developed by channel 2. Simulations reported by Bullock and Grossberg (1988b, 1989) showed that Renshaw-mediated compensation could virtually eliminate undesired joint rotations associated with variations in *P*, for any given choice of (A_1, A_2). In our theory, then, Renshaw cells play a key role in ensuring the invariance principle formalized by Equation 9.6. The dramatic difference made by adding Renshaw cells can be seen by comparing the noninvariant operating characteristic shown in Figure 9.7b with the near-invariant operating characteristic of Figure 9.7d, which summarizes the results of simulations of the Renshaw-augmented neural network schematized in Figure 9.8. Figure 9.7c shows that the near-invariant relation between $A_1 - A_2$ and $M_1 - M_2$ under changes of *P* shown in 9.7d coexists with a near-linear relationship between *P* and $M_1 + M_2$, as required by Equation 9.6. A closely related thesis regarding linearization has been independently advanced by Akazawa and Kato (1990), who, however, restricted their discussion to α-*MNs* and Renshaws in one channel.

Because of the historical status of stretch receptors as preeminent sources of error compensation, it is worth noting that the Renshaw cells, unlike stretch receptors, respond at neural rates with miniscule lags as the net signal to muscle evolves, without having to wait for error to appear after a lag introduced by the relative sluggishness of muscle, and without having to send news of the error back over a long cable. Thus, unlike stretch receptors, the Renshaws can correct for size-principle-based distortions *before* they can cause significant joint rotations. They act preemptively, whereas a stretch reflex—properly speaking—always compensates reactively. This is why we believe it is a very serious mistake, both as a matter of linguistic usage and as a matter of conceptualization, to attempt, as some have done, to subsume virtually the entire motor apparatus under the rubric of the "stretch reflex."

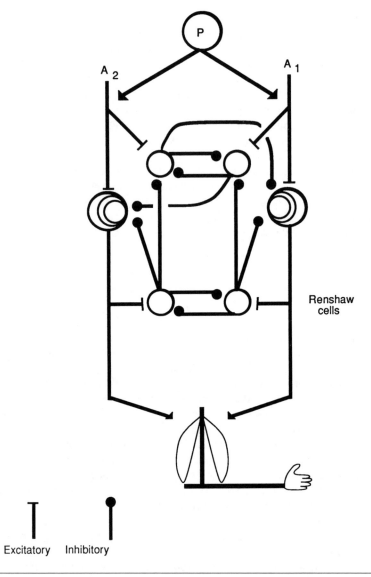

Figure 9.8   Renshaw cells "tap the cables" running from α-*MN* pools to muscles. Their negative feedback to α-*MNs* can compensate for distortion introduced by the size principle.

FATIGUE AND BIAS COMPENSATION BY A FORCE FEEDBACK PATHWAY

Another threat to invariant realization of motor intentions arises from muscle fatigue. Although less catastrophic than muscle yielding, fatigue similarly involves a reduction in the capacity of muscle to produce force. Also as in yielding, fatigue typically affects the opponent system asymmetrically: That muscle working against the greater load fatigues more, which without neural compensation creates an unplanned imbalance that disrupts accurate limb positioning.

In vivo, it has been shown experimentally that a force feedback exists and that it has properties appropriate for fatigue compensation (Kirsch & Rymer, 1987). The pathway carrying the force feedback emerges from the Golgi tendon organs embedded in the border zone between the contractile and the tendinous regions of muscle tissue. As shown in Figure 9.9, these organs return fibers that excite *Ib* interneurons (hereafter *IbINs*) in the associated muscle control channel. *IbINs* in turn inhibit both α-MNs in their own channel and *IbINs* in the opposing channel. This signed connectivity is reminiscent of that of Renshaw cells but is used to funnel measurements taken one stage farther downstream in the outflow channel.

Whereas Renshaw cells respond to *expected* force because they tap the lines from α-*MNs* to the force generators, *IbINs* respond to *actual* force. Otherwise the similarity in their connectivity to that of Renshaws suggests that, in addition to providing fatigue compensation, *IbINs* may be able to assist the Renshaws by compensating for correlates of the size principle that reveal themselves only after the neuromuscular transduction. In fact, our simulations have shown that when there is residual positioning variability due to changes in *P* after the Renshaw subsystem is added to the model, action of the Golgi-*IbIN* system tends to recenter the range of variation at the angle the limb assumes when *P* = 0. In short, the Golgi-*IbIN* system helps reduce bias or constant error even when fatigue is not a significant factor. In keeping with our theme that the important properties of this system are not reducible to mere corollaries of a stretch reflex, we note that force shifts occur prior to rotations and length changes during voluntary limb repositioning, so the Golgi-*IbIN* feedback begins early enough to provide some preemptive reduction of positioning errors.

REGULATING FORCE IMBALANCES WITHOUT SUPPRESSING COCONTRACTION: ONE ROLE FOR MUTUAL INHIBITION AMONG INHIBITORY INTERNEURONS

We have argued that a key design feature of the biological system schematized by the FLETE model is assurance of independent control of a large force range at any desired muscle length. A key design constraint here is that the development of force imbalances $F_1 - F_2$ be regulated in a way that does not interfere with control of the overall force level $F_1 + F_2$. In this regard, note that all the inhibitory interneurons that project to α-*MNs*—*IaINs*, Renshaw cells, and *IbINs*—are themselves subjected to inhibition from the corresponding cellular pool in the opponent muscle control channel. This mutual inhibition tends to reduce the inhibitory suppression of α-*MN*

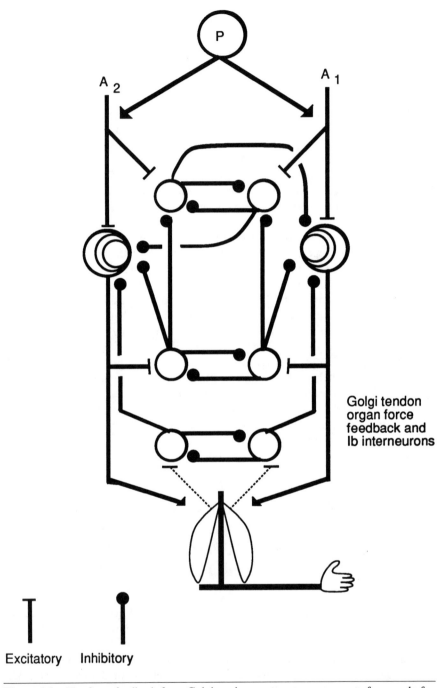

Figure 9.9 The force feedback from Golgi tendon organs can compensate for muscle fatigue as well as correlates of the size principle that reveal themselves only after the neuromuscular transduction.

activity toward the minimum suppression needed to restore a balance of forces at the desired joint angle.[1]

For example, consider the case of Renshaw pool R_1 under the circumstances of Figure 9.6b. Rather than inhibitorily feeding back—and thereby canceling the effect of!—the total increment in α-MN_1 activity induced by change in P, the effective R_1 inhibitory feedback signal corresponds more closely to only that part of the α-MN_1 activity increment not mirrored in the α-MN_2's (the opponent channel's) activity increment due to change in P. Numerical simulations confirm that α-MN activity rises more quickly as a function of signal P with mutual Renshaw inhibition than without.

COMPENSATION FOR VARIATIONS IN MECHANICAL ADVANTAGE AND AGGREGATE ADAPTIVE COMPENSATION FOR RESIDUAL ERRORS OF POSITIONING

It might be thought that the force transduction monitored by the Golgi tendon organs was the last step at which *intrinsic* sources of variability threaten plan realization. However, this is not true even for the idealized single-joint rotations we have been discussing. In particular, we now need to acknowledge that the actual rotary force, or torque, exerted by a muscle depends on both its force F_i and its mechanical advantage. Whenever a muscle is inserted as shown in Figure 9.1, the mechanical advantage of the force exerted by the muscle varies with joint angle. This would not pose a threat to invariant realization of motor intentions if changes in mechanical advantage were the same for both muscles in an opponent system. However, the effect is often asymmetrical, with one muscle gaining advantage relative to its opponent during a rotation (e.g., Hasan & Enoka, 1985).

Muscle torques are related to muscle forces by the equation

$$T_i = F_i D_i , \tag{9.17}$$

where D_i is the perpendicular distance between the axis of joint rotation and the line of force of muscle i. The distance D_i is a function of joint angle, which changes rapidly during movement, and of muscle insertion geometry, which can change slowly during skeletal growth. To compensate for the untoward effects of asymmetric variations in mechanical advantage, we might expect the nervous system to have

---

[1]McCrea (1990) reports that mutual inhibition has been clearly demonstrated for *IbINs* (e.g., Brink, Jankowska, McCrea, & Skoog, 1983), but it is rarely mentioned in textbook discussions. Schwindt (1981) noted that the inverse myotatic response of Lloyd, in which agonist *IbIN* stimulation causes antagonist muscle activation, has both di- and trisynaptic facilitatory components. One of these components is likely to be due to disinhibition of the antagonist, mediated by agonist *IbIN* inhibition of antagonist *IbIN*. The other facilitatory component is usually represented as resulting from an excitatory link from agonist *Ib* fiber to antagonist α-MN, usually mediated by an excitatory interneuron. In the present context, we note that both pathways from agonist *Ib* fiber activation to antagonist activation work to oppose undesirable force imbalances without suppressing desired cocontractions.

followed the same strategy illustrated so many times above. To implement this strategy, the nervous system would measure either the variation-causing factor or a close correlate of it, then use its measurements as the basis for a compensatory flow of signals capable of appropriately adjusting the balance across the α-MN pools. As far as we know, no mechanism exists for direct measurement of D_i, so we must consider correlates. For example, as already noted, D_i is a function of joint angle θ. So feedback from joint receptors sensitive to θ could in principle provide some compensation.[2]

Ultimately, however, a high-performance motor-control system cannot afford to rely solely on the kind of *automatic* compensatory systems so far described. Such systems can greatly reduce error due to evolutionarily stable neuromuscular sources of variablility, but many sources of variability are not of this type, such as the musculoskeletal function $D_i(\theta)$, which depends on happenstances of skeletal growth. Also, accidents or errors of neural growth and development can create compensatory subsystems that are not optimally tuned. Thus residual error is unavoidable. To achieve truly high performance, the movement control system should measure actual limb position errors and use feedback regarding residual positioning error to adaptively retune signal flows within the neural network.

We believe (see also Kuffler & Hunt, 1952; Matthews, 1981) that the parallel neuro-musculo-sensory system comprising gamma-motoneurons, intrafusal muscles or "spindles," and spindle receptors, which has long been studied as a substrate for the stretch reflex, is preeminently nature's solution to the need to *measure* residual positioning errors. Figure 9.10 shows our composite FLETE model network with this parallel system added. Note that the net descending signal $A_i + P$ is now also delivered to the gamma-MN pool (hereafter γ-*MN*) in Channel i. This γ-*MN* pool in turn activates intrafusal muscles situated in parallel with main (or to use the contrasting term, "extrafusal") muscle. This parallelism can be exploited to measure positioning errors if, unlike main muscle (whose contractile state changes partly depend on contraction-opposing loads), intrafusal muscle contractile state changes depend wholly on their level of innervation from γ-*MN*s. Then whenever their contractile state changes are not precisely canceled by concurrent whole-muscle length changes, there will be a deviation from the baseline tension level exerted on the spindle's receptor elements, and fluctuations around the baseline firing rates of these receptors can serve as a measure of the direction and magnitude of positioning errors.

The existence of a baseline spindle tension and an accompanying baseline spindle receptor activation indicates that the spindle feedback signals emerging from both

---

[2]Indeed, it is known that activation of joint receptors by full extension excites extensors while inhibiting flexors. In a recent review, Tracey (1985) remarked that "it is difficult to know how this pattern, a kind of positive feedback, might function in the reflex control of muscle" (p. 179). However, the fact that joint receptors become active only as joint angles become more extreme, and that their action is to excite motoneurons, is perfectly comprehensible if seen as a compensation for the often-severe loss of mechanical advantage experienced near the ends of a joint's range of motion. In such a context, what initially appears to be a positive feedback would in fact be a partial compensation for an undesired, but physically unavoidable, negative feedback: the rapid erosion of mechanical advantage as the limb nears the end of its range of operation.

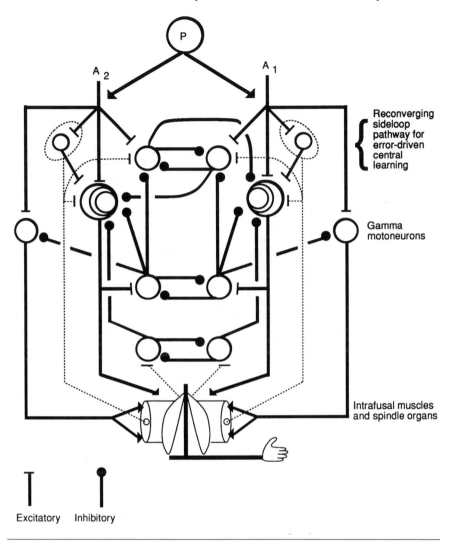

Figure 9.10 A parallel neuromuscular system comprising γ-*MNs*, intrafusal muscles, and spindle receptors allows measurement of residual positioning errors. Spindle feedback signals act locally via the stretch reflex, but also project to the higher brain, where they may guide recalibration of descending commands.

opponent muscles may be large, and this raises the specter of saturative loss of sensitivity to the difference between the opposing channels' spindle feedbacks at whatever stages receive these signals. Because the sign and magnitude of this difference is critical for the process of compensating residual positioning errors, we can expect the same sort of neural provisions for processing spindle feedback signals as described in the section on varying joint stiffness independently of joint angle (to ensure sensitivity to the difference $A_1 - A_2$.) We now note that at the *spinal* stage receiving these signals, as shown in Figure 9.10, these provisions are *identical*: The ascending spindle feedbacks, E_i, like the descending signals $A_i + P$, project both directly to same-channel α-*MNs* and indirectly to opponent-channel α-*MNs* via the *IaINs* (Baldissera, Cavallari, Fournier, Pierrot-Deseilligny, & Shindo, 1987). In light of our prior analysis (in the section on how joint stiffness can vary independently of joint angle), it is easy to see that this routing guarantees that $M_1 - M_2$ will remain sensitive to the difference between spindle feedback signals $E_1 - E_2$ even if both the E_i are very large.

The spinomuscular circuitry just described is that which mediates the classical stretch reflex. But we emphasize that within our theory the immediate, partial, reactive compensation for positioning errors provided by the reflex is not the sole motivation for the parallel muscle system. Following earlier demonstrations by Grossberg and Kuperstein (1986, 1989) and Kawato, Furukawa, and Suzuki (1987), in Bullock and Grossberg (1990, 1991) we summarized how a *central* adaptive process sensitive to spindle feedback signals could learn an intended-angle-dependent, preemptive compensation for angle-dependent variations in mechanical advantage. This construction illustrated that *stretch feedback* can be fully motivated within an adaptive sensory-motor control system without reference to a *stretch reflex*. This is quite important from the perspective of biological theory, because it helps us comprehend sensory-motor systems like that governing the eyes—where stretch receptors have evolved in the absence of any stretch reflex pathways. The construction also further supports our general thesis that the motor control system cannot be fruitfully analyzed as an epicycle of the stretch reflex: Even stretch receptors must not be thought of reductively, as mere servants of a *reactive* compensatory process. We believe they also serve a proactive, error-preempting, adaptive system that is essential for gradual acquisition of skill over repeated performance trials.

LOCUS OF SIGNAL CONVERGENCE AND RENSHAW INHIBITION OF γ-*MNs*

In the version of the FLETE model explored in Bullock and Grossberg (1990, 1991), we assumed that signals A_i and P converged at the α-*MN* stage, and that signal P was not sent to the γ-*MNs* or *IaINs*. Nevertheless, we discovered that preemptive moment-arm compensation required a central site, within each muscle control channel, for a convergence of A_i and P. This total signal was then modified by the adaptive central process mentioned above before being relayed to α-*MNs*. The model presented in this paper illustrates that independent control of length and tension is also achievable if α-*MNs*, *IaINs*, and γ-*MNs* alike receive the total signal $A_i + P$ whose central existence was implicated in our prior study.

A second reason for convergence of these signals at *IaINs* was elaborated in the section on how joint stiffness can vary independently of joint angle. A second reason for convergence of signals A_i and P at γ-*MNs* may now be mentioned. To the extent that spindles respond to *local* contraction within the muscle body as well as to whole-muscle shortening, some spindles could be "unloaded" at high cocontraction levels in the absence of a whole-muscle length change. Such unloading can degrade the pooled length-error signal by reducing the sample size, or number of loaded spindle receptors, on which it is based. Partial unloading might also reduce the pooled signal's proportionality to stretch by causing spindle receptors to operate in a less linear part of their range. Finally, partial unloading would also lead to loss of one tonic source of α-MN excitation, which would work against joint stiffening. All these potential problems are circumvented if signal P activates *gamma-MNs* as well as α-MNs.

The proposal that γ-MNs receive $A_i + P$ rather than A_i alone, helps rationalize a known component of spinal network connectivity not included in our earlier version of the model. This omission was the inhibitory feedback from Renshaw cells to γ-MNs in their channel (see Pompeiano, 1984). Without this feedback, γ-MN pools excited by $A_i + P$ are in danger of saturating when $A_i + P$ becomes large. Such saturation can be avoided if inhibition to γ-MNs increases with P. This is guaranteed by an inhibitory feedback from Renshaws to γ-MNs. However, as just noted, it is equally important that the strength of this Renshaw inhibition not so suppress γ-MN activity that spindle receptors become unloaded when cocontraction is high. Measurements made in animals confirm that γ-MNs are not silenced by the Renshaw inhibition to which they are subjected (Pompeiano, 1984).

SUMMARY OF EQUILIBRIUM STATES OF THE FLETE MODEL WITHOUT STRETCH FEEDBACK AND AFTER SIMULATED LESIONS OF INTERNEURONAL PATHWAYS

Having built up the FLETE model piecewise in order to fully rationalize the known biological circuitry it formalizes, we now summarize its properties by plotting the equilibrium values reached by its state variables after the circuit is activated by a full range of combinations of the net positioning signal $A_1 - A_2$ and the stiffening signal P. To highlight components other than the stretch reflex, we let $D_i = 1$ (i.e., we assume no model moment-arm variations), and we turn off all model spindle feedback signals. The resulting mathematical system is fully specified in Appendix A.[3]

Figure 9.11, a-j, shows equilibrium values of FLETE model state variables. Figure 9.11a shows that agonist muscle length is a linear function of $A_1 - A_2$ with only a small residual band of variability due to variations in P. The width of this band would be reduced further with the stretch reflex reactivated. Figure 9.11d shows that γ-MN activity is also linear in $A_1 - A_2$, which helps provide a reliable basis for registration of length errors by muscle spindles. Figure 9.12e shows, as argued in the previous section, that this linearity would break down in the absence of inhibitory feedback from Renshaw cells to γ-MNs.

Figure 9.11, b, c, e, and f, shows that the linear function relating agonist muscle length and γ-MN activity to $A_1 - A_2$ can coexist with highly nonlinear relations

---

[3]Note that in this version of the model, force feedback is linear, and there is no *IbIN* equation as such, hence also no *IbIN* mutual inhibition. As noted previously, adding such would improve force range.

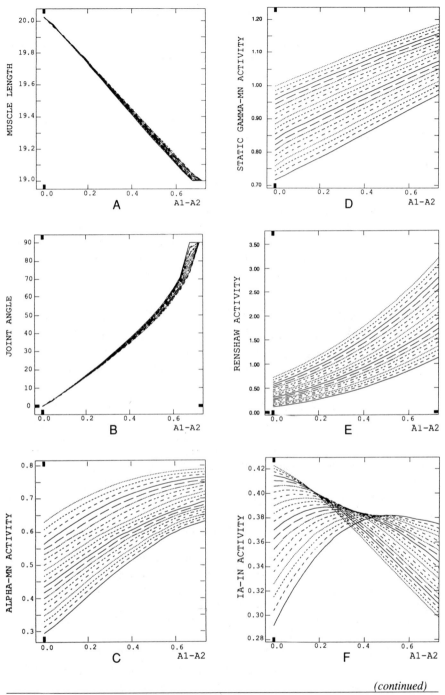

(continued)

Figure 9.11   Equilibrium values of FLETE model state variables. (See legend in Figure 9.12.)

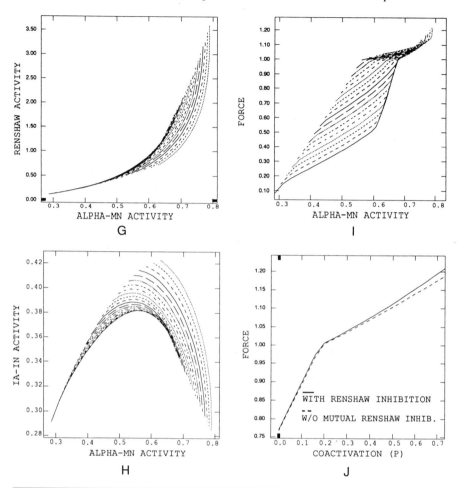

Figure 9.11   *(continued)*

between $A_1 - A_2$ and joint angle, $\alpha\text{-}MN$ activity, Renshaw activity, and *IaIN* activity, respectively.

Figure 9.11. g-i, illustrates the highly nonlinear relations between agonist $\alpha\text{-}MN$ activity and other key agonist channel variables. Figure 9.11j shows that the invariance property specified in Equation 9.6 and demonstrated in Figure 9.11, a and b, can coexist with an abrupt change in the slope of the function relating muscle force to signal P (a change in this case introduced by the threshold Γ_F in Equation 9.15 under the assumption that the opponent muscle's force can indeed cause yielding). Figure 9.10j also shows the slower growth of force that occurs when there is no inhibition between Renshaw cells in opposing channels.

Figure 9.12, a-e, illustrates the severe breakdown of independent control of muscle length in the absence of (a) model force feedback such as arises in vivo from Golgi tendon organs (Figure 9.12a), (b) model Renshaw inhibition (Figure 9.12b), or

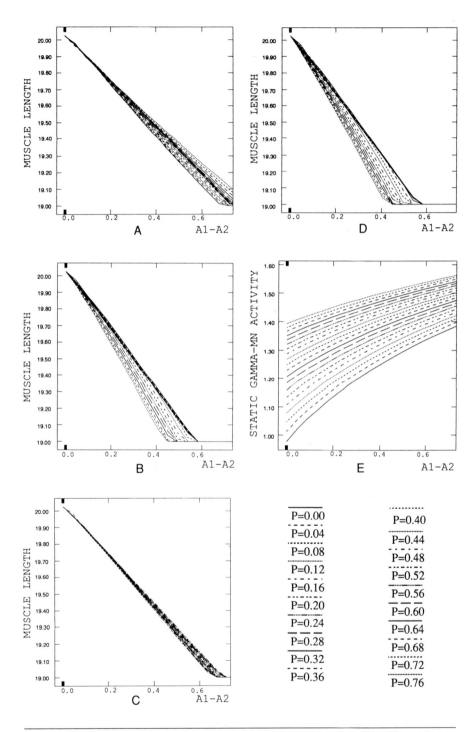

Figure 9.12 Results of simulated lesions on equilibrium states of the FLETE model.

(c) both (Figure 9.12d). Note that the Renshaw lesion produces the more severe degradation: There is greater dispersion at equal $A_1 - A_2$ settings, and a loss of sensitivity to $A_1 - A_2$ even at low P values. In particular, as shown in Figure 9.12b, the joint goes to full flexion at an $A_1 - A_2$ value that otherwise codes a less extreme joint angle. This effect is reminiscent of the tetanus observed in vivo when Renshaw cells are poisoned with strychnine.

Figure 9.12c illustrates that independent control of muscle length survives simultaneous removal of mutual inhibition between opponent Renshaw pools and between opponent $IaIN$ pools. Note, though, that these simulated lesions leave intact the opponent, disinhibitory, compensatory pathways

$$R_i \rightarrow IaIN_i \rightarrow \alpha\text{-}MN_j\,,$$

where $i,j = \{1,2\}$.

Earlier we noted that the Renshaw cells, unlike a stretch reflex, can act preemptively to compensate for unequal amplifications of signal P by the size principle. Ideally, such preemptive action would preclude even transient deviations from the equilibrium state variables shown in Figure 9.11, a and b. Figure 9.13a shows three waveforms, $P_1(t)$, $P_2(t)$, and $P_3(t)$, corresponding to three different schedules for first increasing and then decreasing signal P. Figure 9.13b shows muscle length as a function of time, for three different initial muscle lengths (corresponding to three distinct constant settings of $A_1 - A_2$), during schedule $P_1(t)$. Here we see that a slow change in coactivation produces negligible transient deviations from the equilibrium values. Figure 9.13c shows the response to higher rate, larger amplitude changes in coactivation (schedule $P_2(t)$ in 9.13a), and Figure 9.13d shows the response to square wave changes in signal P (schedule $P_3(t)$ in 9.13a). Remarkably, only small transient deviations from the equilibrium values given in Figure 9.11a emerge under any of these regimes.

JOINT STIFFNESS CONTROL AND THE LENGTH-FORCE FUNCTION FOR ACTIVE MUSCLE

Equation 9.1 represents the effect of activating a muscle as a pure shift in the threshold length for force development, and states that force is a linear function of suprathreshold length. In fact, however, muscle activation causes both a shift in the threshold length for force development and a change in the slope of the length-force curve. Such effects can be approximated by replacing Equation 9.1 with a faster-than-linear function such as

$$F_i = k([L_i - (\Gamma_i - C_i)]^+)^2. \tag{9.18}$$

Figure 9.14, a-j, shows operating characteristics of the FLETE model when Equation 9.18 serves as the force development law. As can be seen in Figure 9.14, a and b, the key invariance property stated in Equation 9.6 continues to hold.

With Equation 9.18, the magnitudes of the muscle force changes $\Delta F_1, \Delta F_2$, induced by a given angular change $\Delta\theta$ imposed on the limb by an external torque change

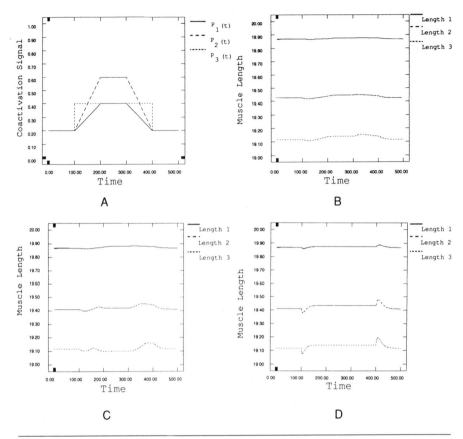

Figure 9.13 (A) Three regimes for varying the coactivation signal P through time. (B) Muscle length during regime $P_1(t)$ of (A). (C) Muscle length during regime $P_2(t)$ of (A). (D) Muscle length during regime $P_3(t)$ of (A).

ΔT_e depend on the states of activation of the muscles. In particular, the higher the activation, the steeper the decline in force by the muscle compressed by the external torque, and the steeper the growth of force by the muscle stretched by the external torque. Thus a given external torque meets greater resistance from a joint whose muscles are more highly activated if Equation 9.18 describes the force development process. Figure 9.15 shows representative results of pertinent FLETE model simulations. With Equation 9.18 serving as the force development law, joint stiffness is an increasing function of coactivating signal P. For the simulation plotted, we reinstated the moment-arm effect and activated the stretch reflex with a low stretch-dependent feedback coefficient ($G_p = 0.1$; see Appendix A). We also started the simulation with the joint 20° flexed. As can be seen, there is an asymmetry in the response of the system to torques rotating the limb to more extreme versus less extreme angles. This asymmetry is due to the spinal reflex's "ignorance" of the moment-arm effect. If this asymmetry is absent in vivo, this presumably indicates

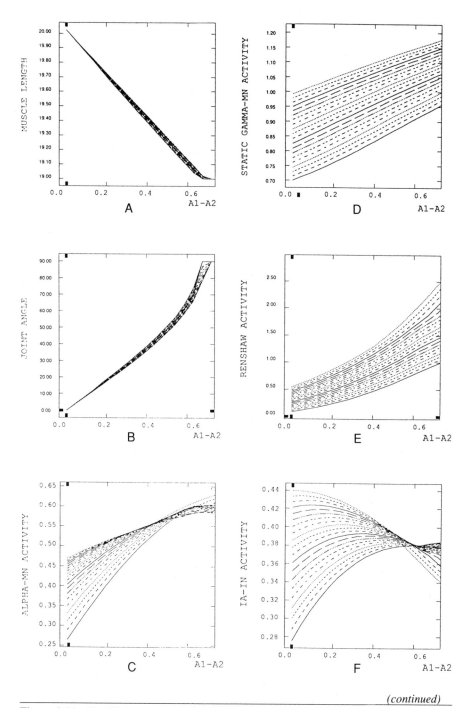

Figure 9.14 Equilibrium states of the FLETE model with a quadratic length-force function. (See legend in Figure 9.12.)

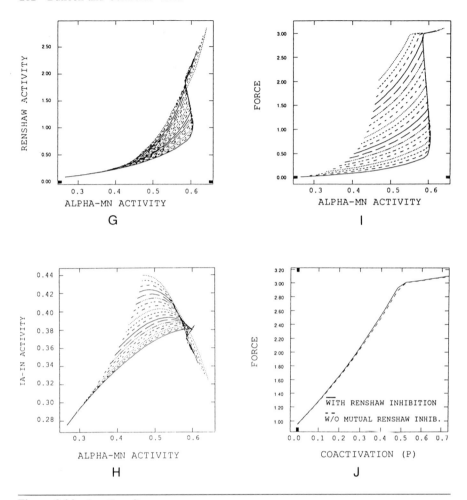

Figure 9.14   *(continued)*

an additional compensation (e.g., via joint receptors or cortically mediated stretch reflexes).

This demonstration completes our conceptual reconstruction of the peripheral neuromuscular system as a module that affords independent control of muscle length and joint stiffness. Table 9.1 summarizes the experimental evidence for all the cell types and connections (including sign) assumed in the model.

CONCLUDING REMARKS: EXPANDED TREATMENT OF DYNAMICS

With the exception of the simulation summarized in Figure 9.13, this paper has focused on equilibrium states of the FLETE system for steady-state values of

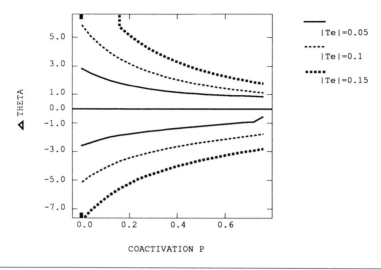

Figure 9.15   Illustration of stiffness control. Angular displacements caused by external torques are inverse functions of cocontraction signal P. For the simulation shown, the initial angle (zero point on plot) was 20° flexed, and the stretch reflex was activated with a low stretch-dependent feedback coefficient ($G_p = 0.1$) or gain, as exists in vivo.

descending signals. A brief prior report (Bullock & Grossberg, 1992) showed that the model's *transient* behavior during rapid shifts of $A_1 - A_2$ is also compatible with physiological observations. In particular, we showed that the FLETE circuit can generate a triphasic EMG burst pattern similar to those frequently observed in vivo (e.g., Lestienne, 1979). Because these burst patterns emerged in the model from the dynamics of the neuromuscular system in interaction with a load, our study proves that such patterns need not be preformed in the descending command. This demonstration is therefore pertinent to long-standing debates (Bullock & Grossberg, 1988a; Feldman, 1986; Feldman, Adamovich, Ostry, & Flanagan, 1990) about whether such burst patterns have their *ultimate* origin in a central motor program or in a dynamical system governed by organism-environment interactions. We are now preparing a series of reports that treat in a much more comprehensive way the dynamics of FLETE modules in interaction with each other, multijoint limbs, and more central neural networks. In the first (Bullock, Contreras-Vidal, & Grossberg, 1992) we incorporate additional known descending pathways as well as the improved competence associated with distinct "static" and "dynamic" populations of γ-*MNs* and resultant distinct components of the spindle feedback signals. We also show that parameter ϕ (see Appendix A, Equations A.11, A.13-A.15) can be eliminated, while improving system performance, if the membrane equations are made more realistic by allowing negative potentials. The second report (Contreras-Vidal & Bullock, 1992) will treat multijoint dynamics, including mechanical, muscular, and neural cross-coupling. The third report (Bullock, Contreras-Vidal, & Grossberg, 1992) shows how cerebellar learning provides feedforward control and improved timing of phasic inputs (to α-*MNs*) originally generated reactively at the periphery.

Table 9.1   Evidence for Connectivity and Physiology Incorporated in the FLETE Model

Connection type	Citations
1. Excitatory $\alpha\text{-}MN_i \rightarrow R_i$	Renshaw (1941, 1946) Eccles, Fatt, & Koketsu (1954)
2. Inhibitory $R_i \rightarrow \alpha\text{-}MN_i$	Renshaw (1946) Eccles, Fatt, & Koketsu (1954)
3. Inhibitory $R_i \rightarrow IaIN_i$	Hultborn, Jankowska, & Lindström (1971a)
4. Inhibitory $R_i \rightarrow \gamma\text{-}MN_i$	Ellaway (1968) Ellaway & Murphy (1980)
5. Inhibitory $R_i \rightarrow R_j$	Ryall (1970) Ryall & Piercey (1971)
6. Inhibitory $IaIN_i \rightarrow \alpha\text{-}MN_j$	Eccles & Lundberg (1958) Araki, Eccles, & Ito (1960)
7. Inhibitory $IaIN_i \rightarrow IaIN_j$	Eccles & Lundberg (1958) Hultborn, Jankowska, & Lindström (1971b) Hultborn, Illert, & Santini (1976) Baldissera, Cavallari, Fournier, Pierrot-Deseilligny, & Shindo (1987)
8. Excitatory Ia_i fiber $\rightarrow IaIN_i$	Hultborn, Jankowska, & Lindström (1971a) Baldiserra et al. (1987)
9. Excitatory Ia_i fiber $\rightarrow \alpha\text{-}MN_i$	Lloyd (1943)
10. Inhibitory $IbIN_i \rightarrow \alpha\text{-}MN_i$	Laporte & Lloyd (1952) Eccles, Eccles, & Lundberg (1957) Kirsch & Rymer (1987)
11. Excitatory $IbIN_i \rightarrow \alpha\text{-}MNs$	Laporte & Lloyd (1952) Eccles, Eccles, & Lundberg (1957)
12. Inhibitory $IbIN_i \rightarrow IbIN_j$	Laporte & Lloyd (1952) Eccles, Eccles, & Lundberg (1957) Brink, Jankowska, McCrea, & Skoog (1983)
13. Nonspecific $P \rightarrow$ spinal motor pools	Humphrey & Reed (1983) DeLuca (1985)

Finally, all these projects make use of the VITE model (Bullock & Grossberg, 1988a, 1991; Gaudiano & Grossberg, 1991) of the central neural network responsible for planned, variable speed shifts between initial and final values of the net positioning command $A_1 - A_2$. In Bullock and Grossberg (1988a), we showed that this model was consistent with, and offered a partial explanation of, Fitts's famous speed-accuracy trade-off function for point-to-point movements with strict spatial accuracy requirements. A forthcoming paper (Bullock & Ross, 1992) extends the VITE model, in a way consistent with known cortical and basal-ganglia circuitry, to propose a possible neural basis for automatically adjusting a central movement-time control signal to expected movement distance, so as to assure a desired level of movement accuracy. Because such an adjustment compensates for distance-dependent error variability intrinsic to the VITE design, it would exemplify central preemptive compensation similar to that seen in the peripheral circuitry studied in earlier sections of this paper. If the extended VITE model and the FLETE model both survive empirical test, they will buttress our initial claim that much of the structure of the neural networks used to realize movement plans can be understood only as adaptations that compensate for intrinsic sources of motor variability.

APPENDIX A

FLETE Model Equations

Force-Length Relationship

$$F_i = k \times g([L_i - \Gamma_i + C_i]^+) \tag{A.1}$$

where

$i = (1,2) = $ (agonist,antagonist)
$k = 0.5$
$F_i = $ force of muscle i
$L_i = $ length of muscle i
$\Gamma_i = 20.9 = $ resting muscle length
$C_i = $ contractile state of muscle i
$g(\omega) = \omega^2$ or $g(\omega) = \omega$

Contraction Rate

$$\frac{d}{dt}C_i = \beta_i[(B_i - C_i)M_i - \delta C_i] - [F_i - \Gamma_F]^+ \tag{A.2}$$

$0 < \beta_i < 1$
$\Gamma_F = 1.0$
$M_i = $ agonist α-MN pool
$\delta = 1 = $ fiber relaxation rate
$B_i = $ number of contractile fibers

Origin-to-Insertion Muscle Lengths

$$L_1 = \sqrt{(\cos\theta)^2 + (20 - \sin\theta)^2} \tag{A.3}$$

$$L_2 = \sqrt{(\cos\theta)^2 + (20 + \sin\theta)^2} \tag{A.4}$$

$\theta = $ joint angle

Limb Dynamics

$$\frac{d^2}{dt^2}\theta = \frac{1}{I_m}(F_1 - F_2 - n\frac{d}{dt}\theta)$$ (A.5)

or

$$\frac{d^2}{dt^2}\theta = \frac{1}{I_m}(T_1 - T_2 + T_e - n\frac{d}{dt}\theta)$$ (A.6)

$\frac{d}{dt}\theta$ = angular velocity

I_m = moment of inertia, which is proportional to mass
n = viscosity coefficient
T_e = external torque
T_i = torque associated with muscle i
$T_i = D_i F_i$

Moment-Arm of Force F_i

$$D_1 = \frac{20}{\sqrt{\left(\frac{sin\ \theta - 20}{cos\ \theta}\right)^2 + 1}}$$ (A.7)

$$D_2 = \frac{20}{\sqrt{\left(\frac{sin\ \theta + 20}{cos\ \theta}\right)^2 + 1}}$$ (A.8)

Contraction Rate

$$\beta_i = 0.10 + 0.05(A_i + P + \chi \times E_i)$$ (A.9)

A_i = present position command
P = coactivation signal
$\chi = 1$
E_i = stretch feedback

Number of Contractile Fibers

$$B_i = 0.3 + 3.0(A_i + P + \chi \times E_i)$$ (A.10)

Renshaw Population

$$\frac{d}{dt}R_i = \phi(\lambda B_i - R_i)z_i M_i - R_i(1 + R_j)$$ (A.11)

$$z_i = 0.2 + 0.8(A_i + P)$$ (A.12)

$\lambda = 5.0$
$\phi = 0.2$

Alpha MN Population

$$\frac{d}{dt}M_i = \phi[(\lambda B_i - M_i)(A_i + P + \chi E_i)] - M_i(\delta_i + \Omega R_i + \rho F_i + I_j) \qquad \text{(A.13)}$$

$\Omega = 1$
$0 < \delta_i < 1$
$\rho = 1$

IaIN Population

$$\frac{d}{dt}I_i = \phi(10 - I_i)(A_i + P + \chi E_i) - I_i(1 + \Omega R_i + I_j) \qquad \text{(A.14)}$$

Static Gamma MN

$$\frac{d}{dt}N_i = \phi[(2 - N_i)(A_i + P)] - N_i(0.2 + \nu f(R_i)) \qquad \text{(A.15)}$$

$\nu = 0.3$
$$f(\omega) = \frac{\omega}{0.1 + \omega}$$

Intrafusal Static Gamma Muscle Contraction

$$\frac{d}{dt}U_i = 4N_i - U_i \qquad \text{(A.16)}$$

Spindle Organ Response

$$\frac{d}{dt}W_i = [U_i + L_i - \Gamma_i]^+ - W_i \qquad \text{(A.17)}$$

Stretch Feedback

$$E_i = G_p \times W_i \qquad \text{(A.18)}$$

G_p = stretch gain

REFERENCES

Akazawa, K., & Kato, K. (1990). Neural network model for control of muscle force based on the size principle of motor unit. *Proceedings of the IEEE*, **78**, 1531-1535.

Araki, T., Eccles, J.C., & Ito, M. (1960). Correlation of the inhibitory postsynaptic potential of motoneurones with the latency and time course of inhibition of monosynaptic reflexes. *Journal of Physiology* (London), **154**, 354-377.

Baldissera, F., Cavallari, P., Fournier, E., Pierrot-Deseilligny, E., & Shindo, M. (1987). Evidence for mutual inhibition of opposite *Ia* interneurones in the human upper limb. *Experimental Brain Research*, **66**, 106-114.

Brink, E., Jankowska, E., McCrea, D.A., & Skoog, B. (1983). Inhibitory interactions between interneurones in reflex pathways from group *Ia* and group *Ib* afferents in the cat. *Journal of Physiology* (London), **343**, 361-373.

Bullock, D., Contreras-Vidal, J.L., & Grossberg, S. (1992). Equilibria and dynamics of a neural network model for opponent muscle control. In G. Bekey & K. Goldberg (Eds.), *Neural networks in robotics*. Norwell, MA: Kluwer Academic.

Bullock, D., & Grossberg, S. (1988a). Neural dynamics of planned arm movements: Emergent invariants and speed-accuracy properties during trajectory formation. *Psychological Review*, **95**(1), 49-90.

Bullock, D., & Grossberg, S. (1988b). Neuromuscular realization of planned trajectories. *Neural Networks*, **1** (Suppl. 1), 329.

Bullock, D., & Grossberg, S. (1989). VITE and FLETE: Neural modules for trajectory formation and tension control. In W. Hershberger (Ed.), *Volitional action*. Amsterdam: North-Holland, 253-297.

Bullock, D., & Grossberg, S. (1990). Spinal network computations enable independent control of muscle length and joint compliance. In R. Eckmiller (Ed.), *Advanced neural computers* (pp. 349-356). Amsterdam: Elsevier.

Bullock, D., & Grossberg, S. (1991). Adaptive neural networks for control of movement trajectories invariant under speed and force rescaling. *Human Movement Science*, **10**, 1-51.

Bullock, D., & Grossberg, S. (1992). Emergence of tri-phasic muscle activation from the nonlinear interactions of central and spinal neural network circuits. *Human Movement Science*, **11**, 157-167.

Bullock, D., Contreras-Vidal, J., & Grossberg, S. (1992). *A new model of cerebellar learning by opponent error processing: Application to rapid joint rotations*. Manuscript in preparation.

Bullock, D., & Ross, W. (1992). *Amplitude-dependent auto-compensation in a vector integration to endpoint (VITE) model of trajectory formation*. Manuscript in preparation.

Contreras-Vidal, J.L., & Bullock, D. (1992). *Modeling mechanical, muscular, and neural interactions in a 2 d.o.f. planar arm subjected to load perturbations*. Manuscript in preparation.

Cooke, J.D. (1980). The organization of simple, skilled movements. In G.E. Stelmach & J. Requin (Eds.), *Tutorials in motor behavior* (pp. 199-212). Amsterdam: North-Holland.

Cullheim, S., & Kellerth, J.O. (1978). A morphological study of the axons and recurrent axon collaterals of cat α-motoneurones supplying different functional types of muscle unit. *Journal of Physiology* (London), **281**, 301-313.

DeLuca, C.J. (1985). Control properties of motor units. *Journal of Experimental Biology*, **115**, 125-136.

Eccles, J.C., Eccles, R.M., & Lundberg, A. (1957). Synaptic actions on motoneurons caused by impulses in Golgi tendon organ afferents. *Journal of Physiology* (London), **138**, 227-252.

Eccles, J.C., Fatt, P., & Koketsu, K. (1954). Cholinergic and inhibitory synapses in a pathway from motor-axon collaterals to motoneurones. *Journal of Physiology* (London), **126**, 524-562.

Eccles, R.M., & Lundberg, A. (1958). Integrative pattern of *Ia* synaptic actions on motoneurons of hip and knee muscles. *Journal of Physiology* (London), **144**, 271-298.

Ellaway, P.H. (1968). Antidromic inhibition of fusimotor neurones. *Journal of Physiology* (London), **198**, 39P-40P.

Ellaway, P.H., & Murphy, P.R. (1980). A quantitative comparison of recurrent inhibition of α- and γ-motoneurones in the cat. *Journal of Physiology* (London), **315**, 43-58.

Feldman, A.G. (1986). Once more on the equilibrium-point hypothesis (λ model) for motor control. *Journal of Motor Behavior*, **18**, 17-54.

Feldman, A.G., Adamovich, S.V., Ostry, D.J., & Flanagan, J.R. (1990). The origin of electromyograms—explanations based on the equilibrium point hypothesis. In J.M. Winters & S.L.-Y. Woo (Eds.), *Multiple muscle systems: Biomechanics and movement organization*. New York: Springer-Verlag.

Fitts, P.M. (1954). The information capacity of the human motor system in controlling the amplitude of movement. *Journal of Experimental Psychology*, **47**(6), 381-391.

Gaudiano, P., & Grossberg, S. (1991). Vector associative maps: Unsupervised real-time error-based learning and control of movement trajectories. *Neural Networks*, **4**, 147-183.

Grossberg, S. (1973). Contour enhancement, short-term memory, and constancies in reverberating neural networks. *Studies in Applied Mathematics*, **52**, 217-257.

Grossberg, S. (1982). *Studies of mind and brain: Neural principles of learning, perception, development, cognition, and motor control*. Boston: Reidel Press.

Grossberg, S., & Kuperstein, M. (1986). *Neural dynamics of adaptive sensory-motor control: Ballistic eye movements*. Amsterdam: Elsevier/North-Holland.

Grossberg, S., & Kuperstein, M. (1989). *Neural dynamics of adaptive sensory-motor control* (expanded ed.). New York: Pergamon Press.

Hasan, Z., & Enoka, R.M. (1985). Isometric torque-angle relationship and movement-related activity of human elbow flexors: Implications for the equilibrium-point hypothesis. *Experimental Brain Research*, **59**, 441-450.

Henneman, E. (1957). Relation between size of neurons and their susceptibility to discharge. *Science*, **26**, 1345-1347.

Henneman, E. (1985). The size-principle: A deterministic output emerges from a set of probabilistic connections. *Journal of Experimental Biology*, **115**, 105-112.

Hodgkin, A.L., & Huxley, A.F. (1952). A quantitative description of membrane current and its applications to conduction and excitation in nerve. *Journal of Physiology*, **117**, 500-544.

Hongo, T., Jankowska, E., & Lundberg, A. (1969). The rubrospinal tract. II. Facilitation of interneuronal transmission in reflex paths to motoneurones. *Experimental Brain Research*, **7**, 365-391.

Hultborn, H., Illert, M., & Santini, M. (1976). Convergence on interneurones mediating the reciprocal *Ia* inhibition of motoneurones *I*. Disynaptic *Ia* inhibition of *Ia* inhibitory interneurones. *Acta Physiology Scandinavia*, **96**, 193-201.

Hultborn, H., Jankowska, E., & Lindström, S. (1971a). Recurrent inhibition of interneurones monosynaptically activated from group *Ia* afferents. *Journal of Physiology* (London), **215**, 613-636.

Hultborn, H., Jankowska, E., & Lindström, S. (1971b). Relative contribution from different nerves to recurrent depression of Ia IPSPs in motoneurones. *Journal of Physiology* (London), **215**, 637-664.

Hultborn, H., Lindström, S., & Wigström, H. (1979). On the function of recurrent inhibition in the spinal cord. *Experimental Brain Research*, **37**, 399-403.

Humphrey, D.R., & Reed, D.J. (1983). Separate cortical systems for control of joint movement and joint stiffness: Reciprocal activation and coactivation of antagonist muscles. In J.E. Desmedt (Ed.), *Motor control mechanisms in health and disease* (pp. 347-372). New York: Raven Press.

Kawato, M., Furukawa, K., & Suzuki, R. (1987). A hierarchical neural-network model for control and learning of voluntary movement. *Biological Cybernetics*, **57**, 169-185.

Kiehn, O. (1991). Plateau potentials and active integration in the "final common pathway" for motor behaviour. *Trends in Neuroscience*, **14**, 68-73.

Kirsch, R.F., & Rymer, W.Z. (1987). Neural compensation for muscular fatigue: Evidence for significant force regulation in man. *Journal of Neurophysiology*, **57**, 1893-1910.

Kuffler, S.W., & Hunt, C.C. (1952). The mammalian small-nerve fibers: A system for efferent nervous regulation of muscle spindle discharge. In P. Bard (Ed.), *Patterns of organization in the central nervous system* (pp. 24-47). Baltimore: Williams & Wilkins.

Laporte, Y., & Lloyd, D.P.C. (1952). Nature and significance of the reflex connections established by large afferent fibers of muscular origin. *American Journal of Physiology*, **169**, 609-621.

Lestienne, F. (1979). Effects of inertial load and velocity on the braking process of voluntary limb movements. *Experimental Brain Research*, **35**, 407-418.

Lloyd, D.P.C. (1943). Conduction and synaptic transmission of the reflex response to stretch in spinal cats. *Journal of Neurophysiology*, **6**, 317-326.

Matthews, P.B.C. (1981). Proprioceptors and the regulation of movement. In A.L. Towe & E.S. Luschei (Eds.), *Handbook of behavioral neurobiology: Vol. 5. Motor coordination* (pp. 93-138). New York: Plenum.

McCrea, D.A. (1990). *Can sense be made of spinal interneuron circuits?* Paper presented at the Controversies in Neuroscience I: Movement Control conference. R.S. Dow Neurological Sciences Institute, Portland, OR, August 25-26, 1990.

Nichols, T.R. (1984). Velocity sensitivity of yielding during stretch in the cat soleus muscle. In G.H. Pollack & H. Sugi (Eds.), *Contractile mechanisms in muscle* (pp. 753-755). New York: Plenum.

Polit, A., & Bizzi, E. (1979). Characteristics of motor programs underlying arm movements in monkeys. *Journal of Neurophysiology*, **42**, 183-194.

Pompeiano, O. (1984). Recurrent inhibition. In R.A. Davidoff (Ed.), *Handbook of the spinal cord: Vols. 2 and 3. Anatomy and physiology*. New York: Marcel Dekker.

Pompeiano, O., & and Wand, P. (1976). The relative sensitivity of Renshaw cells to static and dynamic changes in muscle length. *Progress in Brain Research*, **44**, 199-222.

Rack, P.H.M., & Westbury, D.R. (1969). The effect of length and stimulus rate on the tension in the isometric cat soleus muscle. *Journal of Physiology*, **204**, 443-460.

Renshaw, B. (1941). Influence of discharge of motoneurons upon excitation of neighboring motoneurons. *Journal of Neurophysiology*, **4**, 167-183.

Renshaw, B. (1946). Central effects of centripetal impulses in axons of spinal ventral roots. *Journal of Neurophysiology*, **9**, 191-204.

Ryall, R.W. (1970). Renshaw cell mediated inhibition of Renshaw cells: Patterns of excitation and inhibition from impulses in motor axon collaterals. *Journal of Neurophysiology*, **33**, 257-270.

Ryall, R.W., & Piercey, M. (1971). Excitation and inhibition of Renshaw cells by impulses in peripheral afferent nerve fibers. *Journal of Neurophysiology*, **34**, 242-251.

Schwindt, P.C. (1981). Control of motoneuron output by pathways descending from the brain. In A.L. Towe & E.S. Luschei (Eds.), *Handbook of behavioral neurobiology: Vol. 5. Motor coordination* (pp. 139-229). New York: Plenum.

Shepherd, G. (1990). *The synaptic organization of the brain*. New York: Oxford University Press.

Sherrington, C.S. (1906). *The integrative action of the nervous system*. New Haven: Yale University Press.

Tracey, D.J. (1985). Joint receptors and the control of movement. In E.V. Evarts, S.P. Wise, & D. Bousfield (Eds.), *The motor system in neurobiology*. Amsterdam: Elsevier Press.

Wand, P., & Pompeiano, O. (1979). Contribution of different size motoneurons to Renshaw cell discharge during stretch vibration reflexes. *Progress in Brain Research*, **50**, 45-60.

Wilson, V.J., Talbot, W.H., & Kato, M. (1964). Inhibitory convergence upon Renshaw cells. *Journal of Neurophysiology*, **27**, 1063-1079.

Acknowledgments

This study was supported in part by NSF grants IRI-87-16960 and IRI-90-24877 and in part by Instituto Tecnológico y de Estudios Superiores de Monterrey. The authors wish to thank Kelly Dumont, Cynthia Bradford, and Diana Meyers for their valuable assistance in the preparation of the manuscript.

PART III

DYNAMICAL SYSTEMS APPROACHES TO MOVEMENT VARIABILITY

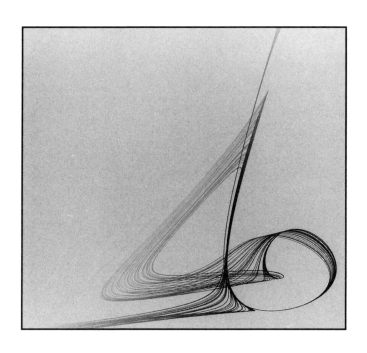

Chapter 10

In Search of a Unified Theory of Biological Organization: What Does the Motor System of a Sea Slug Tell Us About Human Motor Integration?

George J. Mpitsos
Oregon State University and
The Mark O. Hatfield Marine Science Center, Newport, Oregon

Seppo Soinila
University of Helsinki, Helsinki, Finland

Much of our discussion here will address the functional meaning of divergence and convergence of connections among neurons. At the simplest level, both are anatomically definable: Divergence occurs when a single neuron sends synaptic[1] projections to many other neurons, and convergence occurs when many neurons send projections onto a common follower neuron. A more functional definition is to say that divergence distributes information, whereas convergence produces sharing of information. The consequence of divergence is to increase the size of the cofunctional group of neurons, but this alone would only produce a set of independent processors. In parallel programming, the programmer breaks down a problem into different components and then assigns each component to a different processor; the programmer distributes the components, but the processors act independently. Similarly, there may be multiple sites of learning, perhaps arising from divergence of input-stimulus pathways onto many different cells, and each site may involve different cellular mechanisms, but unless there is some interaction or convergence, each site processes information independently. Because of its potential for sharing information, convergence forces many neural sites to work interdependently. Thus, convergence lies at the heart of our definition of parallel processing in biological systems (Mpitsos & Cohan, 1986b; Mpitsos, Collins, & McClellan, 1978), as it does in simple connectionist neural networks (Vemuri, 1988) that have little resemblance to biological ones.

---

[1]We use the terms *synaptic projections* and *connections* to refer both to well-defined pre- and postsynaptic structures involving localized transmitter release and to morphologically indistinct structure involving diffuse transmitter release.

GRAND UNIFICATION THEORIES

In attempting to understand the functional implications of divergence and convergence even in small networks, Pribram's (1971) analogy to holography for distributed memory storage seemed a possibility (Mpitsos et al., 1978), particularly, as Mpitsos and Cohan (1986b) later reported, because some networks are able to reorganize similar motor output patterns of activity after neurons are removed that appear to control the pattern of activity going to motor neurons. In these studies, the neuron was removed from taking part in the motor pattern by hyperpolarizing it below its firing threshold. This produced two types of errors: cessation of firing in the motor neurons that it controlled, and cessation of all motor activity. Eventually the original pattern recovered, even though the hyperpolarized neuron, and the motor neuron(s) it drove, did not take part in the reformed motor pattern. Because the overall firing pattern in the reformed activity in the motor roots appeared similar to the original pattern, it seems reasonable that the error was somehow distributed throughout the generator network. By analogy to holography, the "picture" of activity emerging from the memory distributed among the pattern-generating neurons exhibited graininess when bits of information were lost, rather than exhibiting holes or gaps in some regions while retaining high resolution in others, as would occur in some neural networks (Mpitsos & Burton, 1992). We use "graininess" here because fewer neurons became involved in the reformed pattern than in the original one, yet the overall structure of the pattern seemed the same. There are problems even with the notion of holography, and in carrying the analogy of graininess too far, but for present purposes the real question that these studies point to is one of memory storage and control in high-dimensional systems. The high-dimensionality that we refer to is not just in the number of interacting components. It also includes, as we shall discuss, the storage of different forms of information within the same set of synapses and nonlinear ways of addressing it.

Although it is easy to see high-dimensionality, and the consequences of it, in the human cortex, it has not been so easy to admit that it exists in animals that neuroscience persists in calling "simple." A worldview that polarizes animals into simple and complex (into generalizations relating to invertebrate and vertebrate phyla) emerged (e.g., see comment in Edelman, 1978, p. 89). A wide variety of factors, including the technology of intracellular microelectrode recordings (Ling & Gerard, 1949), the ability to use these recording methods on cells that can be identified in different experimental preparations, findings showing that activity is encoded within the central nervous system itself for generating patterned motor activity (Wilson, 1961), the importation of the ethologist's (Lorenz, 1981) fixed-action pattern (FAP), identification of functional types of cells such as command neurons that control central pattern generators and stereotyped behaviors or FAPs (Delcomyn, 1980; Kupferman & Weiss, 1978), and the related findings showing that much of this activity is genetically encoded (Bentley & Konishi, 1978), worked together to entrench reductionism. Though each finding remains useful in its own right, concepts developed from reductionist single-neuron methods have proved inadequate to understand distributed, multifunctional, and variable systems.

It is an interesting discovery that many biological systems, being potentially high-dimensional, may generate complex behavior that is governed by relatively low-dimensional dynamics.[2] Choatic systems fall into this category and, because of their complex response dynamics, have been a subject of considerable attention over the past 10 years (Mpitsos, Creech, Cohan, & Mendelson, 1988; Osovets et al., 1984; Rapp, Zimmerman, Albano, Deguzman, & Greenbaun, 1985; Skarda & Freeman, 1987). We shall summarize some of these efforts. But rather than dealing with the verifiability of chaos itself or of any dynamic process, which has already been addressed sufficiently elsewhere (Mpitsos, 1989), what we wish to do here is address common features of all nervous systems that give rise to, or exclude the ability of the systems to produce, particular response dynamics. This is to say that the important features are not so much whether repetitive activity, as one example, is generated by limit-cycle or chaotic dynamics, as it is of the system characteristics that permit different activities to arise.

It may be useful to forewarn the reader that our own perspective on brain function, or on the function of systems composed of aggregates of nonlinearly interacting components, has two parts, the one experimental and the other philosophical. It is essential, of course, that the philosophy or theory one holds about the actions of a system have a foundation on hard biological fact. However, problems arise when doing only that. Take just one example: All visual systems use on-responses to respond to the onset of light, and off-responses to respond to the off-set of light. But knowing the cellular and physiological mechanisms that generate off-responses in some molluscs would lead one completely astray about the mechanisms that produce them in vertebrate animals (Mpitsos, 1973; Werblin & Dowling, 1969). Evolutionary selection mechanisms tend to optimize the adaptive[3] mechanisms in each organism. Thus, owing to diversification and optimization, it is often difficult to determine what features permit generalization across organisms or, for that matter, across integrative systems within an organism, because the various systems may have developed under different evolutionary constraints. It is possible to argue in favor of comments one might find in print that go something like this: Owing to the observation that evolution conserves mechanisms, what we understand of mechanisms of learning in a simple animal such as a sea slug will allow us to understand the mechanism of learning in humans. But to take that argument is to forget the equally important fact that diversification is a crucially important driving

---

[2]There is often no need to go beyond the definition of dynamics simply as time-dependent variations of activity, though there are different forms of dynamics. Rather than presenting a formal definition, we shall introduce various ideas that modify our standard working definition as they arise in the course of the discussion.

[3]The term *adaptive* implies some conformation of a system (biological or computational) that allows it to survive in its environment. The process of conforming, as we shall discuss in detail in the "Does a Theory Exist?" section, may represent a gradient descent in the error of the response with respect to the response required for survival, or in the energy required to generate the response. That there may be local minima in such conformations indicates that there may be nonoptimal ways of responding, and, conversely, it indicates that there may also be an absolute minimum representing some optimal way that the system might respond for a given environmental demand. Though local minima may be sufficient for survival.

force in biological evolution, not only through variations arising from random factors, but also through deterministic low-dimensional factors whose dynamics gives them a life of their own.

As neurobiologists, we are interested in the integrative mechanisms of sea slugs, crayfish, insects, leeches, lampreys, or humans. But from a broader perspective, we wish to ask whether there are scale-independent principles, namely, ones that apply to different levels of organization, from chemical processes to cellular, organismal, and social ones. The question is: Can we identify unifying principles, as one might say of the attempts to establish grand unification theories (GUTs) in physics? Unfortunately, biological systems are too complex and uncontrollable to permit such a synthesis presently, as we shall try to show in the present paper. One possibility is to conduct computer simulations of models that reduce a particular biological system within the bounds of definable characteristics. This may give insight into mechanisms pertaining to that system, but it may not provide much insight into general principles.

An alternative simulation approach is to use biological information as "points of departure" to conduct computer simulations that do not necessarily attempt to replicate the structure or function of any particular biological system. We go further to suggest that it might be useful to use simulation systems that are actually extreme caricatures of biology but that nonetheless might give insight into biology generally. What we hope to do eventually is to obtain some idea about how network architecture incorporates various linear and nonlinear interactions between neurons to allow the network as a whole to generate different types of response dynamics. We want also to understand how these fundamental network principles become sculpted selectively to produce the neural responses observed in individual animals. The neural architecture in individual organisms may retain more or less of these primal features, as required or permitted by the tasks presented for adaptive fitness. Thus, by seeking to identify common principles from which different mechanisms may emerge, we are joining a call to reconsider the importance of comparative biology (Bullock, 1984), a subject that has suffered as research has become entrenched in animal-specific encampments. But, as we hope will become apparent, our efforts will not be to determine, for example, whether command processes are the same in different animals or to define the command process more exactly. As important as such issues are, we shall nonetheless aim to address comparisons at a broader or more abstract level. Much of our discussion here will center on making analogies through commonality in dynamical principles rather than in mechanisms.

There are, of course, many people who, in one way or another, have addressed the question of how cooperative action arises among groups of intercommunicating individuals. The works of Grossberg, for example, on neural networks and the mathematical foundation of many psychological phenomena are too numerous even to summarize adequately (Grossberg, 1980; Grossberg & Kuperstein, 1986). It is a theme of studies of modern neural network connectionism (Nadel, Cooper, Culicover, & Harnish, 1989), of chemical dynamics (Andrade, Nuño, Moran, Montero, & Mpitsos, 1992; Eigen & Schuster, 1979; Schnabl, Stadler, Frost, & Schuster, 1991), and the mammalian nervous system (Skarda & Freeman, 1987). In many biological aspects, it can be traced back to Darwin (1859) and Aristotle (Thom, 1990). Such works notwithstanding, we shall attempt to show in the present discussion that a unifying theory of how neurons (or individuals of any type) act cooperatively within a group is presently lacking. Along the way we shall also attempt to identify ways for continuing the search for unifying principles.

FINDINGS IN A SEA SLUG

The idea that neural function arises through changing contexts of interactions in neuronal assemblies in our experimental animal, the gastropod mollusc *Pleurobranchaea californica*, has arisen from a series of interrelated behavioral and neurophysiological experiments. We focus here on behaviors that the animal can generate using its mouth and jaws and identify groups of neurons that may provide insight into some of the dynamical processes that give rise to changing contexts of activity among these neurons. In the course of this paper we shall first describe the behavioral, physiological, and immunohistochemical studies in our experimental system, the sea slug *Pleurobranchaea*, and then compare these results to those obtained in other invertebrate animals and in vertebrates. Another gastropod mollusc, the sea slug *Aplysia*, has been the focus of reductionist researches in many laboratories that have attempted to explain animal behavior and associative learning in terms of definable reflexes. In the section dealing with reductionism, we examine these findings, show the difficulties that have arisen, and then reassess them from the point of view of parallel distributed processing. Given growing interest in nonlinear dynamics in mathematical and physical models, we examine the viability of applying tools arising from these studies to biological systems. In the section on computer simulations, we suggest computer methods that might give some insight into how the integrated activity of large numbers of neurons might arise from interactions occurring locally between individual neurons.

The ensuing discussion also relies on the term *behavior* and identifies a number of behaviors within the repertoire of what the animal can do. For the moment, we use *behavior* to refer specifically to a definable response of the animal, or generically to some unspecified but potentially identifiable response. We shall see, however, that the definition of behavior, of behavioral repertoire, and of behaviorally multi-behavioral or multifunctional systems (ones that can produce different behaviors using the same sets of neurons) needs to be revised to take into account the consequences of variation in "contexts" of neuronal group action.

Following René Thom (1990), we use a call from Aristotle (330 B.C.) to summarize the intent of our own work begun 2 decades ago: ἄλλην ἀρχην ἀρξάμενοι, namely, "Now let us make a fresh start," at least to point out what it is that traditional thinking in neurobiology does not address sufficiently, and what the problems are in progressing further. Just as change was at the center of Aristotle's physics, so it is that change (multifunctionality and variation) is at the center of our view of biological organization.

Behavior

Pleurobranchaea is a large sea slug, a member of the opisthobranch gastropod molluscs, ranging in size from a few millimeters to 10s of centimeters, depending on its age. In general body features it resembles a snail, though like land slugs it has no shell (see photographs in Mpitsos & Collins, 1975; Mpitsos et al., 1978; Mpitsos & Davis, 1973; Mpitsos, Murray, Creech, & Barker, 1988). The animal exhibits a relatively large repertoire of behaviors, including righting when turned upside down, defensive withdrawal, mating, egg laying, and swimming. The jaws, radula (analogous to a tongue), mouth, and lips of the animal generate many different and variable behaviors (Mpitsos & Cohan, 1986a). These include several components of feeding, regurgitation, and defensive biting, among others (Cohan & Mpitsos, 1983b; McClellan, 1978, 1979, 1980, 1982a; Mpitsos & Cohan, 1986a). The animal also exhibits self- and interanimal gill grooming (Mpitsos, unpublished observations), but we presently have no way to evoke gill-grooming behavior reliably. However, of all its behaviors, interanimal gill grooming is particularly interesting because *Pleurobranchaea* is cannibalistic, which raises questions into the mechanisms that turn carnivorous feeding mouth, radula, and jaw movements into cleaning movements.

Feeding behavior usually has dominance over the other behaviors. For example, the animals normally withdraw from tactile stimuli applied to their head regions, but in the presence of food, withdrawal responses are suppressed in feeding-motivated animals (Davis, Mpitsos, & Pinneo, 1974; Mpitsos & Davis, 1973). The most obvious feature of the feeding behavior is the rapid bite-strike response, in which the entire jaw structure comprising the proboscis is rapidly thrust out to bite at a food object and then rapidly withdrawn. Feeding also consists of bite-ingestion movements, in which food is grasped and then sequentially drawn into the mouth cavity largely through cyclical inward and outward movements of the radula and coordinated movements of the anterior regions of the jaws and mouth. A third stage of feeding consists of swallowing movements, in which food is passed from the buccal cavity through the esophagus and then into the stomach. The bite-ingestion and swallow components of feeding (Davis & Mpitsos, 1971) are excellent for neurophysiological work because of their oscillatory characteristics, much as might happen in humans during opening and closing of the jaws and related movements of the tongue. Because of the sequence of oscillations, the behavior persists and is amenable to analysis, whereas single-shot behaviors such as withdrawal are more difficult to analyze. However, as in humans, the number of cycles that the animal may exhibit during a single bout of bite-ingestion-and-swallow is often short and the statistical characteristics of the activity may be nonstationary, both of which, as discussed below, pose difficult problems in studies aimed at understanding the dynamics of the behavior.

Key Features of All Mouth-Related Behaviors Can Be Examined Through a Small Population of Neurons, the BCNs

The cerebropleural ganglion ("brain") of *Pleurobranchaea* innervates the mouth and anterior head regions, whereas the buccal ganglion innervates the muscles that move the jaws and radula. Thus, coordination of buccal-oral behaviors, namely, ones that involve both the buccal structures and the mouth and lips, must happen through these ganglia.

The only way buccal-oral behaviors can happen is through the buccal-cerebral neurons (BCNs), of which there are approximately 15 to 20 in each half of the two buccal hemiganglia. The BCNs are unique because they are (a) the only cells in the buccal ganglion that project to the brain, except for two bilaterally paired giant neurons whose function is presently unknown and (b) either directly involved in generating the central pattern generator for the buccal behaviors or intimately involved in controlling it (Cohan & Mpitsos, 1983a, 1983b; Mpitsos & Cohan, 1986b). There may be other oscillators located in the brain, but their effects are weak in comparison to the effects of the BCN oscillator. The BCNs and the two giant cells are the only sources of information to the brain about processes in the buccal ganglion. All of the behaviors involving movements of the mouth and lips in coordination with the tongue and jaws must act through BCNs, and because the BCNs are part of the central pattern generator, they do more than perform coordination of the different motor centers.

Although the various mouth-related behaviors may involve thousands of neurons, key features of the information required to generate these behaviors may be obtained from much smaller subsets of neurons consisting primarily of the BCNs and some of the neurons with which they interconnect. Thus, the BCNs, acting individually and as a group, are *multifunctional*, because they must generate activity pertaining to multiple behaviors.

Figure 10.1 summarizes the BCN connections. The evidence for these connections has been described in several publications (Cohan, 1980; Cohan & Mpitsos, 1983a, 1983b; Mpitsos & Cohan, 1986b). The present evidence indicates that they connect with one another primarily polysynaptically, as indicated by the interneurons in Figure 10.1; however, many of these polysynaptic connections may be through other BCNs. In a few cases there may be mutual inhibitory connections between the BCNs, but the exact connectivity, if it can be defined, remains for further study. As indicated schematically in Figure 10.1, many BCNs converge onto the same target motor neurons, and individual BCNs diverge onto different motor neurons. In turn, the motor neurons feed back to the BCNs that drive them. An identified group of neurons in the brain, the paracerebral neurons (PCNs), converge onto the BCNs, and the BCNs feed back to the PCNs (Gillette, Kovac, & Davis, 1978, 1982; Mpitsos & Cohan, 1986b).

The actual biological network is much larger and more interconnected than shown in Figure 10.1. For example, there are different pools of neurons that send axons out of the brain through the various motor roots, of which there are approximately a dozen on each side of the brain, though some motor neurons send axons out different roots. Additionally, it is necessary to consider that there are numerous pools of interneurons. Thus, the number of converging and diverging connections in the brain and buccal ganglion is quite large. Moreover, just as there are interactions between the brain and buccal ganglion, there are interconnections between the brain and other ganglia. Therefore, the extended network consisting of neurons affecting the BCNs, and ones that the BCNs affect, involves hundreds of neurons.

What we hope to achieve in our present line of work is to add neuron pools to the core model shown in Figure 10.1. We want especially to obtain the temporal relationships in the firing of as many of the neurons as possible, partly to use the data to reassess the conclusions we have already reached, and partly to use it to

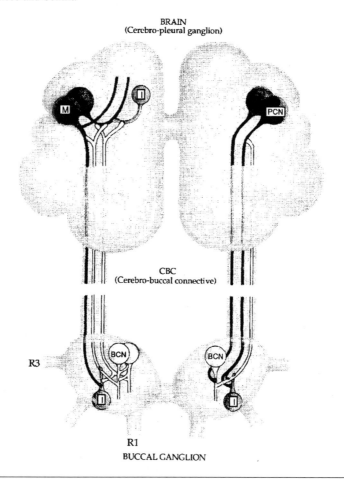

BRAIN
(Cerebro-pleural ganglion)

CBC
(Cerebro-buccal connective)

R3

R1

BUCCAL GANGLION

Figure 10.1 Central features of converging and diverging connections in the *Pleurobranchaea* nervous system. BCN = buccal-cerebral neurons; I = interneuron; M = motor neuron; PCN = paracerebral neuron. The size of each of these pools of neurons is about 10 to 20 units each. There are many more motor neuron pools, one for each motor root; some cells send axons out multiple roots. R1 is the motor root that innervates muscles for opening the jaws; R3 is the motor root for closing the jaws. Motor roots of the brain are not shown. For clarity of presentation, the BCN–motor neuron connections are shown on the left, and BCN-PCN connections are shown on the right.

Note. From "Evidence for Chaos in Spike Trains of Neurons That Generate Rhythmic Motor Patterns" by G.J. Mpitsos, R.M. Burton, H.C. Creech, and S.O. Soinila, 1988, *Brain Research Bulletin*, **21**, pp. 529-538. Copyright 1988 by Pergamon Press Ltd. Adapted by permisison.

obtain some insight into how such large numbers of neurons interact with one another. The time of firing of all BCNs and PCNs is being extracted from multiple recordings conducted simultaneously at different extracellular sites along the nerves that connect the brain and buccal ganglia (the cerebrobuccal connectives, CBCs). Because activity occurs in both directions in the CBC, the multiple recording sites allow us to determine the direction of propagation of firing in different nerve fibers, and thereby to distinguish between the BCNs and other neurons. It is only a matter of extended labor to include the time of firing of motor neurons in the different motor roots.

The point of all of this work, however, is not to obtain a complete network, but to use the data to assure that our computer simulations of different model assumptions will provide activity that reflects the activity in the biological system. A particularly important aspect of this work will be to obtain an indication of the types of variations and motor pattern blending that the system generates. Owing to similarities in the structure of the nervous systems in slugs and snails, the principles obtained in studies on *Pleurobranchaea* may hold for other animals. It is likely, though not demonstrated sufficiently, that neurons analogous to the BCNs in *Pleurobranchaea* may have similar functions in all snails and slugs. But it is not clear presently whether other snails and slugs generate as many mouth-related behaviors as *Pleurobranchaea* do, or whether the behaviors in these other animals are as variable.

Distributed Function, Multifunctionality, and Variation

Our initial aim for studying this ''simple'' sea slug was to understand the cellular basis of learning. The many control experiments in the studies of Mpitsos and Collins (1975) and Mpitsos, Collins, and McClellan (1978) were the first to demonstrate that sea slugs are capable of Pavlovian and avoidance associative learning, and even earlier work, though not as extensively controlled, promised that associative learning could be examined in isolated nervous systems (Mpitsos & Davis, 1973). However, work begun in the mid '70s closely examined the motor patterns and behaviors and showed that networks are multifunctional in being capable of more than generating different behaviors and that similar motor patterns can yield different behaviors (Cohan, 1980; McClellan, 1978, 1979, 1980, 1982a; Mpitsos & Cohan, 1986a, 1986b). More importantly the motor patterns of different behaviors often blend with one another and the underlying motor patterns of neural and muscular activity are quite variable (Cohan, 1980; Mpitsos & Cohan, 1986a, 1986b). As discussed subsequently, rather than there being a definable reflex system, it seemed possible that networks of neurons work by flexible contexts of action. The variations in the contexts might involve linear regroupings or might arise from nonlinearities that cause rapid shifts or bifurcations in the patterns of activity generated by the network. It became apparent that attempts to attribute specific function to a given neuron, or to locate the engram of a learned behavior at a particular synapse, could fail.

Rationale for Change in Conceptual Framework: From Single Cells to Contextual Groups

Consequently, we had to backtrack, to reassess how it is that even innate or ''un-learned'' motor patterns arise in such systems, before we could address the problem of how newly learned information is incorporated into the network. Although we continued to conduct learning studies after the observations made in the mid to late '70s, our rationale for doing them has not been to find the locus of learning at specific synapses, but to determine whether learning could actually be identified in the responses of reduced preparations (Mpitsos & Cohan, 1986a, 1986c, 1986d). Additionally, given the indication that information may be distributed over many neurons, it was necessary to develop the technology for identifying populations of neurons that are involved in specific aspects of learning among which we could examine how learning affected cooperative actions among neurons in the population (Mpitsos, Murray, Creech, & Barber, 1988; Murray & Mpitsos, 1988; Murray, Mpitsos, Siebenaller, & Barker, 1985; Soinila & Mpitsos, 1991).

The idea of cooperativity, which Freeman and co-workers (Skarda & Freeman, 1987) have used to advantage in their studies of the rabbit olfactory bulb, resembles what we refer to as "contexts" in neuronal group function. Much of the discussion in this paper will attempt to present our understanding of functional contexts. Early in the development of the idea of command neurons (cells that evoke stereotypic behaviors), Davis and Kennedy (1972a, 1972b) showed that each command neuron of the lobster swimmeret system produces characteristically different effects and selectively controls different motor neurons, indicating that the command process arises from group action in which each command neuron performs specific subtasks of the command process and activates a particular set of motor neurons. Later work, such as the finding in *Pleurobranchaea* that command neurons receive feedback connections from the motor network that they drive (Gillette et al., 1978), blurred functional distinctions that may be attributed to single neurons because function seemed to be shared. Davis (1976) used the term *consensus* to refer to the emergent actions that might arise among groups of interacting neurons. In studies on locust walking, Kien (1983, 1990b) used *consensus* to refer to variable activity in ensembles of neurons. Our thinking on the ability of groups of neurons to act contextually includes variation in the effects produced by individual neurons, by the group as a whole, and in the neurons that constitute the group. For the present discussion we use the idea of "contexts" interchangeably with "consensus." However, although there are similarities between our use of the terms *contexts* and *consensus* and Davis's and Kien's uses of *consensus*, there are also some important differences that we shall address. Our definition relies on many factors other than the number of neurons that become active. Therefore, we hold off on giving a definition (see the section "Definition of Contexts in Group Action") until we have first presented behavioral examples and provided discussions of principles relating to variation, dynamics, and nonlinear function.

Context of Neuronal Group Action: Inferences From Behavioral Choice

The following example may help to explain our use of the term *consensus* (or *context*): One of the original purposes for studying *Pleurobranchaea* was to examine how animals "choose" to perform a particular behavior when confronted simultaneously by many stimuli that often require conflicting responses, as might occur in the natural environment (Davis & Mpitsos, 1971). For example, turning an animal upside down evokes righting behavior having a definable duration. Presenting food to the animal produces several components of feeding behavior at definable thresholds. When the animal is turned upside down and presented food simultaneously, righting times significantly increase, but feeding thresholds remain constant. By such simultaneous presentations of different stimuli to evoke pairs of behaviors, it is possible to define a behavioral hierarchy (Davis et al., 1974) and to view the process of establishing the hierarchy as a reflex system where one behavior inhibits another (Kovac & Davis, 1980).

It is necessary, however, to go one step further. Early studies on behavioral choice (Mpitsos, unpublished observations) indicated that some behaviors seem to blend into one another, as Kirsti Bellman (1979) has shown in lizards. In *Pleurobranchaea*, for example, the anterior portion of the foot may start to twist in order to right, but, at the same time, it may begin to cup around the descending solution of the food stimulus. The anterior foot appears to be attempting to perform two contradictory behaviors at the same time. Even when righting behavior starts, it is slowed because the foot's motor system is still receiving conflicting activities, one for righting and one for feeding. We do not deny that reflexes involving inhibition can be found, but focusing on that alone places one's concepts on the side of the razor's edge in which behavior and the underlying neurointegrations are viewed as set and repeatedly definable structures.

The important issue to us is the *process* of forming the behavioral choice during the time that the animal is presented multiple stimuli, rather than a stereotyped behavioral hierarchy. The two approaches speak about the same behaviors but give different explanations. The contextual approach views behavior as arising fluidity among many different and blendable behaviors. The reflex approach views the animal as a generator of a set of fixed-action patterns (FAPs; e.g., Gillette, 1986), each relating to definable and repeatedly identifiable responses in the animal. The definition of behavioral hierarchy forces one to think of behaving animals as concatenations of reflexes or FAPs that are repeatedly definable. In the extreme situation in which an inverted animal lies motionless, neither feeding nor righting, the definition of behavioral hierarchy would lead one to develop experiments showing inhibition between feeding and righting sensory-motor systems, as shown for the interaction between feeding and withdrawal (Kovac & Davis, 1980). It would also lead one to identify a particular locus in the nervous system at which such inhibition takes place. The variability of activity in *Pleurobranchaea*, and the high degree of converging and diverging connections in its nervous system, lead us to believe that such localization of mechanism may be misleading. By contrast, when taking these factors into account, one's focus is directed to dynamically shifting contexts of activity in which the identity and location of the underlying mechanism for a behavior is not fixed, just as the behavior may not be fixed and always distinguishable from others. One is more apt to think of variably emerging networks rather than switchboard reflexes.

Thus, although the definition of behavioral hierarchy is useful for categorization, and although it is defined using the behavioral-choice paradigm, it dangerously excludes the dynamics within choice-making processes. To be sure, reflex actions are indications of a process, but the reflex approach leads one to examine the structure of the network itself, whereas an approach that deals with the dynamics of interactions leads one to examine principles of interaction from which networks emerge not only variably but also nonlinearly, as we shall try to illustrate in the section dealing with reductionism and in the section dealing with computer simulations. Inhibitory interactions between motor systems may be used by both explanations, but the dynamical approach uses inhibition either as a potential explanation that may or may not actually take place, or as a participating variable in a system that expresses the dynamics. In either of these nonreflex explanations, the role of inhibition may not be discernible from the structure of the network itself, though dynamical explanations must also account for conditions that actually express reflexes.

Context of Action in the Buccal-Oral System

The buccal-oral system of *Pleurobranchaea*, consisting of the lips, mouth, radula, and jaws, seems to magnify variation and behavioral blending, because, as noted above, it is capable of generating many different behaviors and variants within individual behaviors. Moreover, blending happens among the various mouth-related behaviors themselves, as well as with behaviors produced by other motor systems. A number of studies have provided criteria for identifying motor patterns relating to particular buccal-oral behaviors. McClellan (1978, 1980, 1982a, 1982b) and Croll and Davis (1981, 1982) have established specific motor pattern differences in electrical recordings made from muscles and nerves to distinguish between feeding, regurgitation, and rejection behaviors, but even McClellan's studies demonstrated that different behaviors can be generated by similar motor patterns.

Having observed considerable motor pattern variations, Mpitsos and Cohan (1986a, 1986c, 1986d) devised a series of associative learning experiments to determine whether a learned response persisted in even minimally dissected animals. The results clearly showed that the behaviors of the undissected and dissected behaving animals were identical, as determined by direct observation of what the animals did in response to the applied experimental and control stimuli that were used in training. However, from the electromyographic data alone, obtained simultaneously while observing the behaviors, it was not possible to identify consistent differences in the firing patterns of muscles during feeding, regurgitation, and rejection. The information had to reside within these patterns, but the information itself could not be read simply by examining the temporal orchestration of activity in the recorded motor patterns. An alternative explanation is that the information resides in the dynamics of the neuromuscular system as a whole; that is, in the combination of interactions between the motor output, in the nonlinear loading presented by the muscles and mouth and jaw structures, and in the effect of sensory feedback to the central nervous systems. Such systems may have qualities similar to damped-driven oscillators whose dynamics are sensitive to changes in parameter constants that control the effects of different variables (e.g., see the description of the Duffing oscillator in Thompson & Stewart, 1986). Not inconsistent with this is that the animal can perform a given behavioral effect successfully using a combination of patterns. In neural activity, it may be sufficient to have reached an approximating and variable consensus or context of action; an explicit stereotyped pattern may not be required.

The neural sources of some of this variation were identified in studies of isolated nervous systems that were used in order to remove the influence of sensory perturbations. For example, neural patterns reemerge even when BCNs that were initially responsible for generating patterned activity are reversibly removed from the coactive networks (Mpitsos & Cohan, 1986b, Figure 5), showing that different combinations of neurons generate similar responses. Similarly, the firing of some BCNs shifts variably between completely opposite phases of the cycle of opening and closing of the jaws (Mpitsos & Cohan, 1986b, Figure 16). Graded intermediates may occur as the nervous system generates patterns of rhythmic activity and spontaneously shifts into another pattern.

Dynamics in Distributed Function

Graded, intermediate, and variable forms of activity may give crucial information about integrative mechanisms. Of particular interest are mechanisms that allow high-dimensional or complex neural structures to generate low-dimensional activity.

Attractors as Dissipative Structures:
Low Dimensionality in High-Dimensional Systems

An intuitive definition of *attractor* may be given by examining the property of attraction. Suppose for the moment that we are dealing with a process governed by three variables. The state of the system at any given time is represented by the values of these variables. The progression of these values over time defines the parameter state-space of the activity of the system. Plots of these variables, one variable in each coordinate of three-dimensional space, define the phase space. The flow or trajectory from one point to another provides a view of the phase portrait of the dynamics of the activity. For continuous periodic activity, the trajectory is a closed loop. A brief external perturbation, applied to one, or any combination, of the variables will move the state of the system away from the closed loop. If the trajectory then collapses asymptotically back toward the closed loop, the system may be considered to be governed by an attractor. The set of all possible perturbations, and subsequent dissipative responses shown by the asymptotic recovery, define the inset to the attractor, or its *basin of attraction*. In the case of periodic activity the attractor may be a *limit cycle*. The activity could also be generated by *chaotic attractors* whose trajectories are represented not by a limit set either before or after perturbations, but by an attracting set. An indication of this set may be viewed through the geometry of the topological manifold in which the trajectories mix. Examples of the mixing geometry of attractors in *Pleurobranchaea* responses and model systems in our own work may be found in Mpitsos, Burton, Creech, and Soinila (1988), Mpitsos, Creech, Cohan, & Mendelson, (1988), and Andrade et al. (1992). Though we have used phase portraits to obtain an intuitive view of attractors, a single dynamical system may have phase portraits containing multiple, competing attractors (Thompson & Stewart, 1986).

The above-cited work from our laboratory also discusses a variety of geometrical and computational tools that may be used to determine whether the activity is generated by limit-cycle or chaotic attractors. In either case, the most useful means for determining whether the system is generated by an attractor is to conduct the perturbation experiments already described, which is a major focus of our present efforts in both biological and model systems. Much experimental work needs to be done in this way, but it is quite likely that attractors underlie much biological function, as is shown, for example, by perturbation experiments designed to test for the resetting of the phase of oscillatory activity. (An example of an externally applied current pulses to one of the BCNs in *Pleurobranchaea* is shown in Mpitsos & Cohan, 1986b, Figure 3.)

As the system evolves to dissipate perturbations, one would observe that the ensemble of points in state space decreases over time; that is, that there is volume contraction. Volume contraction simplifies the topology of the structure defined by the trajectories, and, as pointed out by Thompson and Stewart (1986, p. 1), "this can often mean that a complex dynamical system with even infinite-dimensional phase space . . . can settle to final behavior in a subspace of only a few dimensions."

This phenomenon is particularly important in biological systems, because they are inherently high-dimensional. A single cell in the visual cortex of the mouse, for example, receives inputs from approximately 5,000 other cells (Braitenberg, 1989), each of which may be a controlling variable. Numerical analyses of spontaneous cortical neuron activity (Rapp et al., 1985), of EEGs in olfactory bulb (Skarda & Freeman, 1987), of cortex (Albano, Mees, Muench, Rapp, & Schwartz, 1988; Babloy- antz, Salazar, & Nicolis, 1985; Skinner, Mitra, & Fulton, 1989), and of motor patterns in *Pleurobranchaea* (Mpitsos, Burton, Creech, & Soinila, 1988; Mpitsos, Creech, Cohan, & Mendelson, 1988) all indicate that the activity is generated by relatively few variables. One of the tasks facing work in animals such as *Pleurobran- chaea*, and of correlative computer simulations, is to identify the variables, out of the many available, that become active in low-dimensional activity, and to identify the conditions among these variables that permit low-dimensionality to arise. Part of the goal of our computer simulation is to define minimal structures that permit the generation of different types of attractors, and to determine how different at- tractors might arise at different times within the same high-dimensional space. An interesting possibility is that what determines which subspace is occupied may simply be a matter of what attractor becomes established first. In a sense, there may be a type of competition such that the same behavior at some different times may be generated by a somewhat different attractors arising from variable subsets of the available high-dimensional possibilities.

Turbulence, Attracting States, and Self-Organizing Criticality

Thus, given weak connections, which are common in the *Pleurobranchaea* nervous system (Mpitsos & Cohan, 1986b), it is not inconceivable that different limit-cycle and chaotic attractors may emerge simultaneously within the same network, moving and blending in space and time, giving rise to the blending seen in whole-animal behavior (Mpitsos & Cohan, 1986a) and in some motor patterns (Mpitsos & Cohan, 1986b). These conditions may provide the opportunity for analogs of *turbulence* to occur (Mpitsos, 1989). As discussed in our section on computer studies, we believe that large groups of neurons need not all act in a coordinated fashion, particularly when a large number of relatively weak synapses are distributed throughout the network. The statistical properties of the network and the effect of weak coupling may permit conditions under which different subsets of the extended network are able to begin acting cooperatively within themselves. Yet owing to extensive convergence and divergence of the underlying connectivity, one subset of neurons may influence the coordinated firing of other subsets. In this way, small foci of coordinated firing may move spatially, blend, or separate into different foci, much as one might envision of vortices in hydrodynamic turbulence. Instructive examples of such phenomena in physical models have been presented in laboratory simulations (Sommeria, Meyers, & Swinney, 1988) and in computer simulations of the formation of the large red spot of Jupiter (Marcus, 1988). Videotapes showing the evolution of vortices in the hydrodynamic model and in the computer simulations were seminal in solidifying our own intuition about what may happen in neural systems (Professor Harold Swinney, Department of Physics, University of Texas, Austin, personal communication). In considering the possibility of turbulence in neural systems, our own feeling is that the definition of *attractor* in such cases may not be as suitable as in more definable spatiotemporal structures. We prefer to use the term *attracting states*.

Attracting states may have some resemblance to mechanisms of *self-organizing criticality* (SOC) proposed by Bak and co-workers (Bak, 1990a, 1990b, 1990c; Bak & Tang, 1988; Chen & Bak, 1989; Weissenfeld, Tang, & Bak, 1989). The ideas have been applied to models of turbulence in forest fires (Bak, Chen, & Tang, 1990) and the production of unpredictable avalanches that occur when attempting to build mounds of sand by piling one grain of sand over another (Bak & Chen, 1991). Local effects are deterministic and easily observed, but the global effects are not predictable from such local information, and, partly for these reasons, systems governed by SOC seem to be acting near the "border of chaos" (Bak, 1990a). To our knowledge, SOC has not been applied to nervous systems. We envision that conditions that would allow SOC to take place would retain the deterministic character of monosynaptic actions between neurons but, given weak interactions, would also permit statistical or random spatiotemporal long-range effects through polysynaptic action.

Chaos and Other Forms of Variation

Variation in the responses of single neurons and in the interactions between neurons may arise from a variety of mechanisms. For example, it is well established neurophysiologically that the responses of output neurons in multisynaptic pathways have variable latencies with respect to the timing of activity in input neurons. In general, there are two forms of variations, deterministic ones that arise from simple changes in the parametric setting of constants in the neuron or network, and nondeterministic ones such as random noise. We examine examples of these and suggest possible uses that they may have.

Bifurcation Parameters and Chaos.   We shall examine bifurcation parameters in more detail in the section on computer simulation. It is sufficient to state briefly that they are parameter constants that control how a system (or its defining set of equations) expresses its nonlinear characteristics. When the system is far from critical points, changes in bifurcation constants have relatively little effect on the dynamics of the system. At or near critical points, small changes in bifurcation parameters produce rapid changes (bifurcations) in the response of the system. Within certain ranges in the values of these parameters the system may exhibit rapid shifts between different types of periodic activity and chaos as the parameter is successively changed (Andrade et al., 1992; Thompson & Stewart, 1986).

The simplest definition of chaos is that chaos is completely deterministic at each step of its temporal evolution, yet over the long-term, its response is not predictable. An example we shall discuss later is the logistic equation, given by $X_{n+1} = R(1 - X_n)X_n$, where R is the bifurcation constant. This equation has no random factor in it, yet for certain values of R it is not possible to predict the evolution of the time series several iterations into the future, given some initial starting value. Despite its long-term equivalence to random noise, the organized geometry in plots of X_n versus X_{n+1} clearly show the deterministic, nonrandom character of chaos (May, 1976; Mpitsos & Burton, 1992; Thompson & Stewart, 1986).

It is difficult to demonstrate that biological systems generate chaotic attractors, owing primarily to their short-lived and apparently nonstationary behavior (Mpitsos, 1989). However, computer simulations have clearly shown that Hodgkin-Huxley membranes (Chay, 1985; Chay & Rinzel, 1985) and the parabolic burster neuron, R_{15}, in the abdominal ganglion of *Aplysia* (Canavier, Clark, & Byrne, 1990) may be capable of bifurcating into a broad spectrum of simple periodic and chaotic activity. Our previous studies on the implications of attractors and variation, and of their implication in the generation of contexts of interrelated firing in groups of neurons, have been discussed in behavioral and neurophysiological studies (Mpitsos & Cohan, 1986a, 1986b; Mpitsos & Lukowiak, 1985). And there is some evidence for chaos in the responses of individual BCNs and motor neurons in *Pleurobranchaea* (Mpitsos, Burton, Creech, & Soinila, 1988; Mpitsos, Creech, Cohan, & Mendelson, 1988). Other activity of single neurons is more consistent with noisy limit cycles.

The lessons to be gained from chaos are these:

1. As illustrated by the logistic equation, variations arising from chaos are not noise superimposed on the information-carrying signal, they themselves represent the information.
2. The information in chaotic systems is always increasing with respect to information available at a given initial time. This is to say that if chaos is to represent behavior, it is necessary to use the long-term phase-space geometry of the attractor driving the system to gain a view of what the behavior is like. Given equal noise-free conditions, the behavior represented by periodic activity can be defined in a single orbit.
3. Periodic or limit-cycle activity dissipates perturbations differently than do chaotic systems. As pointed out by Conrad (1986), limit cycles in biological motor systems dissipate perturbations in ways equivalent to heat loss through the body structures innervated by the neural system in question, whereas chaotic attractors dissipate the perturbations by generating new variations. Limit-cycle attractors always return to doing behaviors in the same stereotyped ways. Chaotic attractors generate new variations naturally in response to perturbations, because their sensitivity to initial conditions always forces them to generate the behaviors in different ways, which is to say that behaviors are always different in chaotic systems.

4. Mpitsos and Burton (1992) have shown that chaotic discrete processes, much as might occur in spike trains communicating between networks, allow simple networks to perform complicated tasks that would require considerably more complex networks to perform if the signals were generated by nonchaotic discrete processes or by continuous periodic or continuous chaotic processes.

5. It was also shown that the inherent variations of chaotic discrete processes permit networks that receive such signals to optimize their responses either in transmitting the signal one-for-one or in performing computations on them. That is, the deterministic character of chaotic discrete processes allows them to convey information, yet their long-term randomness provides sufficient variation to allow the responding network to learn rapidly.

As we shall discuss subsequently, random noise may be used advantageously to perform such optimizations. But random noise has the disadvantage of being high-dimensional, and high-dimensional processes are difficult to generate, because they must represent many degrees of freedom. Chaotic processes are the long-term equivalent to random noise, yet the expression of chaos can be easily controlled using low-dimensional systems and simple adjustments to a single control parameter, as in the logistic equation. In multibehavioral systems such as *Pleurobranchaea*, the combined informational content and variation of chaos may be useful in accessing the different response possibilities (Mpitsos & Burton, 1992).

Bifurcation-Induced Variations. Another form of low-dimensional variation arises when systems approach bifurcation points. An intuitive understanding of this may be given by recalling the above discussion on the demonstration of attractors lying in three-dimensional space, and using this example to understand what happens to Lyapunov exponents as the system approaches bifurcation points. In a system governed by three variables, there are three exponents. (A useful discussion of Lyapunov exponents and numerical methods for estimating them is presented in Wolf, Swift, Swinney, & Vastano, 1985). A negative Lyapunov exponent indicates that there is contraction in a given direction in phase space. If all three exponents were negative, the flow of points in phase space would collapse in all directions into a single point. For continuous, bounded systems not at a fixed point, at which the system remains at equilibrium at some nonchanging parameter state (see the definition in Thompson & Stewart, 1986, p. 194), Haken (1983) has shown that one of the exponents must be zero. In a simple limit cycle governed by three variables, the remaining exponents must be negative. The negativity in the sum of the exponents assures that there is an overall contraction in the flow of points in phase space to keep the system bounded. The summed negativity also assures that the system will dissipate perturbations if they are not so large as to push the state beyond the attractor's basin of attraction. Bifurcations into chaos introduce a positive exponent but retain the criteria that there be one zero-valued exponent and that the sum of the exponents be negative. The positive exponent shows that the state of the system in the corresponding dimension of phase space is always expanding. Having a zero-valued Lyapunov exponent indicates that the growth in phase space is neither contracting nor expanding over time. Thus, the rate of growth of a three-variable[4] system in phase space is given by $2^{(\lambda_1 + \lambda_2 + \lambda_3)t}$, where λ_1, λ_2, and λ_3 are the corresponding Lyapunov exponents for growth in each direction of phase space, and t is time. Because the exponential change is given as base 2, the exponents express the rate of change of growth in phase space as information in bits per second. Thus limit cycles lose information as they evolve with respect to some initial state, whereas chaotic systems gain information.

As a system approaches bifurcation points, some of the Lyapunov exponents approach zero values, as we show herein for the catalytic network model of Andrade et al. (in press; Mpitsos, in preparation). Setting the bifurcation parameter, μ, to a value of .02, generates a one-period limit cycle far from a bifurcation point, and λ_1, λ_2, and λ_3, have values, respectively, of 0, −2.8, and −43. Adjusting μ to .0125, well past the bifurcation into a two-period limit cycle, the exponents have values of 0, −3.6, and −43. However, setting μ to .0149, which is near the bifurcation point, the exponents are 0, −.05, and −46; λ_2 vanishes. Thus, as the system approaches bifurcation points, a greater number of Lyapunov exponents approach zero than when the system is farther way from these points. Perturbations in directions of phase space governed by exponents having small negative values would be dissipated slowly. Even in model systems having no extraneous injected noise, transient variations are often difficult to remove when one is attempting to locate bifurcation points.

Kelso, Schlultz, and Schöner (1986) have used the term *critical fluctuations* for the variations observed in human finger movements during phase transitions, or, in our terminology, at critical bifurcation conditions. We have observed similar fluctuations in our own studies using sinusoidal current to drive individual neurons in *Pleurobranchaea* and *Aplysia* (Glanzman & Mpitsos, unpublished observations). Moreover, because the *Pleurobranchaea* buccal-oral system appears to sit metastably near transitions into different patterns of activity (as shown, for example, by frequent spontaneous transitions of activity in isolated nervous systems; e.g., see Mpitsos, Creech, Cohan, & Mendelson, 1988), we should expect to see variations in activity simply because of the tendency of the system to pass through bifurcation conditions. In model networks, it is possible to generate activity in the system long enough to

---

[4]The need for three variables in continuous systems that can generate chaos may be viewed intuitively by examining the flow of trajectories in phase space and their ability to mix as they course through the attractor surface; a typical trajectory will visit every vicinity. Evidence for mixing can be obtained by cutting a Poincaré section through the phase portrait and noting the interrelated positions of the crossings of the trajectory through the section (Thompson & Stewart, 1986). If one places a string on a flat surface defined by two variables, it is possible to conform the shape of the string to flow to a fixed point, to form a variety of self-similar spirals (Schroeder, 1991, p. 89), or to connect the two ends of the string to form limit cycles or other periodic activity. (Also see a discussion of the Jordan curve theorem and the theorem of Poincaré-Bendixson in Hofbauer & Sigmund [1988, p. 149].) However, it is not possible to have nearby lengths of the string diverge from one another and eventually mix in their interrelated positions without causing the string to cross on itself somewhere, unless the trajectories flow into a third dimension and then fold back onto a thickened plane; that is, however imperceptible, there must be a thickness to the surface of the attractor, composed of countless layers arising from continuous stretching and folding, which brings distant trajectories close together. Discrete processes, on the other hand, can generate chaos in a single dimension, as shown by the logistic equation.

get rid of transients. But biological systems, which generally do not have such long-term luxury, should exhibit considerable variation simply because of bifurcation effects, unless they lie far from critical points.

A rather interesting problem of bifurcation-induced variations occurs in regions of the controlling parameter that cause chaos. Such regions are filled with subregions that lead to periodic activity, as can easily be demonstrated by examining the bifurcation parameter of the logistic equation at expanded scales (Thompson & Stewart, 1986, p. 162). Therefore, small changes in a control parameter may actually lead to rapid shifts between chaos and periodicity, with each state being accompanied by transient variations. Clearly, there is a need to understand how biological systems cope with the sensitivity in the adjustment of bifurcation parameters and with the different forms of variations that arise from such adjustments. One possibility may be that the large number of converging and diverging connections among neurons may buffer unwanted bifurcation conditions by lifting the controlling effect from residing in a single neuron or a few of them and distributing it over a large number of neurons. In this way, the bifurcation conditions emerge from group action, though individual neurons may exhibit near critical behavior. This may also be a reason for the observation of the wide distribution and convergence of neurotransmitters and modulators.

Random Noise: Variation-Dependent Optimization.   Other variability in *Pleurobranchaea* seems to be high-dimensional, or even random, as shown by the response of a single neuron in Figure 1 of Mpitsos (in press) and by the analysis of electromyograms in Mpitsos and Cohan (1986a). It has long been known that a little random noise may help systems to avoid local minima, which may be defined for the present purposes as nonoptimal responses (see Figure 8 in Burton & Mpitsos, 1992, for a diagrammatic demonstration of local minima). The physicochemical properties of DNA provide an example of one use of noise in biological studies (Doty, Marmur, Eigner, & Schildkraut, 1960). Heating solutions of DNA (injecting noise into the system) breaks the two complementary strands apart. If the solution is cooled too rapidly, the original complementary bonds between base pairs is not completely restored; that is, the system has fallen into a local minimum. If the solution is cooled slowly, the strands recombine optimally, forming the absolute minimum. Thus, the terms *local minima* and *absolute minimum* may be used to refer to a number of characteristics, such as information storage, reconstruction of an original template, and energy level.

Such processes of noise control are time dependent and usually control noise by decreasing it exponentially. The method is referred to as *simulated annealing*. Kirkpatrick, Gelatt, and Becchi (1983) discuss simulated annealing and apply it to several optimization problems, including the placement of computer chips on a circuit board, in which the goal is to minimize wire length and bends, and the traveling salesman problem, in which the goal is to minimize the distance traveled between cities if each city is visited only once. Simulated annealing is time dependent because it requires the noise in the system to have a decay rate, and once the noise has died out, it is necessary to introduce noise into the system again in order for it to be ready to respond to a new situation. Biological systems are generally event dependent, not time dependent. It may be difficult or impossible to determine in advance when the next challenge to survival will occur or what it will be, and when to reinject noise into the system. Once a challenge has presented itself, there may not be enough time to adjust the rate of decay of noise.

As a step in determining how random noise might be used in adaptive systems, Burton and Mpitsos (1992) devised time-independent noise algorithms (TINA) that control noise through the response of the system, as would occur in natural environments, rather than through predefined time schedules. To demonstrate the algorithm, Burton and Mpitsos used simple nonbiological neural networks that were required to learn to transmit or manipulate chaotic input signals, much as might occur if networks communicated with one another with chaotic spike trains. Networks were trained using an error-backpropagation algorithm (Rumelhart, McClelland, & PDP Group, 1986). Random noise was added to the learning-induced changes in synaptic weights and thresholds, but the level of the injected noise was adjusted on the basis of the amount of error generated each time the network responded to an input event. By such adjustments it was possible to avoid local minima and speed the process of reaching maximal levels of learning.

Thus, random noise, chaos, and possibly variations arising from bifurcation conditions may provide conditions leading to two different methods of optimizations. The effect of chaotic discrete processes was shown under conditions in which chaos would act as a transmitter of information between networks, whereas the effect of noise was shown when it was added to changes in synaptic weights and thresholds during learning when the network had to respond to the chaotic signal. However, chaos is only short-term deterministic. The long-term statistics of chaotic processes, as might occur in spike trains, are identical to random noise. For systems such as *Pleurobranchaea* or the mammalian olfactory bulb (Skarda & Freeman, 1987) that are multifunctional or contain multiple information within the same set of connections, variations that allow the system to search for one of many attractors or attracting states may be essential.

The three types of variation mentioned above involve different search strategies and control methods. Chaos has a deterministic search strategy and can be controlled through bifurcation parameters in membrane dynamics (Canavier et al., 1990; Chay, 1985), synaptic release (see the interesting suggestion in Kriebel, Vautrin, & Holsapple, 1990), and, as we shall discuss in the computer simulations section, in synaptic strengths. Neural systems may be able to approximate randomness simply by using weak synapses and by taking advantage of the large number of connections among cells. For example, connections among 10 to 100 neurons may provide sufficient degrees of freedom to approximate the high-dimensionality of Gaussian noise. A number of activity-dependent changes in synaptic strengths or in the probability of transmitter release (Korn & Faber, 1987) might provide methods to control noise naturally and in time-independent ways. Some of the noise or variations that occur near bifurcation points are deterministic and self-controlled because they are transients that die out asymptotically as the activity evolves over time. Decreases in the value of Lyapunov exponents near bifurcation points would also allow random effects to become amplified, but as the system passes through bifurcation, both the transient effects and the random variations diminish.

Variation, Not Chaos.   The point, then, in thinking about adaptive mechanisms is to understand the use of a spectrum of variational types. Owing to its interesting phase-space geometry and its long-term unpredictability, chaos has received much press. The important issue, however, is not chaos, but variation and its control, and the way variation affects the ability of the system to access different dynamical states. The neural architectures that support the generation of these variabilities and ones that lead to control are unexplored. We provide suggestions in the computer simulations section.

Error as an Integrative Principle.   A system that has evolved to meet only one adaptive need can be highly tuned to perform that task well, but when confronted with new adaptive needs, such systems may prove extremely fragile. Alternatively, if the system is naturally variable, the output may never be exactly "right" for a given task, but it may be right enough for the system to adapt successfully to different situations. Moreover, given a limited number of neurons, a greater range of outputs may be possible when the system has variable and blendable outputs than when the system contains a rigidly fixed number of output patterns.

Error may not only be a product of system dynamics; it may also be influential in the establishing the dynamics. The first indication of this was in studies of catalytic networks originally devised to account for the first steps in chemical or prebiotic evolution (Eigen & Schuster, 1979; Küppers, 1983). Schnabl et al. (1991) recently showed that error, expressed as mutual intermutation between reactive molecular species, significantly affects the ability of a system to bifurcate into complex, chaotic oscillations. Andrade et al. (1992) provide a more biologically plausible model of error utilization in catalytic networks that may be modifiable for application to studies of neural networks. In this model, error arises from faulty replication; that is, in mutual intermutation the error is transformed into information contained in another reactant species, whereas in faulty replication information is simply removed from the system. Although the generation of complex (chaotic) behavior in this latter model is less sensitive to changes in error than the mutual-intermutation model, analysis of both models using the level of error as the bifurcation parameter shows that error plays a role in the dynamics occurring among the catalytic interactions.

Definitions

Given the preceding discussion of variability in distributed networks, we extend here some of our working definitions.

Dynamics, Behavior, and Multifunctionality.   The preceding discussions provide the background for us to present several working definitions. In the most general terms, we take the term *dynamics* to imply the generation of cooperative activity among a group of interacting components of a system. There may be many different dynamical mechanisms: Linear shifts in the aggregates of coactive components, bifurcations, limit-cycle and chaotic attractors, attracting states, turbulence, and self-organizing criticalities are just a few examples that we mentioned. As we shall attempt to illustrate further in the computer simulations section, our definition of neurocircuits relies heavily on dynamics rather than network architecture.

In much of the preceding discussion, we have used the term *behavior* in the sense that behaviors are distinctly different from one another, as if feeding, regurgitation, righting, and other behaviors in the animal's repertoire were definable. Indeed, the notion of a repertoire seems to indicate that they are definable. However, our above discussion of contexts and consensuses shows that we do not believe that behaviors need be repeatedly the same. For example, the animal ingests food, it may regurgitate it, and it may right when inverted. Yet the animal may perform these behavioral effects in many different ways. If we are correct in our assessment of variations in neural activity and contexts, it is possible that the kinematics of the behavioral effect are always changing. Given this blurring of what the term *behavior* may mean, it is obvious that systems capable of generating many different behaviors using the same neurons must be defined in ways that include variation. Therefore, the concept of multifunctional networks to us implies patterns of activity and behavioral effects that lead variably from one effect to another as well as the generation of distinctly different behaviors.

Contexts in Group Action: Linear and Nonlinear Organization.   In a previous section (Dynamics in Distributed Function) we referred to studies that proposed that the appropriate behavior arises when a large number of neurons, or perhaps all or most of them, become active (Davis, 1976; Kien, 1983, 1990a, 1990b; Wetzel & Stuart, 1976). This is part of what we mean by contexts and consensuses. Linear summations such as those implied by the expression ''large number'' do not address two important problems. First, if attractors or other nonlinear phenomena arise, it is not necessary for the majority, or a large number, of neurons to become active. That is, coherent activity may take place among a minority of neurons, but we believe that if the coherence is strong enough, its effect may override activity that is less strongly organized, though both coherent and noncoherent activity probably affect the actual expression of the resultant behavior. The question, then, is not how many neurons become active, but how strong the coherent activity is above a noise level. Second, even if the interactions are linearly related, or if robust, stable attractors have not organized, adaptive responses may still take place, though the effect may not be as strong as in cases when the majority of neurons act together or when there are strong attractors.

FINDINGS IN OTHER ANIMALS

The expression of variable activity in neural assemblies is not unique to *Pleurobranchaea*. For comparison to our findings in *Pleurobranchaea*, we examine here a few examples taken from studies in other invertebrates and in mammals.

Multifunctionality and Variability in Invertebrates

Taking advantage of well-defined connections among four identifiable cells in the buccal ganglion of *Aplysia*, Gardner (1990) has shown that synaptic effects between identified neurons vary widely from animal to animal. Drawing an analogy to connectionist neural networks, Gardner points out that the importance of a network is not so much in what its synaptic strengths are but rather in what the set of synapses together can do in expressing the information in an *algorithmic process*. The difference between biological networks and neural networks is that the temporal interrelationships in the firing of neurons may shift, and that the same network may be able to generate different patterns of activity (Mpitsos & Cohan, 1986a, 1986b). Thus, in Gardner's terms, a set of connections may contain the information for many different algorithms. Our modification to this is that one must not consider the algorithm as being repeatedly the same; that is, the algorithm is itself variably expressed.

Recent findings in the sea slug *Aplysia* (Leonard, Edstrom, & Lukowiak, 1989; Zečević et al., 1989) and in lobsters (Cardi, Nagy, Cazalets, & Moulins, 1990; Heinzel, 1988a, 1988b; Heinzel & Selverston, 1988; Leonard et al., 1989; Leonard, Martinez-Padron, Edstrom, & Lukowiak, 1990) are consistent with the notion that the same networks can produce activity relating to different behaviors (i.e., they are multifunctional), as is the work on yet another sea slug, *Tritonia* (Getting, 1989; Getting & Dekin, 1985), although only the work on *Aplysia* has taken notice of variation (Wu, Falk, Höpp, & Cohen, 1989). An important paper describes leech locomotion and asks what it is that the "central pattern generator" really mediates, because a variety of variable behaviors were observed (Ayers, Carpenter, Currie, & Kinch, 1983). Kien (1983, 1990a, 1990b) has published a series of insightful papers on locust walking and has addressed the notion of variation through observations indicating that different groups of neurons become active to produce a behavior. Variability has also been reported in walking motor patterns in cockroaches (Delcomyn & Cocatre-Zilgien, 1988).

By the late 1970s the notion that "hard-wired" networks can explain behavior had received strong support from studies on genetically inherited ability in many animals to generate patterned activity (Bentley & Konishi, 1978). Nonetheless, 10 years later, Getting (1989, pp. 186-187) voiced the following interesting conclusion from his work in *Tritonia*: "Networks with similar connections can produce dramatically different motor patterns, and, conversely, similar motor patterns can be produced by dramatically different networks." Similarly, one can read from the work in *Pleurobranchaea* (Mpitsos & Cohan, 1986a, p. 513) that "organized activity emerges or self-organizes such that different contexts of the same coactive neurons become involved in generating the same or different motor pattern." Much evidence in neurobiology has shown that it is possible to ascribe particular function to identified neurons, and criteria of how to do that have been extensively discussed (Delcomyn, 1980; Kupferman & Weiss, 1978, 1986; Rosen, Teyke, Miller, Weiss, & Kupferman, 1991). Some of the same researchers have also recently put forth the contrasting notion that conditions might exist under which it may not be possible to ascribe function to particular neurons (Kupferman, Deodhar, & Weiss, 1991).

Thus, although the classical perspective still seems to hold, and much evidence exists to support it, there is a growing awareness of alternative possibilities. Our feeling is that it may be difficult to make direct comparisons between animals, even if there seem to be many similarities, as there are, for example, in the general neuroanatomical features of the nervous systems in snails and slugs, indicating that their nervous systems contain neurons like the BCNs in *Pleurobranchaea*. It may be, for example, that feeding systems in animals that evolved to utilize relatively stable and predictable food sources may be less variable than ones having to cope with unpredictable ones. One might envision such a comparison between certain herbivores and carnivores, though the defining experiments have not been done.

What is most important in all of this is that people have begun to address the issues, and quite likely the most illuminating comparisons will be ones that involve different response dynamics. Our bias is that variation should be a common observation. In cases not exhibiting variation, the question then has to do with the mechanisms that control variation.

Bifurcation and Response Modality
in the Lobster Stomatogastric System

The recent discovery of the ability of the stomatogastric ganglion in lobsters to generate different behaviors (Heinzel, 1988a, 1988b; Heinzel & Selverston, 1988) shows clearly that one must not assume that even the simplest networks produce only single responses. The findings of Cardi et al. (1990) are worth casting in our frame of reference relating to bifurcation. The stomatogastric ganglion in lobsters contains a subset of 14 neurons that comprise the pyloric network, which acts as a central pattern generator. Of particular interest in this network is a further subset of three pacemaker neurons that form the oscillator. Another oscillator lying in the commissural ganglion sends projections to the stomatogastric ganglion. By using sucrose-block techniques on the nerve interconnecting the two ganglia, it was possible to reversibly interrupt the connections between the two oscillators. When the projections were blocked, systematic injection of depolarizing and hyperpolarizing current into one of the three pyloric pacemaker neurons resulted in continuous variation in the period of oscillatory bursts of activity in the pyloric rhythm. But when these projections were not interrupted, the period varied discontinuously, and, for some ranges of the injected current, two modes of oscillation emerged at a particular level of injected current. Overall the results show that the timing between the two oscillators affected the modes of integration in the pyloric network, and that the commissural projections also exerted neuromodulatory control over the pyloric network.

There are two ways to look at this data. The first is that there is some reflex circuit change that alters the oscillations in the pyloric network when the connection between the two pattern generators is intact. This seems reasonable if one considers that neuromodulation may be capable of adjusting which neurons participate in the oscillatory interactions or their interrelated timing (e.g., Marder, 1988; Marder & Hooper, 1985; Marder, Hooper, & Eisen, 1987). Using John's (1972) terminology, the network may use switchboard factors to control whether the network produces unimodal or bimodal firing in its burst patterns.

A broader perspective holds that the role of transmitters and modulators is to raise the network closer to a critical point for bifurcation. Small, systematic adjustments in the current injected into one of the three pattern-generating neurons push the system beyond the critical point, allowing the network as a whole to oscillate in two modes, or to jump discontinuously from one period to another. When the transmitter or transmitters are not present, as when the connections between the oscillators are interrupted, the system settles into a state that is far from the bifurcation point. In this case, no amount of injected current will push the network close enough to the critical point to permit bifurcation to take place. What does happen is that the period varies continuously as a function in the strength of the injected current. This is precisely what happens when one varies the bifurcation parameter in a system that is far from a critical point (e.g., Andrade et al., 1992; Thompson & Stewart, 1986). There are two potential bifurcation parameters in the study of Cardi et al. (1990). The way the experiments were conducted uses the polarization state (the amount of injected current) of one of the pattern-generating neurons as the bifurcation parameter. However, if there were sufficient knowledge of the cells in the commissural ganglion that project to the pyloric ganglion, their level of firing could be used as the bifurcation parameter for each level of applied polarization in the pattern-generating neuron.

The advantage of using bifurcation analysis may not be appreciated in studies of most experimental biological systems, because of their complexity and the difficulties they pose in permitting selective control of a single parameter. The utility of the analysis becomes more obvious, however, in computer simulations. Not the least utility of bifurcation analysis is that it may provide some predictability. For example, Feigenbaum (1983) observed that the succession of period-doubling bifurcations occurs in a universally predictable way. The ratio of differences in successive bifurcations is given by $\mathcal{F}_i = (\mu_i - \mu_{i+1})/(\mu_{i+1} - \mu_{i+2})$, where μ is the value of the bifurcation parameter in the sequence of bifurcations from $i = 1, \ldots, \infty$. For many bifurcation maps, $\mathcal{F}_i$ quickly converges to 4.6692 to the fourth decimal place. The pyloric network may be small enough to permit the use of computational methods. The major task will be to determine what parameter to control, though information from neurohumoral experiments may point to candidate factors. Different bifurcation states may use the underlying network architecture in different ways. The way the network expresses the various firing patterns among its constituent neurons is not predictable from knowledge of the bifurcation parameter itself or of the anatomy of the neuronal connections. Predictability of these functional or emergent networks is even more difficult in large networks or if variability is a factor. If there are many weak synapses, there may be insufficient synaptic power to control how the activity traverses the connections among the neurons. Previous activity in the network may alter how the neurons participate in the future to produce similar overall patterns of activity. Both factors have been observed in *Pleurobranchaea* (Mpitsos & Cohan, 1986b) and may affect how the network responds during bifurcation.

Mammals

The importance of variation in brain function was, to our knowledge, noted first in mammalian studies. The work of Adey and co-workers (see summary in Adey, 1972) done over 20 years ago, on the chimpanzee and human electroencephalogram (EEG) and on firing of cortical neurons in cats, clearly expressed the need to consider that noise may have a crucial role in the organization of brain function. Adey noted that although information must be contained in structure, the way the information is expressed quite likely is not obtainable from knowing the connections of structure itself. At about the same time, John (1972) discussed the problem of considering cortical structure as statistical rather than as switchboard circuits that can be deciphered simply by examining the connections. The ideas expressed by Adey and

John were seminal in solidifying reservations in our own laboratory about the viability of ascribing whole-animal behavioral phenomena to simple neurocircuits (Mpitsos et al., 1978). Wetzel and Stuart (1976) clearly favored a variable neuronal group hypothesis to account for vertebrate walking. More recently, Braitenberg (1989) examined the connectivity of visual cortex and suggested that activity flowing through it may resemble a random walk. Rapp et al. (1985) analyzed spontaneous firing in cortical neurons and suggested that the variations observed in cortical neurons may not be random, but rather may arise from deterministic low-dimensional mechanisms such as chaos. Variation appears to be an important avenue for self-organization of cooperative activity occurring simultaneously over the entire surface of the olfactory bulb, as Skarda and Freeman (1987) have proposed that the dynamical state of the bulb shifts from chaotic baseline variations into memory-specific limit cycles that are evoked when the animal inhales odors. Lindsey et al. (1992) have shown that the interrelated firing of neurons in cat respiratory brain stem reconfigures during successive respiratory oscillations.

All of these findings are consistent with our own findings in *Pleurobranchaea*, and, in turn, our findings suggest that the different variational types may provide for response optimization into different attractors. Although the work in *Pleurobranchaea* represents the first demonstration that chaotic activity underlies adaptive responses in animals, it is necessary to take the evidence extremely cautiously, as has been pointed out (Mpitsos, 1989; Mpitsos, Burton, Creech, & Soinila, 1988; Mpitsos, Creech, Cohan, & Mendelson, 1988). To be sure, the responses are often variable, and that is the more important issue than chaos itself. However, to the extent that chaos does hold to be the case in *Pleurobranchaea* and in the various observations described above in mammals, it may prove a general principle to pursue further that the variations may not only convey information for a behavior but also provide for one of the methods for response optimization discussed in the previous section.

Divisions of the Mammalian Motor System: Relationship to Divergence and Convergence

Mammalian motor behavior may be classified as involving the pyramidal system (PS) or the extrapyramidal system (EPS). According to the classical view, execution of all voluntary movement in mammals is initiated by motor cortex acting through the PS, which constitutes a two-neuron chain. The upper motor neuron descends from the cortex and synapses in the spinal cord with the lower motor neuron, which innervates the muscle. Going backward, each muscle fiber is innervated by a single lower motor neuron, which is contacted by only a few upper motor neurons, perhaps only a single one. So, each skeletal muscle of the body has a topical representation in a specific zone of the motor cortex. Stimulation of a specific region results in a stereotype response, which, if the stimulus is focal, includes one muscle fiber only. A given cortical neuron can act in two different states, depending on the context defined by preceding impulses from the associative cortex (Tanji & Evarts, 1976). This seems much like a switchboard, showing a precise structure-function correspondence. It can function as such, but the result is not the kind of movement we would like to perform. We get an idea of what kind of movements the PS can produce by itself by watching patients with dysfunction of the cerebellum or the basal ganglia, as in the case of Parkinson's disease. Their movements are coarse, as if the limb moving is not quite sure of the goal. They have often heavy tremor, suggesting an imbalance of muscular tone at rest. Similar imbalance during movement is indicated by rigidity, suggesting that processing of the sensory information about continuously altered position is not occurring fast enough or precisely enough. We might say that the PS does not tolerate nearly as much error as the EPS. It is interesting to emphasize that in cases of cerebellar infarcts or in Parkinson's disease the spinal cord with all its reflexes is supposed to be intact and functioning the best it can perform. Therefore, the PS may exhibit considerably less convergence of overlapping information and less distributed action. The one-to-one mapping allows the PS to execute precise control of movement but may make it extremely error prone should a particular line fail—whereas the EPS may exhibit less precise control yet may be less error prone when its components fail.

Although the physiological finding that given muscular responses can only be obtained by stimulation of certain cortical neurons indicates that there is little convergence, histochemical data suggest that multiple transmitter systems, presumably from the EPS and spinal cord, converge onto the lower motor neuron. The substances involved include dopamine, noradrenaline, serotonin, histamine, substance P, and thyrotropis releasing hormone (TRH) (Hunt, 1983). The upper motor neuron shows some degree of divergence, because its collaterals contact with EPS neurons and spinal cord interneurons before synapsing with the lower motor neuron.

Classically, anything regulating motor functions other than the PS is defined collectively as the EPS. The EPS includes the basal ganglia, the vestibular system, and the cerebellum, and it is thought to be responsible for coordination of movements. Its components connect indirectly with the PS at both cortical and spinal cord levels. The components of EPS are highly interconnected, although the precise circuitry is incompletely known. High degrees of convergence and divergence are likely to occur in the EPS, as suggested by the morphology of, for instance, the cerebellar Purkinje cells and basket cells. By contrast, the PS has significantly fewer connections among its constituent neurons.

This distinction between PS and EPS, however, may not be immutable, as indicated by motor learning. Consider a musician learning a new piece or a jongleur learning a new number. Initially, the motor pattern is established under cortical control. This always happens relatively slowly, and once it gets fast enough, the cortex cannot handle it and may even inhibit the pattern. Where is the pattern transferred to? It must be some subcortical level that takes over the pattern. All we know is that the control levels must be above the lower motor neuron, which is the final common pathway, and that the pattern must be processed by the EPS. Control can be switched back and forth between the different levels, but the PS and EPS seem almost to have switched their functional categorization. To be sure, learning may model EPS to conform to convergence architectures that exhibit less convergence and variation, as discussed below in relation to Figure 10.4.

The diffuse reticular activating system (RAS) is perhaps most apropos to discussions of convergence and divergence and adds a control factor that must be considered with all somatic motor functions. We know from everyday experience that rather sophisticated motor activity can take place at the lowest states of activation (sleep-walking) or rather gross errors may occur, if the state of activation is overly high. The structure classically thought to be related to state of activation is the RAS of the brain stem. Interestingly, this is not really a structure in the same sense as the nuclei or the cortex. Rather, its neurons are diffusely spread over a large proportion of the brain stem. Considering the anatomical fact that most of the vital regulation centers are located in that region over a very small space, the RAS must be in contact with just about everything. It has been thought that the RAS controls mainly autonomic vital functions. However, it has turned out that a reticular system is found all over the spinal cord as well. So it is reasonable to expect that the RAS is intimately involved with motor functions too. (Our guess is that we would find that the RAS extends over all the cortex as well, if we only had markers to identify the cell types.) Thus, a better understanding of differences in the connectivity and function of the PS, EPS, and RAS, and their interactions, may shed some light on the functional significance of convergence and divergence.

NEUROMODULATION: CONVERGENCE AND DIVERGENCE OF NEUROTRANSMITTER SYSTEMS

In the classic view, experimental manipulation of individual neuromodulators often generates predictable effects, as has long been demonstrated in other animals (Lukowiak, Goldberg, Colmers, & Edstrom, 1986; Marder, 1984, 1988; Murphy, Lukowiak, & Stell, 1985). Our own work began with a similar intention: to identify behavior-specific neurotransmitter evidence relating to associative learning. There is good pharmacological evidence for the classically defined type of cholinergic muscarinic receptors (and of a new form of these) in *Pleurobranchaea* (Murray & Mpitsos, 1988). Behavioral evidence shows that muscarinic receptors have a role in associative learning (Mpitsos, Murray, Creech, & Barker, 1988). Development of immunofluorescence methods for detecting the transmitter for these receptors, acetylcholine (ACH), have allowed us to identify the location of presynaptic cholinergic neurons (Soinila & Mpitsos, 1991, 1992). Using complete serial histological sections to examine the full extent of the projections led us to the finding that we should have expected from our physiological work but that, interestingly, we had not. The histology showed that a relatively few cells diverge perfusely throughout the nervous system, hardly leaving any portion of the neuropil untouched.

This led us to examine the distribution of over a dozen putative neurotransmitters in complete serial sections of all ganglia in both *Aplysia* and *Pleurobranchaea* (e.g., Soinila & Mpitsos, 1991, 1992; Soinila, Mpitsos, & Panula, 1990). Examples of these findings are shown in Figure 10.2, a-f, for *Aplysia* and in Figure 10.2, g-i, for *Pleurobranchaea*. Each transmitter we examined involved a few neurons that diverged and converged extensively over the same target areas of the neuropil and on individual neurons. The alternative possibility that neurotransmitters projected selectively onto different areas was seldom seen. Our present working hypothesis, which is being examined physiologically, is that there may be little motor specificity in the projection of neuromodulators, though there may be differences in their

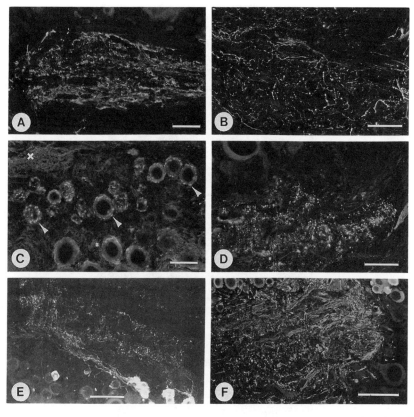

Figure 10.2 (A-F) Photomicrographs of the neuropil region of the *Aplysia* buccal ganglion, showing immunoreactivity for (A) histamine, (B) serotonin, (C) ACH, (D) GABA (gamma-aminobutyric acid), (E) VIP (vasoactive intestinal peptide), and (F) FMRFamide (Phe-Met-Arg-Phe-NH$_2$). The cross in (C) indicates the immunoreactive neuropil, and the arrowhead shows immunoreactive terminals around nonreactive neurons. Bar = 100 μm (A,D,E,F) or 50 μm (B,C). (G-I) Photomicrographs of the neuropil region of the *Pleurobranchaea* buccal ganglion, showing immunoreactivity for (G) histamine, (H) GABA, (I) FMRFamide. Bar = 100 μm. Note the extensiveness of the immunoreactive coverage throughout the neuropil in all tissues from both animals. Positive immunoreactivity is indicated by the white profiles that are extensively distributed over the black, nonreactive areas. For reference, in Figure 10.2I, FMRF-amide covers the entire neuropil of the buccal ganglion. The large cell at the right is the buccal giant, and the commissure leading to the left half of the buccal ganglion is at the left margin. The anterior margin of the ganglion is delineated by the row of dimly stained cells at the top of the micrograph, and the posterior margin is shown at the bottom edge of the neuropil. The area between the neuropil and the row of dimly stained cells contains cell bodies that are not seen because they contain no immunoreactivity.

Note. From "Immunohistochemistry of Converging and Diverging Neurotransmitter Systems in Mollusks" by S. Soinila and G.J. Mpitsos, *Biological Bulletin*, **181**, pp. 484-499. Copyright 1992 by Lancaster Press. Adapted by permission.

253

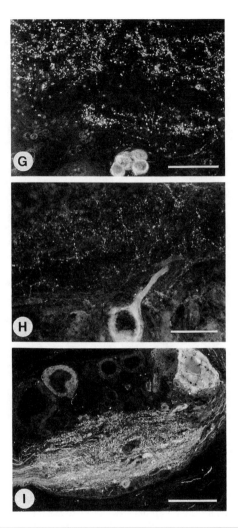

Figure 10.2 *(continued)*

actions. Recent physiological findings in *Aplysia* (Morgan, 1991) support this hypothesis, because individual bath-applied transmitters and neuromodulators appear to affect all motor systems examined.

Given the physiological finding of the extensive convergence and divergence in *Pleurobranchaea* (Mpitsos & Cohan, 1986b), and the corollary finding in *Aplysia* that sensory stimulation activates perhaps the majority of neurons in a ganglion (Zečević et al., 1989), the interesting possibility arises that conditions may often arise when many or possibly all neurotransmitters may become active at the same time. In this case, the classic view of neuromodulation that has been generated using selective applications of single transmitters may not provide adequate insight into the physiological effects produced under normal behavioral conditions. The classic view comes, we believe, dangerously close to making an unstated assumption that the effects of the individual transmitters on common target neurons sum linearly. But if conditions arise when the interactions are nonlinear, the classic experimental approach provides us with little insight into how neuromodulation acts to control network function in normally behaving animals.

A similar situation seems to exist in mammalian nervous systems. Extensive innervation by nerve fiber staining for a large number of transmitters, such as ACH, dopamine, serotonin, histamine, GABA, taurine, glutamate, enkephalin, angiotensin, cholecystokinin, TRH, and vasoactive intestinal polypeptide, has been described in the mammalian striatum (Graybiel & Ragsdale, 1983). Likewise, multiple transmitters (ACH, serotonin, noradrenaline, glutamate, GABA) have been localized throughout the cerebellar cortex (Schulman, 1983). The wulst ("bulge") is a structure in the avian brain that resembles the mammalian neocortex. It is bipartite and runs the length of the dorsomedial portion of the hemisphere. A medial portion is similar to the mammalian hippocampus (wulst regio hippcampalis, Wrh), and a lateral portion is similar to regions of the somatosensory neocortex (wulst regio hyperstriatica, Whs). Both structures are laminated, permitting experiments that can determine whether neurotransmitters are differentially distributed between and within laminae. Shimizu and Karten (1991) examined the immunohistochemical location of cell bodies and fibers containing serotonin, ACH (through localization of choline acetyltransferase, ChAT, and nicotinic ACH receptors, nAChR), catecholamine (through localization of the enzyme tyrosine hydroxylase), GABA (through localization of the enzyme glutamic acid decarboxylase, GAD, and the $GABA_A$ receptor), and the neuropeptides substance-P (SP), leucine-enkephalin (L-ENK), neuropeptide Y (NPY), neurotensin (NT), somatostatin releasing-inhibiting factor (SRIF), corticotropin releasing-factor (CRF), vasoactive intestinal polypeptide (VIP), and cholecystokinin (CCK). Although these substances exhibited laminar specificity, evidence was obtained showing that many regions of the Whs contained overlapping transmitters and neuromodulators. For example, in some portions of a large region, the hyperstriaticum accessorium, evidence was obtained for all substances except CCK, though the density of distribution for each substance was different.

An ideal structure to use for such purposes in vertebrate animals is the retina, because of its well known function and neuroarchitecture and the ease with which its various cell types can be identified (Dowling, 1987; Werblin & Dowling, 1969). Present findings indicate that many neurotransmitters and neuromodulators are located in the various cells of the retina (Karten, Keyser, & Brecha, 1990), although the methods do not show clearly enough how much divergence and convergence occurs among the cells in the retina or wulst and how much occurs from the retinal ganglion cells onto other brain areas. A better method of analysis is to use evidence from the location and distribution of transmitter receptors. Progress in the laboratory of Professor Harvey J. Karten (personal communication) at the Department of Neuroscience, University of California at San Diego, indicates that individual retinal cells contain receptors for many different neurohumoral factors and that many cells stain for the same receptors, indicating that there is extensive convergence and divergence of neurotransmission and neuromodulation. Because of its experimental approachability and well-known function, the retina may provide a rich experimental source for understanding how multiple converging factors interact to control neuronal function.

In human physiology, Parkinson's disease is probably the best known example of a transmitter-specific defect in human motor function. Its cause is considered to be a decrease in the activity of the dopaminergic nigrostriatal tract. Clinical neurology has established that when the amount of dopamine is too low, the action of the dopamine antagonist, the cholinergic system of the basal ganglia, becomes too strong. The treatment, L-dopa, increases dopamine levels to retain the balance between the two systems. However, there is nothing in here to prove that the action of the dopamine-ACH system is necessarily based on fixed circuits and that it acts individually in normal brain function. Although dopamine is found in a specific tract, we do not how much divergence or convergence is involved in that system or what the effects may be when many neurons and transmitters act together.

Although the pituitary is not a classically definable motor organ, it provides an excellent example of multihumoral control. The intermediate lobe is a morphologically homogeneous group of cells that all contain the same hormones, melanocyte-stimulating hormone and beta-endorphin. The question is why are so many different transmitters needed for the simple regulation of inhibition-excitation. Stimulatory (serotonin, ACH) and inhibitory (dopamine, opioids, probably GABA) actions have been described for one substance at a time, but we have no idea how these substances act together. Because the output (hormone secretion) is so simple and easily measurable, this tissue may provide a model for studying the implications of divergence and convergence of multiple neurotransmitter inputs.

Figure 10.3, a-e, summarizes some of our findings in rat pituitary. The data clearly support the possibility of high convergence onto the same target areas, but because there is presently no morphometric evidence of how many neurons provide the innervation, we cannot presently provide an estimate of the ratios of convergence and divergence. The pituitary is particularly interesting, because the output of the system in response to converging actions is neurohumoral rather than electrical.

We suggest that the properties of nonlinearity, distributed function, variability, multifunctionality, convergence/divergence, and the likelihood that the system is error prone, all of which we have attributed to the electrical neurocircuit, may also be ascribable to neuromodulation. It may be possible to obtain repeatable effects when controlling certain transmitters, but what the effects may be or how to conceptualize the interaction of many transmitters (acting at very low concentrations) is presently unclear. If the dynamics of target processes (electrical or chemical) are far from bifurcation points, the nonlinearities (or any effect) may not be observable. But given that the bifurcation points are accessible, the number of possible effects arising from electrical nonlinearities and from the effects of transmitters, cotransmitters, and neurohormones becomes enormous. If we are to believe that neurohumoral agents act variably and in concert, then we must envision further that the subcellular mechanisms that each of these receptors and channels activates may lead to converging and diverging nonlinear actions within the cell itself. Thus, it is conceivable that the clarity of the mechanisms presented for a single neurotransmitter or a single second-messenger system may be somewhat misleading. The point that needs to be examined further is that there may be many different sites of converging interactions in biological systems that process the same information in parallel, and perhaps in

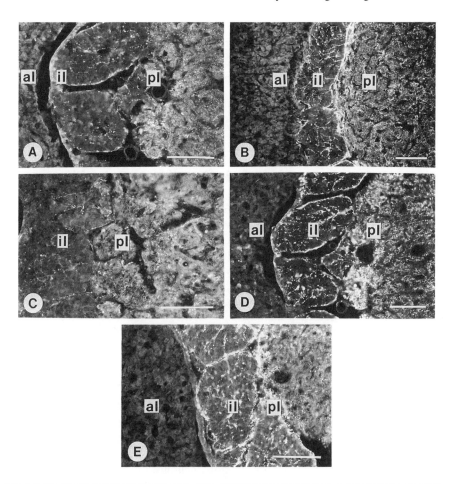

Figure 10.3 Photomicrograph of rat pituitary. al = anterior lobe; il = intermediate lobe; pl = posterior lobe. (A) Acetylcholine. (B) MEAGL (Met^5-enkephalin-ARG^6-GLY^7-LEU^8. (C) Serotonin. (D) GABA. (E) Tyrosine hydroxylase, the dopamine-synthesizing enzyme. Note the convergence of these substances onto similar areas of the intermediate and posterior lobes, as shown in Figure 10.2 in neural tissues of *Aplysia* and *Pleurobranchaea*.

different ways, but may be capable of sharing the results of such processing. Thus, systems may exist in which it may not be possible to ascribe unique function to any motor, cellular, or subcellular process.

REDUCTION AND EMERGENCE
IN CONTROL MECHANISMS

How are these widely distributed physiological and neurohumoral processes controlled? We suggest that many are not, at least not explicitly. It would be too costly, for the same reasons that it would be too costly to devise neurocircuits for each behavior. It seems better to allow the system to be error prone. As discussed in studies on *Pleurobranchaea* (Mpitsos, 1989; Mpitsos & Cohan, 1986a, 1986b) some looseness may actually be beneficial, because systems needing to be highly tuned to specific tasks may prove to be brittle in variable, unpredictable environments. Put differently, it seems better to allow the interaction between the organism and the environment to determine the behavior than to "hard-wire" encode all of the behaviors that an animal can perform.

Transmitters Control Network Function and Architecture

There are, of course, demonstrable control mechanisms that we need to remember that show hard-wiring. For example, as we have mentioned previously, it has been shown that selective application of neurotransmitters evokes different patterns of activity in simple ganglia (Dickinson, Mecsas, & Marder, 1990; Lipton & Kater, 1989; Marder, 1988), just as there is a vast textbook literature showing evidence of the classical "neurocircuit," (Kandel, 1979). Most published evidence weighs heavily in this direction. Thus, good evidence exists to show that "each neurotransmitter or neurotransmitter system may . . . be able to elicit, from the same neuronal circuit, a characteristic and different 'operational state.' In this way it would be possible to obtain a wide range of stable neuronal outputs from a single circuit" (Marder et al., 1987, p. 322).

A remarkable series of experiments by Kater and co-workers (e.g., Kater & Mills, 1990; Lipton & Kater, 1989), begun initially in the freshwater snail *Helisoma* and now extended to mammalian neural tissues, shows the ability of transmitter receptors to control neuronal growth, plasticity, and even survival of neurons. The work has examined a spectrum of neurotransmitters and neuromodulators, including ACH, GABA, dopamine, glutamate, norepinepherin, serotonin, somatostatin, and VIP. Taking advantage of cell culture of identified neurons, the work has been able to provide a strong basis of control experiments. As one example in *Helisoma*, serotonin retards neurite outgrowth, whereas the addition of ACH prevents the serotonin-induced inhibition. The transmitters work through the depolarization state of the cell. For example, presenting an excitatory transmitter alone, such as serotonin, retards the normal neurite outgrowth, but superimposing hyperpolarizing current on transmitter-induced excitation allows the neurite to resume its normal growth rate. The transmitters may act through either voltage- or receptor-activated channels on a common intracellular messenger, calcium. As Lipton and Kater (1989) summarize, neuronal architectures (and therefore neurocircuits) are determined by a fine *balance* in the activation of these two types of channels through an interplay of excitatory and inhibitory transmitters (though different mechanisms may be used in other neural systems; see Garyantes & Regehr, 1992).

The term *balance* clearly indicates that Lipton and Kater are aware that control in natural biological systems may be high-dimensional, because neural tissues are known to contain many transmitters. The problem, then, is to determine how the high-dimensionality is expressed. One possibility is that there is simple linear summation of the effects produced by the various transmitter. However, it is well known that the electrogenic properties of the postsynaptic cell can easily change a simple synaptic input into a nonlinear response. Twenty years ago, Wilson and Cowan (1972) conducted computer simulations on a population model to illustrate that groups of cells intercommunicating through excitatory and inhibitory connections exhibit damped oscillations, multiple stable states, and, under certain constraints, stable limit-cycle oscillations in the number of excitatory and inhibitory neurons firing per unit time. A rather interesting feature of the model is that local interactions were essentially random, yet the long-range effects were quite organized. Another interesting feature of the model that is pertinent to the present discussion is that the populations of excitatory and inhibitory cells were homogeneous; differences arose statistically through use and refractory periods. In even simpler networks, one-shot activation between converging inputs to a common neuron can lead to linear and nonlinear effects in the postsynaptic cell (Andersen, 1987; Koch & Poggio, 1987). In single neurons, it may be possible to generate many different periodic and aperiodic firing patterns by means of fine adjustments to a single ion channel (Chay, 1985). Chay (1985) also showed that intracellular calcium concentration may fluctuate differentially and nonlinearly in each dynamical state. Therefore, the controlling balance between converging transmitters and neuromodulators that affect neuronal structure need not be a simple linear affair. What may seem a linear balance under some parameter ranges of the neurohumoral state can easily switch to drastically different conditions at critical bifurcation conditions.

The dynamics of interactions arising in population of cells need not employ the full high-dimensional space. Going back to our notion of attractors, the different dynamics that a network will allow determine the characteristics of temporal visitation of activity at any given neuron in the coactive group; that is, a set of connections will be activated differently by the types of attractors that it can sustain. Although a developing network at some primitive state may exhibit different dynamical capabilities than a finely tuned, mature one, the same questions of nonlinear conditions arise in both. Finally, if attractors arise either in the responses of single neurons or in networks of them, the high-dimensionality we see in the number of transmitters present may not necessarily be expressed as a high-dimensional process. It is an interesting possibility, raised by numerical studies, that coordinated activity in potentially high-dimensional systems often results in low-dimensional attractors (Mpitsos, Creech, Cohan, & Mendelson, 1988; Skarda & Freeman, 1987). From a simple listing of the number of transmitters resulting from experiments in which transmitters are applied one at a time or in pairs, it is not evident how the system dynamically collapses into low-dimensional control or which of the transmitters become involved in generating low-dimensional activity. Even in small model networks in which all of the driving differential equations are known, it is not obvious from the equations themselves, nor presently from the connectivity, how it is that a lower dimensionality arises from a larger possible set of available variables, unless the system is examined after activating it (Andrade et al., 1992).

Given a linear system, it may be possible to say that neurotransmitters are architects of neural structure. But as we shall discuss later when dealing with bifurcation in minimal networks, conditions may arise in which the activity itself is what fine-tunes a network and, in turn, the network redefines the type of activity that can emerge. There is a dialectical interplay between the two elements, and this dialect, we believe, can act as an architect of neurons and circuits. The chain of events that we might envision of the events that control cell structure is as follows: The dynamics of firing in individual neurons and in networks of them acts on structure through transmitters; the transmitters act on the cell through calcium. The dynamics of changes in intracellular calcium sets up a chain of events that affect cell growth. But cell growth redetermines what the dynamics will be, and so forth recursively. Other factors may contribute, such as synaptic competition. If the notion that many neurons act in close temporal association, or in coordination, is correct, then we must add the complication that the system as a whole is extremely high-dimensional and that many types of nonlinearities may occur. As we shall speak of subsequently regarding the locus of learning, there may be no sine qua non balance of neuro-humoral agents for a given architecture to appear. Although there may be many systems in which there is always a precise connection between a balance between a particular set of chemical elements and structure, understanding these systems gives little insight into others in which variability is an issue.

Thus, although the scientific method at our disposal provides elegant connections between cause and effect, much as Descartes and Euclid would like us to believe, the possibility of high-dimensional space, of nonlinearities, and of the dialectic between structure and dynamics indicates that our view of complex systems may be too simple. However, the scientific methods as they are, are nonetheless the only ones we have. Therefore, our concern is not that the methods and conclusions are simplistic, but rather that they do not address fundamental questions that need to be asked. Moreover, the clarity of some of these reductionistic methods and the importance of the resulting findings have overshadowed the need to go beyond them and to develop methods of data collection that may be useful in taking that step.

Synapse-Specific Control of Whole-Animal Behavior: Complications in the Reductionist Explanation of Learning in *Aplysia*

A tradition in invertebrate neurobiology holds that an advantage of using invertebrate animals is that once a behavior is identified with a particular motor pattern, the same behavior can then be studied neurophysiologically in the motor patterns of isolated nervous systems. As discussed briefly in our earlier section on the context of action in the buccal-oral system, this is quite difficult to do in *Pleurobranchaea* (Mpitsos & Cohan, 1986a). However, the most elegant example of such reductionist approaches has been the identification of site-specific learning in the gill-withdrawal response in *Aplysia* (see reviews in Carew & Sahley, 1986; Farley & Alkon, 1986; Kandel & Schwartz, 1982; Mpitsos & Lukowiak, 1985). A long series of studies has attempted to show how changes at monosynaptic sites between sensory neurons and motor neurons can explain whole-animal phenomena such as sensitization, dishabituation, and associative learning. The mechanism involves serotonin as a neurotransmitter in the reinforcing pathway. The original series of experiments showed that activation of serotonin receptors on sensory neurons leads to a chain of events involving adenosin 3',5'-monophosphate (cyclic AMP) that depress a potassium current when the cell fires. This exposes an inward calcium current that broadens the action potential and, owing to the increase in intracellular calcium, leads to increased transmitter release onto the follower motor neuron. A group of sensory cells, referred to as the LE-neurons, which are usually activated electrically in isolated ganglia, provides the input to identified motor neurons, of which neuron L_7 is perhaps the most important in terms of its effect on the movement of the gill. A group of cells referred to as L_{29} provides the serotonergic input.

A number of important extensions and problems have arisen that both greatly illuminate and complicate this simple model system. We cite only a few examples:

1. *Peripheral nervous system.* From the beginning of work in the late 1960s, evidence has existed indicating that emergent effects may involve the peripheral nervous system, which is distributed within the gill itself. Indeed, in many cases the abdominal ganglion seems not to be necessary for generating robust gill-withdrawal responses and simple forms of learning (Mpitsos & Lukowiak, 1985).

2. *Complex behavior.* The once-presumed-simple withdrawal reflex has turned out not to be so simple, and it consists of several different types of movements (Leonard et al., 1989).

3. *Neuronal function.* Some of the major identifiable motor neurons have variable function within the same experimental preparation within the same behavior (Leonard et al., 1990). This raises strong questions in *Aplysia* as to the accuracy of assuming that identified neurons have consistently the same role in a given behavior, much as Mpitsos and Cohan (1986b) have raised regarding the function of neurons in *Pleurobranchaea*.

4. *Complex network.* Small, well-localized sensory taps activate perhaps half of the cells in the abdominal ganglion, showing that there is extensive divergence of sensory and possibly other effects (Zečević et al., 1989).

5. *Nonconstant activity.* Cells partaking in successive taps are variable (Wu et al., 1989), suggesting that localization of the network may be difficult or impossible.

6. *Source of serotonergic control is unidentified.* Activating L_{29} produces enhanced transmitter release. Serotonin applied experimentally produces the same effect. But L_{29}, which was thought to provide the serotonergic enhancement, apparently does not contain serotonin (Kistler et al., 1985; Ono & McCaman, 1984).

7. *Multiple neurohumoral factors enhance synaptic release.* We now know that at least two other transmitters, small cardioactive peptides A and B (SCP_A, SCP_B), broaden action potentials in LE cells and produce synaptic facilitation on their follower motor neurons (Abrams, Castellucci, Camardo, Kandel, & Lloyd, 1984), but apparently they are not located in L_{29} (Kupferman, Mahon, Scheller, Weiss, & Lloyd, 1984). Interestingly, SCP_B produces spike broadening but not facilitation of transmitter release in depressed sensory neurons (Pieroni & Byrne, 1989), which . may relate to mobilization of the transmitter.

8. *Multiple subcellular processes.* There may be diverging cyclic AMP-dependent processes in different forms of synaptic facilitation (Goldsmith & Abrams, 1989). Conversely, in both the gill-withdrawal system and the analogous tail-withdrawal system, cyclic AMP-dependent and cyclic AMP-independent subcellular processes may converge onto the same spike-broadening mechanisms in both the gill- (Klein,

Bratha, Dale, & Kandel, 1989) and tail-sensory neurons (Sugita, Baxter, & Byrne, 1991).

9. *More than one group of sensory inputs.* The possibility has been raised that under some conditions, novel sensory neurons may be involved in modification of a siphon-withdrawal response whose behavioral modification has been thought to be controlled by changes in the LE sensory neurons (Wright, Marcus, & Carew, 1989).

10. *LE cell activity lacks the timing to be the primary site of learning.* Most importantly, it now appears that there is a second group of sensory cells that have lower thresholds than the LE cells (Cohen, Henzi, Kandel, & Hawkins, 1991) and are probably more likely than the LE cells to activate during training of the gill-withdrawal response itself. It has now been reported (Cohen et al., 1991) that the latency of responses in mechanoactivated LE cells in all of the 32 preparations that were tested always occurred *after* the initiation of the discharge in the motor neurons. Their timing in the behavioral reflex has been difficult to determine (Byrne, Castellucci, & Kandel, 1978). The problem, then, is, if the *cellular basis of behavior* relies on the LE cells as the site of facilitated transmitter release, the responses of the LE cells must occur *before* the initiation of motor output for that behavior, but the recent findings show clearly that they do not.

Emergent Control of *Aplysia* Behavior: Don't Worry, Be Happy

It might be tempting to some interpreters of the above-mentioned complications in *Aplysia* to disparage the original conclusions about site-specific learning. We believe, however, that that would be a mistake. To dismiss the original conclusions would be to fall to the temptation that has faced previous work on learning in *Aplysia*, and of most such attempts in other animals, that there is in fact some other reducible locus of learning, or some reducibly identifiable neurocircuit as the generator of behavior. But by making the dismissal, one would miss the more important issue that emerges from the findings, namely, that the data may be influential in redirecting the focus from reductionism to a higher level of analysis. It is not just that behavior may be different on different occasions. A general scheme appears to have emerged in all of the work on *Aplysia* that is not inconsistent with the findings we have obtained in our attempts to understand the integrative processes that generate behaviors in *Pleurobranchaea*. This scheme relates to our discussion above of parallel processing arising from the extensive distribution and sharing of information, as we will now summarize.

The Locus of Learning May Not Be at a Unique Cellular Site

The evidence cited in our list of complications may be reinterpreted according to the following general scheme: Different sites in the nervous system are capable of generating similar components of the same behavior, and each site is capable of affecting the other; that is, there is apparently extensive convergence and divergence among different sensory and motor centers. Within a given sensorimotor system, divergence is an inherent effect of even small, highly localized stimulation. At the same time, different sensory pathways converge on the same motor neurons. Similar convergence occurs among neurohumoral systems and their subcellular effects (see also Soinila & Mpitsos, 1991). Thus, mounting evidence indicates a cascade of diverging and converging chemical interactions that distribute sensory and motor effects widely in innate responses and in responses arising from different forms of learning.

Evidence exists that supports these possibilities. For example, we know that weak, highly localized tactile stimulations, as used in training experiments to show learning, activate large numbers of neurons (Zečević et al., 1989); that is, that divergence distributes information over many cellular loci. We also know that learning occurs in both the peripheral and the central components of the nervous system of *Aplysia* (see the review in Mpitsos & Lukowiak, 1985). We also know from studies in isolated nervous systems and from more intact preparations that conditioning-related changes occur on LE sensory neurons that synapse on different gill motor neurons. Training-induced changes may occur at the neuromuscular junction (Jacklet & Rine, 1977). Additionally, changes may occur during training that follow all of the criteria established for associative learning but that do not take place between the sensory neurons and their follower neurons. For example, Lukowiak and Colebrook (1986) have obtained evidence of associative conditioning that excludes the major gill motor neurons. The conditioned stimulus (CS) consisted of weak tactile stimulation of the siphon skin. The unconditioned stimulus (UCS), in one set of experiments, consisted of strong electrical stimulation of the pedal nerve, which connects the brain with the foot, and in another set of experiments it consisted of strong tactile stimuli to the gill itself. During training, dual intracellular recordings were made from sensory neurons and major identifiable gill motor neurons (L_7, LDG_1, LDG_2, L_9). The movement of the gill itself was also monitored. In the course of training, the CS produced gill-withdrawal movements that increased as a function of the number of training trials, and the efficiency of the sensory-to-motor neuron synapses increased. Appropriate control experiments showed that the effects were consistent with associative conditioning. However, the number of action potentials produced in the motor neuron in response to the CS correlated well with the actual movement of the gill only during the initial stages of training. But most of the amplitude changes in the gill-withdrawal response were not correlated with any changes in the number of action potentials generated in the motor neurons. In another set of experiments, designed to mimic associative learning observed in whole-animal studies, evidence was obtained for associative learning in a significant number of reduced preparations in which there was an increase in the number of action potentials produced in the motor neurons, but there was no change in the amplitude of the gill-withdrawal response.

Findings such as these show that associative learning, and simpler forms of learning such as sensitization and habituation, may take place at many different loci. Thus, as regards the 10th complication on our list, it is not too big a jump to realize that learning could also happen in classes of sensory neurons other than the LE cells, and eventually to discover that learning-related physiological changes may also be shown postsynaptically in the motor neurons themselves, not just presynaptically in the sensory neurons. Additionally, as Mpitsos et al. (1978) have pointed out in detailed control studies of associative learning in *Pleurobranchaea*, let us not be wedded dogmatically to a definition of associative learning that forces physiology to comply with a particular protocol of stimulus presentations applied by the experimenter to whole animals. Single-trial training in this study showed that, for short intervals between the CS and the UCS, backward conditioning produced almost as strong conditioning as forward conditioning. Mpitsos et al. (1978) pointed out that what may be temporally controllable experimentally in the application of sensory inputs may not hold physiologically. The same set of subcellular mechanisms producing learning-related changes in forward between the CS and the UCS (which is required by the definition of associative learning) may exist to some extent when the stimuli are presented in close temporal pairing but in reverse order. To us, changes arising from both the forward and the backward temporal relationships between the CS and the UCS can represent associative learning (though this does not exclude arguments for different mechanisms, should they occur, to account for backward conditioning). For these reasons, it also may not be too big a jump to accept the fact that learning may still take place in the LE neurons of *Aplysia*, even if their responses arising from stimulation of sensory skin do not occur until after the motor neurons are activated by other sensory neurons.

Thus, although it is possible that a unique "locus of learning," the engram in *Aplysia*, might still be found, the data indicate strongly that *the system seems to consist of many parallel, redundant, and possibly interacting components, none of which may be the* sine qua non *element in the learning process or in the generation of the motor responses, irrespective of whether or not they involve learning.*

The Neurocircuit May Not Be Definable

Another tradition of reductionism in neurobiology, particularly in invertebrate studies, has been the notion that cells and their functions are repeatedly identifiable. We have already mentioned some of the problems in identifying function in *Aplysia* (Gardner, 1990; Leonard et al., 1990). The recent computer simulations of simple neural networks relating to the feeding system of *Aplysia* have led to a similar conclusion, that "tests done on individual neurons can provide misleading information on the actual role of the neuron in generating behavior" (Kupferman et al., 1991). Compare this quote with one from Mpitsos and Cohan (1986b, p. 538): "These findings indicate that the classic technique of driving a particular neuron in order to assess its effect in evoking activity or a behavior may be an insufficient criterion for identifying its functional role." That is, a given neuron's function depends on the context of activity in which it takes part. But given variability in the activity in the firing patterns within such contexts or "mobile consensuses," even this might be an insufficient definition (Cohan, 1980; Mpitsos, 1989; Mpitsos & Cohan, 1986b).

The neurocircuit for a behavior is misrepresented by even the most complete mappings of identified neurons that we see in publications. Studies using voltage-sensitive dyes show that weak, localized stimulation of sensory skin of the siphon produces massive and variable activation of neurons in the abdominal ganglion of *Aplysia* (Wu et al., 1989; Zečević et al., 1989). As we have discussed regarding the simplified networks shown in Figure 10.1 for *Pleurobranchaea*, the connectivity of the actual circuit of interacting neurons is quite large. The larger the overall pool, and the greater the number of weak synapses, the greater the possibility that the actual network generating a behavior will be variable and undefinable.

Different Levels of Learning
Within Definable Sets of Synapses

Let us assume for the moment that a small group of neurons can be isolated functionally from the effects of other groups of cells. Can we then obtain sufficient information about the network to define it completely by looking at the network and knowing all of the connection parameters? We think not. Consider just one example relating only to the strength of synapses. In our own neural network simulations, the data indicate that synapses contain different forms of information (Burton & Mpitsos, 1992; Mpitsos, 1991). One form of information ("knowledge") is task-specific, relating to the computations of one or more functions that network must perform. Another form ("metaknowledge") has to do with the process by which that task was learned; it does not affect the network performance on the specific tasks, but becomes evident only when the network is confronted with new tasks. These conclusions were drawn from experiments that compared learning performance in networks that used random noise to optimize changes in synaptic weights against networks that were not exposed to noise. Both types of networks were allowed to reach the same level of learning on a given task, but the noise-exposed networks learned a subsequent task faster, even when noise was not included during training of the second task, than networks that did not use noise. Starting networks at different initial synaptic strengths at the beginning of a training session yields different final synaptic settings, but all final networks perform the same learned task equally well. Because of this, Burton and Mpitsos initialized networks using different synaptic strengths and thresholds. Examination of a large number of networks at the end of the first training session revealed that the two types of training methods did not generate statistically significant differences in the means and standard deviations of the synaptic weight settings. Both types of networks contained the same information for generating equally accurate computations relating to the first task, but networks that were exposed to noise contained further information that permitted them to perform well on a second task. Each task has a particular error landscape associated with it (see Burton & Mpitsos, 1992, Figure 8, and Mpitsos & Burton, 1992, Figure 13, for examples of error landscapes and volumes). Burton and Mpitsos suggest that noise-exposed networks sample these error structures more completely than networks that were not exposed to noise. Thus, when confronted with new tasks having any similarity in their error structures to the first task, the synaptic settings of networks exposed to noise already contain information about the new task and are able to navigate its error fields rapidly. By contrast, because networks that are not exposed to noise contain less of such information, they are not able to navigate as rapidly through the new error structure.

The implication of these findings for the present discussions is that one may look for changes relating to a given task, but depending on the conditions under which that task has been learned, the aggregate of synapses within a pool of neurons may contain different types of information, where one type pertains specifically to one or more tasks that have been learned, and the second type pertains to more general conditions that do not affect the accuracy of the first, but nonetheless may camouflage the results that the experimenter is seeking to identify. The rabbit olfactory bulb (Skarda & Freeman, 1987) may be a useful example to contrast our findings. In this structure, odor-specific information is stored spatiotemporally, but apparently all neurons take part in expressing the code for each odor. Our simulation networks can also be constructed to encode information relating to multiple tasks (Mpitsos & Burton, 1992), but the noise-induced changes in the network represent an informational abstraction that goes beyond the information need specifically to perform well on previously learned tasks. Therefore, if our computer simulations of connectionist neural networks have analogues in biological systems, the understanding of synaptic modification and the information that the synapses contain cannot be deciphered simply by examining the synapses themselves as they relate to only one task. In their studies of Mauthner neurons, Faber, Korn, and Lin (1991, p. 295) raise the related caveat, but for different reasons, that "although it is possible to derive generalized rules of the operation of synapses, their variants may exert a major role in shaping the behavior of complex circuits."

Problems analogous to those described above and in the preceding two subsections may have beset Lashley (1950), whose unsuccessful attempts to identify the loci of stored memories (engrams) in the cortex have been more inspiring and illuminating, at least to us, than were he to have found them. It is interesting that much of neuroscience has followed the same course as Lashley, but on the cellular level in attempting to identify behavioral phenomena in terms of single synapses and single neurons, and in the process has generated equally instructive findings. It is also interesting that Pavlov, before Lashley, was apparently discontent with the possibility that learning could be localized to particular areas of the cortex, because learning persisted in his animals even after they had suffered brain damage (Boakes, 1984, pp. 127-128).

"Fuzzy" Control

Thus, the "control" we seek to define for the physiological and neurohumoral aspects of the nervous system is oblique and emergent rather than being crisply Euclidean in postulating particular causes and effects as would be expected of reflexes. One feature of such emergence is that there may be many ways to do the same thing, and even gradations between these ways. We know, for example, that under some conditions removal of a neuron from acting in a motor pattern can be compensated by shifts in the activity of other neurons (Mpitsos & Cohan, 1986b). Redundancy, arising from information sharing among convergent pathways, compensates for error or failure in some of its components, even if these components originally generated strong control over the other members of the coactive group. Are neurohumoral systems equally redundant, or does each of the ever-growing number of neurotransmitters being identified daily have a unique task? Our own work leans heavily toward the first of these possibilities (Soinila & Mpitsos, 1991). In the same sense that there may be "lazy" synapses in neural networks (Mpitsos &

Burton, 1992), whose presence is required only under some conditions, are there "lazy" or even unnecessary transmitters? Some of what we see in a given system may represent baggage of evolutionary or developmental processes. This, however, provides for yet another form of variation that permits possible adventitious incorporation into further evolution or behavior.

Is Our View Holistic?

Is our view holistic? No. Being concerned with mechanisms that generate global behavior is not necessarily being holistic. In our approach, global behavior depends on local rules followed by individuals acting within a large group. It is these rules that we seek to identify, though there may be different rules that relate to global behavior directly. Even in simple processes such as the building of sand-grain mounds (Bak & Chen, 1991) and affine transformations (Barnsley, 1988), the global consequences of local behavior are not predictable. Nevertheless, emergent function need not be a property of large groups of neurons.

It is interesting, however, that one of the best examples of work in artificial intelligence in many decades employed a top-down analysis in which a principle obtained from studies on the behavior of whole animals was used to gain insight into how that behavior might have emerged from individual neuronal units. The work we refer to is Klopf's (1988) drive-reinforcement model of associative learning, which extends Hebb's (1949, p. 62) rule to account for Pavlovian conditioning. Hebb's rule states, "When an axon of cell A is near enough to excite cell B and repeatedly and persistently takes part in firing it, some growth or metabolic change takes place in one or both cells such that A's efficiency as one of the cells firing B is increased." Before Klopf's model, computer simulations of Hebb's rule in simple networks were not successful in demonstrating learning that mimicked findings in biological systems.

Hebb's rule may be interpreted as a three-cell network (Mpitsos et al., 1978), one input cell for the CS and one input cell for the UCS, both of which synapse on a common follower cell (Cell B). Klopf (1988) made the following crucial modifications to the rule to make it work in such a simple system: (a) Temporal delay was introduced between the onset of the CS and UCS. (b) Synaptic modification was made proportional to the rate of change in the CS and UCS. (c) The follower cell (B) itself expressed a form of behavior analogous to tendencies that may be observed in whole animals: Whole animals seek to optimize some quality of their environment, such as avoiding pain and enhancing pleasure. Klopf (1982) made the simple, but crucial, analogous assumption that cells tend to optimize excitation and reduce inhibition. Additionally, to account for excitation and inhibition, the follower cell received excitatory and inhibitory terminals in its CS input pathway.

The methodology for training the network is the same as for training a whole animal. In each training trial, a pulse is presented to the CS input, which initially produces little effect, and after a short delay, a pulse is presented to the UCS input. The only parameter that is arbitrarily set in the model is the constant for the rate of learning. Amazingly, training-induced changes in the synaptic effect of the CS input on the follower cell reproduced all of the known Pavlovian conditioning phenomena in experimental animals and in humans (backward conditioning, CS alone, UCS alone, trace conditioning, second-order conditioning, foreshadowing, blocking, conditioned inhibition, etc.).

The model has now been extended to account for instrumental conditioning (Morgan, Patterson, & Klopf, 1990). The work also made progress in resolving the long-standing debate relating to the theoretical relationship between Pavlovian and instrumental conditioning, because the instrumental conditioning effects in the model emerge from Pavlovian conditioning. Thus, computational methods may have resolved what psychological debate and experimentation in biological systems have not been able to resolve. The studies discussed below in the section dealing with computer simulations pursue the same rationale of using simple rules to lead to understanding of global effects.

DOES A THEORY EXIST?

At least three important principles have emerged from dynamical systems studies that are important to biologists: (a) The notion that distributed networks can generate attractors; (b) that a considerable amount of information about a system can be gained from bifurcation analysis; and (c) that an understanding of the dynamics of a system can be obtained from the phase-space geometry of such attractors. By these methods, it is possible to discover much about a system without having to resort to the difficult if not impossible task of uncovering the sets of equations that actually run the system.

A long history of work has developed these ideas, from Poincaré to Lorenz, Crutchfield, Farmer, Packard, Rössler, Ruelle, Takens, Swinney, Shaw, Yorke, and others of the many recent contributors to the knowledge of nonlinear dynamics (see the reviews in Abraham & Shaw, 1983; Thompson & Stewart, 1986). There are many theorems in the field of nonlinear dynamics, and there are many discussions of how to handle the nonlinearities (e.g., Grossberg, 1980; Grossberg & Kuperstein, 1986; Seydel, 1988), beautiful demonstrations of attractor topologies, bifurcations, and stability analyses, when these are in fact available. As important as these are, they do not constitute a unified theory, at least not as it might apply to brain function, though Bak and co-workers suggest that their mathematics or models of self-organizing criticalities (Bak, 1990a, 1990b, 1990c; Bak & Chen, 1991; Bak & Tang, 1988; Chen & Bak, 1989; Weissenfeld et al., 1989), which apparently account well for many physical and biological phenomena, may provide an encompassing dynamical theory.

One way to get around the theoretical problems, as is often suggested by physiologists and nonphysiologists alike, is to perform computer simulations on systems whose state space is completely defined and parameterized; that is, to determine all of the connections among neurons, membrane properties, neurotransmitters, firing thresholds, and the like. However, one look at the complexity of the connections and at the wide divergence and convergence occurring in even "simple" systems should provide convincing evidence that this approach is hopeless (Cohen, Hopp, Wu, Xaio, & London, 1989; Mpitsos & Cohan, 1986b; Soinila & Mpitsos, 1991; Zečević et al., 1989). Moreover, as discussed above, the reductionist neurocircuits that have been developed over the years to account for behaviors are but a caricature of the actual "networks" that generate the behaviors in intact animals.

The possibility might also be suggested that insight into the integrative principles might be obtained from the mathematics describing the biological systems. This also seems an unlikely possibility at present, even in relatively small systems. Even in well-defined experimental systems, the first evidence of dynamical states and their bifurcations came from direct observations. One such example is the Belousov-Zhabotinsky reaction, which consists of about 30 chemical constituents in which malonic acid is oxidized in an acidic bromate solution (Roux, 1983; Roux, Simoyi, & Swinney, 1983). Although it may be possible to define the various reactant species and list the reactions, it has not been possible, to our knowledge, to predict the dynamics of the system using the mathematics of the reactions. Another example is the demonstration of different dynamical states in yeast glycolysis (Markus, Kuschmitz, & Hess, 1985). And, for yet another example, near the turn of the century Duffing extensively studied damped-driven oscillators, yet the full force of the dynamics in his simple model system was not uncovered until recently, using computer simulations (Thompson & Stewart, 1986; Ueda, 1980). Lorenz's landmark paper (Lorenz, 1963) showing the first instance of persistent chaos in a simple mathematical model of fluid convection was found accidentally in computer simulations, not theory.

Finally, even the application of extant dynamical systems tools to time series of experimental data provides little recourse (Mpitsos, 1989). These tools have largely been developed using simple models whose responses can be generated sufficiently long to obtain an indication of their dynamics. Biological responses, by contrast, are often extremely short-lived. For example, chewing and swallowing behaviors in humans, as in *Pleurobranchaea*, may be generated by robust attractors, but so few cycles are generated that characterization of their dynamics, whether they be limit-cycle or chaotic attractors, is not possible. Even in ideal systems, a certain amount of guesswork needs to be done. For example, the Grassberger-Procaccia algorithm can significantly overestimate the attractor dimension of limit cycles and underestimate it for chaotic systems, particularly as the dimension increases, even for model systems such as the Rössler hyperchaos (Rössler, 1979).

The positive side of all of these problems is that biology stands on an exciting, albeit difficult, threshold of growth in theories and concepts. And it is biology that will force further development of dynamical tools. The work of Ellner and co-workers on nonparametric methods to calculate Lyapunov exponents is an example (Ellner, 1988, 1991; Ellner, Gallant, McCaffery, & Nychka, 1991).

COMPUTER SIMULATIONS: MINIMAL MULTIFUNCTIONAL NETWORKS

Computational analogies may provide insight where theory is lacking. Lorenz's (1963) work on convection provides an excellent example of how computer simulations may spark insight into new methods for handling complex systems. The work of Klopf and co-workers (Klopf, 1988; Morgan et al., 1990), which was discussed earlier in the section on reductionism, is another example in which computational methods have proved decisive in addressing an important problem in the theory of learning. In Lorenz's case, the outcome was unexpected. In Klopf's case, the outcome was planned because of the equivalence of the statement of drive reinforcement at both the unit and the global levels. Both of these examples show that certain statements or assumptions about interacting systems can be used to address complex behavior through computational methods without having first to develop a proved theory about the global system. Put differently, given certain assumptions about local events, it may be possible to allow the system to generate coherent global activity. In the following discussion we examine two subjects, bifurcation dynamics and attractors, taken from dynamical systems studies that may help to illustrate some of the features of self-organizing activity that emerge from local interactions.

Bifurcation in Simple Mathematical Models: The Rössler and Logistic Equations

Nonlinearities are easy to see in simple models such as the Rössler system (Rössler, 1976) of coupled ordinary differential equations that generate complex chaotic dynamics:

$$dX/dt = -Y - Z \quad dY/dt = X + aY \quad dZ/dt = b + ZX - cZ, \quad (10.1)$$

where a, b, and c are constants. Here X is a function of Y and Z, Y is a function of X and itself, and Z is a nonlinear function of itself and X. Each of these variables is expressed nonlinearly through the others. The variables themselves provide the nonlinear drive for the system, but the constants determine how the drive is expressed. For example, setting a = .019, b = .2, c = 10 (and setting X = .65, Y = 2.12, and Z = 4.7 in order to initialize the equations) causes the system to generate predictable limit-cycle activity. Plotting X against Y, for example, produces a perfect circle. However, by simply setting a = .15, the system generates chaos (Rössler, 1976). That is, the activity is completely deterministic, but the long-term evolution of X, Y, and Z is unpredictable, as can be seen by the complex though pleasing movement of the trajectories in X-Y plots. The constant a acts as a bifurcation parameter that controls how the system functions.

   The logistic equation, $X_{n+1} = R(1 - X_n)X_n$, is an even simpler example, where the new value on the left is generated by the nonlinear drive of the previous value on the right (initialized between 0 and 1) and is then reintroduced into the system to generate the subsequent number. For values of the constant R between 0 and about 3.55, the process of nonlinear action followed by recursive folding back into the equation produces periodic sequences of numbers, but for R greater than 3.55 the system generates chaotic sequences (May, 1976). Successive, linear adjustments to a constant such as R may produce only minor changes in the system over a large portion of R's allowable range. But at critical points, very small alterations in R produce nonlinear shifts (bifurcations) in the sequence of numbers. At low R-scale resolutions, regions are observed at which only chaos appears to occur. By expanding the R-scale, one observes that chaotic regions contain periodic regimes.

Bifurcation in Hodgkin-Huxley Membrane

Teresa Chay's (1985) seminal paper examined a three-variable Hodgkin-Huxley membrane precisely as described above for the logistic equation. The time variation of voltage in the model is given by

$$dV/dt = g_I^* m_\infty^3 h_\infty (V_I - V) + g_{K,V}^* n^4 (V_K - V) + g_{K,C}^* \frac{C}{1 + C}(V_K - V) + g_L^*(V_L - V),$$

$$(10.2)$$

where I is mixed inward currents (sodium, calcium); K,V is the voltage-sensitive potassium channel; C is the internal calcium concentration; K,C is the calcium-sensitive potassium current; L is leakage; n is the probability of opening K,V; m is the probability of activation; h is the probability of inhibition; and g^* is the maximal conductance divided by capacitance. The three variables in the system are (a) membrane potential, V; (b) n, the probability of opening the voltage-dependent potassium channel; and (c) the intracellular concentration of calcium. Intracellular calcium is voltage-dependent, as are sodium, one of the potassium channels, n, m, and h. It can be easily seen mathematically that all of these variables affect one another through voltage (as a consequence of their effects on currents), and that the system of such interactions is highly nonlinear, although examination of the equations would not necessarily give immediate insight into which parameters to use to control bifurcations. The bifurcation parameter is the calcium-dependent potassium conductance $g_{K,C}$, and, as described earlier for the logistic equation, the membrane produces many different firing patterns when this conductance is systematically changed.

Relationship Between Bifurcation Dynamics and Network Architecture

To illustrate the difficulties encountered in attempting to understand the dynamical capabilities of network architectures and the direction we have taken in some of our computer studies, let us consider the (overly) simplified cartoons in Figure 10.4,

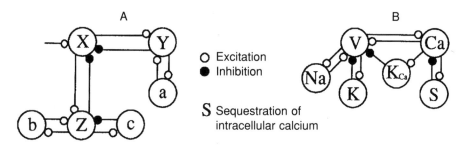

Figure 10.4 "Minimal" neurocircuit transpositions of the three-variable Rössler system of coupled differential equations (A) and of the Chay's three-variable Hodgkin-Huxley membrane (B).

a and b, that transpose the Rössler system and the Chay membrane into "realistic" analogues of neuronal networks. "Realistic" might include voltage-sensitive ion channels, calcium-dependent ones, transmitter-release dynamics, transmitter re-uptake, second-messenger systems, and other processes one might want to include of an experimental system.

Given tonic excitatory input to X in Figure 10.4a, and making X capable of postinhibitory rebound, it may be possible for X and Y, and for X and Z, to oscillate in opposition if there is sufficient accommodation in the firing of Z and/or Y. Figure 10.4b shows a network cartoon of a subset of the variables in the Chay membrane. Given Chay's simulations, it might be predicted that the synapse of K_{Ca} onto V would provide access to bifurcation dynamics. The nonlinearities in the Rössler and Chay systems are easily identifiable in the differential equations that compose them. And it is possible to see how the calcium-dependent potassium conductance can influence the dynamics of the Chay model. But it is considerably more difficult to identify analogous nonlinearities and bifurcation conditions in neuronal networks. It has long been established that synaptic activation of neurons leads to nonlinear responses because of the firing threshold in the driven neuron. It is also known how to simulate individual synapses using digital integration, by describing the kinetics mathematically, or by examining nonlinear interactions between different types of synapses (Koch & Poggio, 1987). But the dynamical implications of different net-work architectures and of the synapse characteristics that affect the dynamics of regenerative electrical activity of neurons in these networks are problems that remain largely untapped.

Along this line, present efforts in our laboratory are aimed at understanding what types of converging and diverging centers in minimal networks are required for bifurcations to occur. In the same way as Chay used the calcium-dependent potassium conductance to control the bifurcations, our efforts are to determine whether synaptic strengths can also be used as bifurcation parameters. The problem facing us in dealing with the biological system is much more difficult than that which faced Chay, because (a) our system has many more degrees of freedom; (b) our system is not as smoothly continuous as the Hodgkin-Huxley membrane—that is, the mem-brane responses may seem continuous, but cells usually receive information in short pulses or bursts; and (c) there are no previous network examples for us to follow in which bifurcation has been demonstrated. Interestingly, the types of convergence centers that have proved capable of bifurcating into variable activity in our prelimi-nary computer simulations are ones having structures similar to the one shown in Figure 10.4b.

As our knowledge grows of the connectivity among the BCNs and of their connections with other neuronal groups, we shall construct computer simulations of networks having increasing sizes. We shall then progressively introduce the effects of the many converging neurotransmitter systems. Additionally, by imple-menting early behavioral evidence of synaptic competition during learning in *Pleuro-branchaea* (Mpitsos et al., 1978), and the evidence for synaptic competition in mammalian cortex (Merzenich et al., 1983), we expect to see our networks remodel their connections over time. Interactive groups may actually grow or shrink in time; large populations may split into subsets; the spatial boundaries between coactive groups may move in time; and network architectures may emerge that affect the amount of variation occurring in the network.

Continuous Versus Discrete Processes

The Rössler and Chay models are both three-variable systems, as required of any continuous bounded system that is capable of generating chaos. We summarized the reasons behind the need for three variables, using mixing of trajectories in three-space and an examination of Lyapunov exponents, in the section on bifurcation-induced variations. By contrast, discrete processes can generate chaos in one dimension, as in the case of the logistic equation, and coupled discrete processes can generate chaos in two-space, as shown by the Hénon system, where $X_{n+1} = 1 - aX_n^2 + Y_n$ and $Y_{n+1} = bX_n$ (Hénon & Pomeau, 1975). Recall also that the issue is not whether a system generates chaos, but its ability to exhibit both simple and complex behaviors, depending on its bifurcation conditions arising from simple quantitative alterations rather than from qualitative changes in network structure. Moreover, if the bifurcation parameter is the driving frequency of an input signal, it is not necessary even for quantitative changes to occur in the network for simple and complex dynamics to appear.

The difference between continuous and discrete processes is of significance to neurobiologists. The neural networks studies of Mpitsos and Burton (1992) indicate that when signals between networks are chaotic discrete processes, simple networks are able to perform difficult tasks on these signals that would otherwise require more complex networks to perform if the mode of transmission used continuous periodic or even continuous chaotic processes. Continuous processes are used in neural integration (e.g., Werblin & Dowling, 1969), but the usual mode of information transfer is through trains of action potentials. Trains of action potentials in pacemaker firing cells are generated by continuous fluctuations in membrane potentials and in the dynamics of ionic species. Examples may be found in computer simulations of the parabolic burster neuron R_{15} in *Aplysia* (Canavier et al., 1990) and in the Chay model described earlier. The information in these spike trains, though generated by continuous processes, is in a pulse code. Therefore, there are a number of questions that need examination. For example, is there a difference between the information contained in the dynamics of spike trains and the information contained in the continuous membrane processes that generate them? What happens in postsynaptic cells when they receive such spike trains, and when are we to consider the dynamics in the postsynaptic cells as continuous processes or as analogues of discrete processes? The membrane potentials of these follower cells may appear continuous, but they are driven by discontinuous input events.

The differences between discrete and continuous processes pose problems in numerical analyses. Experimental data usually consists of the time series of one or several dependent variables, but the methods provide little knowledge of the number of dependent variables that actually drive the system. Numerical methods provide some help. For example, it is possible to conduct phase-space analyses that give information about the topological dimension of attractors and about the number of dependent variables (embedding space) that may be involved in generating the attractors (Mpitsos, Burton, Creech, & Soinila, 1988; Mpitsos, Creech, Cohan, & Mendelson, 1988). The evidence provides some justification supporting chaotic attractors and low-dimensional embedding space.

However, some of the calculated attractor dimensions were lower than 2, posing some difficulties in interpretation of what the dynamics is. Continuous systems must have at least three Lyapunov exponents; there must be at least two nonnegative ones, one being positive, as required for chaos, and one having zero value, as required by Haken's (1983) theorem. Given two nonnegative exponents, by calculations using the Kaplan-Yorke (1979) conjecture it should be expected that the lowest attractor dimension for continuous chaotic systems be greater than 2 (examples are given in Andrade et al., 1992; Wolf, 1985). One-variable, such as the logistic equation, have dimensions less than 1. Two-variable discrete processes have dimensions between 1 and 2; our own estimate of the Hénon system gives a dimension of about 1.36. Knowing the mathematical representation of a system allows one to place such numbers in appropriate context, but experimental data leaves numerical results ambiguous. Do we assume that attractor dimensions of less than 2 are coupled discrete processes, or is there a problem with the analytical methods? Of the latter possibility, the available tools, whether using time series of a single variable or all variables, calculations of attractor dimensions are difficult to obtain even for model systems (Andrade et al., 1992).

Answers to questions such as the one just given are necessary, because they provide an indication about how information is processed and encoded. We are presently addressing them using numerical analyses of data from computer simulations of membrane patches and of responses of cells in networks where we have access to all parameters and variables of the system. Comparison of analyses on the data from measurements of continuous variables and from spike trains may yield some insight into implications relating to continuous and discrete processes.

Attractors and Gradients in Biological Systems: Comparison Through Analogy in Principles, Not in Identity of Mechanisms

The potential consequences of the identity between attractors and optimization are rather interesting. Consider the following situations. In attempts to simplify computer simulations, it is often difficult to determine exactly where to limit the characterization of the biology. For example, the connectionist methods of error backpropagation are usually faulted because of their obvious nonbiological nature. But the answers that come from the use of such networks depend on the principles that are actually being simulated. The major driving element of error backpropagation is that the system must follow a negative error gradient between a teacher function and the output of the system (Rumelhart, Hinton, & Williams, 1986). If the question being addressed has to do with the principle of error reduction rather than, say, what second messengers might be involved in a cellular process or how feedback actually occurs in a real nervous system, the backpropagation method might give some insight into how gradient-seeking systems store information in their distributed elements.

Response Thresholds

Following the rationale just described, Mpitsos and Burton (1992) obtained a number of results that might have relevance to biological systems. They found, for example, that the computational capabilities of networks are severely limited when only trainable synaptic strengths are used. Adding trainable thresholds significantly expands the computational power of the networks. In invertebrate learning studies, thresholds (as might be inferred from membrane changes in postsynaptic cells) have either not been observed at the cellular level and or have not been generally attended to (Mpitsos & Lukowiak, 1985). Studies on long-term potentiation (LTP) in rats have, however, provided evidence implicating response thresholds through changes in synaptically induced changes in the ratio of excitation and inhibition rather than changes in membrane impedance (Barrionuevo, Kelso, Johnston, & Brown, 1986; Chavez-Noriega, Halliwell, & Bliss, 1990). Heretofore, the methods used to test LTP have focused neither on assessing the computational implications of threshold adjustments nor on the technical conditions to extend the findings, but it would be extremely interesting to determine whether adjustments in the ratio of excitation to inhibition were set differently for each cell, as might occur in gradient-descent adjustments in thresholds during learning in neural networks.

Network Size May Be Self-Limiting

An unexpected finding in the studies of Mpitsos and Burton (1992) was that increasing the number of neurons in a hidden layer or interneuronal layer beyond a certain point slows, and eventually causes the system to cease, learning; that is, group size may be self-limiting. Limitation of group size has been enforced algorithmically in simulations of mammalian cortex through synaptic competition and inhibitory synapses (Edelman, 1987; Pearson, Finkel, & Edelman, 1987). It is also conceivable, however, that group size may be additionally limited by the gradient tendencies of attractors. If the findings of Mpitsos and Burton hold biologically, the slower organizational times of large networks may be superseded by smaller subsets of neurons as they form attractors. Once sufficiently formed, the attractors themselves may restrict group size, partly by their gradient processes and partly by learning-related synaptic competition.

Metaknowledge and Lazy Synapses

Metaknowledge represents that ability of networks to store different forms of information (Burton & Mpitsos, 1992). We discussed it earlier in dealing with reductionism, and we believe that it may be a consequence of gradient tendencies. Our computational studies also found that although networks set their synaptic weights and thresholds at optimum levels, many of the synaptic weights produce little effect when removed from the network; that is, they are "lazy." Mpitsos and Burton (1992) discuss a number of uses for such synapses. One of the most interesting possibilities comes from somewhat different studies by Warren (Warren, 1989), who showed that certain synapses may be deleted after training without significantly affecting network performance on a previously learned task, but that networks were unable to learn the task if they started with the reduced number of synapses in the first place. This poses interesting problems to biologists, because weak connections are often observed among the interactive components of their experimental systems. The tendency in the past has been to dismiss such connections, or to presume that they would be "pruned" away if not used. Our findings, along with Warren's, indicate that these synapses may be crucial for learning new tasks. By analogy to computers, they might be considered as temporary registers that permit gradient descent, but once gradient descent has been reached, they are no longer needed for that task.

Attractors, From Sea Slugs to Bees

It is easy to see how attractors arise in simple mathematical models, in cell membranes, and even in networks of neurons. Varients of the Volterra-Lotka equation, which is used in studies of population dynamics, for example, indicate that attractors also arise through the interactions of individuals within a population of animals (Cohen & Grossberg, 1983; Hofbauer & Sigmund, 1988; Andrade et al., in press). Perhaps it is also possible that attractors exist in the interaction between an animal and the resources it has available to it in its environment. Real (1991) has shown recently that bees are able to adjust their behavior so as to optimize the use of food resources. Whether or not this involves gradients and attractors has not been addressed. The idea is consistent with the possibility that biological networks (and biological systems generally) may exhibit behavior that tends to minimize some gradient factor (as error or energy) through the ability of attractors to dissipate energy (Mpitsos, 1991; Mpitsos, in preparation). Attractors pull in any phase-space trajectory that falls within their basin of attraction. Thus, for example, in limit cycles, externally applied perturbations move the trajectory of the system away from the limit set, but if the state of the trajectory remains within the attractor's basin of insets, the trajectory will fall asymptotically back into the limit set. Chaotic attractors also attract nearby states but dissipate perturbations over their entire surface. We might say that attractors minimize energy or error (Mpitsos, in preparation). Put differently, attractors optimize the match between their attracting set and activity that falls near it. In either case, the action may be considered a minimization process. On the behavioral level, bees are able to control their foraging techniques so as to optimize the use of food resources (Real, 1991). Whether this involves attractors and gradient seeking is presently not known but is potentially testable.

Local Error Minima in Biological Adaptation

The idea that a system tends to optimize its behavior has a somewhat different expression in biological systems than it might have in computer simulations of connectionist neural networks. We can envision that, with enough time and stable environmental conditions, evolutionary competition among organisms will produce changes that best adapt the species to the environment. One might think of the process as reaching an absolute error minimum between the response of the organism and the best possible response under the imposed conditions. Any response that is not optimal represents a local minimum. In neural networks, methods have been developed (see the section on random noise) to avoid local minima, using, for example, simulated annealing (Kirkpatrick et al., 1983) and time-invariant noise algorithms (TINA) (Burton & Mpitsos, 1992). Simulated annealing usually involves exponential decay of noise over time. TINA adjusts noise as a function of the amount of error that is produced when a system responds to its input stimuli. This method, however, was chosen only as a vehicle to demonstrate the idea of TINA. Other methods, not necessarily directly related to error feedback, may also be used that retain time invariance. For example, our present attempts to implement TINA in networks consisting of neurons having biologically realistic characteristics is to adjust the probabilistic release of the transmitter (Korn & Faber, 1987) or to use short-term activity-dependent learning rules such as sensitization (Mpitsos & Lukowiak, 1985) to maintain the flow in a given part of the network. Our goal is to assign certain facilitatory responses to classes of neurons and then to allow the actual pathway to emerge dynamically. Low error would be represented by activity recurring through a particular part of the network. As error increases, diffusely distributed feedback onto the network would disrupt such preferentially frequented pathways, permitting others to emerge. If these new pathways lead to low error, feedback decreases, allowing the flow through the pathway to continue. If attractors self-organize, the preferential pathways would then be further entrenched, because, as discussed above, the basin of insets to the attractor itself may represent an energy- or error-minimizing process.

This process does not require that the tendency to follow a gradient actually reaches an optimal minimum or, equivalently, that the attractor be spatiotemporally a robust, stable structure. Biologically, both in the daily behavior of organisms and in their evolutionary succession, local minima are extremely important in generating adaptive responses. Whatever works is sufficient, whether the response is optimal or not. Thus, our notion of an adaptive system is one that can generate different minima that can be addressed rapidly, and exited rapidly if they do not meet the need. Indeed, we believe that it is from the ability to generate many local minima that multibehavioral networks may have evolved.

Part of understanding the generation of local minima will be to see how multibehavioral networks generate different attractors in computer simulations. Transitions between different attractors may yield labile intermediate forms that only partially resemble more stable ones. The most difficult problem that we face here is to determine how best to visualize temporal activity graphically for spike trains (Mpitsos, Burton, Creech, & Soinila, 1988; Mpitsos, Creech, Cohan, & Mendelson, 1988). Continuous nonspiking processes pose less of a problem (Andrade et al., in press). Part of the answer may also come from an understanding of spatiotemporal dynamics.

Visualization of Spatiotemporal Dynamics

The more we study biology, the more it seems that we must somehow leave it to gain a feel for what may be happening there. Put simply, biological systems are too complex and uncontrollable even to perform such experiments as those represented by Figure 10.4. We must imbue these simulation networks with as much biological information as needed to obtain activity that somehow resembles the activity of the biological system. But complete state-space parameterization of the biological system is beyond hope, as one glimpse of the complexity in Figures 10.2 and 10.3 will show. At the level at which we can attribute realistic biological characteristics to a network, the system becomes intractable even for simple analyses of steady states (see the example analysis of a simple model system in Andrade et al., 1992).

Given the growing power of computer graphics and the increasingly easier access to supercomputers, the recourse for biologists interested in the emergence of group dynamics is to conduct the type of experiments shown in Figure 10.4 and, especially, to visualize the spatiotemporal flow of activity in large-scale simulations involving many interacting units. An understanding of such spatiotemporal flows is, we believe, one of the central questions facing neuroscience. Walter Freeman and co-workers were perhaps the first to begin a detailed account of spatially distributed recordings, in their studies of rabbit olfactory bulb (Skarda & Freeman, 1987). But even in these studies, the analysis of the temporal flow is of the time series of single recording sites. Perhaps the major lesson in dynamical systems work over the decade has been the fact that much can be learned about the activity of a system by the analysis of its phase-space geometry. Up to four variables can be analyzed simultaneously using time-series analysis (e.g., see Andrade et al., 1992, Figures 8 to 11; Mpitsos & Burton, 1992, Figure 13). We need to do the same for many variables, both spatially and temporally.

By such methods it may be possible to examine the possibility of limit cycles, chaotic attractors, SOCs and turbulence, the coexistence of multiple attractors, and the movement of these attractors spatially and possibly even their blending into one another. It may also be possible to determine how particular circuit structures emerge, how variability appears to be controlled by particular circuit characteristics. In the long term it will be important to ask how such structures are affected by systemwide factors. If we are to believe our neurochemical findings, it is quite likely that bifurcation parameters may be defined more accurately as being distributed over a large number of cells than as, for example, being in the conductance modification of a single cell. The first possibility may explain the fact that some systems are relatively insensitive to changes in only a few of their components.

CONCLUSION

In answer to the title of this paper, we have actually said little about what sea slugs can tell us explicitly about the neurointegration of specific human movement. But we believe that the findings tell us a considerable amount about what must be addressed in order to gain a unified perspective on biological integration that might eventually affect how we view human movement. We understand that much has been said appropriately by others about coordination of limbs in invertebrates and vertebrates, the rightful importance of FAPs, and selective control of individual neurotransmitters on pattern generation and in the formation of network structure, and that such findings may be applicable to human motor behavior. Perhaps most of the time all of these studies provide the best answers, just as most of the time Newtonian physics provides the right answers in daily engineering problems. Perhaps, also, the neurointegrative processes in *Pleurobranchaea* and *Aplysia* follow the same predictabilities most of the time.

The instances that are not explainable by traditional neurocircuit perspectives might be dismissed as biological aberrances. Alternatively, owing to the fact that the animal seems to function well enough with them, they may be pursued as being of adaptive significance. We have followed the latter route and have been forced into a perspective that is more statistical-mechanical and dynamical than classically switchboard. Lorenz (1974) voiced the long-held view that all biological information is stored in structure. We hardly disagree with that. But the question is, How do we read that information, and is much of it redundant and even of nonsense or accidental value? The latter possibilities may actually provide certain adaptive value adventitiously in ever-changing and unpredictable environments. In reaching a new theoretical perspective that addresses these issues, our view is that there are two levels of solution: the special case, relating to the switchboard neurocircuit, and the general solution, which must be reducible to the special case but must also provide a general theoretical foundation that is extensible to many other cases.

The shift to dynamics, or at least away from answering all questions by using reflexes, marks a shift away from mechanism to organization. Although each biological level of organization may express the dynamics in its own processes, the dynamical principles may be applicable to all levels of organization. The central question in all of these systems is, How does the individual influence the group, and, in turn, how does the group influence the actions of the individual? We have tried as much as possible to couch our ideas in biological findings, though much more data needs to be gathered (and regathered) before we feel more comfortable. If what we have discussed is accurate, then, as Barbara McClintock envisioned, "we are going to have a new realization of the relationship of things to each other" (Keller, 1983, p. 207).

REFERENCES

Abraham, R.H., & Shaw, C.D. (1983). *Dynamics—the geometry of behavior (parts 1-4)*. Santa Cruz: Aerial Press.

Abrams, T.W., Castellucci, V.F., Camardo, J.S., Kandel, E.R., & Lloyd, P.E. (1984). Two endogenous neuropeptides modulate the gill and siphon withdrawal reflex in *Aplysia* by presynaptic facilitation involving cAMP-dependent closure of a serotonin-sensitive potassium channel. *Proceedings of the National Academy of Science USA*, **81**, 7956-7960.

Adey, R.W. (1972). Organization of brain tissue; is the brain a noisy processor? *International Journal of Neuroscience*, **3**, 271-284.

Albano, A.M., Mees, A.I., Muench, J., Rapp, P.E., & Schwartz, C. (1988). Singular-value decomposition and the Grassberger-Procaccia algorithm. *Physical Review A*, **38**(6), 3017-3026.

Andersen, P.O. (1987). Properties of hippocampal synapses of importance for integration and memory. In G. Edelman, W.E. Gall, & W.M. Cowan (Eds.), *Synaptic function* (pp. 403-429). New York: Wiley.

Andrade, M.A.,Nuño, J.C., Moran, F., Montero, F., & Mpitsos, G.J. (in press). Complex dynamics of a catalytic network having faulty replication into an error-species. *Physica D.*

Aristotle (330 BC). *Physica I* (9), 192b, 6. Athens, The Lyceum: Tsipouro Press, Mpitsopoulos & Sons.

Ayers, J., Carpenter, G., Currie, S., & Kinch, J. (1983). Which behavior does the lamprey central motor program mediate? *Science*, **221**, 1312-1315.

Babloyantz, A., Salazar, J.M., & Nicolis, C. (1985). Evidence of chaotic dynamics of brain activity during the sleep cycle. *Physics Letters A*, **111**, 152-155.

Bak, P. (1990a). Is the world at the border of chaos? *Annals of the New York Academy of Science*, **581**, 110-118.

Bak, P. (1990b). Self-organized criticality. *Physica A*, **163**, 403-409.

Bak, P. (1990c). Simulation of self-organized criticality. *Physica Scripta*, **T33**, 9-10.

Bak, P., & Chen, K. (1991). Self-organized criticality. *Scientific American*, **264**, 46-53.

Bak, P., Chen, K., & Tang, C. (1990). A forest-fire model and some thoughts on turbulence. *Physics Letters A*, **147**, 297-300.

Bak, P., & Tang, C. (1988). Self-organized criticality. *Physical Review A*, **38**, 364-374.

Barnsley, M. (1988). *Fractals everywhere*. San Diego: Academic Press.

Barrionuevo, G., Kelso, S.R., Johnston, D., & Brown, T.H. (1986). Conductance mechanism responsible for long-term potentiation in monosynaptic and isolated excitatory synaptic inputs to hippocampus. *Journal of Neurophysiology*, **55**, 540-550.

Bellman, K.L. (1979) *The conflict behavior of the lizard,* Sceloporus occidentalis, *and its implication for the organization of motor behavior.* Unpublished doctoral dissertation, University of California, San Diego.

Bentley, D., & Konishi, M. (1978). Neural control of behavior. *Annual Review of Neuroscience*, **1**, 35-59.

Boakes, R. (1984). *From Darwin to behaviorism*. Cambridge: Cambridge University Press.

Braitenberg, V. (1989). Some arguments for a theory of cell assemblies in the cerebral cortex. In L. Nadel, L.A. Cooper, P. Culicover, & R.M. Harnish (Eds.), *Neural connections, mental computations* (pp. 137-145). Cambridge, MA: MIT Press.

Bullock, T.H. (1984). Comparative neuroscience holds promise for quiet revolution. *Science*, **222**, 473-478.

Burton, R.M., & Mpitsos, G.J. (1992). Event-dependent control of noise enhances learning in neural networks. *Neural Networks*, **5**, 627-637.

Byrne, J.H., Castellucci, V.F., & Kandel, E.R. (1978). Contribution of individual mechanoreceptor sensory neurons to defensive gill-withdrawal reflex in *Aplysia. Journal of Neurophysiology*, **41**, 418-431.

Canavier, C., Clark, J.W., & Byrne, J.H. (1990). Routes to chaos in a model of a bursting neuron. *Biophysical Journal*, **57**, 1245-1251.

Cardi, P., Nagy, F., Cazalets, J.-R., & Moulins, M. (1990). Multimodal distribution of discontinuous variation in period of interacting oscillators in the crustacean stomatogastric nervous system. *Journal of Comparative Physiology A*, **167**, 23-41.

Carew, T.J., & Sahley, C.L. (1986). Invertebrate learning and memory: From behavior to molecules. *Annual Review of Neuroscience*, **9**, 435-487.

Chavez-Noriega, L.E., Halliwell, J.V., & Bliss, T.V. (1990). A decrease in firing threshold observed after induction of EPSP-Spike (E-S) component in rat hippocampal slices. *Experimental Brain Research*, **79**, 633-641.

Chay, T.R. (1985). Chaos in a three-variable model of an excitable cell. *Physica D*, **16**, 233-242.

Chay, T.R., & Rinzel, J. (1985). Bursting, beating, and chaos in an excitable membrane model. *Biophysical Journal*, **47**, 357-366.

Chen, K., & Bak, P. (1989). Is the universe operating at a self-organized critical state? *Physics Letters A*, **140**, 299-302.

Cohan, C.S. (1980) *Centralized control of distributed motor networks in* Pleurobranchaea californica. Unpublished doctoral dissertation, Case Western Reserve University, Cleveland, OH.

Cohan, C.S., & Mpitsos, G.J. (1983a). The generation of rhythmic activity in a distributed motor system. *Journal of Experimental Biology*, **102**, 25-42.

Cohan, C.S., & Mpitsos, G.J. (1983b). Selective recruitment of interganglionic interneurons during different motor patterns in *Pleurobranchaea*. *Journal of Experimental Biology*, **102**, 43-58.

Cohen, L., Hopp, H.P., Wu, J.Y., Xaio, C., & London, J. (1989). Optical measurement of action potential activity in invertebrate ganglia. *Annual Review of Physiology*, **51**, 527-541.

Cohen, T.E., Henzi, V., Kandel, E.R., & Hawkins, R.D. (1991). Further behavioral and cellular studies of dishabituation and sensitization in *Aplysia*. *Society for Neuroscience Abstracts*, **17**, 1302.

Conrad, M. (1986). What is the use of chaos? In A V. Holden (Ed.), *Chaos* (pp. 3-14). Princeton, NJ: Princeton University Press.

Croll, R.P., & Davis, W.J. (1981). Motor program switching in *Pleurobranchaea*. I. Behavioral and electromyographic study of ingestion and egestion in intact specimens. *Journal of Comparative Physiology*, **145**, 277-287.

Croll, R.P., & Davis, W.J. (1982). Motor program switching in *Pleurobranchaea*. II. Ingestion and egestion in the reduced preparation. *Journal of Comparative Physiology*, **147**, 143-153.

Darwin, C. (1859). *The origin of species*. London: John Murray.

Davis, W.J. (1976). Organizational concepts in the central motor networks of invertebrates. In R.M. Herman, S. Grillner, P.S.G. Stein, & D.G. Stuart (Eds.), *Neural control of locomotion* (pp. 265-292). New York: Plenum Press.

Davis, W.J., & Kennedy, D. (1972a). Command interneurons controlling swimmeret movements in the lobster. I. Types of effects on motoneurons. *Journal of Neurophysiology*, **35**, 1-12.

Davis, W.J., & Kennedy, D. (1972b). Command interneurons controlling swimmeret movements in the lobster. II. Interaction of effects on motoneurons. *Journal of Neurophysiology*, **35**, 13-19.

Davis, W.J., & Mpitsos, G.J. (1971). Behavioral choice and habituation in the marine mollusk *Pleurobranchaea californica. Zeitschrift für vergleichende Physiologie*, **75**, 207-232.

Davis, W.J., Mpitsos, G.J., & Pinneo, J.M. (1974). The behavioral hierarchy of the mollusc *Pleurobranchaea*. I. The dominant position of the feeding behavior. *Journal of Comparative Physiology*, **90**, 207-224.

Delcomyn, F. (1980). Neural control of movement. *Science*, **210**, 492-498.

Delcomyn, F., & Cocatre-Zilgien, J.H. (1988). Individual differences and variability in the timing of motor activity during walking in insects. *Biological Cybernetics*, **59**, 379-384.

Dickinson, P.S., Mecsas, C., & Marder, E. (1990). Neuropeptide fusion of two motor-pattern generator circuits. *Nature*, **344**, 155-158.

Doty, P., Marmur, J., Eigner, J., & Schildkraut, C. (1960). Strand separation and specific recombination in deoxyribonucleic acids: Physical chemical studies. *Proceedings of the National Academy of Science USA*, **46**, 461-476.

Dowling, J.E. (1987). *The retina: An approachable part of the brain*. Cambridge, MA: Harvard University Press.

Edelman, G.M. (1978). Group selection and phasic reentry signalling: A theory of higher brain function. In G.M. Edelman & V.B. Mountcastle (Eds.), *The mindful brain* (pp. 55-110). Cambridge, MA: MIT Press.

Edelman, G.M. (1987). *Neural Darwinism: The theory of neuronal group selection*. New York: Basic Books.

Eigen, M., & Schuster, P. (1979). *The hypercycle. A principle of natural self-organization*. New York: Springer-Verlag.

Ellner, S. (1988). Estimating attractor dimensions from limited data: A new method with error-estimates. *Phys. Lett. A*, **133**, 128-133.

Ellner, S. (1991). Detecting low-dimensional chaos in population dynamics data: A critical review. In J. Logan & F. Hain (Eds.), *Does chaos exist in ecological systems?* Charlottesville: University of Virginia Press.

Ellner, S., Gallant, A.R., McCaffery, D., & Nychka, D. (1991). Convergence rates and data requirements for Jacobian-based estimates of Lyapunov exponents from data. *Physics Letters A*, **153**, 357-363.

Faber, D.S., Korn, H., & Lin, J.-W. (1991). Role of medullary networks and post-synaptic membrane properties regulating Mauthner cell responsiveness to sensory excitation. *Brain, Behavior, and Evolution*, **37**, 286-297.

Farley, J., & Alkon, D.L. (1986). Cellular analysis of gastropod learning. In A.J. Greenberg (Ed.), *Cell receptors and cell communication in learning* (pp. 220-266). Basel: S. Karger.

Feigenbaum, M.J. (1983). Universal behavior in nonlinear systems. *Physica D*, **7**, 16-39.

Gardner, D. (1990). Paired individual and mean postsynaptic currents recorded in 4-cell networks of *Aplysia. Journal of Neurophysiology*, **63**, 1226-1240.

Garyantes, T.K., & Regehr, W.G. (1992). Electrical activity increases growth cone calcium but fails to inhibit neurite outgrowth from rat sympathetic neurons. *Journal of Neuroscience*, **12**, 96-103.

Getting, P.A. (1989). Emerging principles governing the operation of neural networks. *Annual Review of Neuroscience*, **12**, 185-204.

Getting, P.A., & Dekin, M.S. (1985). *Tritonia* swimming: A model system for integration within rhythmic motor systems. In A.I. Selverston (Ed.), *Model neural networks and behavior* (pp. 3-20). New York: Plenum Press.

Gillette, R. (1986). Command neurons-FAP. *Behavioral Brain Sciences*, **9**, 727-729.

Gillette, R., Kovac, M., & Davis, W.J. (1978). Command neurons in *Pleurobranchaea* receive synaptic feedback from the motor network they excite. *Science*, **199**, 798-801.

Gillette, R., Kovac, M., & Davis, W.J. (1982). Control of feeding motor output by paracerebral neurons in the brain of *Pleurobranchaea*. *Journal of Neurophysiology*, **47**, 885-908.

Goldsmith, B.A., & Abrams, T.W. (1989). Role of adenylate cyclase in several forms of synaptic facilitation in *Aplysia* sensory neurons. *Society for Neuroscience Abstracts*, **15**, 1624.

Grassberger, P., & Procaccia, I. (1983). Characterization of strange attractors. *Physics Review Letters*, **50**, 346-349.

Graybiel, A.M., & Ragsdale, C.W. (1983). Biochemical anatomy of the striatum. In P.C. Emson (Ed.), *Chemical neuroanatomy* (pp. 427-504). New York: Raven Press.

Grossberg, S. (1980). *Studies of mind and brain*. Boston: Reidel.

Grossberg, S., & Kuperstein, M. (1986). *Neural dynamics of adaptive sensory-motor control*. Amsterdam: North-Holland.

Haken, H. (1983). At least one Lyapunov exponent vanishes if the trajectory of an attractor does not contain a fixed point. *Physics Letters A*, **94**, 71-72.

Hebb, D.O. (1949). *Organization of behavior*. New York: Wiley.

Heinzel, H.G. (1988a). Gastric mill activity in the lobster. 1. Spontaneous modes of chewing. *Journal of Neurophysiology*, **59**, 528-550.

Heinzel, H.G. (1988b). Gastric mill activity in the lobster. 2. Proctolin and octopamine initiate and modulate chewing. *Journal of Neurophysiology*, **59**, 551-565.

Heinzel, H.G., & Selverston, A.I. (1988). Gastric mill activity in the lobster. 3. Effects of proctolin on isolated central pattern generator. *Journal of Neurophysiology*, **59**, 565-585.

Hénon, M., & Pomeau, Y. (1975). *Two strange attractors with a simple structure*. New York: Springer-Verlag.

Hofbauer, J., & Sigmund, K. (1988). *The theory of evolution dynamics and dynamical systems*. Cambridge, MA: Cambridge University Press.

Hunt, S.P. (1983). Cytochemistry of the spinal cord. In P.C. Emson (Ed.), *Chemical neuroanatomy* (pp. 53-84). New York: Raven Press.

Jacklet, J.W., & Rine, J. (1977). Facilitation at the neuromuscular junction: Contribution to habituation and dishabituation of the *Aplysia* gill withdrawal reflex. *Proceedings of the National Academy of Science USA*, **74**, 1267-1271.

John, E.R. (1972). Switchboard versus statistical theories of learning and memory. *Science*, **177**, 850-864.

Kandel, E.R. (1979). *Behavioral biology of* Aplysia. San Francisco: Freeman.

Kandel, E.R., & Schwartz, J.H. (1982). Molecular biology of learning: Modulation of transmitter release. *Science*, **218**, 433-443.

Kaplan, J., & Yorke, J. (1979). Chaotic behavior of multidimensional difference equations. In H.-O. Peitgen & H.-O. Walther (Eds.), *Lecture notes in mathematics: Functional difference equations and approximation of fixed points* (pp. 204-227). New York: Springer.

Karten, H.J., Keyser, K.T., & Brecha, N.C. (1990). Biochemical and morphological heterogeneity of retinal ganglion cells. In B. Cohen & I. Bodis-Woliner (Eds.), *Vision and the brain*. New York: Raven Press.

Kater, S.B., & Mills, L.R. (1990). Neurotransmitter activation of second messenger pathways for the control of growth cone behaviors. In J.M. Lauder (Ed.), *Molecular aspects of development and aging of the nervous system* (pp. 217-225). New York: Plenum Press.

Keller, E.F. (1983). *A feeling for the organism: The life and work of Barbara McClintock*. New York: Freeman.

Kelso, J.A.S., Scholz, J.P., & Schöner, G. (1986). Nonequilibrium phase transitions in coordinated biological motion: Critical fluctuations. *Physics Letters A*, **118**, 279-284.

Kien, J. (1983). The initiation and maintenance of walking in the locust: An alternative to the command hypothesis. *Proceedings of the Royal Society of London B*, **219**, 137-174.

Kien, J. (1990a). Neuronal activity during spontaneous walking. I. Starting and stopping. *Comparative Biochemistry and Physiology A*, **95**, 607-621.

Kien, J. (1990b). Neuronal activity during spontaneous walking. II. Correlation with stepping. *Comparative Biochemistry and Physiology A*, **95**, 623-638.

Kirkpatrick, S., Gelatt, C.D., & Becchi, M.P. (1983). Optimization by simulated annealing. *Science*, **220**, 671-680.

Kistler, H.B., Hawkins, R.D., Koester, H.W., Steinbusch, W.M., Kandel, E.R., & Schwartz, J.H. (1985). Distribution of serotonin: Immunoreactive cell bodies and processes in the abdominal ganglion of mature *Aplysia*. *Journal of Neuroscience*, **5**, 72-80.

Klein, M., Bratha, O., Dale, N., & Kandel, E.R. (1989). Analysis of a newly described cellular process contributing to facilitation at depressed neuron synapses. *Society for Neuroscience Abstracts*, **15**, 1264.

Klopf, A.H. (1982). *The hedonistic neuron: A theory of memory, learning, and intelligence*. New York: Hemisphere.

Klopf, A.H. (1988). A neuronal model of classical conditioning. *Psychobiology*, **16**, 85-125.

Koch, C., & Poggio, T. (1987). Biophysics of computation: Neurons, synapses, and membranes. In G. Edelman, W.E. Gall, & W.M. Cowan (Eds.), *Synaptic function* (pp. 637-697). New York: Wiley.

Korn, H., & Faber, D.S. (1987). Regulation and significance of probabilistic release mechanisms at central synapses. In G.M. Edelman, E.W. Gall, & W.M. Cowan (Eds.), *Synaptic function* (pp. 57-108). New York: Wiley.

Kovac, M.P., & Davis, W.J. (1980). Reciprocal inhibition between feeding and withdrawal behaviors in *Pleurobrachaea*. *Journal of Comparative Physiology*, **139**, 77-86.

Kriebel, M.E., Vautrin, J., & Holsapple, J. (1990). Transmitter release: Prepackaging and random mechanism or dynamic and deterministic process? *Brain Research Reviews*, **15**, 167-178.

Kupferman, I., Deodhar, D., & Weiss, K.R. (1991). Simple neural network models provide heuristic tools for understanding the possible role of command-like neurons controlling behaviors in *Aplysia*. *Society for Neuroscience Abstracts*, **17**, 1591.

Kupferman, I., Mahon, A., Scheller, R., Weiss, K.R., & Lloyd, P.E. (1984). Immunocytochemical study of the distribution of small cardioactive peptide (SCP$_b$) in *Aplysia*. *Society for Neuroscience Abstracts*, **10**, 153.

Kupferman, I., & Weiss, K.R. (1978). The command neuron concept. *Behavaioral Brain Sciences*, **1**, 3-39.

Kupferman, I., & Weiss, K.R. (1986). Command performance. *Behavioral Brain Sciences*, **9**, 736-739.

Küppers, B.O. (1983). *Molecular theory of evolution*. Berlin: Springer-Verlag.

Lashley, K. (1950). In search of the engram. *Symposia of the Society for Experimental Biology*, **4**, 454-482.

Leonard, J.L., Edstrom, J., & Lukowiak, K. (1989). A re-examination of the ''gill withdrawal reflex'' of *Aplysia californica* Cooper (Gastropoda; Opisthobranchia). *Behavioral Neuroscience*, **103**, 585-604.

Leonard, J.L., Martinez-Padron, M., Edstrom, J.P., & Lukowiak, K. (1990). Does altering identified gill motor neuron activity alter gill behavior in *Aplysia*? In K.S. Kits, H.H. Boer, & J. Joose (Eds.), *Molluscan neurobiology* (pp. 30-37). Amsterdam: North-Holland.

Lindsey, G.B., Hernandez, Y.M., Morris, K.F., Shannon, R., & Gerstein, G. (1992). Dynamic reconfiguration of brain stem neural assemblies: Respiratory phase-dependent synchrony versus modulation of firing rates. *Journal of Neurophysiology*, **67**, 923-930.

Ling, G., & Gerard, R.W. (1949). The normal membrane potential of frog sartorius fibers. *Journal of Cellular and Comparative Physiology*, **34**, 383-396.

Lipton, S.A., & Kater, S.B. (1989). Neurotransmitter regulation of neuronal outgrowth, plasticity, and survival. *Trends in Neuroscience*, **12**, 265-270.

Lorenz, E.N. (1963). Deterministic non-periodic flows. *Journal of Atmospheric Science*, **20**, 130-141.

Lorenz, K.Z. (1974). Analogy as a source of knowledge. *Science*, **185**, 229-234.

Lorenz, K.Z. (1981). *The foundations of ethology*. New York: Simon & Schuster.

Lukowiak, K., & Colebrook, E. (1986). Classical conditioning of *in vitro Aplysia* preparations: Multiple sites of neuronal changes. In H.H. Boer, W.P.M. Geraerts, & J. Joose (Eds.), *Neurobiology of molluscan models* (pp. 320-325). New York: North-Holland.

Lukowiak, K., Goldberg, J., Colmers, W.F., & Edstrom, J.P. (1986). Peptide modulation of neuronal activity and behavior in *Aplysia*. In G.B. Stephano (Ed.), *CRC handbook of comparative opioid and related neuropeptide mechanisms*. Boca Raton, FL: CRC Press.

Marcus, P.S. (1988). Numerical simulation of Jupiter's great red spot. *Nature*, **331**, 693-696.

Marder, E. (1984). Mechanisms underlying neurotransmitter modulation of a neuronal circuit. *Trends in Neuroscience*, **7**, 48-53.

Marder, E. (1988). Pattern generators: Modulating a neural network. *Nature*, **335**, 296-297.

Marder, E., & Hooper, S.L. (1985). Neurotransmitter modulation of the stomatogastric ganglion of decapod crustaceans. In A.I. Selverston (Ed.), *Model neural networks and behavior* (pp. 319-338). New York: Plenum Press.

Marder, E.E., Hooper, S.L., & Eisen, J.S. (1987). Multiple neurotransmitters provide a mechanism for the production of multiple outputs from a single neuronal circuit. In G. Edelman, W.E. Gall, & W.M. Cowan (Eds.), *Synaptic function* (pp. 305-327). New York: Wiley.

Markus, M., Kuschmitz, D., & Hess, B. (1985). Properties of strange attractors in yeast glycolysis. *Biophysical Chemistry*, **22**, 95-105.

May, R.M. (1976). Simple mathematical models with very complicated dynamics. *Nature*, **261**, 459-467.

McClellan, A.D. (1978). Feeding and rejection in *Pleurobranchaea*: Comparison of two behaviors using some of the same musculature. *Society for Neuroscience Abstracts*, **4**, 201.

McClellan, A.D. (1979). Swallowing and regurgitation in the isolated nervous system of *Pleurobranchaea*: Distinguishing features and higher order control. *Society for Neuroscience Abstracts*, **5**, 253.

McClellan, A.D. (1980) *Feeding and regurgitation in* Pleurobranchaea californica: *Multibehavioral organization of pattern generation and higher order control.* Unpublished doctoral dissertation, Case Western Reserve University, Cleveland, OH.

McClellan, A.D. (1982a). Movements and motor patterns of the buccal mass of *Pleurobranchaea* during feeding, regurgitation, and rejection. *Journal of Experimental Biology*, **98**, 195-211.

McClellan, A.D. (1982b). Re-examination of presumed feeding motor activity in the isolated nervous system of *Pleurobranchaea*. *Journal of Experimental Biology*, **98**, 212-228.

Merzenich, M.M., Kaas, J.H., Wall, J.T., Nelson, R.J., Sur, M., & Felleman, D.J. (1983). Topographic reorganization of somatosensory cortical areas 3b and 1 in adult monkeys following restricted deafferentation. *Neuroscience*, **8**, 33-55.

Morgan, J.L.M. (1991). *Peptidergic regulation of visceral motor circuits in the sea hare,* Aplysia californica. Unpublished doctoral dissertation, Oregon State University, Corvalis, OR.

Morgan, J.S., Patterson, E.C., & Klopf, A.H. (1990). A drive-reinforcement model of simple instrumental conditioning. *Proceedings of the International Joint Conference on Neural Networks*, **2**, 227-232.

Mpitsos, G.J. (1973). Physiology of vision in the file clam *Lima scabra*. *Journal of Neurophysiology*, **367**, 371-383.

Mpitsos, G.J. (1989). Chaos in brain function and the problem of nonstationarity: A commentary. In E. Basar & T.H. Bullock (Eds.), *Dynamics of sensory and cognitive processing by the brain* (pp. 521-535). New York: Springer-Verlag.

Mpitsos, G.J. (1991). Neural network error surfaces: Limitation of network size, input signal dynamics, and metaknowledge in memory storage. *Society for Neuroscience Abstracts*, **17**, 484.

Mpitsos, G.J. Attractors provide mechanism for gradient descent in biological organization. (Manuscript in preparation.)

Mpitsos, G.J., & Burton, R.M. (1992). Convergence and divergence in neural networks: Processing of chaos and biological analogy. *Neural Networks*, **5**, 605-625.

Mpitsos, G.J., Burton, R.M., Creech, H.C., & Soinila, S.O. (1988). Evidence for chaos in spike trains of neurons that generate rhythmic motor patterns. *Brain Research Bulletin*, **21**, 529-538.

Mpitsos, G.J., & Cohan, C.S. (1986a). Comparison of differential Pavlovian conditioning in whole animals and physiological preparations of *Pleurobranchaea*: Implications of motor pattern variability. *Journal of Neurobiology*, **17**, 498-516.

Mpitsos, G.J., & Cohan, C.S. (1986b). Convergence in a distributed motor system: Parallel processing and self-organization. *Journal of Neurobiology*, **17**, 517-545.

Mpitsos, G.J., & Cohan, C.S. (1986c). Differential Pavlovian conditioning in the mollusk *Pleurobranchaea*. *Journal of Neurobiology*, **17**, 487-497.

Mpitsos, G.J., & Cohan, C.S. (1986d). Discriminative behavior and Pavlovian conditioning in the mollusc *Pleurobranchaea*. *Journal of Neurobiology*, **17**, 469-486.

Mpitsos, G.J., & Collins, S.D. (1975). Learning: Rapid aversive conditioning in the gastropod mollusc *Pleurobranchaea*. *Science*, **188**, 954-957.

Mpitsos, G.J., Collins, S.D., & McClellan, A.D. (1978). Learning: A model system for physiological studies. *Science*, **199**, 497-506.

Mpitsos, G.J., Creech, H.C., Cohan, C.S., & Mendelson, M. (1988). Variability and chaos: Neurointegrative principles in self-organization of motor patterns. In J.A.S. Kelso, A.J. Mandell, & M.F. Shlesinger (Eds.), *Dynamic patterns in complex systems* (pp. 162-190). Singapore: World Scientific.

Mpitsos, G.J., & Davis, W.J. (1973). Learning: Classical and avoidance conditioning in the mollusk *Pleurobranchaea*. *Science*, **180**, 317-320.

Mpitsos, G.J., & Lukowiak, K. (1985). Learning in gastropod molluscs. In A.O.D. Willows (Eds.), *The mollusca* (pp. 95-267). New York: Academic Press.

Mpitsos, G.J., Murray, T.F., Creech, H.C., & Barker, D.L. (1988). Muscarinic antagonist enhances one-trial food-aversion learning in *Pleurobranchaea*. *Brain Research Bulletin*, **21**, 169-179.

Murphy, A.D., Lukowiak, K., & Stell, W.K. (1985). Peptidergic modulation of patterned motor activity in identified neurons of *Helisoma*. *Proceedings of the National Academy of Science USA*, **82**, 7140-7144.

Murray, T.F., & Mpitsos, G.J. (1988). Evidence for heterogeneity of muscarinic receptors in the mollusc *Pleurobranchaea*. *Brain Research Bulletin*, **21**, 181-190.

Murray, T.F., Mpitsos, G.J., Siebenaller, J.F., & Barker, D.L. (1985). Stereoselective L-[^{3}H] quinuclidinyl benzilate-binding sites in nervous tissue of *Aplysia californica*: Evidence for muscarinic receptors. *Journal of Neuroscience*, **5**(12), 3184-3188.

Nadel, L., Cooper, L.A., Culicover, P., & Harnish, R.M. (Eds.). (1989). *Neural connections, mental computation*. Cambridge, MA: MIT Press.

Ono, J., & McCaman, R.E. (1984). Immunocytochemical localization and direct assay of serotonin-containing neurons in *Aplysia californica*. *Neuroscience*, **11**, 549-560.

Osovets, S.M., Ginzburg, D.-A., Gurfinkel, V.S., Zenkov, L.P., Latash, L.P., Malkin, V.B., Melnichuk, P.V., & Pasternak, E.B. (1984). Electrical activity of the brain: Mechanisms and interpretation. *Soviet Physics Uspehki*, **26**, 801-828.

Pearson, J.C., Finkel, L.H., & Edelman, G.M. (1987). Plasticity in the organization of adult cerebral cortical maps: A computer simulation based on neuronal group selection. *Journal of Neuroscience*, **7**, 4209-4223.

Pieroni, J.P., & Byrne, J.H. (1989). Differential effects of serotonin, SCP_B, and FMRFamide on processes contributing to presynaptic facilitation in sensory neurons of *Aplysia*. *Society for Neuroscience Abstracts*, **15**, 1284.

Pribram, C. (1971). *Languages of the brain*. Englewood Cliffs, NJ: Prentice-Hall.

Rapp, P.E., Zimmerman, I.D., Albano, A.M., Deguzman, G.C., & Greenbaun, N.N. (1985). Dynamics of spontaneous neural activity in the simian motor cortex: The dimension of chaotic neurons. *Physics Letters A*, **110**, 335-338.

Real, L.A. (1991). Animal choice behavior and the evolution of cognitive architecture. *Science*, **253**, 980-986.

Rosen, S.C., Teyke, T., Miller, M.W., Weiss, K.R., & Kupferman, I. (1991). Identification and characterization of cerebral-to-buccal interneurons implicated in the control of motor programs associated with feeding in *Aplysia*. *Journal of Neuroscience*, **11**, 3630-3655.

Rössler, O. (1976). An equation for continuous chaos. *Physics Letters A*, **57**, 397-398.

Rössler, O. (1979). An equation for hyperchaos. *Physics Letters A*, **71**, 155-157.

Roux, J.-C. (1983). Experimental studies of bifurcations leading to chaos in the Belousof-Zhabotinsky reaction. *Physica D*, **7**, 57-68.

Roux, J.C., Simoyi, R.H., & Swinney, H.L. (1983). Observation of a strange attractor. *Physica D*, **8**, 257-266.

Rumelhart, D.E., Hinton, G.E., & Williams, R.J. (1986). Learning internal representations by error propagation. In D.E. Rumelhart & J.L. McClelland (Eds.), *Parallel distributed processing: Explorations in the microstructure of cognition. Vol 1. Foundations* (pp. 318-362). Cambridge, MA: MIT Press.

Rumelhart, D.E., McClelland, J.L., & PDP Group (Eds.). (1986). *Parallel distributed processing: Explorations in the microstructure of cognition. Vol 1. Foundations.* Cambridge, MA: MIT Press.

Schnabl, W., Stadler, P.F., Frost, C., & Schuster, P. (1991). Full characterization of a strange attractor: Chaotic dynamics in low-dimensional replicator systems. *Physica D*, **48**, 65-90.

Schroeder, M. (1991). *Fractals, chaos, power laws—minutes from an infinite paradise*. New York: Freeman.

Schulman, J.A. (1983). Chemical neuroanatomy of the cerebellar cortex. In P.C. Emson (Ed.), *Chemical neuroanatomy* (pp. 209-228). New York: Raven Press.

Seydel, R. (1988). *From equilibrium to chaos. Practical bifurcation and stability analysis*. New York: Elvesier.

Shimizu, T., & Karten, H.J. (1991). Immunohistochemical analysis of the visual wulst of the pigeon (*Columba livia*). *Journal of Comparative Neurology*, **300**, 346-369.

Skarda, C.A., & Freeman, W.J. (1987). How brains make chaos in order to make sense of the world. *Behavioral Brain Sciences*, **10**, 161-195.

Skinner, J.E., Mitra, M., & Fulton, K.W. (1989). Low-dimensional chaos in a simple biological model of neocortex: Implications for cardiovascular and cognitive disorders. In J.G. Carlson & A.R. Seifer (Eds.), *An international perspective on self-regulation and health*. New York: Plenum Press.

Soinila, S., & Mpitsos, G.J. (1991). Immunohistochemistry of diverging and converging neurotransmitter systems in mollusks. *Biological Bulletin*, **181**, 484-499.

Soinila, S., & Mpitsos, G.J. (1992). *Distribution of acetylcholine in the nervous system of* Aplysia *and* Pleurobranchaea. Manuscript in preparation.

Soinila, S., Mpitsos, G.J., & Panula, P. (1990). Comparative study of histamine immunoreactivity in nervous systems of *Aplysia* and *Pleurobranchaea*. *Journal of Comparative Neurology*, **298**, 83-96.

Sommeria, J., Meyers, S.D., & Swinney, H.L. (1988). Laboratory simulation of Jupiter's great red spot. *Nature*, **331**, 689-693.

Sperry, R.W. (1981). Changing priorities. *Annual Review of Neuroscience*, **4**, 1-15.

Sugita, S., Baxter, D.A., & Byrne, J.H. (1991). Serotonin- and PKC-induced spike broadening in tail sensory neurons of *Aplysia*. *Society for Neuroscience Abstracts*, **17**, 1590.

Tanji, J., & Evarts, E.V. (1976). Anticipatory activity of motor cortex neurons in relation to direction of an intended movement. *Journal of Neurophysiology*, **39**, 1062-1068.

Thom, R. (1990). *Semio physics: A sketch of Aristotelian physics and catastrophe theory*. Redwood City, CA: Addison-Wesley.

Thompson, J.M.T., & Stewart, H.B. (1986). *Nonlinear dynamics and chaos*. New York: Wiley .

Ueda, Y. (1980). Steady motions exhibited by Duffing's equation: A picture book of regular and chaotic motion. In P.J. Holmes (Ed.), *New approaches to nonlinear problems in dynamics* (pp. 311-322). Philadelphia: SIAM.

Vemuri, V. (1988). *Artificial neural networks: Theoretical concepts*. Washington, DC: Computer Society Press of the IEEE.

Warren, A.H. (1989). *An investigation in size reduction in neural networks*. Unpublished masters thesis, Oregon State University, Corvalis, OR.

Weissenfeld, K., Tang, C., & Bak, P. (1989). A physicist's sandbox. *Journal of Statistical Physics*, **54**, 1441-1458.

Werblin, F.S., & Dowling, J.E. (1969). Organization of the retina of the mud puppy, *Necturus maculosus*. *Journal of Neurophysiology*, **32**, 339-355.

Wetzel, M.C., & Stuart, D.G. (1976). Ensemble characteristics of cat locomotion and its neuronal control. *Progress in Neurobiology*, **7**, 1-98.

Wilson, D.M. (1961). The central nervous control of flight in a locust. *Journal of Experimental Biology*, **38**, 471-490.

Wilson, H.R., & Cowan, J.D. (1972). Excitatory and inhibitory interactions in localized populations of model neurons. *Biophysical Journal*, **12**, 1-24.

Wolf, A., Swift, J.B., Swinney, H.L., & Vastano, J.A. (1985). Determining Lyapunov exponents from a time series. *Physica D*, **16**, 285-317.

Wright, W.G., Marcus, E.A., & Carew, T.J. (1989). Dissociation of monosynaptic and polysynaptic contributions to dishabituation, sensitization, and inhibition in *Aplysia*. *Society for Neuroscience Abstracts*, **15**, 1265.

Wu, J., Falk, C.X., Höpp, H., & Cohen, L.B. (1989). Trial-to-trial variability in the neuronal response to siphon touch in the *Aplysia* abdominal ganglion. *Society for Neuroscience Abstracts*, **15**, 1264.

Zečević, D., Wu, J., Cohen, L.B., London, J.A., Höpp, H., & Falk, C.X. (1989). Hundreds of neurons in the *Aplysia* abdominal ganglion are active during the gill-withdrawal reflex. *Journal of Neuroscience*, **9**, 3681-3689.

Acknowledgments

George J. Mpitsos was supported by the Air Force Office of Scientific Research (89-0262) and by a grant from the PGR Fund. Seppo Soinila was supported by a grant from the Finnish Cultural Foundation. The authors wish to thank Dr. Janet Leonard for her critical reading of a previous version of this manuscript, and Professor Lavern Weber, Director of the Mark O. Hatfield Marine Science Center, for making space available to us and for his continuing encouragement.

Chapter 11

Fluctuations, Intermittency, and Controllable Chaos in Biological Coordination

J.A.S. Kelso, M. Ding
Florida Atlantic University, Boca Raton, Florida

When confronted with the issue of "variability" in the workings of organisms, it is tempting to turn the issue upside down and ask, instead, what the consequences of "constancy" are. The constancy of perception, for example, demands a variable world: Were this not the case, there would be no need to talk about perceptual constancy. Without variety of stimulation the brain makes up things that are not a function of sensory experience. For example, play the acoustic stimulus for |ba| into a person's ear repetitively, and the perception of |ba| will undergo a progressive series of (lawfully related) perceptual transformations (Tuller, Kelso, & Mandell, 1992; Warren, 1961). Sometimes such phenomena are called "errors" or "hallucinations." But another possibility is that when we turn off variety in the world, we see the expanding, intrinsically creative workings of the brain.

In a similar vein, when we require a skilled individual to repeatedly perform the same task (a constant goal), it should not be so surprising that we observe variability in the produced trajectories. For without this variability we would have no cause to talk about "motor equivalence." The very existence of so-called motor equivalence implies creative, variable solutions to repeatedly solving the same task demand. Such variability should hardly be viewed as error (except, perhaps, under the strange conditions we sometimes impose on subjects), but should be viewed, rather, as a quite essential feature of behavior.

Two points are in order:

1. Under constant task conditions the trajectories of a multi-degree-of-freedom system, though never exactly identical, are nevertheless very similar. It is possible, therefore, to conceive of the entire trajectory as an attractor, such as a piece of a limit cycle in the case of a discrete movement (e.g., Kelso & Schöner, 1988; Schöner, 1990) or as a fixed point of the relative phasing among interacting components (e.g., Haken, Kelso, & Bunz, 1985). In the next section of this paper we will demonstrate the conceptual and empirical importance of fluctuations around attractive states of the system's coordination dynamics. Such variability is crucial to understanding the stability of movement patterns and the creation of new coordinative patterns.

2. The spontaneously creative and variable solutions to motor tasks suggest that an "infinite" number of trajectories are available to the system. In the concluding part of this paper we will provide a theoretical mechanism that makes it possible to select from an infinite set of trajectories (or orbits) and to stabilize the one selected. The key idea is to build flexibility into the coordination dynamics, thereby allowing the system access to any trajectory.

The core of the present paper, however, follows the background material briefly presented in the next section. Then we present some recent results and theoretical considerations that provide an understanding of von Holst's (1973) absolute and relative coordination. In particular, we will show that the same mechanism underlies the distributional behavior of a system whose components are interacting absolutely (i.e., in a phase- and frequency-locked fashion) or partially. This mechanism arises when the symmetry of the coordination dynamics is broken (DeGuzman & Kelso, in press; Kelso, DelColle, & Schöner, 1990), producing statistical distributions of phasing behavior that have been seen in a very large number of different systems, ranging from the wingbeat patterns of the locust, to respiratory-heart coordination, to interacting cell populations in mammalian visual cortex, to two people visually coordinating their movements, and so on (for these and many other examples, see Haken & Köpchen, 1991). Although essential aspects of this theory have been published elsewhere (e.g., DeGuzman & Kelso, 1992; Kelso & DeGuzman, 1991; Kelso, DeGuzman, & Holroyd, 1991a, b), the focus here, in line with the theme of this book, is on the relative roles of stochastic and deterministic sources of variability in motor coordination.

THE "CLASSICAL" PICTURE: VARIABILITY AND NOISE

A key concept in the dynamic-pattern approach to coordination (e.g., Kelso & Schöner, 1987; Schöner & Kelso, 1988; see also Beek, 1989; Schmidt, Carello, & Turvey, 1990; Turvey, 1990) is that of a *collective variable* (or *order parameter*; cf. Haken, 1983) that captures the system's coordination activity on a chosen level of description. Collective variables define stable and reproducible relationships among the system's interacting components: They can be found most easily around phase transitions or bifurcations where different coordination patterns are clearly distinguishable (Kelso, 1984). The system's coordination dynamics, that is, equations of motion that govern its coordination activity, are defined in terms of the dynamics of these (empirically identified) collective variables. Examples of such collective variables are relative phases that capture patterns of coordination in the nervous system and behavior (for a review, see Kelso, 1991). It should be emphasized that the coordination dynamics do not typically refer to the physical mechanism of moving masses as, say, in the field of biomechanics. In the present context they refer to the equations of motion that generate the time course of (collective) states of the central nervous system (CNS) itself. What does it mean for biological systems to be subject to coordination dynamics? A major consequence of this viewpoint is that *stability* plays a central role, most dramatically evident when stability is lost. Theoretically, loss of stability leads to changes in coordination; predictions regarding loss of stability can be (and have been) tested even before such change occurs. So how does one determine the stability of coordination patterns?

In the "classical picture" (e.g., as described in Kelso, Schöner, Scholz, & Haken, 1987, and Schöner, Haken, & Kelso, 1986), it is assumed that stochastic forces act on the collective variable(s), thereby rendering *fluctuations* of fundamental and practical importance. In the self-organizing, pattern-forming systems of synergetics (Haken, 1983, chaps. 6 and 7), stochastic forces act as continuously applied perturbations producing deviations from the attractor state. The size of these fluctuations as measured, for example, by the variance or standard deviation (SD) of a collective variable, x, around its attractor state is a measure of the stability of this state. The more stable the attractor, the smaller the deviation from the attractor state for a given strength of the stochastic force.[1] A simple, illustrative example is provided in Figure 11.1, a-c.

Still another measure of stability is the relaxation time τ_{rel}, which is a local measure of the stability of the attractor state. If a small perturbation is applied to the system, driving it away from its stationary fixed point state, x_f, the time for the system to return to its stationary state is essentially independent of the size of the perturbation but strongly dependent on the stability of the attractor state. The smaller τ_{rel} is, the more stable is the attractor. The case $\tau_{rel} \rightarrow \infty$ corresponds to a loss of stability. Figure 11.2, a-d, illustrates these relationships. Without elaborating on the details, it is worth mentioning that τ_{rel} may also be determined from fluctuation measures (e.g., by measuring the line width of the spectrum of collective variable fluctuations; see Kelso et al., 1987). This measure has recently been employed by Schmidt, Beek, Treffner, and Turvey (1991) in an effort to uncover scaling properties of the fluctuations.

Changes in coordination, and in general spontaneous pattern formation processes, may arise due to *instability*. Such instabilities are, of course, reflected in stability measures. For example, in Figure 11.3, a-e, we provide a schematic example showing that as the "attractor layout" is deformed by changing parameters, a previously stable minimum becomes shallower and shallower (Fig. 11.3b). At a critical point, the minimum becomes completely flat, due to its collision with a neighboring maximum, and then completely disappears (Fig. 11.3c). The shallower, less articulated minimum has a less restraining influence on fluctuations producing so-called *fluctuation enhancement* or *critical fluctuations*, as seen by a broadening of the local probability distribution (dotted line). After the transition, a new or different coordination pattern appears; the probability distribution is narrowed again, compared to its pretransition value. Similarly, relaxation time increases as the minimum flattens out (compare Figs. 11.3a and 11.3b) and then, as switching occurs, decreases to correspond to the new coordinative pattern. The strong increase in relaxation time, τ_{rel}, is called *critical slowing down*. Since their discovery in bimanual coordination (Kelso & Scholz,

---

[1]A more complete explanation and formalization of these concepts is given in Kelso, Ding, & Schöner (1992). Here we stress the basic ideas with minimal mathematical formalism.

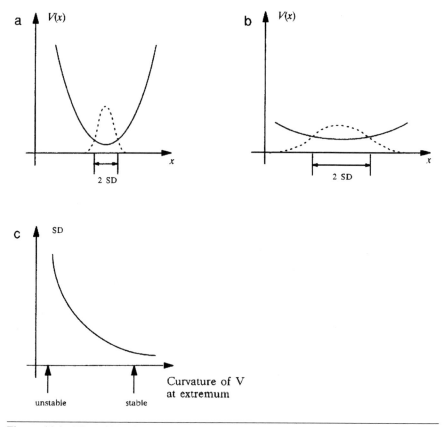

Figure 11.1   The width of the probability distribution (dashed line), as measured by the standard deviation (SD), is a measure of stability. It is smaller for a more pronounced minimum of the potential (a) than for a shallower potential (b). If one varies the shape of the potential experimentally, the SD exhibits the corrresponding change in stability (c).

1985), theoretically predicted critical fluctuations (Figure 11.3e) and critical slowing down (Figure 11.3d) have been experimentally detected in an increasing number of different experimental systems (e.g., Schmidt et al., 1990; Wimmers, Beek, & Van Wieringen, 1992).

In summary, stochastic fluctuations provide a source of variability that has been demonstrated to be conceptually and technically important. From a theoretical viewpoint, fluctuations have enabled the complete specification of coordination dynamics. Not only have patterns of coordination been shown to correspond to attractor states, but switching between attractors is caused by loss of stability. Fluctuations probe the stability of collective states and, by doing so, enable the system to discover new states. We see then the *constructive* role of variability in self-organizing systems, with variability providing an essential source of flexibility. A similar view is expressed elegantly by Mpitsos (1989) in the context of neural and behavioral patterns. An important point for motor behavior is that parametrically

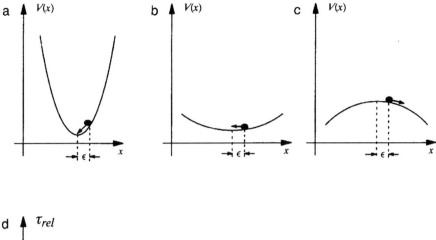

Figure 11.2   (a) In a steep potential, the system relaxes quickly from a small perturbation of size ε. (b) In a shallower potential (smaller curvature at the minimum), the relaxation after the same perturbation as in (a) takes longer due to the smaller restoring force exerted. (c) When the shape of the potential is changed by varying a parameter, the stability of the stationary state is also changed. (d) τ_{rel} reflects changes in stability.

induced patterns of variance reveal underlying mechanisms. Such systematic variability, indicative of upcoming instability, should not be confused with, or bundled into, the concept of error. Naturally enough, the analysis of error scores tends to dominate in fields such as learning. Yet even there, recent evidence shows that instabilities in the coordination dynamics may underlie learning and transfer (Zanone & Kelso, 1991, 1992).[2]

---

[2]There is a message here also for experimental design (Kelso, 1990). In conventional designs that aim to detect significant differences, treatment levels must be randomly assigned in order to homogenize unwanted sources of variance. Failure to randomly sample treatment levels proves, in the conventional approach, to be a serious limitation in terms of generalizability of the results. The present perspective advocates the study of parameter-dependent patterns of variance as crucial to understanding the system's underlying dynamics. Universality (generalizability) emerges when features such as critical fluctuations and critical slowing down are found in very different systems.

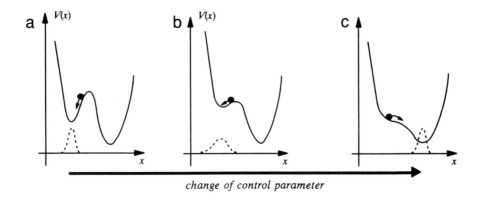

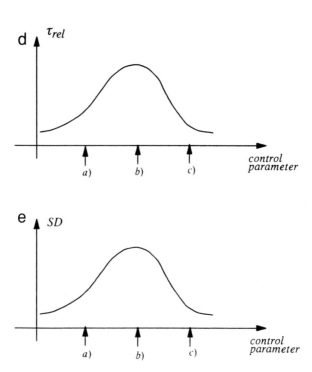

Figure 11.3 (a) A bistable potential (solid line) with a local probability distribution (dashed line). Local relaxation is fast. (b) One minimum has flattened out as a control parameter is changed to a critical value. Local relaxation is slow, and the local probability distribution is very wide. (c) As the control parameter is scaled beyond the transition, the system switches to another available stable stationary state. (d) Relaxation time is plotted as a function of the control parameter (the approximate locations of the situations a, b, and c are indicated). The maximum, as the system goes through the transition, indicates critical slowing down. (e) The standard deviation as a function of the control parameter, the maximum of which reveals critical fluctuations.

THE DYNAMICS
OF ABSOLUTE AND RELATIVE COORDINATION

The complexity of central processes is based not upon the number of qualitatively different forces but upon the wide variability of interaction, i.e., on the rich variety of possible dynamic equilibria. (von Holst, 1973, p. 134)

The behavioral physiologist E. von Holst was able to classify a wide range of coordination phenomena into two types: One, *absolute coordination* involved phase relationships between two or more interacting components that were essentially constant. The other, *relative coordination* involved a *statistical rule* in which every possible phase relation between interacting components occurred, even though a "common phase" characteristic of absolute coordination was still present. Figure 11.4, a and b, shows examples of the distributional behavior of the two kinds of coordination, and the time series from which these statistical distributions were derived. Von Holst (1973) proposed that the reason for the statistical rule of (relative) coordination was probably the same as that which produced absolute coordination, but that some kind of "counter-influence" hampered the production of fixed phase relationships. Here we want to demonstrate that the distributional behavior of absolute and relative coordination can be produced by the same *coordination dynamics*. It is not immediately intuitive that such a view is correct. For example, the phase distribution in Figure 11.4a (relative coordination) might simply be created by adding noise to Figure 11.4b (absolute coordination), where the term *noise* is used in the same sense as discussed in the previous section—that is, as statistical fluctuations of, say, a given strength and unit variance. No doubt noise, which is present in all real systems, is present also in experimental systems that display absolute and relative coordination. But there is no a priori reason why more noise should be present in relative than in absolute coordination.

a   b

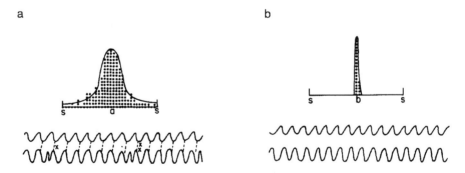

Figure 11.4   (a) Relative coordination. The frequency histogram of possible phase relations between two signals (top) and the time series from which these phase relations are extracted (bottom). The signals come from pectoral (upper curve) and dorsal (lower curve) fin rhythms. (b) Absolute coordination. Histogram of phase relations and corresponding time series as in (a).

Another reason why relative coordination is not simply absolute coordination plus noise is that, as von Holst (1973) clearly showed, relative coordination is a specific phenomenon in its own right. In particular, the phase slippage between two interacting components that occurs in relative coordination is a *progressive* and *systematic* process, not a random process. Many examples of this directed-phase-slippage effect can be found in von Holst's studies of medulla-operated fish. In a series of experiments on human multifrequency (e.g., Kelso & DeGuzman, 1988; Kelso, DeGuzman, & Holroyd, 1991b) and sensorimotor coordination (Kelso et al., 1990), phase slippage and the addition of extra cycles are found to be quite typical. Unlike absolute coordination, in which two or more components move at the same frequency and maintain a fixed phase relation, this less rigid, more flexible form of relative coordination refers to conditions where the *tendency* for phase attraction persists even though the individual components are allowed to express their inherent spatial and temporal variation. Relative coordination may be viewed as a manifestation of flexibility in biological systems despite, or rather because of, competing tendencies (Kelso et al., 1991b). Von Holst (1973) recognized that the shift from absolute to relative coordination may be produced in two ways: either through weakening the so-called "M-effect" (or "coupling strength" between the components) or through changing the frequency of one component relative to the other.

Elsewhere (Kelso et al., 1990, 1991b; Kelso & DeGuzman, 1991) it has been shown that the fundamental reason for the switch from absolute to relative coordination is due to *symmetry breaking* of the following coordination dynamics:

$$\dot{\phi} = -a \sin \phi - 2b \sin 2\phi \qquad (11.1)$$

where ϕ is the relative phase, the order parameter in this case. Equation 11.1 was proposed by Haken et al. (1985) for the experimental model system studied by Kelso and colleagues (see also, e.g., Schmidt et al., 1990, who use Equation 11.1 for an entirely different system), and it provides a good description of absolute coordination and switching between different modes of coordination. For example, a phase transition or pitchfork bifurcation from bistability ($\phi = 0$ and $\phi = \pm\pi$) to monostability ($\phi = 0$) occurs at a critical ratio of the control parameter $|a/4b| = 1$, corresponding to the frequency of movement in experiments. The stochastic version of Equation 11.1 also contains additional properties (discussed previously) that have been studied in considerable detail, especially the behavior of the system as it evolves toward the transition (for review, see Schöner & Kelso, 1988).

But what does it mean to break the symmetry of the coordination dynamics? The functional form of Equation 11.1 is constrained by symmetry (similar components, same frequency), which explains the position of the natural phases 0 and π. Symmetry breaking amounts to a simple correction of the basic coordination dynamics specified in Equation 11.1, but now the relative phase dynamics are no longer invariant under the operation $\phi \rightarrow -\phi$. The reason is that a term $\delta\Omega$ corresponding to the eigenfrequency difference between interacting components is introduced into Equation 11.1 (Kelso et al., 1990):

$$\dot{\phi} = \delta\Omega - a \sin \phi - 2b \sin 2\phi. \qquad (11.2)$$

In the theoretical model described by Equation 11.2, *any* influence that causes differences in the eigenfrequencies of the components may act as a source of symmetry breaking, for instance, handedness, hemispheric asymmetry, and differential loading of identical limbs as in Turvey et al. (chapter 14 of this book). Such effects have been observed experimentally and modeled theoretically under conditions in which human subjects coordinate their actions with a periodic environmental signal (Kelso et al., 1990) or when humans coordinate their arms with their legs. Due to the inherent, biophysical asymmetry between the components, systematic drift in the relative phase in the direction of the upcoming pattern is observed. And strong directional biases occur in the transition regime itself; for example, the arm preferentially changes its phase relative to the leg (Kelso & Jeka, 1992). The sources of symmetry breaking are manifold and open up the study of many interesting pattern-formation phenomena in behavior and cognition (for review, see Kelso, 1990). Figure 11.5, a-c, plots the vector field of the dynamics (Equation 11.2) $\dot{\phi}$ versus ϕ for different parameter values. The system (Equation 11.2) contains stationary patterns or fixed points of ϕ where the time derivative of ϕ ($\dot{\phi}$) crosses the ϕ-axis. When the slope of $\dot{\phi}$ is negative, the fixed points are stable coordinated (phase- and frequency-synchronized) states; when the slope is positive, the fixed points are unstable (repellers). Arrows in Figure 11.5 indicate the direction of the flow.

It can be easily verified that for small values of $\delta\Omega$ one can obtain phase locking, but the fixed points are slightly shifted away from the pure inphase and antiphase

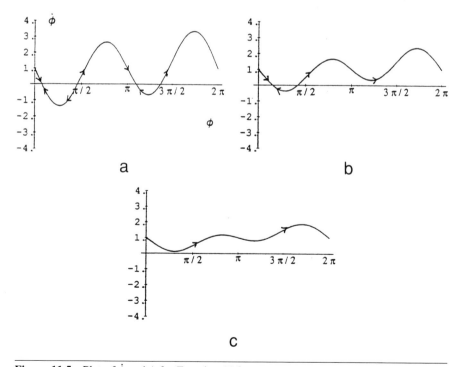

Figure 11.5   Plot of $\dot{\phi}$ and ϕ for Equation 11.2.

patterns of absolute coordination (Figure 11.5a). As the control parameter changes, lifting the curve up (Figure 11.5b), the fixed point of ϕ near zero drifts along the x-axis toward $\pi/2$. Likewise the unstable fixed point moves leftward (Figure 11.5b). Slowly drifting fixed points signal upcoming *tangent* or *saddle-node* bifurcations where the stable and unstable fixed points coalesce, and one obtains a fixed point that is neither an attractor nor a repeller. As the curve is lifted further, no fixed points exist in the system (Figure 11.5c), but, as shown in Figure 11.5c, there is still *attraction* to certain phase relations (where the fixed point used to be), even though the relative phase itself is no longer an attractor ($\dot{\phi} > 0$). The system hovers around this "remnant of the attractor" for a long period of time and exits along the repelling direction. One can readily intuit that were one to plot a statistical distribution of the relative phase in this *intermittent* regime of the coordination dynamics, it would contain all possible phase values but in a concentration around a single, preferred phase, just like the relative coordination shown in Figure 11.4a. Is this, in fact, the case?

To examine this question closely we choose a discrete form of the coordination dynamics rather than the flows of Equations 11.1 and 11.2. The motivation for using discrete dynamical systems comes from both theoretical and experimental considerations (DeGuzman & Kelso, 1992; Kelso & DeGuzman, 1991). As before, we study the evolution of the collective variable, relative phase but now use the point estimate, ϕ_n, instead of the continuous estimate, $\phi(t)$. The dynamics of the relative phase can then be inferred from the return map of ϕ_{n+1} versus ϕ_n. In the return map, for 1:1 coordination, phase-locked modes appear as fixed points, whereas for non-1:1 coordination, they are stable orbits consisting of a discrete set of points. The fixed point properties of the basic coordination dynamics (Equations 11.1 and 11.2) are the main requirement for the map. These are accommodated in the so-called phase-attractive circle map that has been used to model multifrequency coordination (DeGuzman & Kelso, 1991; Kelso & DeGuzman, 1988):

$$\phi_{n+1} = f(\phi_n) = \phi_n + \Omega - \frac{K}{2\pi}(1 + A \cos 2\pi\phi_n) \sin 2\pi\phi_n, \qquad (11.3)$$

where the relative phase, ϕ, is normalized to the interval [0,1]. The way this works is shown schematically in Figure 11.6, which is pretty much self-explanatory. For given parameter values and some initial phase ϕ_n as input, Equation 11.3 is computed yielding an output, ϕ_{n+1}, which is then fed back into the function, and so on.

The meaning of the parameters in Equation 11.3 is inferred partly from general properties of the well-studied circle map (see, e.g., Glazier & Libchaber, 1988, and references therein). K is the strength of coupling between the components, Ω is the ratio between the natural frequencies of the uncoupled components, and the parameter A, by analogy with Equations 11.1 and 11.2, is a measure of the relative importance of the intrinsic phase states, $\phi = 0$ and π.

The detailed analysis of Equation 11.3 has been carried out elsewhere (DeGuzman & Kelso, 1991; see also Kelso et al., 1991a, b, for additional features). Furthermore, the connection between intermittency in the map and relative coordination has been established (Kelso et al., 1991a, b; Kelso & DeGuzman, 1991), and several experimentally testable predictions proposed (DeGuzman & Kelso, 1992).

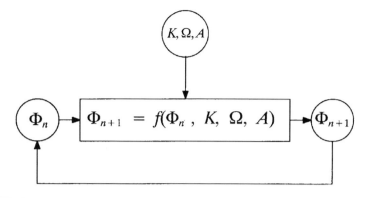

Figure 11.6   Illustration of iterative dynamics of the phase attractive circle map (Equation 11.3).

Entirely analogous to Equation 11.2 and Figure 11.5, intermittency occurs near a *tangent* or *saddle-node* bifurcation in the map. We show an example in Figure 11.7, a-c, in which we vary the values of Ω near the period 1 boundary, that is, near a 1:1 phase- and frequency-synchronized state. The boundary in this case ($A = 0$) is defined by $K = 2\pi\Omega$. For $K = 0.6$, the saddle-node bifurcation occurs at $\Omega_c = 0.6/2\pi \approx 0.0455$. Figure 11.7a shows the function $f(\phi)$ intersecting the diagonal line at two points: ϕ^- and ϕ^+ where ϕ^- is a fixed point attractor and ϕ^+ is a fixed point repeller ($\Omega = \Omega_c - 0.03$ in Figure 11.7a). Initial conditions other than exactly $\phi = \phi^+$ converge to ϕ^- as $n \to \infty$. As Ω increases, ϕ^- and ϕ^+ approach each other and coalesce when $\Omega = \Omega_c$ (Figure 11.7b). For $\Omega = \Omega_c + .01$ beyond the boundary of the Arnold tongue, ϕ^- and ϕ^+ cease to exist (Figure 11.7c), and the system exhibits either periodic orbits of higher period or quasiperiodic orbits, depending on the exact location of Ω. If Ω is decreased, then the reverse sequence of events is observed.

The narrow channel between the function $f(\phi)$ and the diagonal line in Figure 11.7c induces so-called *type I intermittency* (Pomeau & Manneville, 1980). The dynamical behavior is as follows: Inside the channel, iterates of the map move very slowly (Figure 11.7c), giving rise to the impression that the fixed point attractor was already in place (from the point of view of decreasing Ω). After exiting the channel, the trajectory takes large strides for a number of times before reentering the channel. That is, *phase slippage* occurs, and there is no longer any mode locking, because the fixed points have disappeared. The appearance of phase slippage means that between two channel crossings one of the oscillators gains a period: exactly the phenomenon of *relative coordination* (see again Figure 11.4a). This is just like the situation of a father walking along with his small child: Because of their intrinsically different cycle periods, either the one (the father) must slow down, or the other (the child) must skip steps, if they are to keep pace with each other. They are poised near the ghost of mode-locked states (fixed points), relatively but not absolutely coordinated.

Using the map Equation 11.3, it is easy to plot the relative phase distributions corresponding to mode-locked and intermittent regimes of the dynamics. Study of

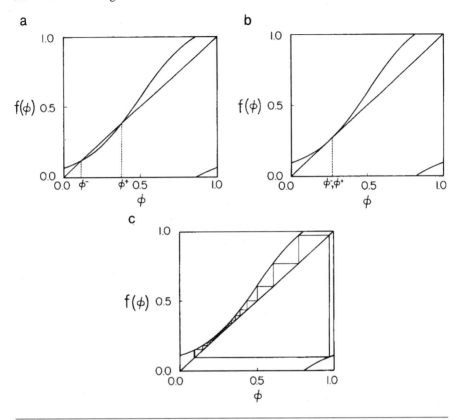

Figure 11.7   Function $f(\phi)$ for three values of Ω. (a) $\Omega = \Omega_c - 0.03$, (b) $\Omega = \Omega_c$, and (c) $\Omega = \Omega_c + 0.01$.

these distributions strengthens the proposal that absolute and relative coordination correspond to mode-locking and intermittent regimes of the coordination dynamics, whether defined as a flow (Equation 11.2) or in discrete form (Equation 11.3). The reader can prove this for him- or herself by iterating Equation 11.3 on a computer and plotting the phase histogram. We present the relevant comparison in Figures 11.8 and 11.9, a and b. The dotted distribution in Figure 11.8 is produced by adding a small Gaussian noise to the phase when the parameters are set well inside the Arnold tongue for 1:1 mode locking. The map is iterated 10,000 times after removal of any initial transients. Figure 11.9a shows the corresponding return map; the large "blob" means the phase is concentrated around the mode-locked state, with the size of the blob corresponding to the magnitude of the noise.

The solid-line distribution in Figure 11.8 shows the system's behavior just outside the boundary of the Arnold tongue for 1:1 synchronization. Note that the distribution is much broader, with a long tail stretched throughout the interval. For this reason, given the same number of iterates ($N = 10,000$), the peak is smaller than its dotted counterpart. Figure 11.9b shows the dynamics of Equation 11.3 in this intermittent regime. Now one sees that the phase is concentrated *near* the mode-locked state,

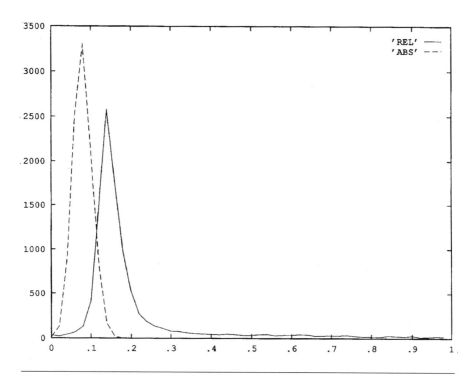

Figure 11.8 Histogram of relative phase distributions in the intermittent regime (relative coordination, solid line) and the mode-locked regime (absolute coordination, dashed line).

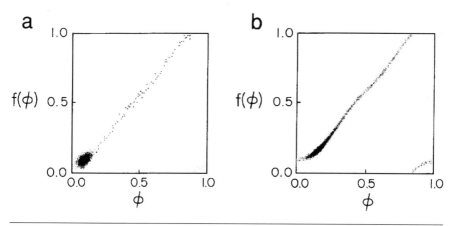

Figure 11.9 (a) Iteration of the map (Equation 11.3) in the 1:1 mode-locked regime, with noise added. (b) Iteration of the map (Equation 11.3) near the 1:1 mode-locked regime, i.e., intermittent dynamics. The noise added in (a) is twice as great as in (b), but the system stays mode-locked in an absolutely coordinated state.

shifted a little compared to Figure 11.9a and spread out over the whole interval. *Even though the noise strength is reduced by a factor of two in the intermittent (Figure 11.9b) case, compared to the mode-locked case (Figure 11.9a) the distribution is far broader in the former than in the latter.*

In summary, we have demonstrated conclusively how the distributional states of both absolute and relative coordination may be produced by the same coordination dynamics. A key insight lies in the identification of relative coordination with the intermittent regime of the dynamics (Kelso et al., 1990; Kelso & DeGuzman, 1991). This more flexible and varied form of coordination arises when the symmetry of the basic coordination dynamics Equation 11.1 is broken. All of the previous concepts have been quite precisely established through theory and experiment, and further tests (e.g., regarding the dwell time, namely, how long a system can stay nearly mode locked in relative coordination before drifting away or switching) are open to test.

VARIABILITY, FLEXIBILITY, AND DETERMINISTIC CHAOS

In this section we explore the connection between variability and flexibility in a somewhat more hypothetical fashion, namely, in systems that exhibit chaotic dynamics. What relevance might chaotic dynamical systems have for the field of motor control? The answer is, we do not really know. Nevertheless, certain recalcitrant problems in biological coordination point in the direction of chaotic dynamics, aspects of which may contain part of the solution. The three problems we have in mind are:

1. How can the motor control system generate an apparently infinite number of trajectories? That is, no single movement trajectory is ever exactly the same, even though an ensemble of such trajectories might show sufficient similarity as to suggest an invariant of motion.
2. Relatedly, why does the motor control system appear to be so sensitive to its initial conditions (ask any golfer), which is, after all, the hallmark of chaotic dynamics?
3. How does one conceptualize the selection of a desired trajectory, given an "infinite" number to choose from?

In an attempt to address these questions we make the key observation that *embedded within a chaotic attractor is an infinite number of unstable periodic orbits.* Each of these orbits may represent a different behavior of the system: Being chaotic endows the system with dynamical access to any orbit or trajectory.

In physical applications it has been demonstrated that motion on a chaotic attractor can be converted to any desired orbit by turning on a small time-dependent perturbation of some accessible system parameters (Ditto, Rauseo, & Spano, 1990; Ott, Grebogi, & Yorke, 1990; Singer, Wang, & Bau, 1991). This area of research is now referred to as controlling chaos.[3] Depending on the purpose, it is possible to

---

[3]There are other types of methods to control a chaotic process. We refer the interested reader to the following papers: A. Hubler, *Helv. Phys. Acta,* **62** (1989), 343; T.B. Fowler, *IEEE Trans. Autom. Control,* **34** (1989), 201; B.A. Huberman and E. Lumer, *IEEE Trans. Circuits Syst.* **37** (1990), 547.

select and stabilize a different orbit in response to different circumstances by using a different temporal pattern of parameters. Imagine a system that is not chaotic but instead is, say, periodic, then it is impossible to achieve anything qualitatively different from the periodic state, short of a major change of the system. Thus, being in a chaotic state is actually a great advantage, because chaos allows enormous intrinsic flexibility. In the case of biological systems such as the brain, "building in" multipurpose flexibility and control appears to be particularly important, especially if we think in terms of information storage and information processing (Ding & Kelso, 1991). We argue that if information is associated with unstable periodic orbits, then being chaotic allows the brain to dynamically seek out any one among a host of available periodic orbits in response to specific needs. It is worth noting that the presence of unstable periodic orbits can be viewed as a further indication of deterministic chaos in the system.

In what follows we illustrate the concepts of chaotic attractors and the associated unstable periodic orbits, using the Lorenz system. Then we apply the controlling chaos algorithm proposed by Ott, Grebogi, and Yorke (OGY; Ott et al., 1990) to the Lorenz chaotic attractor to obtain various desired periodic orbits.

The Lorenz system is a set of three coupled nonlinear differential equations:

$$\begin{aligned}
\dot{x} &= -\sigma x + \sigma y, \\
\dot{y} &= -xz + rx - y, \\
\dot{z} &= xy - bz,
\end{aligned} \tag{11.4}$$

where r, σ, b are system parameters. In a now-classic paper, Lorenz showed that when $r = 28$, $\sigma = 10$, and $b = 8/3$, the above system exhibits a chaotic attractor (Lorenz, 1963), which we hereafter refer to as the Lorenz attractor (Figure 11.10, a and b). A typical chaotic trajectory on the Lorenz attractor swings back and forth between the two lobes (see Figure 11.10a) in a very erratic manner. A glimpse of this irregular behavior can be seen from the time series plot x versus t (Figure 11.10b). If we denote the trajectory swing on the positive side of the x-axis as x and that on the negative side of the x-axis as y, then each trajectory on the chaotic attractor can be characterized by an infinite symbol sequence of x's and y's. An important property of chaotic motion is that the occurrence of x or y in the symbol sequence is unpredictable, and the entire sequence is essentially indistinguishable from a random sequence of heads and tails generated by tossing a coin. Besides the apparent random aspects of the chaotic dynamics, there are also very regular features associated with the Lorenz attractor. In particular, embedded within the Lorenz attractor are an infinite number of unstable periodic orbits. Each of these orbits represents a regular periodic behavior of the system. In Figure 11.11, a-d, we show two of such orbits. Figure 11.11a is the "x-y" orbit, where x and y bear the same symbolic meaning as described earlier, and Figure 11.11b is the corresponding time series. Figure 11.11c shows the "x-yyy" orbit, whose corresponding time series is displayed in Figure 11.11d. Because these orbits are unstable, we need very precise initial conditions to produce the pictures shown in Figure 11.11. If we integrate the system long enough (longer than the time used to produce Figure 11.11), the trajectory will move away from the unstable periodic orbit and produce chaotic swings similar to that shown in Figure 11.10.

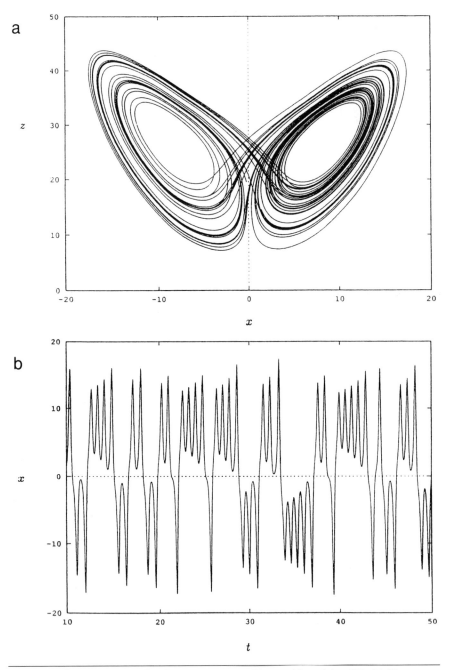

Figure 11.10 The Lorenz chaotic attractor at $r = 28$, $\sigma = 10$, and $b = 8/3$. (a) The x-z projection; (b) the time series x versus t.

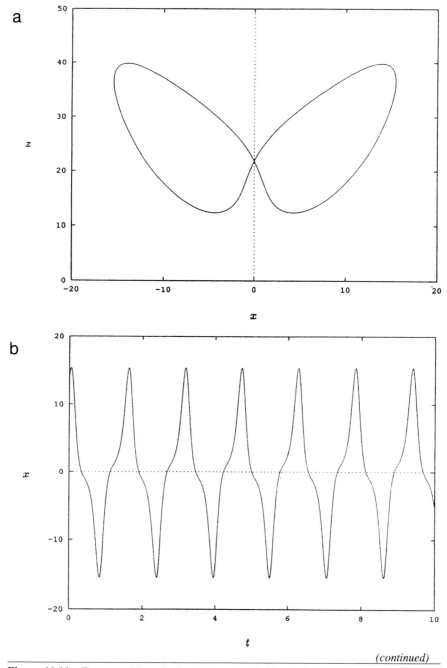

(continued)

Figure 11.11 Two unstable periodic orbits embedded in the Lorenz attractor. (a) The *x-y*
orbit and (b) the corresponding time series; (c) the *x-yyy* orbit and (d) the corresponding
time series.

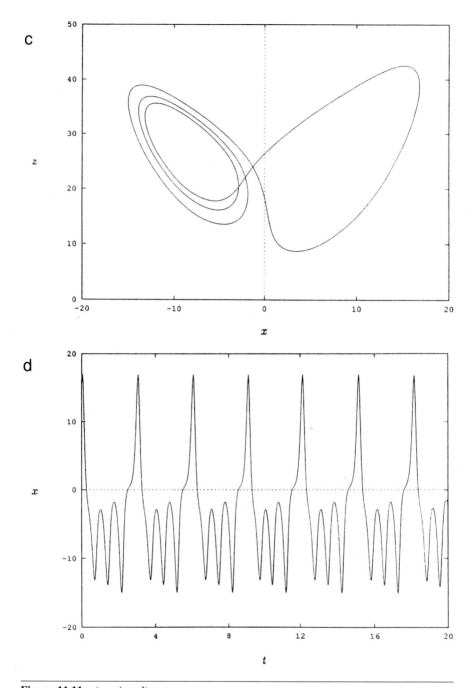

Figure 11.11 *(continued)*

308

As discussed earlier, the dynamics on a chaotic attractor can be controlled by slightly perturbing a system parameter. An explicit algorithm for achieving this end has been proposed by Ott et al. (1990). The original aim of the OGY algorithm is to enhance the performance of a given physical system by utilizing chaos. A particularly appealing aspect of this algorithm is that it does not require a priori analytical knowledge of the system. Thus it may enjoy wider applicability in practice.

In what follows we demonstrate how to control the Lorenz chaotic attractor by using the OGY algorithm. More specifically, we want to convert the chaotic time series shown in Figure 11.10b to either one of the unstable periodic orbits shown in Figure 11.11. In this case, the time-dependent perturbation is applied to the parameter r around its nominal value $r = 28$. The maximum allowable perturbation of r is ± 0.5. This restriction on the range of r stems from the small perturbation nature of the OGY algorithm. Moreover, for small variations of r, we do not envision any qualitative change in the system dynamics.

Figure 11.12 shows a segment of the chaotic time series of the Lorenz attractor, followed by a segment of the controlled "x-y" periodic orbit. The control is turned on at the arrow. The steps we take to produce Figure 11.12 are the following. First, we initialize a trajectory at, say, $x = y = z = 10$ (a point not on the Lorenz attractor). We then integrate the Lorenz equations forward using a 4th-order Runge-Kutta integrator with a fixed step size of 0.001. The resulting trajectory is rapidly pulled

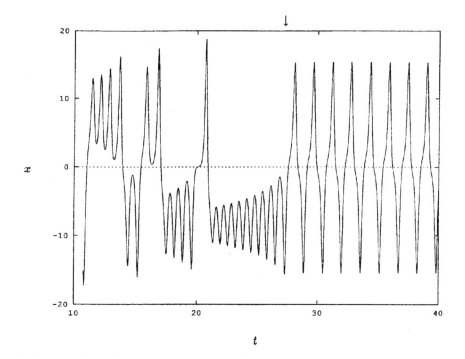

Figure 11.12   The chaotic motion is converted to the x-y periodic behavior via the OGY control algorithm. The control is turned on at the arrow.

toward the attractor. Because the dynamics on the attractor are ergodic, the trajectory will visit the vicinity of every unstable periodic orbit embedded in the attractor. When the trajectory comes close to the ''x-y'' orbit, the perturbation of r is switched on to keep the trajectory in the neighborhood of the ''x-y'' orbit, thereby producing the regular periodic behavior seen in Figure 11.12. The explicit technique for perturbing the parameter r will be illustrated at the end of this section. The time evolution before the onset of control can be viewed as a chaotic transient whose characteristics are discussed in Ott et al. (1990). Figure 11.13 shows the result of controlling the ''x-yyy'' orbit. The steps used here are similar to those used in Figure 11.12.

A very interesting issue concerns the role of external noise. Figure 11.14 shows the result of controlling chaos in the presence of noise. The output we obtain resembles that of intermittency. That is, the control can be successful for a while, then the trajectory is kicked out of the vicinity of the target behavior (''x-y'' orbit in the case of Figure 11.14). After a certain period of chaotic evolution with no control, the trajectory comes back to the neighborhood of the target orbit, due to ergodicity, where the control can be switched on again. This process is repeated intermittently for the entire course of time evolution. We note that in the case of very strong external noise the OGY algorithm will fail to hold the trajectory near the ''x-y'' orbit for any meaningful length of time, hence it will fail to achieve the desired effect.

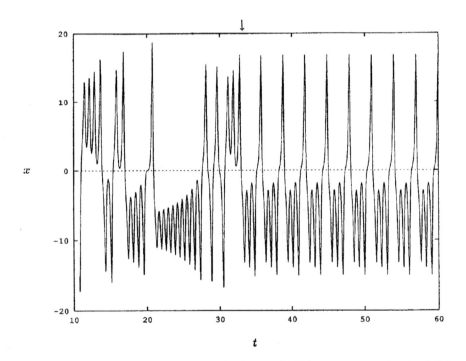

Figure 11.13   The chaotic motion is converted to the *x-yyy* periodic behavior.

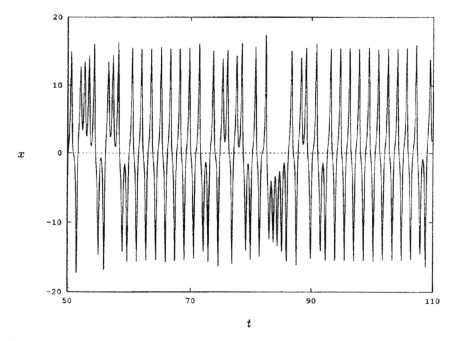

Figure 11.14 Controlling chaos in the presence of external noise. The target behavior in this case is the *x-y* unstable periodic orbit.

To implement the OGY control algorithm in a continuous time system, we need to construct the Poincaré surface of section for the chaotic attractor and observe the behavior of the resulting Poincaré map. For the Lorenz system it is known that a typical orbit on the attractor pierces the plane $z = r - 1$ transversally and recurrently (Lorenz, 1963). (This result is valid for values of r other than $r = 28$.) Therefore the plane $z = 27$ provides a natural location for constructing the Poincaré surface of section for our use. Figure 11.15 shows the Lorenz attractor represented in this surface of section. The horizontal and vertical axes are the x and y coordinates at each upward ($\dot{z} > 0$) piercing of the plane $z = 27$ by a continuous trajectory. In the surface of section, a continuous-time-periodic orbit appears as a discrete-time orbit cycling through a finite set of points. The number of points in this set may be referred to as the period of that discrete orbit. Figure 11.15 shows two such discrete sets of points, denoted by "+" and "*" respectively, where "+" corresponds to the "*x-y*" orbit, and "*" the "*x-yyy*" orbit. It can be seen that these two orbits are embedded in the chaotic attractor with periods of 2 and 4 (same as the number of symbols needed to classify the geometry of the orbits).

Assume that the surface of section can be constructed either theoretically or experimentally, and then in the surface of section the dynamics is described by the following two-dimensional Poincaré map:

$$\xi_{n+1} = F(\xi_n, p), \qquad (11.5)$$

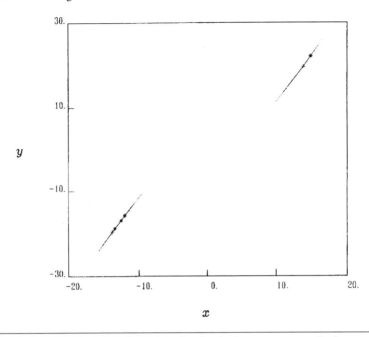

Figure 11.15   The Poincaré surface of section for the Lorenz attractor. In the same surface section the crossings of the *x-y* unstable periodic orbit are represented by ''+'' and that of the *x-yyy* unstable periodic orbit by ''*''.

where ξ_n is the *n*th iterate of the map and p is the system parameter. The essence of this method is to understand the dynamics F in a small neighborhood of the desired orbit. This can be done even in the case where the analytical form of F is unknown (as in the Lorenz system). To control the chaos we attempt to confine the iterates of the map (in the surface of section) in that neighborhood in the following adaptive fashion. When an iterate falls near the desired orbit, we switch on the control by changing the parameter p from its nominal value p_0 by δp. Under the new parameter $p = p_0 + \delta p$ the system forces the next iterate to fall onto the stable manifold of the original orbit for $p = p_0$. Then p is set back to p_0 so that the subsequent iterates can converge to the desired orbit along the stable manifold. When the iterates start to move off the stable manifold of the original orbit, p is adjusted again to bring the iterate back to the stable manifold. The entire process repeats until the required length of the periodic behavior is obtained. Figure 11.16, a-c, pictorially illustrates the control process for the case of a saddle fixed point located at $\xi_F(p_0)$.

This treatment of controlling chaos in a model system raises a number of challenging hypotheses for perceptual-motor coordination. For example, it is well known that voluntary movement ''rides'' on, or is phased with, an underlying physiological tremor (Goodman & Kelso, 1983). Suppose that physiological tremor, a natural process in the 8- to 12-Hz range, is generated by deterministically chaotic dynamics. Could it be, then, that the generation of voluntary movement is related to selecting and

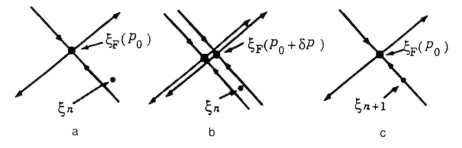

Figure 11.16   Schematic illustration of the OGY algorithm (Ditto et al., 1990). p_0 is the nominal value of the control parameter p. $\xi_F(p_0)$ is the fixed point embedded in the chaotic attractor and is the target behavior. The stable and unstable directions of the fixed point are indicated by the arrows. (a) The nth iterate ξ_n falls near the fixed point $\xi_F(p_0)$. (b) Perturb the parameter p to move the fixed point. (c) The next iterate is forced onto the stable manifold of $\xi_F(p_0)$. Repeat (a), (b), and (c) until the desired length of fixed-point behavior is obtained.

stabilizing an orbit from this underlying process? Similar issues arise in perception (Freeman, 1991), where the percept of an odor, for example, appears to emerge from a chaotic rest state in the olfactory bulb.

EPILOGUE

It is now well established that laws of coordination of the kinds described by Equations 11.1 to 11.3 can be derived in terms of self-organized coupling among the individual components involved. Fluctuations have been demonstrated to be of both fundamental and practical importance to this self-organization. Such fluctuations can be generated by a variable internal or external environment. Fluctuations in macroscopic, collective variables reconcile the stability of coordination with the ability to change behavioral patterns flexibly. Indeed, it is the presence of fluctuations that establishes time-scale relations that governs the switching dynamics among coordinative states.

Variability near stable coordinated states (fixed points) can arise from two sources: stochastic and (deterministically) chaotic. In particular, we have shown that more flexible behavior, called *relative coordination*, corresponds to the intermittent regime of the deterministic dynamics (Equation 11.3). The ubiquity of relative coordination in biology may be tied to the fact that intermittency is a typical feature of dynamical systems near tangent or saddle-node bifurcations. As a state of coordination arising near the boundary of regular and irregular behavior—in the margin of instability, as it were—relative coordination provides an optimal mix of flexibility and stability.

Chaotic dynamics (see the preceding section) may, at first blush, seem too flexible for the purpose of motor control. On the other hand, recent demonstrations that it is possible to control a chaotic process are intriguing, provoking thoughts about some old problems in motor control that have thus far eluded solutions.

REFERENCES

Beek, P.J. (1989). Timing and phase-locking in cascade juggling. *Ecological Psychology*, **1**, 55-96.

DeGuzman, G.C., & Kelso, J.A.S. (1991). Multifrequency behavioral patterns and the phase attractive circle map. *Biological Cybernetics*, **64**, 485-495.

DeGuzman, G.C., & Kelso, J.A.S. (1992). The flexible dynamics of biological coordination: Living in the niche between order and disorder. In J. Mittenthal & A. Baskin (Eds.), *Principles of organization in organisms*. New York: Addison-Wesley.

Ding, M., & Kelso, J.A.S. (1991). Controlling chaos: A selection mechanism for neural information processing? In D. Duke & W. Pritchard (Eds.), *Measuring chaos in the human brain* (pp. 17-31). Singapore: World Scientific.

Ditto, W.L., Rauseo, S.N., & Spano, M.L. (1990). Experimental control of chaos. *Physical Review Letters*, **65**(26), 3211-3214.

Freeman, W.J. (1991, February). The physiology of perception. *Scientific American*, pp. 78-85.

Glazier, J.A., & Libchaber, A. (1988). Quasi-periodicity and dynamical systems. *IEEE Transactions on Circuits and Systems*, **35**, 790-809.

Goodman, D., & Kelso, J.A.S. (1983). Exploring the functional significance of physiological tremor. *Experimental Brain Research*, **49**, 419-431.

Haken, H. (1983). *Synergetics, an introduction: Non-equilibrium phase transitions and self-organization in physics, chemistry and biology*. Berlin: Springer-Verlag.

Haken, H., Kelso, J.A.S., & Bunz, H. (1985). A theoretical model of phase transitions in human hand movements. *Biological Cybernetics*, **51**, 347-356.

Haken, H., & Köpchen, H.-P. (1991). *Physiology of rhythms*. Berlin: Springer-Verlag.

Holst, E. von (1973). Relative coordination as a phenomenon and as a method of analysis of central nervous function. In R. Martin (Ed.), *The collected papers of Erich von Holst* (pp. 33-135). Coral Gables, FL: University of Miami.

Kelso, J.A.S. (1984). Phase transitions and critical behavior in human bimanual coordination. *American Journal of Physiology: Regulatory, Integrative, and Comparative Physiology*, **15**, R1000-R1004.

Kelso, J.A.S. (1990). Phase transitions: Foundations of behavior. In H. Haken (Ed.), *Synergetics of cognition* (pp. 249-268). Berlin: Springer-Verlag.

Kelso, J.A.S. (1991). Behavioral and neural pattern generation: The concept of neurobiological dynamical systems (NBDS). In H.P. Köpchen & T. Huopaniemi (Eds.), *Cardiorespiratory and motor coordination*. Berlin: Springer-Verlag.

Kelso, J.A.S., & DeGuzman, G.C. (1988). Order in time: How cooperation between the hands informs the design of the brain. In H. Haken (Ed.), *Neural and synergetic computers* (pp. 180-196). Berlin: Springer-Verlag.

Kelso, J.A.S., & DeGuzman, G.C. (1991). An intermittency mechanism for coherent and flexible brain and behavioral function. In J. Requin & G. Stelmach (Eds.), *Tutorials in motor neurosciences* (pp. 305-310). Dordrecht: Kluwer.

Kelso, J.A.S., DeGuzman, G.C., & Holroyd, T. (1991a). The self-organized phase attractive dynamics of coordination. In A. Babloyantz (Ed.), *Self-organization, emerging properties, and learning* (pp. 41-62). New York: Plenum Press.

Kelso, J.A.S., DeGuzman, G.C., & Holroyd, T. (1991b). Synergetic dynamics of biological coordination with special reference to phase attraction and intermittency. In H. Haken & H.P. Köpchen (Eds.), *Physiology of rhythms*. Berlin: Springer-Verlag.

Kelso, J.A.S., Delcolle, J.D., & Schöner, G.S. (1990). Action-perception as a pattern formation process. In M. Jeannerod (Ed.), *Attention and performance XIII* (pp. 139-169). Hillsdale, NJ: Erlbaum.

Kelso, J.A.S., Ding, M., & Schöner, G.S. (1992). Dynamic pattern formation: A primer. In J. Mittenthal & A. Baskin (Eds.), *Principles of organization in organisms*. New York: Addison-Wesley.

Kelso, J.A.S., & Jeka, J.J. (1992). Symmetry breaking dynamics in human multilimb coordination. *Journal of Experimental Psychology: Human Perception and Performance*, **18**, 645-668.

Kelso, J.A.S., & Scholz, J.P. (1985). Cooperative phenomena in biological motion. In H. Haken (Ed.), *Complex systems: Operational approaches in neurobiology, physical systems, and computers* (pp. 124-149). Berlin: Springer-Verlag.

Kelso, J.A.S., & Schöner, G.S. (1987). Toward a physical (synergetic) theory of biological coordination. *Springer Proceedings in Physics*, **19**, 224-237.

Kelso, J.A.S., & Schöner, G.S. (1988). Self-organization of coordinative movement patterns. *Human Movement Science*, **7**, 27-46.

Kelso, J.A.S., Schöner, G.S., Scholz, J.P., & Haken, H. (1987). Phase-locked modes, phase transitions, and component oscillators in biological motion. *Physica Scripta*, **35**, 79-87.

Lorenz, E.N. (1963). Deterministic nonperiodic flow. *Journal of Atmospheric Science*, **20**, 130.

Mpitsos, G.J. (1989). Chaos in brain function and the problem of nonstationarity: A commentary. In E. Baser & T. Bullock (Eds.), *Brain dynamics* (pp. 521-535). Berlin: Springer-Verlag.

Ott, E., Grebogi, C., & Yorke, J.A. (1990). Controlling chaos. *Physics Review Letters*, **64**(11), 1196-1199.

Pomeau, Y., & Manneville, P. (1980). Intermittent transitions to turbulence in dissipative dynamical systems. *Communications in Mathematical Physics*, **74**, 189.

Schmidt, R.C., Beek, P.J., Treffner, P.J., & Turvey, M.T. (1991). Dynamical substructure of coordinated rhythmic movements. *Journal of Experimental Psychology: Human Perception and Performance*, **17**(3), 635-651.

Schmidt, R.C., Carello, C., & Turvey, M.T. (1990). Phase transitions and critical fluctuations in the visual coordination of rhythmic movements between people. *Journal of Experimental Psychology: Human Perception and Performance*, **16**(2), 227-247.

Schöner, G.S. (1990). A dynamic theory of coordination of discrete movement. *Biological Cybernetics*, **63**, 257-270.

Schöner, G.S., Haken, H., & Kelso, J.A.S. (1986). A stochastic theory of phase transitions in human hand movement. *Biological Cybernetics*, **53**, 442-452.

Schöner, G.S., & Kelso, J.A.S. (1988). Dynamic pattern generation in behavioral and neural systems. *Science*, **239**, 1513-1520.

Singer, J., Wang, Y.-Z., & Bau, H.H. (1991). Controlling a chaotic system. *Physical Review Letters*, **66**, 1123-1125.

Tuller, B., Kelso, J.A.S., & Mandell, A.J. (1992). Manuscript in preparation.

Turvey, M.T. (1990). Coordination. *American Psychologist*, **45**, 938-953.

Turvey, M.T., Schmidt, R.C., & Beek, P.J. (1991, May). *Fluctuations in interlimb rhythmic coordination*. Paper presented at the conference on variability and motor control, Chicago, IL. (This volume)

Warren, R.M. (1961). Illusory changes of distinct speech upon repetition—the verbal transformation effect. *British Journal of Psychology*, **52**, 249-258.

Wimmers, R.H., Beek, P.J., & Van Wieringen, P.C.W.V. (1992). Phase transitions in rhythmic tracking movement: A case of unilateral coupling. *Human Movement Science*, **11**, 217-226.

Zanone, P.G., & Kelso, J.A.S. (1991). Experimental studies of behavioral attractors and their evolution with learning. In J. Requin & G.E. Stelmach (Eds.), *Tutorial in motor neurosciences* (pp. 121-133). Dordrecht: Kluwer.

Zanone, P.G., & Kelso, J.A.S. (1992). The evolution of behavioral attractors with learning: Nonequilibrium phase transitions. *Journal of Experimental Psychology: Human Perception and Performance*, **18**, 403-421.

Acknowledgments

This work was supported by NIMH (Neurosciences Research Branch) grant MH 42900, BRS grant RR07258, and U.S. Office of Naval Research contract N00014-88-J119.

Chapter 12

Information in Movement Variability About the Qualitative Dynamics of Posture and Orientation

Gary E. Riccio
University of Illinois at Urbana-Champaign

Movement variability is examined within the context of perception and adaptive control of posture and orientation. Variability is viewed as movement that does not serve immediate control objectives. However, it constitutes a pattern of stimulation that provides task-relevant information about the dynamical interaction between an animal and its environment. In principle, information about the animal-environment system facilitates adaptation of control strategies to task-relevant variations in the system. Thus, the movement variability that provides this information may be an essential feature of an adaptive control system rather than irrelevant activity or noise. There are four aspects to a program of research that would demonstrate a functional role for movement variability: (a) The properties of the animal-environment system that are relevant to the task should be identified theoretically and empirically. (b) Perception of task-relevant properties should be demonstrated phenomenologically or behaviorally. (c) Relations should be identified between task-relevant properties and patterns defined by the movement variability. (d) Adaptation of control strategies to variations in the animal-environment system should be related to systematic variations in patterns of movement variability. Examples are presented for each of these research issues.

PERCEPTION AND ACTION

The fundamental premise of this chapter is that the study of perception is necessary in movement science. Perception is important because it is the basis for the essential properties of animate movement. In the present work, *adaptability* and *goal-directedness* are considered to be the essence of animate movement (as opposed to inanimate motion). Perception allows for adaptation of movement patterns to variations in the animal, the environment, and the task so that the goals of the animal are accomplished. The study of perception is the study of how an animal comes to know itself and its environment; more broadly, it is the study of knowledge or epistemology. An ecological epistemology provides the foundation, and the broader context, for this work (E.J. Gibson, 1991; J.J. Gibson, 1979; Lombardo, 1987;

Reed & Jones, 1982; Riccio & Stoffregen, 1988; Stoffregen & Riccio, 1988). It views perception as process rather than product. The perceiver is active, and this activity is both the reason and the "mechanism" for perceiving. Movement is both exploratory and performatory. It provides information and accomplishes overt goals. More generally, the achievement of goals includes the pickup of information or the obtaining of knowledge about oneself or one's environment.

Multiple Methodologies

A variety of scientific methodologies are presented in this chapter. This reflects the fact that there are many methods that can be exploited in the study of perception. One way in which these methods differ is in the process of scientific observation (or data collection). For example, one could introspect or reflect on one's own knowledge of, and experiences in, a particular situation. Such phenomenological methods have not been fashionable, or at least central, in the scientific study of perception for some time (see Boring, 1929/1957; Evans, 1973; Heidbreder, 1933). However, reflection can be of some value in pedagogical contexts, and ultimately it can have an influence on scientific investigations (e.g., Bailey, 1991). Other, more common, methods require that one investigate another person's perception. For example, the investigator may simply watch a person perform an activity without interacting with that person. However, there is generally an interaction between the investigator and the person who is the subject of the investigation, and there can be extended verbal or nonverbal communication between the investigator and the subject (cf. Locke, 1988; Sanderson, James, & Seidler, 1989). In this approach, the subject is a participant in the investigation, in that both the subject and the investigator have some control over the evolving content of, and the process of observation in, the investigation; in other words, they inform each other (cf. Bain, 1989). Because the investigator necessarily has less control of the activities of the participants, the methodology is usually considered to be unacceptable in experimental investigations. However, such loosely constrained interactions provide a way to improve the external validity of experimental manipulations, observations, and analyses. For example, interaction with pilots outside of an experimental context is useful if an investigator is interested in the perception and control of aircraft motion (e.g., Brown, Cardullo, McMillan, Riccio, & Sinacori, 1991); interaction with athletes outside of an experimental context is useful if an investigator is interested in the perception and control of balance in sport (e.g., Bailey, 1991). Such interactions are especially enlightening when the participants include instructors who are skilled in reflecting on, and communicating, their knowledge of a situation.

Experiments in perception require that the investigator have some control over the situation in which perception is studied. Although there is debate over the extent of control that is adequate or appropriate (see, e.g., J.J. Gibson, 1979; Riccio, E.J. Martin, & Stoffregen, 1992), it is usually considered desirable to maximize the control that the experimenter has over the process of observation and the content of what is observed. For example, the experimenter may probe the subject about the phenomenology (conscious experiences) or elicit behavioral responses (overt actions) in a specific set of conditions (e.g., Flach, Riccio, McMillan, & R. Warren,

1986; Riccio & Cress, 1986; Riccio et al., 1992; R. Warren & Riccio, 1985). Verbal or nonverbal responses of subjects are generally described in written or spoken instructions and, thus, are constrained by the accuracy and precision of communication between the experimenter and the subject. Experimenters may attempt to minimize the ambiguity of responses by reducing the task to a binary decision. The decision can be indicated by a verbal response such as yes or no, same or different, or a nonverbal response such as a key press. Thus, the subject indicates whether objects or events are perceivable or not (i.e., above or below threshold) or whether they are differentiable or not (i.e., greater than or less than a just-noticeable difference). However, such decisions can be ambiguous if the instructions are not specific about the aspects of the situation to which the subject should attend and upon which the subject's response should be based. Achieving specificity of instructions is not a straightforward matter in complex (natural) situations. This problem applies equally to verbal and nonverbal responses. The reductionistic solution to this dilemma is to simplify the situation to the point where a passive subject has little, if any, choice about what to attend to and how to respond. This maximizes the experimenter's control over the process of observation and the content of what is observed, but it eliminates a fundamental property of perception: the *obtaining* of stimulation and the *search* for information by an *active* perceiver (see E.J. Gibson, 1991; J.J. Gibson, 1979; Reed & Jones, 1982). In the present work, the activity of perceivers is not sacrificed for experimental control. Management of this trade-off is considered to be an important component of good experimental design (e.g., Riccio et al., 1992).

Multiple Mappings

Verbal descriptions and overt actions provide *intrinsic* measurement systems for that which is perceived. Because they are dependent on the perceiver, such measurement systems have great potential for *meaningfulness* and *representativeness* (see Coombs, Dawes, & Tversky, 1970, for a discussion of these properties of measurement systems). In other words, there is potentially a close correspondence between the descriptions or actions in a particular situation and perceivable properties of the situation. However, the entities and operations (e.g., numerical, algebraic, geometric, or set-theoretic properties) in such loosely constrained measurement systems are generally not well understood. Mathematical properties are more developed in highly constrained *extrinsic* measurement systems such as those provided by standardized laboratory devices and instruments. However, there is little or no reason to believe that such measurement systems are meaningful or representative with respect to that which is perceivable, because they are independent of the perceiver or their relation to the perceiver is not obvious (J.J. Gibson, 1979; Kugler & Turvey, 1987). Because of these complementary properties, the correspondence between intrinsic and extrinsic measurement systems is an important area of investigation. More generally, the correspondence between different measurement systems (intrinsic or extrinsic) is enlightening, because the various methods of scientific observation differ with respect to precision, accuracy, and generality.

The study of perception is replete with investigations of the mappings between measurement systems (Figure 12.1). Scientists are generally committed to particular methodologies and particular measurement systems. Consequently, methodologies, measurement systems, and mappings delineate areas of study and relatively autonomous subsets of the scientific community. The measurement systems and mappings that are relevant to the present work are those that represent purposeful interactions of an animal with its animate and inanimate environment. All other mappings are considered to be nested within this ecological "outer loop." An appreciation of the nested systems of measurement is ultimately required for a comprehensive understanding of perception. However, in the present work, it is assumed that a unified view of the physical, biophysical, neurophysiological, and psychophysiological aspects of perception should be pursued after, and be guided by, ecological

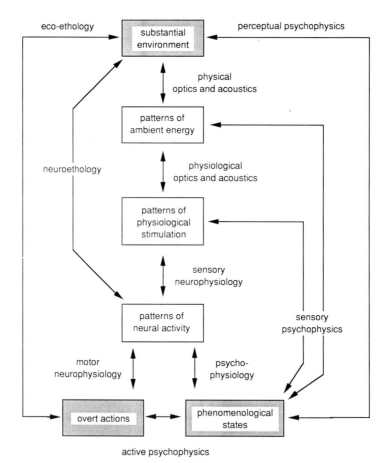

Figure 12.1 Multiple mappings in the study of perception and action. The boxes represent measurable aspects of the animal or the environment. Next to the arrows connecting the boxes are typical names for the corresponding areas of study. The shaded boxes represent observables in the present work.

analyses and investigations. For example, it is assumed that ecological principles will have much greater explanatory value and generative power for neurophysiology than neurophysiological principles will have for the the study of purposeful interactions of an animal with its environment (cf. Kugler & Turvey, 1987).

An Ecological Approach

In ecological approaches to perception and action, interactions between an animal and the environment are considered to be fundamental (J.J. Gibson, 1979; see also Beek & Bingham, 1991; Dainoff & Mark, 1987; Flach, 1990; E.J. Gibson & Schmuckler, 1989; Kondo, 1985; Kugler & Turvey, 1987; Loveland, 1991; Morbeck, Preushoft, & Gomberg, 1979; Newell, 1986; Owings & Coss, 1991; Riccio, in press; Riccio & Stoffregen, 1988; Robinson, 1972; Shaw & Kinsella-Shaw, 1988; Shaw, Kugler, & Kinsella-Shaw, 1990; Tobias, 1982; Valenti & Good, 1991; Vicente & Rasmussen, 1990; W.H. Warren, 1984). This is not to say that an ecological approach focuses on a subset of an animal's overt behavior. The medium, substances, and surfaces that make up the terrestrial environment always have consequences for movement (J.J. Gibson, 1979; see also Stoffregen & Riccio; 1988). Thus, to study movement independently of the environment is to ignore general constraints on movement. An ecological approach assumes that movement science should develop from the most general principles (e.g., Kugler & Turvey, 1987). One way to discover general principles in movement science is to study the most pervasive characteristics of the interaction of an animal with its environment. Postural behavior is arguably the most pervasive aspect of this interaction, and, consequently, it is the starting point of the present work (E.J. Gibson et al., 1987; Riccio & Stoffregen, 1988) and of work in related fields (Kondo, 1985; Morbeck et al., 1979).

A central assumption in the present work is that the most important and pervasive aspects of the animal-environment interaction are the *affordances* of postural behavior for perception and action, that is, the functional consequences that body configuration and stability have for the pickup of information or the achievement of overt goals. It follows that an essential characteristic of postural behavior is the effective maintenance of the orientation and stability of the sensory and motor platforms (e.g., head or shoulders) over variations in the animal, the environment, and the task. Furthermore, it has been argued that this requires perceptual sensitivity to the functional consequences of body configuration and stability; animals should perceive the relation between configuration, stability, and perception or action performance (Riccio & Stoffregen, 1988). It has also been argued that the *topological*, rather than the metrical, properties of these relations are important (Riccio & Stoffregen, 1988; cf. Kugler, Kelso, & Turvey, 1980; McGinnis & Newell, 1982). Body configuration or movement must change adaptively when variations in the animal, environment, or task change this functional topology. However, topological properties may persist over variations in the animal, environment, or task. Body configuration or movement need not change adaptively when there is persistence of the functional topology; that is, body configuration and movement must be *robust*, or insensitive to functionally inconsequential variations in the animal, the environment, and the task (see Riccio & Stoffregen, 1988, 1991).

The animal-environment interaction is a *dynamical system*, and the functional topological characteristics of this interaction are the dynamics of the system (see, e.g., Riccio & Stoffregen, 1988; Riccio, in press). It is important to distinguish the dynamics of a system from the movements of a system. Movements are observable "outputs" of system components; and movement variability is generally defined in terms of particular parameters in these observable outputs. Dynamics are the cause-effect relations within the system (e.g., the relation between the "inputs" to and "outputs" from the components). Inputs to system components can be either information or energy from other system components. If change in one component of a system causes change in another component of the system, there is a dynamical relation between the observable outputs of these components; the outputs of one component are inputs to the other component. Dynamical relations between two components of a system can be complex in that they can be influenced or mediated by other components of the system and by other systems. This is essentially the case when the inputs to one component are based on an "evaluation" of the outputs of another component with respect to particular functional criteria or objectives (see Riccio, in press). The complexity and variability of these multiple and partial relations can frustrate the analysis of adaptive systems. In the present work, phenomenology improves the external validity of the experimental investigations, and it helps isolate essential characteristics of complex systems. It should be noted that there is considerable disagreement about the role of phenomenological methods in ecological psychology and movement science. However, it is an important part of the ecological approach described in this chapter (cf. Runeson, 1977; Ryan, 1938; Thinés, Costall, & Butterworth, 1991).

Exploration and Adaptation

Robustness of postural behavior gives animals the freedom to adopt orientations, configurations, and movements that are not optimal with respect to particular task-relevant criteria (e.g., stability) and, thus, allows them to generate exploratory variation in postural behavior (Riccio & Stoffregen, 1988; cf. McCollum & Leen, 1989; Thelen, 1985). Exploratory behavior generates stimulation, and the obtained stimulation is "textured" by the dynamics of the animal-environment interaction (Riccio, in press). The pickup of information in exploratory behavior (e.g., movement variability) by active perceivers has been demonstrated in a variety of paradigms (Barac-Cikoja & Turvey, 1991; Beek, 1989; Beek, Turvey, & Schmidt, 1992; Bingham, Schmidt, & Rosenblum, 1989; Chan & Turvey, 1991; E.J. Gibson, 1988; E.J. Gibson et al., 1987; Mark, Balliett, Craver, Douglas, & Fox, 1990; Newell, Kugler, van Emmerik, & McDonald, 1989; Newell & McDonald, 1992; Solomon, 1988; Solomon & Turvey, 1988). In control-systems terminology, exploratory behavior provides for *persistent excitation* of the perception-action systems, which is an important characteristic of *adaptive control* (cf. Canudas de Wit, 1988; Chalam, 1987; Narendra, 1986; Riccio & Stoffregen, 1988). Excitation (or stimulation) is persistent (and thus affords adaptation) to the extent that it spans the range of the animal-environment state space in which there is functionally relevant variation (dynamical variability). If stimulation spans the entire range of states over which dynamical variability occurs, then it is *sufficiently rich* to specify these functionally relevant variations and, consequently, to allow for adaptive control (Canudas de Wit, 1988; Chalam, 1987; Narendra, 1986).

The characteristics of the animal-environment state space that are functionally relevant must be identified before hypotheses about informative exploration can be formulated and tested. The present work focuses on (a) the orientations and configurations for which perception and action are optimal (*attractors*), and (b) the limiting orientations and configurations (*separatrices*) within which perception and action can be maintained without a qualitative change in postural behavior (Bailey, 1991; E.J. Martin, 1990; Riccio et al., 1992; Riccio & Stoffregen, 1988, 1991). These directional (i.e., orientation-dependent) constraints on posture and orientation describe what is explored. Functionally relevant variation in these aspects of the animal-environment system should also be identified before exploratory behavior is investigated. The present work is based on a broader program of research on variations in attractors and separatrices that are due to (a) locomotion, (b) objects that are carried by an animal, (c) mechanics of the support surface, and (d) looking at or manipulating objects (Riccio, in press; Riccio & Stoffregen, 1988, 1990, 1991; Stoffregen & Riccio, 1988, 1991; see also Kondo, 1985; Morbeck et al., 1979; Robinson, 1972; Tobias, 1982). This dynamical variability is the reason for exploratory movement variability.

DIRECTIONAL CONSTRAINTS ON BEHAVIOR

The remainder of this chapter presents, and describes the motivation for, four experimental paradigms that have been developed for the ecological study of directional constraints on posture and orientation. Although the research focuses on perception of these constraints, it is also relevant to research in movement variability. The hypotheses in these experiments assume that movement variability provides information about the dynamics of posture and orientation; that is, the characteristics of movement variability are assumed to be lawfully related to the underlying dynamics. This assumption is implicit in research on movement variability conducted within the biomechanics and motor control communities. In these communities, characteristics of movement variability, and factors affecting these characteristics, are generally used to develop models about persistent properties of movement systems: Movement variability informs the scientist about the dynamics of the movement system. In the work presented in the following sections, it is further assumed that movement variability can inform individuals about the dynamics of their own movement systems and, thus, that movement variability can have a functional role in adaptive systems.

Dynamics of Balance

Pendulum dynamics in general, and balance in particular, are pervasive dynamical properties of the interaction between an animal and the terrestrial environment (Kugler & Turvey, 1987; Stoffregen & Riccio, 1988). The direction of balance is generally determined by the vector sum of gravitational force and inertial forces due to acceleration, and it is contraparallel to the direction of this "gravitoinertial force" vector. When the orientation of the body deviates from the direction of balance, torque is produced by the nonalignment of gravitoinertial and support-surface (resistive) forces. Thus, alignment with the direction of balance minimizes the torque or effort required to maintain a particular orientation. However, this does not mean that animals necessarily align with the direction of balance. The goals of perception and action often require that animals achieve other orientations or configurations (Riccio & Stoffregen, 1988). Nevertheless, orientation with respect to the direction of balance always has consequences for control. The magnitude and variations of the torque acting on the body are specific (lawfully related) to the orientation of the body with respect to the direction of balance (Figure 12.2, a-d; cf. Stoffregen & Riccio, 1988). The orientation of the body relative to the direction of balance is specified by patterns of body movement together with the actions required to resist such movements. Thus, the perception of orientation could be based on the *dynamics* of balance, which are always relevant to interactions with the environment.

Classical theories of orientation assume the comparison of body axes to some (usually static) external reference frame such as gravity or anisotropic patterns in optical stimulation (Howard, 1982; Schöne, 1984). This is problematic, because such reference frames are not necessarily relevant to the animal-environment interaction. For example, the direction of gravity is irrelevant to orientation underwater when the body's center of mass is at or near its center of buoyancy. This severely attenuates the effects of gravity on body dynamics; that is, orientation-dependent constraints on control of the body are minimal. In fact, the perception of orientation is notoriously poor underwater, where errors in pointing "up" can be as large as 180°. Perception of orientation by unrestrained animals is much better on land (see Riccio & Stoffregen, 1990). The critical difference on land is that it matters whether or not segments of the body (e.g., head, torso) are balanced with respect to the coupled gravitoinertial and support-surface forces (Riccio & Stoffregen, 1988; Stoffregen & Riccio, 1988). In an ecological approach, the direction of balance is the fundamental referent for the perception and control of orientation (cf. the discussion of the "behavioural vertical" in Roberts, 1978). This does not require the perception of gravity or gravitoinertial force (Riccio & Stoffregen, 1990; Stoffregen & Riccio, 1988).

The Active Perception of Orientation

In general, the direction of balance is primarily determined by the gravitoinertial-force vector. However, with sufficiently powerful laboratory devices, the two directions can be decoupled. Such a device, a roll-axis tracking simulator (RATS), was used to determine whether perceived orientation is influenced by balance dynamics (Riccio et al., 1992). Perception of orientation was evaluated in the context of a task that required subjects to control the roll orientation of a device in which they were seated. They were told to maintain the RATS in an upright orientation while the device was exposed to a continuous disturbance. This is important given the premise that the perceived upright should be intimately related to the act of orienting

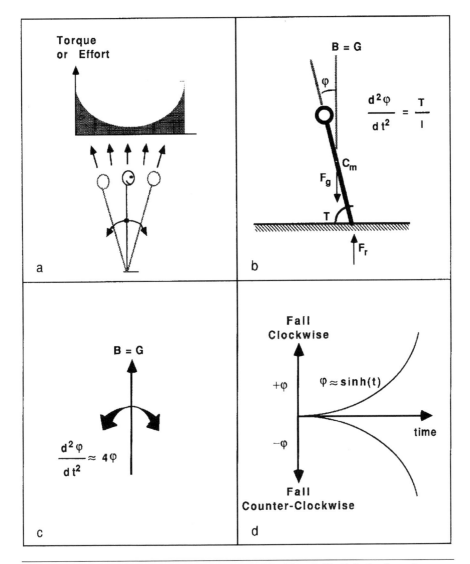

Figure 12.2 The dynamics of falling are depicted in (a) and (b). F_g is the force due to gravity, C_m is the center of mass, I is the moment of inertia of the body, T is the torque on the body, B is the direction of balance, G is the direction of the gravity vector, φ is the orientation of the body with respect to B. The kinematics of falling are depicted in (c) and (d), where t is time and $d^2\varphi/dt^2$ is the second derivative of φ with respect to time. The constant of 4 in the lower left panel is reasonable given the dimensions of the human body and assuming a moderate amount of passive viscous damping in the musculoskeletal system.

Note. From "The Role of Balance Dynamics in the Active Perception of Orientation" by G.E. Riccio, E.J. Martin, and T.A. Stoffregen, in press, *Journal of Experimental Psychology: Human Perception and Performance*, **18**(3), pp. 624-644. Copyright 1992 by the American Psychological Association. Reprinted by permission.

325

(Riccio & Stoffregen, 1990; Stoffregen & Riccio, 1988). The RATS had a balance point and inherently unstable dynamics; this meant that there were orientation-dependent constraints on control. The direction of balance was experimentally manipulated across trials, so that it was independent of the direction of gravity (Figure 12.3, a and b). This experimental manipulation produced dynamical variability (across trials) that was functionally relevant for the control of balance. After each trial, subjects verbally estimated their mean tilt with respect to "upright." Partial correlations between perceived tilt and measured tilts permitted the independent effects of balance and gravity to be assessed. The findings indicate that balance can have the predominant influence on the perception of upright. Information about balance is available only in the movements used to balance the RATS. The informativeness of these movements is inconsistent with classical theories of orientation in which movement relative to an external reference frame (e.g., gravity) would have to be viewed as "noise" that would interfere with the perception of orientation. The findings are more consistent with the assumption that perception is linked to the dynamical interaction of an animal with its environment.

The arguments presented above emphasize that perception of orientation (a direction) is meaningful only when there are orientation-dependent (directional) constraints on control. The experiments of Riccio et al. (1992) suggest that the perception of orientation with respect to balance is important because it supports the maintenance of balance. If control of orientation were precluded, as in most studies of perceived orientation, the perception of orientation would have been irrelevant and unnecessary

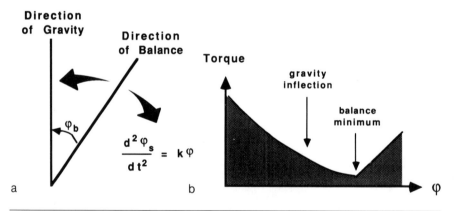

Figure 12.3 (a) The direction of balance was the primary manipulation in this study. φ_b is the angle between the directions of balance and gravity. $d^2\varphi_s/dt^2$ is the angular acceleration of the RATS. The expanding arrows indicate that the dynamics of the RATS were influenced primarily by the balance point. (b) Gravity may have had an effect on any unrestrained parts of the body. This effect is represented by an inflection point in the relation between torque and tilt.

Note. From "The Role of Balance Dynamics in the Active Perception of Orientation" by G.E. Riccio, E.J. Martin, and T.A. Stoffregen, in press, *Journal of Experimental Psychology: Human Perception and Performance*, **18**(3), pp. 624-644. Copyright 1992 by the American Psychological Association. Reprinted by permission.

(Riccio & Stoffregen, 1990). However, there is another reason why it is important to allow subjects (i.e., perception) to be active in ecological investigations. Activity provides persistent excitation of the perception-action systems that may be necessary to pick up information about the dynamics of the animal-environment interaction. Movement variability produces stimulation that is textured by the dynamics of this interaction. For example, the movement or moveability of the animal may become increasingly asymmetrical with increasing tilt from the direction of balance. Such asymmetries could provide information about orientation. This hypothesis has been initially tested by injecting asymmetrical disturbances into the RATS using the paradigm described above (Riccio et al., 1992). Preliminary results indicate that the perception of orientation is systematically related to manipulations of asymmetry (Riccio & E.J. Martin, 1990). In addition, a yoked-control experiment indicated that passive observers had difficulty picking up information about orientation with respect to the direction of balance. Balance dynamics had less influence on perception when passive observers were subjected to the motions that active subjects produced to control the orientation of the RATS (Riccio & E.J. Martin, 1990). This suggests that information in movement variability (i.e., in movement of the RATS) is more accessible if the movement is produced by an active perceiver. In this paradigm, it seems that perception supports action and action supports perception.

Implications for Movement Science

The results just summarized indicate that movement variability can inform individuals about the dynamics of their own movement systems. This suggests caution in the use of models that assume that movement variability is noise in the system. Noise, by definition, is neither informative nor controllable. However, if movement variability is informative, it would be adaptive for animals to modify the characteristics of variability in order to facilitate information pickup. Modification or control of movement variability may be as simple as increasing (or not minimizing) the magnitude of movement so that *patterns* of movement are more salient. In addition, if patterns of movement (e.g., asymmetry) are more salient in particular regions of the state space (e.g., for particular orientations), it may be adaptive to occupy or move toward these regions even if they are not the most energy-efficient states. Evidence for systematic bias away from energy minima was obtained in the experiment described in the previous section (Riccio et al., 1992) and has been obtained in experiments by other investigators (Beek et al., 1992). In these experiments, systematic bias apparently improved the observability of the system dynamics. In any case, the informativeness and controllability of movement variability should be included in models of the movement systems. Moreover, controllability implies that the characteristics of movement variability may be different for two tasks if different information about systems dynamics is required (perhaps implicitly) by these tasks. This suggests caution in generalizing from particular experiments on, and models of, movement variability, unless the task-specific constraints on movement variability are understood.

Maintenance of Orientation

The instructions in the experiments of Riccio et al. (1992) emphasized the perception of upright. The perception of upright was strongly influenced by the direction of balance, which would be an attractor with respect to the minimization of compensatory torque or effort required to maintain control (see Figures 12.2 and 12.3). However, informal observations in the RATS suggested that the threat to balance was more salient than the direction of upright. The threat to balance presumably reflected one's proximity to the limits for which orientation could be maintained (for which perturbations could be reversed). Furthermore, a model for the data on the control of orientation also suggested that the orientation limits had an effect in these experiments (Riccio et al., 1992). More generally, it has been suggested that sensitivity to control limits is important for the robust and adaptive control of posture and orientation (Riccio & Stoffregen, 1988, 1991; E.J. Martin, 1990). Sensitivity to such limits insures that performatory and exploratory behavior remains bounded or stable. Such limits specify the domain of variations in the dynamics of the animal-environment system over which exploratory behavior is sufficiently rich to promote adaptive control. Sensitivity to these limits allows exploratory behavior to be robust and goal directed.

The dynamics of the RATS were qualitatively, or topologically, similar to the dynamics of balance in more common situations such as stance (they were also similar in some subtle quantitative details; see Riccio et al., 1992). The most important similarities are considered to be the *global extrema*, which include the minimum for compensatory torque and the limits for maintaining orientation (see Figures 12.2 and 12.3). For these reasons, it was not surprising that control of this second-order unstable device was learned very quickly; there was apparently transfer of skill from balancing in general to balancing the RATS. However, other investigators might consider this control task to be importantly different, because of the lack of "involuntary" postural "reflexes" that are assumed to play a critical role in the maintenance of balance. In the present work, the action systems that are used to control posture and orientation, and whether or not they are consciously controlled, are not central issues. The central issue is whether or not the effectiveness of these action systems is perceivable. If the effects of a particular action system (e.g., a postural reflex) are not sufficient to achieve a desired goal, and this is perceivable, then a different strategy or action system can be selected. Action systems can be selected (or avoided) whether or not they can be consciously controlled once selected. For example, if a particular postural reflex cannot keep one from falling in a particular situation, one can change the dynamics of the response by changing body configuration (e.g., changing muscular preload by leaning, or reducing the moment of inertia by crouching), bracing oneself with the arms, or even lying down. These are *basic facts* about the adaptive control of posture that should be considered at the outset of any theorizing about postural control.

Stability Limits for Stance

The existing research on postural control has been heavily influenced by neurophysiology (see, e.g., Nashner & McCollum, 1985; Roberts, 1978). Neurophysiological models are ultimately important because they describe the mechanistic instantiation of postural control. However, the appeal to neurophysiology would be considered premature from a control-theoretic perspective. In control theory, the *functions* of a control system (e.g., as represented in a cost functional) should be specified before the components of a control system are selected, designed, or analyzed (see, e.g., Riccio, in press). Functionality either has not been addressed or has not been adequately represented in most research on postural control. An ecological approach starts with a consideration of the functions of postural control (Riccio & Stoffregen, 1988). The simplest and most basic function of postural control is to prevent an animal from falling over. Although this fact is obvious, it is noteworthy because it motivates the novel research on postural control that is described below (E.J. Martin, 1990).

A useful paradigm in posturography is to study the effects of perturbations caused by sudden movements of the support surface (see Nashner & McCollum, 1985). Effects can be measured on body movements (kinematics), reactive forces on the support surface (kinetics), or the activity of muscle groups (electromyography). These measures are interchangeable, to some extent, if one is interested in the correspondence between parameters of the perturbation (e.g., displacement, velocity, and duration) and the strength and latency of the response. Perturbation experiments indicate, for example, that the amount of muscle activity is positively correlated with the displacement and velocity of the support surface; and such data are used to develop hypotheses about the "neural organization" of postural control (Diener, Horak, & Nashner, 1988). However, it is noteworthy that the specification of the inputs (platform displacement and velocity) and the outputs (muscle activation) is not sufficient to provide information about the functional effectiveness of postural responses. For example, neither the inputs nor the outputs nor any combination of them indicate how close subjects come to falling over or changing strategy (e.g., stepping).

The experiment of E.J. Martin (1990) exemplifies a functional (ecological) approach to postural control. The task in this, and other, perturbation experiments is to maintain stance (e.g., not to fall over). For this reason, Martin started with the assumption that the fundamental perceivable was one's spatiotemporal proximity to limits for maintaining orientation (i.e., *stability boundaries*). These limits are essentially the boundaries to the region within which postural perturbations can be reversed (cf. McCollum & Leen, 1989; Riccio & Stoffregen, 1988). The reversal of postural perturbations is dependent on the ability to generate thrust at the support surface. This requires a support surface of adequate extent and rigidity as well as the generation of muscular force (Horak & Nashner, 1986; Riccio & Stoffregen, 1988; Stoffregen & Riccio, 1988). Martin measured this functionally relevant capacity in terms of the distance of the *center of pressure* (i.e., thrust) from the anterior or posterior limits of the base of support (the heels or the toes, for an extensive surface) as this distance decreased during the perturbation (Figure 12.4). Building on the work of Lee (1976), Martin assumed that the most functionally relevant measure of this decreasing distance was the *time to contact* between the center of pressure and the stability boundaries. The meaningfulness of this perceivable would ultimately be based on its relationship to time constants for the various action systems that could be employed to reverse the perturbations (cf. Stoffregen & Riccio,

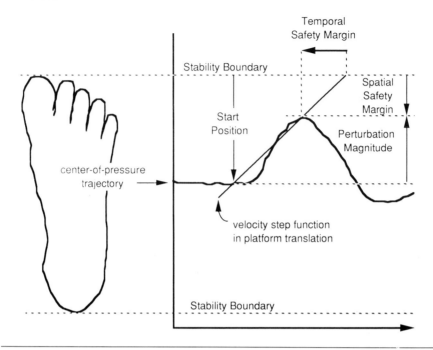

Figure 12.4   Control of orientation with respect to a postural separatrix. A typical trajectory for the center of pressure is depicted. Time is represented along the horizontal. Note that movement of the center of pressure beyond the boundary of a foot requires the use of another action system to maintain balance (e.g., support with the other foot). Derivation of functionally relevant parameters is also represented.

Note. From "An Information-Based Study of Postural Control: The Role of Time-to-Contact With Stability Boundaries" an unpublished master's thesis by E.J. Martin, 1990, University of Illinois at Urbana-Champaign. Reprinted by permission.

1990). If the time to contact were less than the time constant of a particular action (e.g., ankle torque), then another action (e.g., stepping) would have to be used to prevent falling. Because the boundaries of the foot are relevant only to the former (on an extensive surface), they separate the domain of perturbations that can be reversed by ankle torque from the larger domain within which stepping is effective. The boundaries that separate the domains of different action systems are referred to as *separatrices*, and they represent qualitative transitions in the behavior of a system that can be discovered empirically (E.J. Martin, 1990; Riccio & Stoffregen, 1988).

From an ecological perspective, stability of stance is related more closely to the proximity of the center of pressure to stability limits than to the amount of movement of the center of pressure. Thus, stance could be unstable, even with very little sway, if the center of pressure were close to the limits of the base of support. This could occur during stance on a narrow or short support surface or during leaning. Martin chose the latter as an ecologically valid experimental manipulation (subjects leaned to varying degrees until the onset of the perturbation). This allowed Martin to test his a priori hypothesis that displacement and velocity of the support surface mattered only insofar as they influenced the time to contact with stability boundaries. His findings not only supported this hypothesis, but they also showed that leaning could have a larger effect on postural responses than the parameters of platform motion. For example, subjects responded sluggishly when perturbed backward after initially leaning forward, but they responded vigorously when perturbed backward after initially leaning backward. Although this effect may not be particularly surprising, it is not accounted for in other theories and experiments on postural control. Moreover, Martin found that postural responses are organized so that perturbations are reversed at a relatively constant *temporal safety margin* (approximately 300 ms from when the center of pressure would have contacted the stability limits without compensation). The fact that people maintain a margin of safety from a postural brink (cf. J.J. Gibson, 1979) requires significant modifications in traditional assumptions about compensatory postural responses.

Implications for Movement Science

The results summarized above emphasize that measures of movement should be functional and meaningful. The functionality and meaning of movement can only be addressed by considering the context for movement. Systems do not behave in isolation. Adaptability (and intentionality) are possible only if a system is sensitive to those aspects of its surroundings that have consequences for its behavior and if it is sensitive to those aspects of its surroundings for which its behavior has consequences (Riccio, in press). Consider the action systems used to regulate postural sway (see, e.g., Nashner & McCollum, 1985). The functionally relevant surroundings of these systems could include other parts of the body (e.g., the head), objects with which the system is physically coupled (e.g., a skateboard), or objects with which the system is informationally coupled (e.g., an object of visual regard). The characteristics of postural sway have consequences for these aspects of the surroundings or the coupling of the postural system with them (falling would have the most extreme consequences). The surroundings of a particular action system (e.g., ankle-hip synergies used for postural control) also include the complementary action systems within which it is nested (e.g., stepping to avoid falling over). The behavior of a nested action system with respect to the associated separatrix is meaningful because it provides an informational basis for the selection of a more robust action system. In the absence of this ubiquitous functional context, one is left with an isolated analysis of postural sway and, by default, an assumption that the task or function of the system is to minimize sway. Such an assumption is vacuous unless one also assumes that the function of the postural system is to minimize energy expenditure. However, these assumptions are insufficient to explain why an animal would stand instead of sit, crawl, or lie down; thus, they could never lead to an adequate theory of stance. The functional context for stance is discussed in more detail below (see also Kondo, 1985; Morbeck et al., 1979; Riccio & Stoffregen, 1988; Robinson, 1972; Tobias, 1982).

Coordination of Contiguous Segments

The experiments described above suggest that attractors and separatrices play an important role as intrinsic referents for perception. That is, interactions of an animal with its environment are apparently perceived with respect to dynamical attractors and separatrices. This is believed to be a general property of the perception of movement by vestibular, somatosensory, and visual systems. Thus, these hypotheses about perception should apply when attractors, separatrices, or movement trajectories are of higher dimensionality than those in the experiments of Riccio et al. (1992) and E.J. Martin (1990). A simple extension beyond the concepts addressed in these experiments is to consider the properties of a two-segment system (Figure 12.5, a and b; cf. Nashner & McCollum, 1985; Riccio & Stoffregen, 1988).

The emergent properties of multilink systems involve the coordination of contiguous segments. Coordination is necessary because contiguous segments exert forces against one another, and these forces are influenced by the orientation and motion of the segments. That is, the orientation and motion of one segment has consequences for the control of the orientation and motion of a contiguous segment. The torque or effort required to maintain various configurations of a two-segment system is described by an (emergent) interaction between the orientations of each segment (Figure 12.5, a and b; Nashner & McCollum, 1985; cf. Krinskii & Shik, 1964).

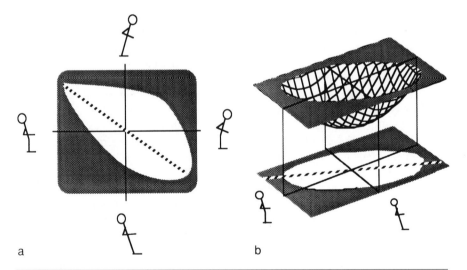

a   b

Figure 12.5   Two-dimensional (a) and three-dimensional (b) representations of a configuration space for postural control. The darkly shaded area represents hip-ankle configurations for which bipedal stance is not possible. The relation between postural configuration and a particular evaluation metric (e.g., torque) is explicitly represented in the three-dimensional space. The dotted line in the two-dimensional projection represents the shallow gradient in torque due to bending without leaning.

Note. From "The Organization of Human Postural Movements: A Formal Basis and Experimental Synthesis" by L.M. Nashner and G. McCollum, 1985, *Behavioral and Brain Sciences*, **8**, pp. 135-172. Copyright 1985 by Cambridge University Press. Adapted by permission.

There is a shallow gradient along the locus of configurations over which torques on the two segments tend to counterbalance each other, whereas there is a steep gradient along an orthogonal locus over which the torques tend to add. The orientations of both segments must be specified in order to describe this functional topology. As in the single-segment system, there is a separatrix for the maintenance of balance in the two-segment system that is partially determined by "contact" between the center of pressure and the limits of the base of support (Riccio & Stoffregen, 1988). The shape of this boundary is also influenced by (emergent) interactions between segments such as the range of motion at the joints. The consequences of movements (change in orientation or configuration) are defined by the topology of the configuration space (Riccio & Stoffregen, 1988). The system is relatively free to change orientation or configuration in the vicinity of the global minimum and, to some extent, along the shallow gradient. Changes in orientation or configuration are much more limited near the separatrix if stability is to be maintained without a change in dimensionality of the system's behavior (e.g., the use of additional segments to balance). Sensitivity to the functional topology of such configuration spaces is required for the stability of performatory and exploratory behavior.

Time Scales and Dual Control

Exploratory behavior is important because of variation in the functional topology of an animal-environment interaction. Common variations include changes in the location, size, and shape of the basin represented in Figure 12.5 caused by variations in velocity of locomotion and by fatigue, injury, or carried objects. (Variations in the functional topology that are caused by variations in the characteristics of the support surface require additional dimensions in the representation; see Riccio & Stoffregen, 1988). Stability can be maintained over such dynamical variations if they are more gradual than the rate at which an animal can adapt to the variations (see Figure 12.6, a-c; see Riccio & Stoffregen, 1991). Adaptation is dependent on the pickup of information. The experiments described earlier indicate that animals can pick up information about the functional topology of the interaction with the environment (i.e., postural attractors and separatrices). This information seems to be picked up primarily through movement that stimulates the proprioceptive systems. This suggests that attractors and separatrices provide texture (or structure) to stimulation. Such structure may include parameters defined over movement variability (Riccio et al., 1992; Stoffregen & Riccio, 1988).

Both muscular action (inputs to an action system) and variations in the animal-environment system (dynamics of the system) can cause changes in orientation and configuration (outputs of the system). For example, one would tend to fall over if (a) one leaned away from the direction of balance while standing (muscular initiation of falling) or (b) one did not lean into a turn during vehicular motion (falling initiated by change in the direction of balance). Information about falling provides feedback about postural actions (e.g., leaning) or variation in the animal-environment system (e.g., the direction of balance). Thus, there is a potential ambiguity in feedback about movement, which would frustrate an attempt to identify variations in the animal-environment system through exploratory behavior (Riccio, in press). The

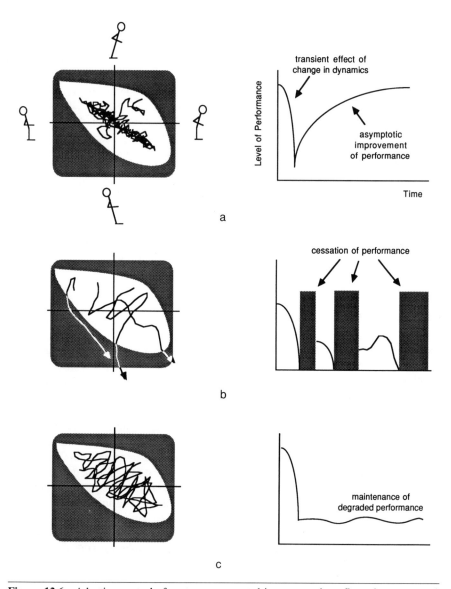

Figure 12.6 Adaptive control of posture represented in a postural configuration space and an associated performance time-history (hypothetical). Performance might be described, for example, in terms of the magnitude of perturbations for which stance can be maintained. Performance drops after a change in the dynamics of interaction with the environment.
(a) The original performance level is approached asymptotically as the animal discovers the control actions that are appropriate for the new dynamics. (b) Performance may be terminated if adaptation is not quick enough (in this case, cessation of performance results in a fall). (c) Performance may be maintained at a degraded level, that is, without adaptation.
Note. From "An Ecological Theory of Motion Sickness and Postural Instability" by G.E. Riccio and T.A. Stoffregen, 1991, *Ecological Psychology*, **3**(3), pp. 195-240. Copyright 1991 by Lawrence Erlbaum Associates. Reprinted by permisison.

ambiguity can be reduced if exploratory behavior (e.g., movement variability) occurs on temporal scales that are smaller than the temporal scale for variations in the animal-environment system (dynamical variability). It is also desirable that exploratory behavior not interfere with performatory behavior; that is, it should not result in loss of control (Figure 12.6; Riccio & Stoffregen, 1991). This *dual-control* problem could be minimized if exploratory behavior occurs on spatial and temporal scales that are smaller (i.e., of lower amplitude and higher frequency) than those characteristic of performatory behavior (cf. Wiener, 1948).

A pervasive source of high-frequency low-amplitude movement variability is muscle tremor (Akamatsu, Hannaford, & Stark, 1986; Stein & Oguztoreli, 1976). Such tremor exists in postural movements in the range of 8 to 12 cycles per second (Lippold, 1970; Mori, 1973), which is an order of magnitude higher than the range of controllable postural frequencies (Johansson, Magnusson, & Akesson, 1988; Maki, 1986). It is possible that low-frequency modulation (e.g., variation in amplitude, frequency, or symmetry) of the high-frequency variability provides information (i.e., feedback) about low-frequency postural dynamics. Because data on high-frequency variability is commonly filtered out in research on postural control, its functional role in such behavior is unknown. A study by Watanabe, Yokoyama, Takata, and Takeuchi (1987) is an interesting exception. The data reveal a negative correlation between the magnitudes of high-frequency and low-frequency variability in the center of pressure during bipedal stance. Because stability is primarily dependent on the relatively high-amplitude sway at low frequencies, the data suggest that tremor (high-frequency variability) promotes stability. An implication of this hypothesis is represented in Figure 12.7. Enhanced postural tremor is hypothesized for orientations or configurations that are close to the separatrix because of the effort in leaning or bending. The tremor could promote stability (reduced sway), where the consequences of high-amplitude sway would pose the greatest threat to balance, by providing informational support for controllable (low-frequency) changes in orientation and configuration. The role of postural tremor in proprioception could be analogous to the role of eye tremor in vision. These sources of movement variability seem to provide ubiquitous exploratory behavior that enhances perception without interfering with the controllable movements that are both larger and slower (cf. J.J. Gibson, 1979; Howard, 1982). Perception of the functional topology for postural control, facilitated by postural tremor, could increase coordination among the segments of the postural system (cf. Newell et al., 1989; Newell & McDonald, 1992).

Multiple Evaluation Functions

The functional topology of the interaction between an animal and the environment has been discussed primarily in terms of the mapping between orientation or configuration and torque or effort (Figures 12.2, 12.3, 12.5). However, it is implicit in the preceding discussion that orientation and configuration can be evaluated with respect to (can be mapped into) other parameters (Figures 12.6, 12.7). For example, orientations and configurations could be evaluated with respect to the likelihood of falling or stepping, the magnitude of movement, or the asymmetry of movement. The

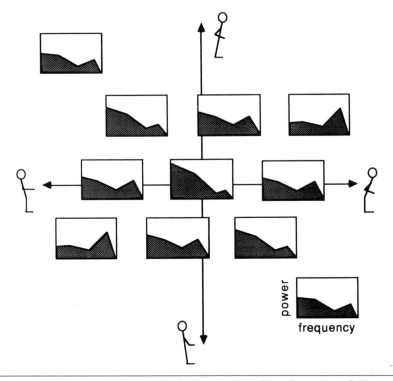

Figure 12.7   Time scales for movement in a postural configuration space (cf. Figures 12.2, 12.3, 12.5, and 12.6). Insets represent hypothetical frequency spectra of postural sway for various orientations and configurations. Bimodal spectral distributions are typical for postural sway. One mode represents controllable low-frequency movements, and the other represents relatively high-frequency physiological tremor.

topology of these mappings (e.g., the shape and location of the basins in a configuration space) could be different for different parameters or *evaluation functions* (Figure 12.8, a-d). The parameters that are relevant are determined by the tasks in which the animal is engaged (Riccio & Stoffregen, 1988). Performance on a particular task (e.g., visual tracking) will be sensitive to some postural parameters (e.g., magnitude of movement) and not others (e.g., likelihood of stepping); a different set of parameters (e.g., effort) might be relevant to a different task (e.g., prolonged standing). This is important because it means that the functional topology for postural control can vary across the plethora of tasks for which postural configuration and stability have functional consequences (Riccio & Stoffregen, 1988, 1991). Thus, variations in the animal, the environment, *and the task* must be considered in a comprehensive approach to postural control (Riccio & Stoffregen, 1988; cf. Beek & Bingham, 1991; Newell, 1986; Saltzman & Kelso, 1987). This dynamical variability can frustrate the analysis of information in movement variability, because it may be difficult to identify the dynamical characteristics to which patterns of movement variability are specific. Phenomenological methods can facilitate the identification

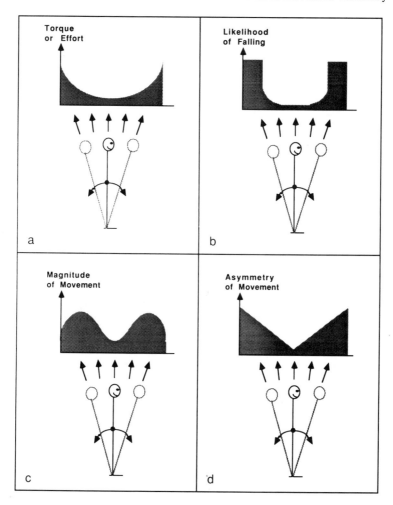

Figure 12.8 Hypothetical relations between orientation and various parameters (a-d) on which orientation could be evaluated. Note that the ''shapes'' of the evaluation functions can vary.

of the appropriate situation-specific evaluation function and the associated functional topology. That is, phenomenology can provide insights about the meaning of particular movement patterns.

Relations Between Configuration and Stability

An ecological approach to the perception and control of posture and orientation has been presented in the preceding sections in the context of theory and experiments. However, one experiment involved highly constrained control of orientation in a laboratory device (the RATS), and the other involved an activity (standing) with which subjects had a lifetime of experience. Although there are strong theoretical reasons for believing that these experiments have considerable generality, it could be argued, on purely empirical grounds, that they are special cases. The power of an ecological approach should also be demonstrable in situations that involve the development of *skilled* behavior (cf. Newell, 1986) and are relatively unencumbered by contrived experimental methods. This has been done in an investigation of skilled and unskilled performance in a cheerleading stunt (Bailey, 1991). In the stunt (the "liberty lift"), one cheerleader stands, on one foot, on the hands of another cheerleader (the other leg is retracted by raising the thigh and flexing the knee). One cheerleader (the "base") holds the other (the "top") directly overhead by supporting the toe of the top's shoe with one hand and the heel of the shoe with the other. One difficulty of investigating such a complex behavior is that the relevant performance measures are not transparent to the investigator. Consequently, it is important to exploit loosely constrained interactions between the investigator and a skilled participant prior to the quantitative phase of the investigation. Bailey (1991) reflected on his own knowledge of, and experiences in, the stunt. He also interviewed other cheerleaders to determine the essential aspects of skilled performance in the stunt. It was generally acknowledged by the elite cheerleaders in this study that the perception and control of the top's configuration and stability was essential for both the base and the top. In addition, the relation between configuration and stability was considered to be critical for skilled performance. Beginners who were able to perform the stunt (but not in a skilled, stable, or aesthetic manner) could not differentiate good and bad configurations, and they did not understand the relation between configuration and stability. It was generally acknowledged that these aspects of the stunt were difficult to teach and to learn. For these reasons, Bailey's investigation focused on the perception and control of configuration and stability.

An ecological approach to postural control provided the theoretical foundation for Bailey's investigation (Riccio & Stoffregen, 1988). An hypothesis was that the optimal configuration of the top, referred to as "hollowing out," was one in which the constellation of forces acting on the top were at *equilibrium*, and that this equilibrium configuration would appear relatively stable and effortless (aesthetic). The optimal configuration is determined primarily by gravity and the forces exerted by the base to maintain balance of the top. The base is responsible for the fine control by pulling down on the heel or the toe of the top. This control is characterized by movements on small spatial and temporal scales. The role of the top is to "stay tight" (be relatively stiff), which presumably helps to suppress fine postural control movements that would interfere with the control by the base (an important but difficult aspect of the stunt for the top is allowing the base to do most of the balancing). However, the top has coarse control of the stunt, through adjustments in configuration, because some configurations are inherently more stable than others. The interviews suggested that stability is evaluated, by both the base and the top, with respect to the movement of the foot (primarily dorsiflexion and plantarflexion). Thus, movement of the foot provides the informational basis for fine control by the base, coarse control by the top, and coordination between the two (i.e., dynamical coupling between the base and the top).

The relation between configuration and stability was justified on both theoretical grounds (Riccio & Stoffregen, 1988) and empirical grounds (reflection, observation, interviews). *Perception* of configuration, stability, and the relation between the two is also suggested. Stronger evidence that these aspects of the stunt are perceived is provided by the *adaptation* in the configuration that allows the stunt to be achieved in novel conditions. Bailey found that configurations changed systematically in both the beginners and the experts when the base modified the dynamics of the stunt by pulling excessively either on the heel or the toe (dynamical variability), although the beginners were more variable in each condition (Figure 12.9, a and b). Bailey hypothesized that configurations were modified by the top so as to achieve stability and equilibrium, and that the perception of stability and equilibrium was based on parameters defined over variability of foot movement (movement variability). In other words, information in movement variability facilitated the adaptation in body configuration that was required by dynamical variability in the stunt. In order to test this hypothesis, the angles of the upper body, the lower body, and the foot with respect to a vertical reference line were measured through frame-by-frame analysis of videotapes. The angles were sampled 10 times per second. The data were then separated into two time scales: variations that were faster than once per second, and variations that were slower than once per second. The high-frequency variations in the foot angle were assumed to reflect movements that were not controllable by the top. Presumably they would arise from muscle tremor or manual control by the base. The low-frequency variations in the upper and lower body angles were assumed to reflect controlled movements by the top. The low-frequency variations in the

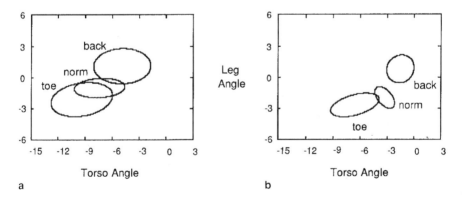

Figure 12.9   Postural configurations observed when toe is pulled down (toe), when heel is pulled down (back), and under normal conditions (norm) in cheerleading stunt (based on actual data). Ellipses represent 50% bivariate confidence regions in three experimental conditions for a beginner (a) and an elite (b) cheerleader. Numbers on axes indicate mean degrees from vertical. Note that posterior tilts (negative numbers) increase to compensate for downward pull on the toe.
Note. From "Topological Dynamics and Performer's Conceptions of Coordination, Control, and Skill in a Cheerleading Stunt," an unpublished honors thesis by M.A. Bailey, 1991, University of Illinois at Urbana-Champaign. Adapted by permisison.

foot angle were assumed to reflect manual control by the base and the effects of variations in configuration (upper and lower body angles).

Stability was defined as the standard deviation of the foot angle for each second of data (i.e., for each set of 10 data points). Although stability was considered to be a characteristic of equilibrium (cf. Riccio & Stoffregen, 1988), equilibrium was also defined with respect to the skewness (asymmetry) of the foot angle for each second of data. Nonequilibrium states (falling forward or backward) would be characterized by movements that were larger or more frequent in plantarflexion or dorsiflexion. The standard deviation and skewness for each second of data were compared to the mean upper body angle and the mean lower body angle for each second of data. Two sets of relations defined the topology of the configuration space. In one, the upper and lower body angles were evaluated in terms of the standard deviation of the foot angles. In the other, the body angles were evaluated in terms of skewness of the foot angles. The relations were described with quadratic *response-surface regression* (see Box & Draper, 1987). The results revealed significant effects of postural configuration on the standard deviation of foot angle: *Multiple R* = .766, $F(5,24) = 6.80$, $p < .001$; and on the skewness of foot angle: *Multiple R* = .603, $F(5,24) = 2.74$, $p = .043$ (quadratic relations are depicted in Figure 12.10, a and b).

The coordinated changes in body angles and the associated changes in parameters of foot movement, within each trial, are represented by the trajectories in Figure 12.10. These trajectories indicated that the cheerleaders *symmetrized* the movements of the foot; that is, the top moved to a configuration for which movements of the foot were symmetrical. The relation between configuration and the standard deviation of foot movement was generally saddle shaped, and trajectories were attracted to the seat of the saddle. This means that cheerleaders did not tend to minimize variability of foot movement. Minimal variability might occur in states, such as leaning, where the body is especially stiff. Such states may not be very robust to perturbations because of the proximity to the separatrix, and it may not be possible to maintain them for very long. The cheerleaders tended to reduce variability to, but not below, a level that was associated with symmetrical movements. This suggests that a certain amount of variability may be necessary to notice an asymmetry in movement. Both the beginners and the elite cheerleaders symmetrized movement, but apparently the beginners required more variability in order to perceive symmetry. Because symmetry is a general property of equilibrium movements (Riccio & Stoffregen, 1988; Stoffregen & Riccio, 1988), it should be a familiar criterion in novel tasks. However, the movements for which symmetry is relevant can be task specific. Beginners may not be as perceptually sensitive as elite cheerleaders to symmetry in movements of the foot. The exploratory data from this study, together with the preliminary data on perception of orientation in the RATS (Riccio & E.J. Martin, 1990), suggest that symmetry of movement variability is an important parameter in the perception and control of orientation and configuration.

Implications for Movement Science

In the experiment of Riccio et al. (1992) the direction of balance was known, a priori, to be a key topological feature (attractor) in the dynamics of the system. In

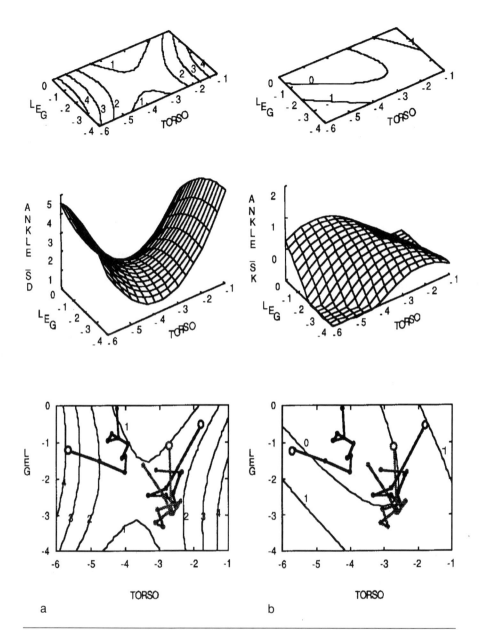

Figure 12.10 Quadratic response surfaces that describe how postural configuration influences movement variability (a) and symmetry (b). Isovariability and isoasymmetry contours are also presented. Trajectories for each 10-s trial are represented in the lower contour plots. The initial configuration for each trial is indicated by an open circle. Negative numbers indicate posterior tilt of body segment.

Note. From "Topological Dynamics and Performer's Conceptions of Coordination, Control, and Skill in a Cheerleading Stunt," an unpublished honors thesis by M.A. Bailey, 1991, University of Illinois at Urbana-Champaign. Adapted by permission.

the experiment of Martin (1990) the anterior and posterior boundaries of the feet were known to be key topological features (separatrices). In other situations, such as the cheerleading stunt described above, the functional topology is largely unknown. Bailey's (1991) exploratory experiment demonstrates that the functional topology of an action system can be revealed in patterns of movement variability. The relation between these patterns and the underlying dynamics is analogous to the relation between patterns in the optic array and the substantial environment. The environment structures the ambient light, and the environment is revealed in this structure (J.J. Gibson, 1979). Similarly, the dynamics of an action system structure the movements of the system, and the dynamics are revealed in this structure.

A variety of analytical techniques can be used to identify structure in the optic array or in movement. Examples include Fourier analysis, autocorrelation, cross-correlation, analysis of dimensionality, and Gestalt methods. Such ad hoc pattern-recognition techniques provide little, if any, information about the sources of structure. A qualitatively different approach in visual science (or movement science) is to identify the source of structure in the optic array (or movement). In visual science, this *informational* approach is exemplified by the identification of the optical structure that reflects the layout of the environment (e.g., J.J. Gibson, 1979). Examples of informational (or *systems identification*) approaches in movement science include identification of movement patterns of a wrist-pendulum system, and the concomitant tissue strains, that relate to the inertia tensor of the pendulum (Solomon & Turvey, 1988); frequency-amplitude patterns that relate to the dynamical coupling between two wrist-pendulum systems (Kugler & Turvey, 1987); and responses of systems to punctate disturbances (Kelso, Tuller, Vatikiotis-Bateson, & Fowler, 1984) or continuous disturbances (Johansson et al., 1988).

An informational approach to the analysis of structure in movement generates hypotheses about specific patterns of movement and their relation to the underlying dynamics. For example, symmetry is an important property of systems in equilibrium (Onsager, 1931), and the magnitude of asymmetry is a plausible metric for distance from equilibrium in near-equilibrium systems (Stoffregen & Riccio, 1988). Bailey's (1991) data suggest that the *effect* of distance from equilibrium on asymmetry of movement is very robust, in that it can be revealed in the *skewness* of a relatively small amount of data. The use of skewness as a dependent measure in movement science is not unprecedented (see Newell & Hancock, 1984), but it may be considered too unreliable for general use. Bailey's (1991) study emphasizes that such general conclusions about skewness are inappropriate. The usefulness of skewness as a dependent measure is determined by the relative sensitivity of skewness to experimental effects and extraneous influences (i.e, sources of *error*); and the relative sensitivity of any measure varies across experiments, because it is determined by experimental design and grounding of the experiment in theory. In Bailey's (1991) study, hypotheses about symmetry and its relation to postural control were well grounded in theory and in the content knowledge of skilled perceiver-actors. The results suggest that the content knowledge of participants in experiments on human movement can facilitate the development of dynamical models for complex movement skills.

Constraints Imposed by Suprapostural Tasks

In Bailey's (1991) study of cheerleading, the task was defined in terms of body configuration and movement. That is, the task of performing the liberty lift was defined by a particular body form that was to be maintained in a relatively stable manner. Performance on many tasks is influenced by body configuration and movement, but a task is not necessarily defined in terms of body configuration and movement. For example, body configuration influences how close the eyes are to potential objects of regard and whether the objects are in the *field of view*. Body configuration also influences whether potential manipulanda are within the functional *reach envelope*. Body movement (i.e., instability) influences the precision of vision and prehension. Together, configuration and stability have consequences for the ease or difficulty of seeing or manipulating objects (Riccio & Stoffregen, 1988, 1991). Thus, visual- or manual-control performance may serve as evaluation functions for body configuration (Figure 12.11, a-d). In the present context, this means that body configuration may be perceived and controlled with respect to performance on suprapostural tasks (i.e., with respect to its affordances).

Postural adjustments may be required for looking at, looking around, and looking through, or for touching, reaching around, and reaching through (cf. J.J. Gibson, 1979). In principle, these task constraints on posture can be described in ways that are commensurate with the description of high-energy constraints (Kugler & Turvey, 1987; Riccio, in press; Riccio & Stoffregen, 1988). This is important, because commensurability of "behavioral" and "physical" constraints is a necessary condition for the development of explanatory models in *ecological mechanics* (cf. Shaw & Kinsella-Shaw, 1988). Explanations that are explicit about the way such disparate constraints combine or interact in the perception and control of behavior should generate predictions about changes in behavior caused by changes in the task. However, the development of such models from first principles (e.g., Beek, 1989; Kugler & Turvey, 1987) is not currently possible for postural control, because very little is known about the constraints imposed by the animal, the environment, and the task in such complex activities. Empirical investigations are needed to examine the effects of these constraints on the perception and control of posture in contexts that are functionally richer than those in which posture is usually studied. For example, with empirical methods analogous to those described above (e.g., Bailey, 1991), the functional topological effects of configuration on the difficulty of manipulating an object could be compared with effects on the compensatory torque required to maintain stance (Figure 12.12, a and b; cf. Riccio & Stoffregen, 1988).

Posture and Manual Performance

The effects of body configuration on performance in a suprapostural task have been examined with techniques similar to those used in Bailey's study of cheerleading (Riccio & van Emmerik, 1991). The goal of this preliminary investigation was to determine whether there were functional topological relations between postural configuration during bipedal stance and performance on a manual control task. The manual control task required that a subject tap at a constant rate (approximately three times per second) and with a constant force on an electronic keyboard. This is a variant of the interval production task in which constant force is generally not a criterion (Michon, 1966). The *variability of intervals* in the interval production task has been shown to be influenced by work load in situations that require

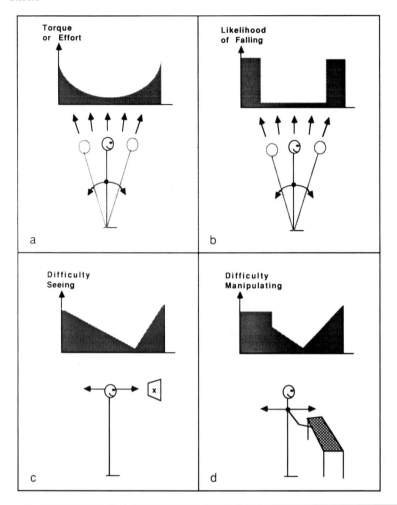

Figure 12.11 Hypothetical relations between orientation and various measures of task performance on which orientation (a-d) could be evaluated. Note that the shapes of and location of global extrema in the evaluation functions can vary (cf. Figure 12.8).

simultaneous control on some other task (e.g., E.A. Martin, McMillan, R. Warren, & Riccio, 1986). Thus, variability of intervals could be sensitive to the effort required to maintain various postural configurations. The constant-force criterion was added so that *force variability* would be a task-relevant measure in this postural control experiment. The hypothesis was that force variability would be sensitive to the instability inherent in various postural configurations (cf. Bailey, 1991). Unlike in Bailey's study of cheerleading, the relations between configuration and task performance were evaluated in a situation where stance is usually very stable: The support surface was static, rigid, flat, extensive, of high friction; the room was well lit; the participant stood on two feet and was not perturbed by external forces.

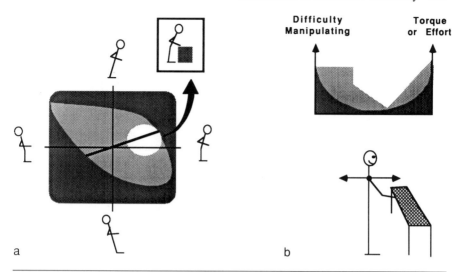

Figure 12.12   (a) Hypothetical effect of a suprapostural task on a postural configuration space. The darkly shaded area represents hip-ankle configurations for which bipedal stance is not possible. The lightly shaded area represents configurations for which stance is possible but for which the task (depicted in the inset) cannot be performed. The unshaded areas represent the configurations for which the task can be performed. (b) The evaluation function for a slice through the configuration space (corresponding to the bold line in [a]).

Performing a manual task while standing was assumed to be a general skill. Consequently, it was also assumed that a person would "know" the relevant topology of postural space and, given that the environmental conditions promoted stability, would adopt a configuration that would be both stable and efficient for the task (see Riccio & Stoffregen, 1988). Under these conditions, it would be difficult to discover a relation between postural configuration and task performance, because observation of an adequate range of configurations would be unlikely. The purpose of investigating this relation was to learn what, presumably, the participant had learned prior to the experiment: that certain postures should be avoided because of their effect on task performance. Consequently, the participant was instructed to adopt a different configuration on every trial. Configuration was defined by the angle of the upper and lower body segments relative to gravity (cf. Figures 12.5, 12.6, 12.7, 12.10, and 12.12). The upper and lower body angles were displayed as orthogonal coordinates of a luminous point on an oscilloscope. The oscilloscope was placed directly in front of the participant and behind the keyboard. Thus, continuous feedback was available to the participant about the current "position" in this configuration space. Auditory feedback was also provided about the force of each tap. Data were collected on the body angles and on the time history for tapping. The *coefficient of variation* was computed for peak tapping force and for intervals between the discrete taps. Two sets of relations defined the functional topology of the configuration space. In one, the upper and lower body angles were evaluated in terms of variability of intervals. In the other, the body angles were evaluated in terms of variability of peak force. The relations were described with quadratic response-surface regression so as to determine the location and shape of attractors and separatrices (cf. Bailey, 1991). In addition, *distance-weighted least squares (DWLS) regression* was used to explore the relation for local variations (Figure 12.13, a and b).

The results reveal a strong quadratic relation between postural configuration and variability of peak force; *Multiple R* = .811, $F(5,18)$ = 6.92, p = .001. The shape of the *manifold* (relation) is a saddle, with the inflection point (seat) close to erect posture. Variability of force was increased by bending and decreased by leaning. Although any interpretation of these preliminary results is necessarily speculative, there are some promising directions for more formal experimentation. The effect of leaning may be due to a decrease in relatively high-amplitude low-frequency sway near the separatrix (see Figure 12.7). Presumably, sway at frequencies near or below three cycles per second (the tapping frequency) could modulate the amplitude of tapping and, thus, be a source of variability in tapping force. The increase in force variability with bending may reflect an instability that can be tolerated (not necessarily consciously), because there was not an apparent threat of falling. The correlation between variability of force and variability of intervals was essentially zero (r = −.007). This suggests that force and timing are influenced by different factors in this experiment. The results failed to reveal a significant quadratic relation between postural configuration and variability of intervals. This may be due to the presence of a more complex manifold for this measure of task performance. DWLS regression indicated that there was very little change in the variability of intervals on the shallow gradient along which torques due to upper and lower body tilt tend to counterbalance each other (note the contour plots in Figure 12.13; cf. Figure 12.5). This is consistent with the expectation that interval variability reflects effortfulness (or energy expenditure). There also is a distinct asymmetry in interval variability with respect to anterior and posterior leaning. This may be due to the fact that posterior leaning was more effortful in this task. The effortfulness of posterior leaning was due to the need to extend the arms in order to reach the keyboard.

The fact that there was not a correlation between force variability and interval variability is consistent with the hypothesis that the former is influenced by postural stability, that the latter is influenced by postural effort, and that stability and effort can vary independently. If this finding holds up, then the constant-force version of the interval production task could be a valuable tool for studying the qualitative dynamics of postural control (see Riccio & Stoffregen, 1988). Such multivariate tasks would be especially useful in situations where there is a competition between information-based attractors and energy-based attractors. For example, bending over may improve the precision with which an object can be manipulated or visually inspected, although this may be very effortful (e.g., in the experiment described above, if the keyboard had been lower or farther in front of the subject). Conversely, it may be impossible to observe or reach an object if one adopts the most comfortable or least effortful orientation or configuration (Figure 12.14, a-c). A principled basis for modeling the interaction of information and energy is possible (e.g., Kugler & Turvey, 1987), although it has not yet been extended to research on posture and orientation. Nevertheless, it is clear that energy extrema alone do not predict the behavior of adaptive systems such as animals interacting with a variable environment.

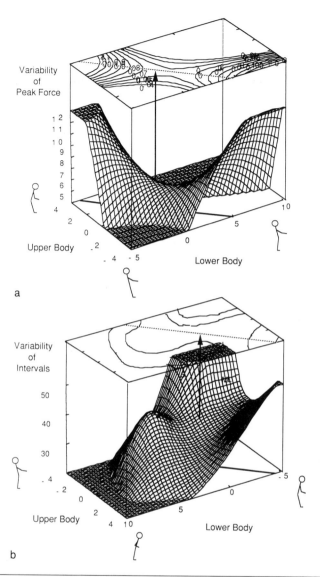

Figure 12.13 Task-relevant postural spaces for performance on the constant-force interval-production task. The vertical axis represents the coefficient of variation for peak force (a) and for intervals between taps (b). Numbers on other axes indicate degrees from vertical. Response surfaces represent the influence of orientation and configuration on performance. The surface in (a) is derived from quadratic regression. The surface in (b) is derived from DWLS regression because the quadratic fit was inadequate. Isoperformance contours are also represented. The bold arrow represents erect stance. The bold line and the dotted line in the two-dimensional projections represent the shallow gradient in torque due to bending without leaning (cf. Figure 12.5). Note that the two postural spaces are viewed from different perspectives.

Note. Compiled from unpublished raw data from Riccio and van Emmerik, 1991.

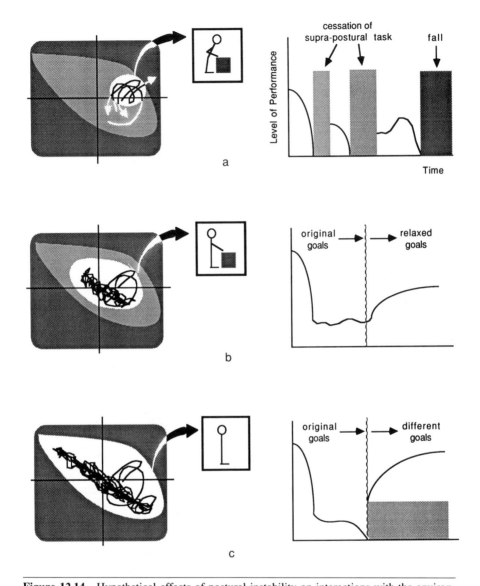

Figure 12.14 Hypothetical effects of postural instability on interactions with the environment. (a) The animal may be incapable of maintaining the configuration and stability that is necessary for the task. Attempted performance on the task may have deleterious consequences for performance on a subordinate task (e.g., maintenance of balance). (b) The animal may accept a reduced level of performance that is within its capabilities for postural control. (c) If performance on a task is impossible without the risk of additional instability, the animal may cease performance on the task but maintain performance on a subordinate task. The shaded area on the right represents cessation of the original goals.

Note. From "An Ecological Theory of Motion Sickness and Postural Instability" by G.E. Riccio and T.A. Stoffregen, 1991, *Ecological Psychology*, 3(3), pp. 195-240. Copyright 1991 by Lawrence Erlbaum Associates. Reprinted by permission.

Attractors and separatrices must be defined on a functional basis that includes both energy exchanges and information pickup (cf. Beek, 1989; Beek et al., 1992; Riccio et al., 1992; Riccio & Stoffregen, 1988).

Implications for Movement Science

The paradigm described above reflects an affordance-based view of task constraints on action systems (see Riccio & Stoffregen, 1988). Affordances are observable interactions between a system and its surroundings (cf. J.J. Gibson, 1979; Riccio, in press). The surroundings of a human action system can be the surfaces, media, and objects in the "natural" environment; human artifacts in the "modified" environment; or other systems and components of the human body. A superordinate system is formed when an action system is coupled with aspects of its surroundings, and this superordinate system may be capable of achieving goals that cannot be achieved with any of the component subsystems. These superordinate goals do not necessarily replace the goals or functions of the subsystems. Instead the goals and systems become nested: The goal-directed behavior of the system constrains the way in which the goals of a component subsystem can be achieved, and vice versa. Although the goal-directed behavior of a system imposes such task constraints on the behavior of component subsystems, the associated coupling among subsystems affords opportunities that may not be possible without the coupling. Intentional systems presumably perceive and act upon these affordances by adaptively coupling with their surroundings in ways that are consistent with the attendant constraints.

The coupling or nesting of systems to achieve nested goals is a pervasive characteristic of human behavior; however, it is generally not addressed in experiments on perception and action (although, see Stark, 1968; Vicente & Rasmussen, 1990). The exploratory investigation of Riccio and van Emmerik is an attempt to study nested action systems with methods that are familiar in movement science (e.g., Nashner & McCollum, 1985; Newell et al., 1989). The investigation is relatively simple, in that only two levels of nesting are considered: postural control (coordination of the upper and lower segments of the body during bipedal stance) and manual control (tapping on a keyboard). Nevertheless, it is sufficiently complex to reveal experimental and theoretical issues that should be generally important in the study of nested action systems. An especially important issue for further investigation is the control of systems and subsystems with respect to nested evaluation functions. For example, the various segments of the body play different roles in postural control and manual control, and these roles are linked to different performance criteria. The coordinated system is heterogeneous with respect to both components and performance criteria (cf. Beek & Bingham, 1991). An intriguing question is whether such systems can be viewed as homogeneous with respect to an integrated set of performance criteria (cf. Beek et al., 1992; Riccio, in press). In the preceding section, postural control and manual control were conceptually integrated by linking manual force variability and timing variability to postural effort and stability, respectively. Further development of the paradigm is required to evaluate this hypothesis.

SUMMARY AND CONCLUSIONS

The approach to ''motor behavior'' in this chapter is unapologetically psychological and phenomenological. Conscious experience is not viewed as an epiphenomenon of behavior. Instead, it is believed that behavioral research should be grounded in the knowledge of interactions with the environment that can be described by investigators and by participants in the research. Although, from this perspective, behavioral research should be grounded in phenomenology, it need not be limited to phenomenology for motivation or methodology. For example, the present work focuses on aspects of behavior that are ecologically pervasive and functionally important as well as phenomenally salient. A number of fundamentally interrelated phenomena meet these criteria: self-motion (Riccio, in press; R. Warren & Riccio, 1985), object approach (Stoffregen & Riccio, 1990), orientation (Riccio et al., 1992; Riccio & Stoffregen, 1990; Stoffregen & Riccio, 1988), postural stability and equilibrium (Bailey, 1991; E.J. Martin, 1990; Riccio & Stoffregen, 1988, 1991), and motion sickness (Riccio & Stoffregen, 1991; Stoffregen, & Riccio, 1991). Investigation of this range of phenomena requires a variety of methodologies and measurement systems. In this sense, the present work is methodologically eclectic. However, it is not theoretically eclectic. All theoretical and empirical developments are based on an ecological approach to behavior (E.J. Gibson, 1991; J.J. Gibson, 1979; see also Lombardo, 1987; Reed & Jones, 1982). In fact, the methodological eclecticism is made possible by, and reveals the power of, the theoretical commitment. This is in distinct contrast to more common scientific approaches for which theoretical eclecticism and a commitment to methodology are apparently acceptable.

The present work should be contrasted with different, but related, research on motor behavior that looks to physics or mathematics for much of its motivation (e.g., Kelso, 1990; Kugler & Turvey, 1987). Such research is important to the extent that it reveals principles that are general for living and nonliving systems. Attempts to identify points of convergence between such research and the present work are considered desirable (e.g., Riccio & Stoffregen, 1988, 1991). However, it should be noted that the present approach is conservative in that it does not assume consistency between research in physics and research in psychology or kinesiology. Consequently, research problems are not derived from physics or mathematics. At least initially, the research problems are unique to perception in the context of adaptable and goal-directed movement; and as stated above, phenomenology plays a central role in identifying *what* phenomena should be studied. Other disciplines are useful in identifying *how* a phenomenon can be studied. Thus, it is useful to adapt methods from psychophysics (e.g., Riccio & Cress, 1986; Riccio et al., 1992), kinesiology (e.g., Nashner & McCollum, 1985; Newell et al., 1989; Watanabe et al., 1987), eco-ethology or comparative behavioral ecology (e.g., E.J. Gibson et al., 1987; Kondo, 1985; Morbeck et al., 1979), systems engineering (e.g., Flach et al., 1986; E.A. Martin et al., 1986; Stark, 1968); and ethnography (e.g., Locke, 1988); and from research on self-organizing systems (e.g., Kelso, 1990; Kugler & Turvey, 1987) and physiological control systems (e.g., Milton, Longtin, Beuter, Mackey, & Glass, 1989).

The research strategy described in this chapter is reminiscent of E.G. Boring's description of the *Zeitgeist* from which scientific psychology emerged: "Phenomenology comes first, even though it does not get far by itself" (Boring, 1929/1957, p. 21). The research, in general, and the view of phenomenology, in particular, are consistent with the *functionalistic* tradition in psychology (see, e.g., James, 1890). The essence of functionalistic phenomenology is captured in the following passage written by Edna Heidbreder: "With James, introspection was the exercise of a natural gift; it consisted in catching the very life of a moment as it passed, in fixing and reporting of the fleeting event as it occurred in its natural setting" (Heidbreder, 1933; p. 171). The present work emphasizes that the adaptive and intentional qualities of animate movement (i.e., its "very life") can be appreciated only by studying movement in a meaningful context (i.e., an "event as it occurred in its natural setting"). This requires that the analysis of the context for a particular movement is commensurate with the analysis of the movement that occurs in that context. If the analysis of movement leads to dynamical models for particular action systems of an animal, then the analysis of context should lead to dynamical models for particular aspects of the animal's surroundings and to models that describe the functional linkage between the animal and its surroundings. The meaningfulness of movement is revealed in this functional dynamical linkage between an animal and its surroundings.

The present work assumes that the basic "observables" in movement science are the qualitative dynamics of the interaction between an animal and its surroundings. The qualitative dynamics (e.g., attractors and separatrices) are determined by the nesting of dynamical systems and the tasks or functions of these systems. Thus, the study of nested dynamical systems should be central in movement science (Riccio, in press). The nesting of dynamical systems varies across situations, and the associated dynamical variability necessitates robust and adaptive control of such systems. Robustness and adaptation to dynamical variability requires information about dynamical variability. This functionally relevant information may be available in movement variability that occurs on relatively small spatial and temporal scales. Small fast movements provide a pattern of variation (persistent excitation) that can be modulated by larger and slower variations in the underlying dynamics of the system. The pickup of information in movement variability has been demonstrated (implicitly or explicitly) in the experiments described above, and, in some cases, the information has been identified with particular patterns of movement (e.g., asymmetry). Such research on *systems identification* should have as important a role in movement science as it does in related research (e.g., Canudas de Wit, 1988; Chalam, 1987; Narendra, 1986; Weiner, 1948). In summary, the nesting of dynamical systems and the informativeness of movement variability appear to be key areas for future research in movement science, and phenomenology can be an important part of the associated research methodology.

REFERENCES

Akamatsu, N., Hannaford, B., & Stark, L. (1986). An intrinsic mechanism for the oscillatory contraction of muscle. *Biological Cybernetics, 53*, 219-227.

Bailey, M.A. (1991). *Topological dynamics and performer's conceptions of coordination, control, and skill in a cheerleading stunt.* Unpublished honors thesis, University of Illinois at Urbana-Champaign.

Bain, L.L. (1989). Interpretive and critical research in sport and physical education. *Research Quarterly for Exercise and Sport,* **60**, 21-24.

Barac-Cikoja, D., & Turvey, M.T. (1991). Perceiving aperture size by striking. *Journal of Experimental Psychology: Human Perception and Performance,* **17**, 330-346.

Beek, P.J. (1989). *Juggling dynamics.* Unpublished doctoral dissertation, Free University, Amsterdam.

Beek, P.J., & Bingham, G.P. (1991). Task-specific dynamics and the study of perception and action: A reaction to von Hofsten (1989). *Ecological Psychology,* **3**, 35-54.

Beek, P.J., Turvey, M.T., & Schmidt, R.C. (1992). Autonomous and nonautonomous dynamics of coordinated rhythmic movements. *Ecological Psychology,* **42**, 65-95.

Bingham, G.P., Schmidt, R.C., & Rosenblum, L.D. (1989). Hefting for a maximum distance throw: A smart perceptual mechanism. *Journal of Experimental Psychology: Human Perception and Performance,* **15**, 507-528.

Boring, E.G. (1957). *A history of experimental psychology.* New York: Appleton-Century-Crofts. (Originally published in 1929)

Brown, Y.J., Cardullo, F.M., McMillan, G.R., Riccio, G.E., & Sinacori, J.B. (1991). *New approaches to motion cuing in flight simulators* (AL-TR-1991-0139). Wright-Patterson AFB, OH: Armstrong Laboratory.

Box, G.E.P., & Draper, N.R. (1987). *Empirical model-building and response surfaces.* New York: Wiley.

Canudas de Wit, C.A. (1988). *Adaptive control for partially known systems: Theory and applications.* Amsterdam: Elsevier.

Chalam, V.V. (1987). *Adaptive control systems: Techniques and applications.* New York: Marcel Dekker.

Chan, T.-C., & Turvey, M.T. (1991). Perceiving the vertical distances of surfaces by means of a hand-held probe. *Journal of Experimental Psychology: Human Perception and Performance,* **17**, 347-358.

Coombs, C.H., Dawes, R.M., & Tversky, A. (1970). *Mathematical psychology.* Englewood Cliffs, NJ: Prentice Hall.

Dainoff, M.J., & Mark, L.S. (1987). Task and the adjustment of ergonomic chairs. In B. Knave & P.G. Wideback (Eds.), *Work with display units 86.* New York: Elsevier.

Diener, H.C., Horak, F.B., & Nashner, L.M. (1988). Influence of stimulus parameters on human postural responses. *Journal of Neurophysiology,* **59**, 1888-1903.

Evans, R.B. (1973). E.B. Titchener and his lost system. In M. Henle, J. Jaynes, & J.J. Sullivan (Eds.), *Historical conceptions of psychology* (pp. 83-97). New York: Springer-Verlag.

Flach, J. (1990). The ecology of human-machine systems. I. Introduction. *Ecological Psychology,* **2**, 191-205.

Flach, J., Riccio, G., McMillan, G., & Warren, R. (1986). Psychophysical methods for equating performance between alternative motion simulators. *Ergonomics,* **29**, 1423-1438.

Gibson, E.J. (1988). Exploratory behavior. *Annual Review of Psychology*, **39**, 1-41.

Gibson, E.J. (1991). *An odyssey in learning and perception*. Cambridge, MA: Massachusetts Institute of Technology.

Gibson, E.J., Riccio, G.E., Schmuckler, M.A., Stoffregen, T.A., Rosenberg, D., & Taormina, J. (1987). Detection of the traversability of surfaces by crawling and walking infants. *Journal of Experimental Psychology: Human Perception and Performance*, **13**, 533-544.

Gibson, E.J., & Schmuckler, M.A. (1989). Going somewhere: An ecological and experimental approach to development of mobility. *Ecological Psychology*, **1**, 3-25.

Gibson, J.J. (1979). *The ecological approach to visual perception*. Boston: Houghton Mifflin.

Heidbreder, E. (1933). *Seven psychologies*. Englewood Cliffs, NJ: Prentice Hall.

Horak, F.B., & Nashner, L.M. (1986). Central programming of postural movements: Adaptation to altered support-surface configurations. *Journal of Neurophysiology*, **55**, 1369-1381.

Howard, I.P. (1982). *Human visual orientation*. New York: Wiley.

James, W. (1890). *The principles of psychology*. New York: Henry Holt.

Johansson, R., Magnusson, M., & Akesson, M. (1988). Identification of human postural dynamics. *IEEE Transactions on Biomedical Engineering*, **35**, 858-869.

Kelso, J.A.S. (1990). Phase transitions: Foundations of behavior. In H. Haken & M. Stadler (Eds.), *Synergetics of cognition* (pp. 249-268). Heidelberg: Springer-Verlag.

Kelso, J.A.S., Tuller, B., Vatikiotis-Bateson, E., & Fowler, C.A. (1984). Functionally specific articulatory cooperation following jaw perturbations during speech: Evidence for coordinative structures. *Journal of Experimental Psychology: Human Perception and Performance*, **10**, 812-832.

Kondo, S. (Ed.) (1985). *Primate morphophysiology, locomotor analyses, and human bipedalism*. Tokyo: University of Tokyo Press.

Krinskii, V.I., & Shik, M.L. (1964). A simple motor task. *Biophysics*, **9**, 661-666.

Kugler, P.N., Kelso, J.A.S., & Turvey, M.T. (1980). On the concept of coordinative structures as dissipative structures. I. Theoretical lines of convergence. In G.E. Stelmach & J. Requin (Eds.), *Tutorials in motor behavior* (pp. 3-47). Amsterdam: North-Holland.

Kugler, P.N., & Turvey, M.T. (1987). *Information, natural law, and the self-assembly of rhythmic movement*. Hillsdale, NJ: Erlbaum.

Lee, D.N. (1976). A theory of visual control of braking based on information about time-to-collision. *Perception*, **5**, 437-459.

Lippold, O. (1970). Oscillation in the stretch reflex arc and the origins of rhythmic 8-12 c/s component of physiological tremor. *Journal of Physiology* (London), **206**, 359-382.

Locke, L.F. (1988). Qualitative research as a form of scientific inquiry in sport and physical education. *Research Quarterly for Exercise and Sport*, **59**, 1-20.

Lombardo, T.J. (1987). *The reciprocity of perceiver and environment: The evolution of James J. Gibson's Ecological Psychology*. Hillsdale, NJ: Erlbaum.

Loveland, K.A. (1991). Social affordances and interaction. II. Autism and the affordances of the human environment. *Ecological Psychology*, **3**, 99-120.

Maki, B.E. (1986). Selection of perturbation parameters for identification of the posture-control system. *Medical and Biological Engineering*, **24**, 561-568.

Mark, L.S., Balliett, J.A., Craver, K.D., Douglas, S.D., & Fox, T. (1990). What an actor must do in order to perceive the affordance for sitting. *Ecological Psychology*, **2**, 325-366.

Martin, E.A., McMillan, G.R., Warren, R., & Riccio, G.E. (1986). A program to investigate requirements for effective flight simulator displays. *International Conference on Advances in Flight Simulation*. London: Royal Aeronautical Society.

Martin, E.J. (1990). *An information-based study of postural control: The role of time-to-contact with stability boundaries*. Unpublished master's thesis, University of Illinois at Urbana-Champaign.

McCollum, G., & Leen, T.K. (1989). Form and exploration of mechanical stability limits in erect stance. *Journal of Motor Behavior*, **21**, 225-244.

McGinnis, P.M., & Newell, K.M. (1982). Topological dynamics: A framework for describing movement and its constraints. *Human Movement Science*, **1**, 289-305.

Michon. J.A. (1966). Tapping regularity as a measure of perceptual motor load. *Ergonomics*, **9**, 401-412.

Milton, J.G., Longtin, A., Beuter, A., Mackey, M.C., & Glass, L. (1989). Complex dynamics and bifurcations in neurology. *Journal of Theoretical Biology*, **138**, 129-147.

Morbeck, M.E., Preushoft, H., & Gomberg, N. (Eds.) (1979). *Environment, behavior, and morphology: Dynamic interaction in primates*. New York: Gustav Fischer.

Mori, S. (1973). Discharge patterns of soleus motor units with associated changes in force exerted by foot during quiet stance in man. *Journal of Neurophysiology*, **36**, 458-471.

Narendra, K.S. (Ed.) (1986). *Adaptive and learning systems: Theory and applications*. New York: Plenum Press.

Nashner, L.M., & McCollum, G. (1985). The organization of human postural movements: A formal basis and experimental synthesis. *Behavioral and Brain Sciences*, **8**, 135-172.

Newell, K.M. (1986). Constraints on the development of coordination. In M.G. Wade & H.T.A. Whiting (Eds.), *Motor development in children: Aspects of coordination and control* (pp. 341-360). Amsterdam: Martinus Nijhoff.

Newell, K.M., & Hancock, P.A. (1984). Forgotten moments: A note on skewness and kurtosis as influential factors in inferences extrapolated from response distributions. *Journal of Motor Behavior*, **16**, 320-335.

Newell, K.M., Kugler, P.N., van Emmerik, R.E.A., & McDonald, P.V. (1989). Search strategies and the acquisition of coordination. In S.A. Wallace (Ed.), *Perspectives on the coordination of movement* (pp. 85-122). Amsterdam: North-Holland.

Newell, K.M., & McDonald, P.V. (1992). Searching for solutions to the coordination function: Learning as exploratory behavior. In G.E. Stelmach & J. Requin (Eds.), *Tutorials in motor behavior II* (pp. 517-532). Amsterdam: North-Holland.

Onsager, L. (1931). Reciprocal relations in irreversible systems. I. *Physical Review*, **37**, 405-426.

Owings, D.H., & Coss, R.G. (1991). Context and animal behavior. I. Introduction and review of theoretical issues. *Ecological Psychology*, **3**, 1-9.

Reed, E., & Jones, R. (Eds.) (1982). *Reasons for realism: Selected essays of James J. Gibson*. Hillsdale, NJ: Erlbaum.

Riccio, G.E. (in press). Coordination of postural and vehicular control: Implications for multimodal perception. In J. Flach, P. Hancock, J. Caird, & K. Vicente (Eds.), *The ecology of human-machine systems*. Hillsdale, NJ: Erlbaum.

Riccio, G.E., & Cress, J. (1986). Frequency response of the visual system to simulated changes in altitude and its relationship to active control. *Proceedings of the 22nd Annual Conference on Manual Control* (pp. 117-134). Wright-Patterson Air Force Base, OH: Aeronautical Systems Division.

Riccio, G.E., & Emmerik, R.E.A. van (1991). [Topological relations between postural configuration and manual control performance]. Unpublished raw data.

Riccio, G.E., & Martin, E.J. (1990). [Further experiments on the active perception of orientation]. Unpublished raw data.

Riccio, G.E., Martin, E.J., & Stoffregen, T.A. (1992). The role of balance dynamics in the active perception of orientation. *Journal of Experimental Psychology: Human Perception and Performance*, **18**, 624-644.

Riccio, G.E., & Stoffregen, T.A. (1988). Affordances as constraints on the control of stance. *Human Movement Science*, **7**, 265-300.

Riccio, G.E., & Stoffregen, T.A. (1990). Gravitoinertial force versus the direction of balance in the perception and control of orientation. *Psychological Review*, **97**, 135-137.

Riccio, G.E., & Stoffregen, T.A. (1991). An ecological theory of motion sickness and postural instability. *Ecological Psychology*, **3**, 195-240.

Roberts, T.D.M. (1978). *Neurophysiology of postural mechanisms*. London: Butterworths.

Robinson, J.T. (1972). *Early hominid posture and evolution*. Chicago: University of Chicago Press.

Runeson, S. (1977). *On visual perception of dynamic events*. Unpublished doctoral dissertation, University of Uppsala, Uppsala, Sweden.

Ryan, T.A. (1938). Dynamic, physiognomic, and other neglected properties of perceived objects: A new approach to comprehending. *American Journal of Psychology*, **51**, 629-650.

Saltzman, E., & Kelso, J.A.S. (1987). Skilled actions: A task-dynamic approach. *Psychological Review*, **94**, 84-106.

Sanderson, P.M., James, J.M., & Seidler, K.S. (1989). SHAPA: An interactive software environment for protocol analysis. *Ergonomics*, **32**, 1271-1308.

Schöne, H. (1984). *Spatial orientation: The spatial control of behavior in animals and man*. (C. Strausfeld, Trans.). Princeton, NJ: Princeton University Press.

Shaw, R.R., & Kinsella-Shaw, J. (1988). Ecological mechanics: A physical geometry for intentional constraints. *Human Movement Science*, **7**, 155-200.

Shaw, R.R., Kugler, P.N., & Kinsella-Shaw, J. (1990). Reciprocities of intentional systems. In R. Warren & A.H. Wertheim (Eds.), *Perception and control of self-motion* (pp. 579-619). Hillsdale, NJ: Erlbaum.

Solomon, H.Y. (1988). Movement produced invariants in haptic explorations: An example of a self-organizing, information-driven, intentional system. *Human Movement Science*, **7**, 201-224.

Solomon, H.Y., & Turvey, M.T. (1988). Haptically perceiving the distances reachable with hand-held objects. *Journal of Experimental Psychology: Human Perception and Performance*, **14**, 404-427.

Stark, L.W. (1968). *Neurological control systems: Studies in bioengineering*. New York: Plenum Press.

Stein, R.B., & Oguztoreli, M.N. (1976). Tremor and other oscillations in neuromuscular systems. *Biological Cybernetics*, **22**, 147-157.

Stoffregen, T.A., & Riccio, G.E. (1988). An ecological theory of orientation and the vestibular system. *Psychological Review*, **95**, 3-14.

Stoffregen, T.A., & Riccio, G.E. (1990). Responses to optical looming in the retinal center and periphery. *Ecological Psychology*, **2**, 251-274.

Stoffregen, T.A., & Riccio, G.E. (1991). A critique of the sensory conflict theory of motion sickness. *Ecological Psychology*, **3**, 159-194.

Thelen, E. (1985). Developmental origins of motor coordination: Leg movements in human infants. *Developmental Psychobiology*, **18**, 1-22.

Thinés, G., Costall, A., & Butterworth, G. (1991). *Michotte's experimental phenomenology of perception*. Hillsdale, NJ: Erlbaum.

Tobias, P.V. (1982). *Man: The tottering biped*. Kensington, New South Wales, Australia: Committee on Postgraduate Medical Education, University of New South Wales.

Valenti, S.S., & Good, J.M.M. (1991). Social affordances and interaction I: Introduction. *Ecological Psychology*, **3**, 77-98.

Vicente, K.J., & Rasmussen, J. (1990). The ecology of human-machine systems II: Mediating "direct" perception in complex work domains. *Ecological Psychology*, **2**, 207-250.

Warren, R., & Riccio, G. (1985). Visual cue dominance hierarchies: Implications for simulator design. *Transactions of the Society for Automotive Engineering*, **6**, 937-951.

Warren, W.H. (1984). Perceiving affordances: Visual guidance of stair climbing. *Journal of Experimental Psychology: Human Perception and Performance*, **10**, 683-703.

Watanabe, Y., Yokoyama, K., Takata, K., & Takeuchi, S. (1987). An evaluation of control mechanisms in a standing posture using velocity and acceleration of body sway. In B. Johnsson (Ed.), *Biomechanics X-B*. Champaign, IL: Human Kinetics.

Wiener, N. (1948). *Cybernetics: Or the control and communication in the animal and the machine*. Cambridge, MA: Massachusetts Institute of Technology.

Acknowledgments

This chapter was written while the author was at Wright-Patterson Air Force Base, Ohio, in a division of a national laboratory (the Armstrong Laboratory) that was formerly the Armstrong Aerospace Medical Research Laboratory. During this time, the author was supported by the Air Force Office of Scientific Research (AFOSR) Summer Research Program. The author is grateful to Thomas Stoffregen, Eric Martin, Peter Beek, Karl Newell, and Barry Hughes for comments on the manuscript.

Chapter 13

Stability of Symmetric and Asymmetric Discrete Bimanual Actions

Charles B. Walter
University of Illinois at Chicago

Stephan P. Swinnen
Catholic University of Leuven, Leuven, Belgium

Elizabeth A. Franz
Purdue University, West Lafayette, Indiana

The skill with which a motor task is performed can be operationalized in a number of ways. Most sets of behavioral criteria include measures of the consistency and efficiency in attaining a goal (e.g., Guthrie, 1935; Welford, 1976). An extremely skilled performer is thus one who achieves a goal consistently and with the minimal possible effort. Similar criteria can be applied to the processes underlying skilled performance. One perspective, for example, holds that once a coordination function is acquired, skill emerges from the ability to consistently assign optimal parameters to the function (Kugler, Kelso, & Turvey, 1980; Newell, 1985). It seems clear that a critical characteristic of both the process of skill production and its behavioral outcome is consistency. An understanding of motor skill is thus inextricably bound to an understanding of the factors that influence consistency through their effects on movement variability.

Two motor variability issues have been the focus of a substantial amount of research. These issues are the relationship between the magnitude of various scalar task parameters and variability in performance, and the effect of practice on performance consistency. Most of this work has been performed using single-limb (often uniarticular) tasks. A third variability issue arises when considering more complex behavior. Actions that require coordination among limbs or limb segments may elicit an interaction among the movements that itself influences variability. This suggests that variability principles derived from uniarticular or single-limb tasks may not entirely account for the variability of multiarticular or multilimb tasks, respectively. This paper addresses the potential effect of bimanual interactions on motor variability, as well as the mediating influence of practice on this effect. The emphasis is on discrete tasks, although relevant findings from continuous movements will be discussed to provide a broad conceptual context.

359

COMPLEMENTARY APPROACHES
TO MOTOR VARIABILITY

Two general approaches have been adopted for the study of variability. Each view has implications for the sources of inconsistent behavior.

Elemental and Dynamical Frameworks

Perhaps the most common method for examining motor variability is to determine a task parameter that, when scaled, yields a change in the variability of either that parameter or of another parameter. Indeed, the variability of virtually any movement parameter (time, distance, force, etc.) appears to be related to the magnitude of the parameter in some manner. The form of the relationship (linear, logarithmic, etc.) and the nature of the independent variable (temporal, kinetic, etc.) determine the theoretical account of the variability. Investigators adopting this approach have typically examined unimanual tasks to reduce the number of potential confounding variables and to aid in movement quantification. Considering the ubiquity of unimanual actions in everyday life, developing variability principles for these tasks is certainly a worthy goal. Some effort has also been made to directly apply principles of unimanual variability to dual-limb performance. This strategy is essentially an "elemental" one, with the single-limb principles serving as elements that can be combined for extrapolation to more complex tasks.[1]

The assumption that elemental principles directly apply to complex behavior is valid if the effects of the principles are essentially additive; that is, interactions among system components must be linear. There is growing evidence from bimanual tasks to suggest that many interlimb interactions are nonlinear, however (Haken, Kelso, & Bunz, 1985; Kelso, Holt, Rubin, & Kugler, 1981). It thus appears to be necessary to directly examine more complex, coordinated actions to access aspects of behavior that arise only through dynamic interactions among system components. Because of this emphasis placed on dynamics, the "dynamical systems" framework recently advanced on a number of fronts (see later) is particularly useful for the study of coordinated motor behavior. Of course, the problem of generalizability is as relevant for this approach as it is for the elemental strategy, but from the transposed perspective. Some principles concerning the variability of complex actions may not apply to simpler tasks. Elemental and dynamical perspectives are thus perhaps best viewed as complementary rather than competing; basic principles of motor variability may emerge through lines of evidence that converge from both approaches.

Sources of Motor Variability

Before discussing the variability of bimanual actions, it is useful to briefly mention two general sources of motor variability that have often been noted in the literature.

---

[1]*Elemental* is presently used in a relative sense and as such is not associated with specific criteria. For example, although the term is used here to distinguish between investigations of single-limb and dual-limb tasks, it could easily be used to contrast uniarticular actions with multiarticular actions or, as noted below, the activity of individual neurons with collectives of neurons.

Movement inconsistency is often attributed to a combination of variability in the organization of the control signal and variability in the effector mechanisms responsible for the action. For example, Hatze (1986) notes that "an observed deviation process is a particular realization of a (approximately normally distributed) controlled variation and a superimposed random excursion" (p. 10). R.A. Schmidt, Zelaznik, Hawkins, Frank, and Quinn (1979) apply the same reasoning to two-handed movements in stating that overt variability is due to "(a) variability in program selection or in the selection of parameters *common* to both limbs and (b) variability in parameters *specific* to the limb or in the recruitment of motor units at the spinal level" (p. 432). The neural variability referred to here would presumably contribute to the "superimposed random excursion" noted by Hatze (1986).

These characterizations of motor variability, however, fail to address the potential effect of the interaction between concurrent movements. The critical question for present purposes is, Does this interaction provide an additional source of motor variability for discrete bimanual actions? The failure of Fitts' law (Fitts, 1954) under bimanual conditions (e.g., Robinson & Kavinsky, 1976) provides one indication that it may. In this case, the particular speed-accuracy trade-off description that is valid for individual limbs is not applicable when the limbs must act in concert. Because inaccuracy arises in part through variability, this evidence suggests that the interaction between bilateral limb movements may provide a third source of motor variability for coordinated actions.

THE DYNAMICAL SYSTEMS FRAMEWORK

The present focus is on the principles associated with the effect of dynamic interactions on variability. Although a number of excellent discussions of basic dynamical systems concepts have been provided, it may be useful to briefly review several notions that are particularly relevant for present purposes.

Background and Related Issues

A "dynamical system" is simply a system that changes over time. Stewart (1990) describes it as "a cloud of points moving around in space" (p. 321). Depending on the level of analysis adopted for examining motor behavior, the points may represent individual neurons, motor units, muscles, synergies, or whole body segments. We will assume that limb movements reflect critical system dynamics for present purposes. A dynamical system can be characterized as a discrete system or as a continuous system. A discrete system is modeled by an iterative process or "mapping." A continuous system can potentially be described by a system of differential equations that reflect the "flow" of the system (Crutchfield, Farmer, Packard, & Shaw, 1986), although natural systems often contain a degree of complexity that initially demands a more topological description. The dynamical perspective is thus partly characterized by a qualitative, generic focus that serves as a precursor to, and sometimes as a replacement of, analytic quantification. It is in this qualitative spirit that we will attempt to apply the dynamical principles developed for oscillatory tasks to the stability of discrete tasks.

The topological description of behavior is generally conveyed in the "state space" of the system in the dynamical framework. This space is a "geometric model for the set of all idealized states" of the system (Abraham & Shaw, 1982, p. 13). The states are termed idealized because they are based on a few variables that represent observable behavior. The variables that *determine* the dynamics of a complex system typically comprise many more dimensions than the state space and are often invisible to macroscopic inspection. For example, the behavioral state of a bird in flight can be described by its location and velocity in each of three orthogonal dimensions, but the individual factors influencing its state constitute a much higher dimensional space. Importantly, there is no a priori method for selecting the appropriate variables to construct a state space for examining an arbitrary system (Packard, Crutchfield, Farmer, & Shaw, 1980).

One method for selecting state-space dimensions for motor behavior is to search for variables that consistently capture critical aspects of the behavior, termed "collective variables" (e.g., Kelso & Schöner, 1988). These variables can be proposed for any number of levels of analysis, but they are most useful if they convey the dynamic properties associated with a specific function of interest. The dynamics of neural activation in response to sensory stimulation can be studied at the cellular level, for example, but it is the dynamics of *collective* neuronal behavior that determines the perception of many stimuli (e.g., Freeman, 1991). Several lines of evidence suggest that the "relative phase" of bimanual movements serves as a useful collective variable for bimanual oscillations (e.g., Haken et al., 1985; Kelso & Scholz, 1985; Turvey, Rosenblum, Schmidt, & Kugler, 1986). The phase of an oscillation is the point of advancement through a single cycle, and the relative phase is the difference between the phase of each limb.

A final essential dynamical systems concept is that of an "attractor." An attractor is the behavior that a system settles into if given the opportunity to do so; that is, the state to which a finite set of trajectories converges in state space (Baker & Gollub, 1990; Crutchfield et al., 1986). The relevant state for present purposes is not that of each individual moving limb, but rather that of the relationship between the movements, or the "coordinated state." The three general forms of attractors that have been identified for purely physical systems are point attractors (static equilibrium in at least one dimension), periodic attractors (repeating waveforms), and strange or chaotic attractors (e.g., Abraham & Shaw, 1982, 1983; Baker & Gollub, 1990). Finally, the strength of attraction toward a specific state can vary. The stronger the attractor, the higher the rate at which the system attains the preferred state (Abraham & Shaw, 1983).

Motor Variability and Dynamic Stability

Within the dynamical systems framework, motor variability is related to the concept of system stability and, consequently, to the notion of an attractor. As noted above, an attractor describes the eventual behavior exhibited by a physical system that is given sufficient time to equilibrate. Behavior that is consistent with the attractor is often stable, although the degree of stability depends on the strength and form of the attractor. Unlike an autonomous physical system, this preferred state may seldom be produced (at a fine level of analysis) by the neuromuscular system, because the behavior is influenced by numerous sources of "noise." But characteristics of the attractor may still be inferred by observing the system's general behavior near the preferred state. The greater the strength of the attractor, for example, the greater the systematic bias of an arbitrary movement pattern toward that state. Knowledge of the general form of the attraction is important, so that actions that require a departure from the preferred state can be identified.

This discussion relates to motor variability in several ways. First, as noted above, any number of mechanisms associated with movement organization and execution serve as sources of motor variability. The overt expression of variability, however, may be constrained by intrinsic attractors. When applied to the present topic, this suggests that discrete bimanual movements that differ from the preferred state of coordination should exhibit greater variability than those that are consistent with it. This is one notion that will be explored here. Secondly, emergent processes attributable to interlimb interactions may themselves serve as additional sources of variability for bimanual tasks. If this is true, then concurrent movements may exhibit greater variability than unimanual actions with identical task requirements. This possibility will also be examined. Finally, the effect of different forms of practice on stability will be examined. Practice that successfully changes intrinsic dynamics should be reflected by an increase in the stability of the new coordinated state (e.g., Zanone & Kelso, in press). This should be reflected both by a reduced bias toward the initially preferred pattern and by reduced variability. Conversely, novel coordinated patterns that are essentially superimposed on a "stationary" dynamical landscape that doesn't intrinsically support the patterns should display greater variability than those that are consistent with the landscape.

DYNAMIC STABILITY OF BIMANUAL ACTIONS

Woodworth (1903) noted long ago that "movements with the left and right hands are easy to execute simultaneously. We need hardly try at all for them to be nearly the same" (cited in R.A. Schmidt, 1988, p. 258). This statement nicely captures a consistent quality of bimanual tasks: Actions that are bilaterally symmetrical are naturally preferred. Stated another way, coordinated actions tend to be attracted toward synchronized (in phase), topologically similar trajectories. This simple principle characterizes one important aspect of the stability of continuous bimanual actions and, we will argue, the stability of discrete bimanual actions as well. We will begin by reviewing the evidence concerning continuous tasks.

Stability of Continuous Bimanual Actions

The coordination of continuous bimanual tasks has been examined by a number of investigators. The task for the subject usually requires bilateral tapping or oscillations of the index fingers or hands. An extremely robust finding for 1:1 bimanual oscillations at slow to moderate rates is that two stable relative phase modes are exhibited (e.g., Kelso, 1984; Yamanishi, Kawato, & Suzuki, 1979). One is at a relative phase of 0° (bilateral, mirror-image symmetry) and the other is at a relative phase of 180°, also termed an "antiphase" relationship. In dynamical terms, these two modes represent attractors. Greater temporal variability in the antiphase mode than the in phase mode indicates a stronger attraction toward the latter (Cohen, 1971; Haken et al., 1985; Kelso & Scholz, 1985; Turvey et al., 1986). This notion is supported by two further observations. First, the antiphase mode loses stability as the frequency of oscillation is increased (Haken et al., 1985; Kelso & Scholz, 1985). This attractor disappears past a "critical frequency," leaving the symmetrical mode as the only stable state. Secondly, the recovery time following a physical perturbation is greater for antiphase coordination than for in phase coordination (Scholz & Kelso, 1989). The delay becomes progressively greater as the frequency increases ("critical slowing") for the antiphase mode.

Other studies have suggested that bimanual tapping tasks with harmonic (integer) ratios that follow the pacing of metronomes are relatively easy to perform (Deutsch, 1983; Klapp, 1979). This evidence initially appears to conflict with the principles noted above; any frequency ratio other than 1:1 requires a constantly changing relative phase and should therefore be quite difficult to perform. It should be noted that these tapping studies, however, have only examined the timing of finger contact. The *trajectories* of oscillations at harmonic frequency ratios can reflect a great deal of interference (e.g., Kelso et al., 1981). Bimanual oscillations at the elbow often exhibit mutual spatiotemporal assimilation for ratios of 2:1 and 3:1, for example (Pollatou, 1991). The limb producing the lower frequency may pause or display small oscillations near the point of reversal while the contralateral limb performs its additional cycle(s). When the limbs do move together, they appear to be very tightly phase-locked, and their amplitudes increase. This evidence indicates that even harmonic bimanual oscillations tend to be drawn toward the same in phase frequency.

Another method for manipulating oscillatory frequency is to physically induce alterations in the individual characteristic frequencies of bilateral "pendulum systems" (Kugler & Turvey, 1987; Rosenblum & Turvey, 1988). This technique precludes the need for subjects to follow external timing sources, as they are simply asked to swing the pendula "comfortably" in synchrony or alternation. Rosenblum and Turvey (1988) examined fluctuations in the period of each wrist-pendulum system when systems with two intrinsically different characteristic frequencies were coupled. A departure from similar bilateral characteristic frequencies was, to a point, again accompanied by increased periodic fluctuations for this task.

Relative phase clearly captures a great deal of the critical behavior displayed by coupled oscillatory systems. But other variables, such as spatial topology, appear to reflect coupling as well. Franz, Zelaznik, and McCabe (1991) examined subjects attempting to concurrently produce an alternating linear trajectory with one arm and a continuous circular trajectory with the contralateral arm. The movements were to be generated with the same frequency, which was determined by a metronome. The relative direction of the movements was selected by the subject, and the majority chose to move the limbs in mirror-image fashion. Most movements were thus executed in phase but required disparate spatial patterns. Systematic deviations from the intended spatial trajectories were noted. Interlimb assimilation appeared, with the intended linear trajectory opening into a narrow elliptical orbit and the intended circular motion compressing into a broad ellipse (although absolute symmetry was rarely observed). This suggests that concurrent trajectories are biased toward a common spatial pattern, much as they are attracted temporally toward a common phase.

Practice can clearly affect the stability of an action that requires a departure from a preferred coordinated state. Zanone and Kelso (in press) instructed subjects to attempt bilateral finger oscillations with similar frequencies but a relative phase offset of 90°. The variability in relative phase significantly decreased across training trials, although individual differences, attributed to different initial intrinsic dynamics, were clearly evident. A systematic "scan" of the stability of various relative phase relationships suggested that a new attractor had appeared at the goal phase offset. Swinnen, De Pooter, and Delrue (1991) also examined bimanual, 1:1 elbow oscillations with a phase offset of 90°. Feedback in the form of displacement-displacement plots was provided after every trial. Subjects were able to reduce both the absolute deviation from the intended relative phase and the variability in relative phase with practice. Learning effects have been noted for a bimanual elbow oscillation task requiring a 2:1 frequency ratio as well (Swinnen, Walter, & Willekens, 1991). Individual period variability was significantly greater for each limb when performing in the bimanual condition than when performing alone, but period variability decreased with practice.

Together, these studies provide convincing evidence for several general dynamical principles of bimanual oscillations. First, attractors initially only appear at in phase and antiphase interlimb relationships, for most subjects performing 1:1 ratios. Although these are oscillatory actions, the attractors can be considered as point attractors in relative phase space, because the *relationship* between the two movements is constant (Schöner & Kelso, 1988). Relative phase space is defined by a single dimension (i.e., a line) with limits of 0° and 360°. "Sinks" can be placed at the attractor points to indicate their respective basins of attraction (Haken et al., 1985). The line can be curved into a circle, so that the phase offsets of 0° and 360° coincide, because they are equivalent. Continuous bimanual movements are also drawn toward similar spatial patterns. This suggests that the specific dimensions of the attractor observed may depend on the dimensions of the task requirements. The frequency of the oscillation appears to directly influence the strength of attraction toward the most preferred state, indicated by the transition in relative phase for tasks that initially alternate. Actions that must be performed away from the preferred relative phase states lose stability, as manifested through an increase in variability. Finally, practice can facilitate bimanual decoupling. Attempts at performing actions that are not initially supported by intrinsic dynamics are generally quite variable, but stability improves over trials. Preliminary evidence suggests that new attractor states may emerge via changes in intrinsic dynamics.

Stability of Discrete Bimanual Actions

One general principle that is confirmed by the evidence discussed thus far is that symmetric actions are extremely easy to produce. Few purposeful actions satisfy this condition, however, rendering a need to examine coordinated actions with disparate bilateral requirements (e.g., Swinnen & Walter, 1988). Movement requirements can differ in two fundamental ways—in their "structural" and in their "metrical" specifications (Kugler et al., 1980; Turvey, Shaw, & Mace, 1978). Briefly, the structural specification refers to the movement pattern or topology, and the metrical specification refers to the magnitude or scale of a given variable. The structure of a given action has also been termed an "essential" characteristic (Gelfand & Tsetlin, 1971). Structural invariance among movements implies that the same trajectory can be obtained by scaling the appropriate variable(s). For the sake of consistency, we will use the term *symmetric* to describe bimanual movements with similar structural and metrical requirements, and *asymmetric* for movements whose structural *or* metrical specifications differ bilaterally.

There are several issues of interest here. Bimanual movements with similar metrical and/or structural requirements will be compared with tasks requiring different bilateral requirements, to examine the effect of movement asymmetry on bimanual stability. Both bias and variability will be examined in this context. Unimanual movements will also be compared with bimanual movements with identical requirements, to examine the effect of bimanual interactions on movement variability. If variability is wholly determined by physical task requirements (speed, force, etc.), then unimanual movements and symmetric bimanual movements should exhibit similar levels of variability. But if interlimb interactions contribute to instability, then the bimanual actions may be more variable. Finally, the effect of practice on the stability of bimanual actions that must diverge from metrical and/or structural symmetry will be discussed. The extent to which the dynamical principles elaborated for oscillatory actions apply to the stability of discrete bimanual actions will be examined through the following four questions:

1. What is the general form of attraction for discrete bimanual actions?
2. What factors influence the strength of attraction toward the preferred coordinated state?
3. How does a forced departure from the preferred coordinated state affect motor variability?
4. How does practice mediate the stability of these actions?

Evidence concerning each question will be addressed separately below.

What Is the General Form of Attraction for Discrete Bimanual Actions?

Identifying attractors for discrete actions poses two unique problems. The first problem arises from the relatively short duration of the actions. As noted above, an attractor is the behavior that a system settles into if given the opportunity to do so. Discrete tasks of relatively short duration, however, provide the system with little settling time. The preferred coordinated state is thus rarely attained, rendering a rigorous, formal identification of the attractors for discrete actions unlikely. The problem can be partially rectified by noting that the stability of a system is determined by examining behavior *near* the attractor. As Stewart (1990) notes, "An unstable state of motion *can* be observed, but only as a transient phenomenon—while the system is *en route* from its original unstable state to wherever it will finally end up'' (pp. 61-62). In the case of a discrete bimanual action, the system may relax toward dynamic equilibrium (i.e., the preferred coordinated state) as the action progresses, but it seldom achieves it. The general form of attraction must be inferred from the direction of the spatiotemporal bias, and its strength from the magnitude and consistency of the bias. Once the general nature of the attractor is identified, the variability of behavior removed from it can be examined.

The second problem of identifying attractors for discrete movements concerns the appropriate variable(s) to characterize the state of the system (Walter, Corcos, & Swinnen, 1990). The state of a system of two oscillating limbs is conveniently described by their relative phase, but a direct application of this concept to discrete bimanual tasks is difficult. If relative phase is defined as the difference between the point of advancement through each of the two discrete movements, then it is entirely determined by the timing of the initiation and termination of the movements. That is, movements that are initated and terminated together would be considered to be in phase throughout the action, regardless of differences in spatial trajectories. This is perhaps not a useful definition, because, as noted earlier and further supported later, spatial topology can reflect bimanual interactions independent of temporal synchrony. A slight modification of the description of interlimb attraction developed for oscillatory actions may be required for discrete actions. This issue is addressed in greater detail in the final section.

A number of studies have examined bimanual movements with the same general topology but different metrical requirements. The tasks have typically consisted of unidirectional aiming movements where parameters such as distance, inertial load, or speed are manipulated (the latter often through different accuracy requirements using the Fitts aiming task). One relevant finding from these studies is that assimilation tends to occur between the limbs for the variable manipulated (Kelso, Southard, & Goodman, 1979; Marteniuk, MacKenzie, & Baba, 1984; Sherwood, 1989, 1990; Sherwood & Canabal, 1988). That is, a fast movement slows down when paired with a slow movement (and vice versa), a long movement shortens when paired with a short movement, and so forth. The actions appear to be drawn toward a common metric (magnitude) of the disparate variable. Bimanual actions that would normally be of different durations when performed separately are typically initiated and terminated simultaneously when performed together (Kelso et al., 1979). This trend toward temporal synchrony for discrete actions is, in a general sense, consistent with the notion of attraction toward a phase offset of $0°$ for continuous actions. If the movement of one limb is physically constrained by its being required to traverse a barrier, mutual assimilation is precluded; that is, the lower trajectory can be raised, but the higher one cannot be lowered, due to the barrier. But attraction toward a similar scale, the greater height in this case, is still often observed (Kelso, Putnam, & Goodman, 1983).

We have been performing a series of studies examining a discrete bimanual task that imposes different structural and metrical requirements on the two limbs (Swinnen, Walter, & Shapiro, 1988). The task is to generate simultaneous movements with different spatiotemporal patterns (Figure 13.1, a-c). One arm is to produce a unidirectional movement (elbow flexion) while the other produces a

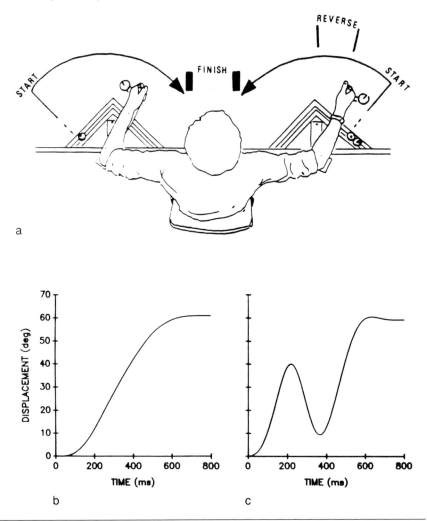

Figure 13.1   (a) Movement apparatus and intended displacement-time traces for (b) unidirectional and (c) reversal movements.

reversal movement (elbow flexion-extension-flexion) of similar duration, typically about 600 ms. The different bilateral metrical requirements arise from the fact that the reversal limb must travel over a greater distance than the unidirectional limb in the same time, requiring a greater magnitude of parameters such as average velocity, average acceleration, and net work. Targets indicate the general location of the reversal and termination points, but spatial accuracy is not stressed, and the timing of each segment of the reversal movement is not specified. The primary objective of the subject is to produce the two patterns together but independent of each other while generally maintaining the goal movement time; that is, to decouple the movements.

A consistent finding emerging from these studies is that the limb movements tend to be drawn to a common spatiotemporal trajectory. The unidirectional limb usually slows, pauses, or reverses direction when the contralateral limb changes direction. The reversal limb is also affected; that is, the amplitude of the middle (extension) segment of the movement is typically smaller than intended during bimanual performance (Swinnen, Walter, Beirinckx, & Meugens, in press). Greater cross-correlations are noted between the acceleration patterns of the movements performed together than when the patterns of the limbs from separate, unimanual trials are correlated (Swinnen, Walter, Pauwels, Meugens, & Beirinckx, 1990). This is one indication of structural symmetry. There is greater similarity between limbs for a number of discrete kinematic measures in the bimanual condition than the unimanual condition as well (Walter & Swinnen, 1990a, 1990b). The net work generated by each limb is much more similar in bimanual conditions than in separate unimanual conditions (Swinnen et al., in press). This tendency for bilateral symmetry is reflected in the electromyographic activity of homologous muscles (Swinnen, Young, Walter, & Serrien, 1991). In summary, a substantial amount of evidence favors the conclusion that at least some discrete bimanual actions are drawn toward a common structure and metric.[2]

What Factors Influence the Strength of Attraction Toward the Preferred State?

It is important to note that the accommodations displayed by each limb during an asymmetric bimanual task are not equivalent in either metrical or structural terms. For example, Marteniuk et al. (1984) reported that the spatial bias (CE) exhibited by the left hand in bimanual aiming conditions with different distance requirements is greater than that shown by the right hand. We have also observed greater coupling for our task when the reversal movement is produced by the nonpreferred left arm than when it is performed by the right arm (Walter & Swinnen, 1990a). These findings suggest a laterality effect, where the strength of interlimb attraction may depend on which limb is producing each of two different movements.

Another form of asymmetry has been noted for the bimanual Fitts task. The limb moving the shorter of disparate bilateral distances for this aiming task exhibits greater temporal bias than the longer distance limb (Corcos, 1984; Kelso et al., 1979). The accommodation noted in our studies is also asymmetric with respect to individual task (limb) requirements. Greater structural and metrical accommodations have been noted in the unidirectional limb than in the reversal limb (Swinnen et al., in press). This is indicated by greater differences in spatial topology and net work between unimanual and bimanual performance for the unidirectional movement than for the reversal movement. Together, these findings indicate that disparate limb

---

[2]Because very few combinations of bimanual spatiotemporal trajectories have been examined, it is perhaps premature to suggest that all discrete bimanual actions are drawn toward metrical and structural symmetry. Other actions may be naturally drawn toward other coordinated states. It does appear, however, that bilateral movements that are fairly similar (but that differ in a few critical respects) do exhibit this form of interlimb attraction. Moreover, the general principles discussed here may well apply to a divergence from *any* preferred state. The authors are indebted to R.A. Schmidt for pointing out this limitation.

movements performed concurrently differentially affect each other both metrically and structurally.

Additional variables affect the strength of interlimb attraction toward a common trajectory. The consistent effect of frequency on attractor strength for oscillatory actions implies that movement kinematics may influence the attraction of discrete movements as well. A recent study has shown that slower movements do exhibit less metrical and structural coupling than faster ones for our discrete task described above (Swinnen, Walter, Vandendriessche, & Serrien, 1991). Faster movements yielded greater similarity in net work produced by each limb and larger cross-correlations of acceleration traces. Another study using this task indicated that initially slowing, then progressively increasing, velocity across practice trials until the criterion speed was reached promoted greater interlimb decoupling than practicing at the faster criterion speed throughout acquisition (Walter & Swinnen, in press). These findings initially suggest that speed and/or movement time may influence the degree of bimanual synchronization.

It is somewhat unclear whether interlimb attraction is mediated kinematically or kinetically, however, because these factors generally covary in studies that intentionally manipulate only the former variable (Walter & Swinnen, 1987). Indeed, an independent manipulation of kinematics and kinetics seems to favor kinetics as a critical factor. Increasing torque requirements by adding an inertial load to the reversal limb increased coupling for our task even when kinematic consistency was maintained across conditions (Figure 13.2; Walter & Swinnen, 1990a, 1990b). The converse manipulation, changing kinematics with relatively constant kinetic levels

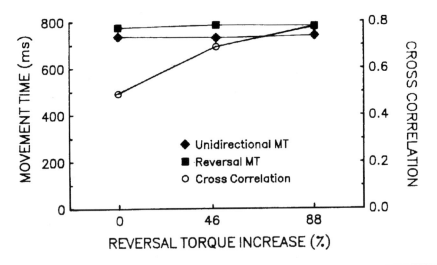

Figure 13.2   Increase in bilateral coupling with increased reversal torque requirements despite temporal constancy across conditions.
Note. From ''Kinetic Attraction During Bimanual Coordination'' by C.B. Walter and S.P. Swinnen, 1990, *Journal of Motor Behavior*, **22**(4), pp. 451-473. Reprinted with permission of the Helen Dwight Reid Educational Foundation. Published by Heldref Publications, 1319 18th Street N.W., Washington, DC 20036-1802. Copyright 1990.

across conditions, failed to elicit this effect (Walter & Swinnen, 1990b). This is not inconsistent with the proposition that frequency is a parameter influencing attraction for oscillatory actions, because increased frequency is often accompanied by increased torque. Moreover, the acceleration pattern for a sinusoidal oscillation is itself a sinusoidal oscillation offset from the position pattern by π. This suggests that torque-time phasing is very similar to position-time phasing for bimanual oscillations. The fact that limb coupling can also be observed between subjects through visual feedback alone, however, suggests that other forms of information influence interlimb attraction as well (R.C. Schmidt, Carello, & Turvey, 1990).

How Does a Forced Departure From the Preferred Coordinated State Affect Movement Variability?

Marteniuk et al. (1984) observed that bimanual spatial variability was always greater than unimanual variability for aiming movements with similar goals. Subjects were instructed to move quickly and accurately and had a spatial goal but not a specific movement time goal. The unimanual/bimanual difference provides preliminary evidence that interlimb interactions themselves contribute to motor variability. They reported mixed findings for the variability of bimanual movements of different distances compared with those of similar length. Sometimes variable error (VE) increased for the divergent condition (i.e., as compared with a movement of similar amplitude in the symmetrical condition), and in other cases it did not. Sherwood and Canabal (1988) did not compare unimanual and bimanual conditions, but they did note that the temporal variability of subjects performing a simultaneous, four-limb lever task was greater than that for subjects moving the limbs in a sequential manner. Sherwood (1989) also observed greater spatial variability for subjects moving different distances concurrently with contralateral limbs (again all four) than for subjects moving the same distance. The majority of this evidence suggests that bimanual actions that are different metrically display greater variability than those that are metrically symmetric.

Swinnen and Walter (1991) noted that temporal VE was greater for subjects performing our bimanual task than for subjects producing each pattern unimanually for the 1st day of practice. The interaction between limb movements again appears to provide a source of variability that is not present for unimanual actions. In a separate study, temporal variability during a bimanually asymmetric condition tended to be greater than that during a symmetric condition (Swinnen, Young, Walter, & Serrien, 1991), particularly for the unidirectional movement. Bilateral movements that require the same spatial and metrical specifications thus appear to be more consistent than those that require different specifications; greater movement variability is associated with a departure from the preferred, symmetric state.

How Does Practice Mediate the Stability of These Actions?

The greater initial variability noted by Sherwood and Canabal (1988) for simultaneous movements than for movements performed sequentially for their four-limb task disappeared over practice trials. The correlation between contralateral movement times (MTs) (adjusted for differences in variability) increased over practice, however, indicating an increase in temporal coupling and suggesting a shift toward the attractor rather than away from it. Sherwood (1989) also reported an increase in correlations between contralateral distances and contralateral MTs, as well as a reduction in bilateral timing differences, over 125 practice trials, when subjects attempted to move different distances. An associated decrease in overall spatial error was also noted, which, together with the increased correlations, again suggests convergence toward the preferred coordinated state. The error in the "different" conditions remained greater than that in the "same" condition, throughout practice. Sherwood (1990) recently noted that the degree of interlimb spatial assimilation decreased for this task over practice trials for three of the four limbs, indicating successful decoupling of the limbs. Although spatial VE decreased with practice for the group moving different distances, it remained greater than the variability for the group moving the same distance with both limbs.

Two of our studies have examined the effect of providing kinematic information feedback on bimanual dissociation. Both studies found that subjects with augmented feedback reduced interlimb coupling over practice trials more than those without it (Figure 13.3a; Swinnen et al., 1990; Swinnen & Walter, 1991). Variable error also decreased for the feedback groups, but it remained higher than the VE displayed by groups that lacked the feedback and were therefore less successful at decoupling the action (e.g., Figure 13.3b). The greater variability in the kinematic feedback group is perhaps due to a slow evolution of the intrinsic dynamics. As a result, the new coordinated state appears to be initially achieved at the expense of stability, although additional practice would presumably further reduce variability.

A somewhat different effect was observed when decoupling was achieved by first reducing, then progressively increasing, movement speed (Walter & Swinnen, in press). This manipulation constitutes "adaptive tuning" of the intrinsic dynamics. The notion here was to initially reduce the strength of interlimb attraction by slowing the action, thereby reducing its torque requirements. This was intended to essentially

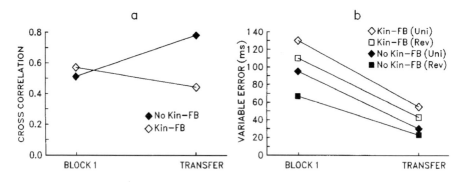

Figure 13.3 (a) Kinematic feedback enhanced bimanual decoupling, but (b) temporal variability remained greater for subjects who received feedback than for the more coupled subjects who did not.

Note. From "The Dissociation of Interlimb Constraints" by S.P. Swinnen, C.B. Walter, J.M. Pauwels, P.F. Meugens, and M.B. Beirinckx, 1990, *Human Performance*, **3**(3), pp. 187-215. Copyright 1990 by Lawrence Erlbaum Associates. Adapted by permission.

flatten the basin of attraction toward symmetry (see Figure 13.5 in the final section) and thereby to facilitate acquisition of the appropriate relative motion pattern. Once this was achieved, speed was progressively increased to the criterion level. Acquisition by this "adaptive" group was compared with that by a "constant" group that practiced at the goal speed throughout training. Both groups received kinematic feedback concerning their performance. The adaptive tuning indeed facilitated metrical and structural decoupling beyond the effect of feedback alone (Figure 13.4a). VE generally decreased over practice trials and, interestingly, remained lower for the adaptive group than for the group constantly practicing at the criterion speed (Figure 13.4b). The concomitant decreases in bias and variability suggest an increase in stability that is perhaps achieved through a change in the intrinsic dynamics of the system (Schöner, 1989; Zanone & Kelso, in press).

Summary

Taken together, these findings support the following preliminary conclusions regarding the four fundamental questions we have posed:

1. It appears that at least some discrete movements are drawn toward both metrical and structural symmetry. This is consistent with the abundant evidence suggesting that continuous movements are strongly attracted toward spatiotemporal synchrony.

2. The strength of attraction depends on both task and laterality factors. Greater accommodation appears for what could loosely be termed the "simpler" of the concurrent movements (shorter, fewer movement segments, etc.), and the coupling is greatest when the more complex movement is performed by the nonpreferred left

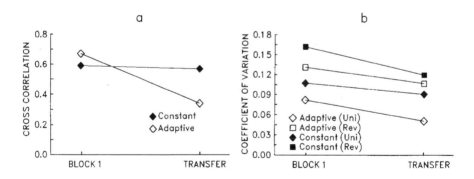

Figure 13.4 Subjects who achieved bimanual independence by a combination of kinematic feedback and adaptive tuning of interlimb attraction displayed lower interlimb coupling (a) and lower variability for each respective limb (b) than subjects receiving kinematic feedback alone. Coefficient of variation was used to adjust for intergroup differences in mean MT.

Note. Figure 13.4 from "Adaptive Tuning of Interlimb Attraction to Facilitate Bimanual Decoupling" by C.B. Walter and S.P. Swinnen, 1992, *Journal of Motor Behavior*, **24**(1), pp. 95-104. Reprinted with permission of the Helen Dwight Reid Educational Foundation. Published by Heldref Publications, 1819 18th Street, N.W., Washington, DC 20036-1802. Copyright 1992.

arm. Attraction toward the preferred state is also directly affected by the kinetic and kinematic requirements of the task. Increased coupling is due not solely to greater neural activation, but perhaps to informational interactions that operate independently of mode as well.

3. Divergence from structural and metrical symmetry produces an increase in variability. In almost every case, bilateral tasks with different movement requirements exhibited greater variability than bilaterally similar actions. The observation that the variability of each limb performing alone is less than that for the same movement performed bimanually also supports the notion that a dynamical interaction between limbs may itself contribute to instability. Together, these findings suggest a source of motor variability that is not observable for single-limb tasks.

4. Subjects who successfully depart from the attractor through practice may initially demonstrate greater variability than those who have more difficulty decoupling the movements. The extent to which a reduced bias and increased consistency are achieved together may, however, depend on the training method used. Preliminary evidence suggests that a combination of kinematic feedback and an initial relaxation of intrinsic dynamical constraints may decrease both coupling and variability more than feedback alone.

Additional work is clearly needed to support these claims, but they provide a point of departure for further investigations concerning the stability of discrete bimanual tasks with disparate metrical and structural requirements.

CONCLUSIONS AND OUTLOOK

The primary goal of this chapter was to determine whether the dynamical principles developed for the stability of bimanual oscillations may apply to discrete coordinated actions. The evidence from discrete tasks indeed appears to generally coincide with that for continuous tasks. Bilateral movements are drawn toward symmetry, analogous to the in phase attractor for continuous actions. A departure from symmetry yields instability in the form of systematic bias and increased motor variability. Finally, the stability of actions that must diverge from the preferred coordinated state generally increases with practice.

The next logical problem is perhaps to empirically map out the basin of attraction for these actions (e.g., Kelso & Schöner, 1988). The appropriate state space in which to convey the map, however, is somewhat unclear for discrete tasks. Relative phase space has proven to be extremely useful for capturing the critical behavior of bilateral oscillations, but the phase of an arbitrary, discrete movement pattern is not formally defined. A potentially useful space to consider for discrete tasks is relative motion space. The attractor in this case is the relative motion trajectory to which behavior is drawn (Walter & Swinnen, 1990b). The trajectory is obtained by plotting the kinematics of one limb as a function of the same variable for the contralateral limb for a given trial. In phase oscillations are represented in relative motion space by a direct linear relationship, and antiphase oscillations exhibit an inverse linear relationship. These trajectories correspond to the two point attractors in one-dimensional relative phase space. The structural attraction toward bilateral symmetry for discrete actions suggests that their preferred relative motion trajectory is linear, and the attraction toward metrical symmetry indicates that the slope of the trajectory is unity. The basin of attraction would presumably take the form of a potential "valley" arising from the minimum of the relative motion potential (Figure 13.5).[3] The effects of torque magnitude and laterality on interlimb coupling indicate that the depth of the potential valley is partially dependent on both of these factors. Finally, the asymmetric accommodation by limbs with dissimilar structural and metrical demands suggests that the valley itself is not symmetric; rather, the preferred trajectory appears to be approached at different rates from different directions.

We will now turn to the effect of practice on the stability of asymmetric bimanual tasks. A useful assumption to adopt for this discussion is that the attractors discussed represent the "natural" behavior of the system at a given point in time. This appears to be a reasonable assumption, because the preferred behavior is often exhibited even when the subject intends to produce a movement with a quite different pattern of coordination. In learning an action that is topologically inconsistent with the natural (in this case, symmetric) pattern, first attempts perhaps represent a compromise between the natural state and the desired state. Movement away from the attractor toward the desired state results initially in a systematic bias and often in an increase in variability. Once the desired coordination is acquired, consistency can be gained through repetitive attempts at the appropriate parameterization (Newell, 1985). The increased consistency perhaps reflects an alteration of intrinsic dynamics (Schöner, 1989; Zanone & Kelso, in press).

Independently examining the bias and variability of movement patterns (and outcome measures) over practice may provide insights into this learning process. Practice accompanied by kinematic feedback increases the correctness of the new

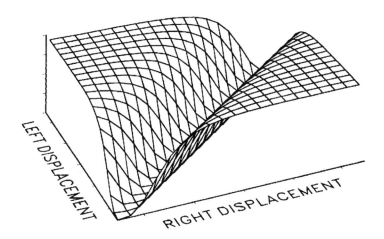

Figure 13.5   Potential "valley" demonstrating attraction toward relative motion symmetry.

---

[3]This figure is provided purely for illustration. The exact form of the valley is not based on empirical evidence and therefore does not explicitly conform to the description in the text.

pattern as revealed by a decreased bias toward the preferred coordinated state and by reduced variability. Motor variability, however, appears to remain at a higher level than that displayed by subjects who are less successful at decoupling (Swinnen et al., 1990). Augmented feedback alone perhaps provides "behavioral information" that contributes to the vector field (basin of attraction) determined by intrinsic dynamics (Schöner, 1990). The greater variability for the decoupled subjects suggests a "struggle" between the intrinsic dynamics and the superimposed information, reminiscent of von Holst's "struggle" between the magnet effect and the natural frequency for coupled oscillators of different endogenous rhythms (Rosenblum & Turvey, 1988; von Holst, 1973). Combining kinematic feedback with adaptive tuning of intrinsic dynamics (Walter & Swinnen, in press) may facilitate learning in a somewhat different way. Tuning the attractor (i.e., decreasing the depth of the potential valley) may allow for a more direct alteration of the dynamics of the system, such that new topological forms are more readily acquired. The intrinsic vector field is perhaps itself altered rather than combined with the additive effect of a superimposed field based on the feedback. Tuning system dynamics, in addition to providing information feedback to facilitate stability during skill acquisition, is a technique that may merit further exploration.

Another present goal was to determine whether bimanual interactions provide a source of variability that is absent for unimanual tasks. The evidence is quite strong that this is the case for both continuous and discrete actions. What might this variability be due to? It is useful to note that many complex systems exhibit apparent variability without the assumption of an ad hoc stochastic process, through deterministic chaos (e.g., Abraham & Shaw, 1983; Gleick, 1987; Stewart, 1990). A relatively simple mapping (the cubic map) that demonstrates symmetry in the form of reflection in the origin, for example, generates a chaotic attractor that might be interpreted as random variability (Stewart, 1990). This example is provided more for heuristic purposes than as a potential account for the present observations. But the general principle of apparently random time-series behavior emerging through a deterministic process is one that may be relevant here, and indeed it is discussed in greater detail elsewhere in this volume. Perhaps one useful goal is to separate the sources of observed motor variability into stochastic and deterministic processes.

We would finally like to reiterate our view that the "elemental" approach noted earlier remains a useful complement to the dynamical framework adopted here in the search for accounts of motor variability. Useful variability principles can indeed be obtained from the study of very simple actions. It appears, however, that the elemental view is not likely to develop a veridical explanation of the stability of complex behavior. The fallibility of such reductionism when applied to complex systems in general is succinctly described by noted evolutionary biologist Stephen J. Gould: "We are tied to historical habits of thought . . . and methods of procedure (reduction to component parts as a mode of explanation, rather than direct study of interaction). When we fail to recognize that these are habits of inertia rather than nature's truths, new paths are closed off" (1987, p. 187).

REFERENCES

Abraham, R.H., & Shaw, C.D. (1982). *Dynamics—The geometry of behavior. Part I. Periodic behavior.* Santa Cruz: Aerial Press.

Abraham, R.H., & Shaw, C.D. (1983). *Dynamics—The geometry of behavior. Part II. Chaotic behavior.* Santa Cruz: Aerial Press.

Baker, G.L., & Gollub, J.P. (1990). *Chaotic dynamics.* Cambridge, England: Cambridge University Press.

Cohen, L. (1971). Synchronous bimanual movements performed by homologous and non-homologous muscles. *Perceptual and Motor Skills, 32,* 639-644.

Corcos, D.M. (1984). Two-handed movement control. *Research Quarterly for Exercise and Sport, 55,* 117-122.

Crutchfield, J.P., Farmer, J.D., Packard, N.H., & Shaw, R.S. (1986, December). Chaos. *Scientific American, 255,* 46-57.

Deutsch, D. (1983). The generation of two isochronous sequences in parallel. *Perception and Psychophysics, 34,* 331-337.

Fitts, P.M. (1954). The information complexity of the human motor system in controlling the amplitude of a movement. *Journal of Experimental Psychology, 47,* 381-391.

Franz, E.A., Zelaznik, H.N., & McCabe, G. (1991). Spatial topological constraints in a bimanual task. *Acta Psychologica, 77,* 137-151.

Freeman, W.J. (1991, February). The physiology of perception. *Scientific American, 264,* 78-85.

Gelfand, I.M., & Tsetlin, M.L. (1971). Mathematical modeling of mechanisms of the central nervous system. In I.M. Gelfand, V.S. Gurfinkel, S.V. Fomin, & M.L. Tsetlin (Eds.), *Models of the structural-functional organization of certain biological systems* (pp. 1-22). Cambridge, MA: MIT Press.

Gleick, J. (1987). *Chaos.* New York: Viking.

Gould, S.J. (1987). *An urchin in the storm.* New York: Norton.

Guthrie, E.R. (1935). *The psychology of learning.* New York: Harper.

Haken, H., Kelso, J.A.S., & Bunz, H. (1985). A theoretical model of phase transitions in human hand movements. *Biological Cybernetics, 51,* 347-356.

Hatze, H. (1986). Motion variability—its definition, quantification, and origin. *Journal of Motor Behavior, 18,* 5-16.

Holst, E. von (1973). *The behavioral physiology of animal and man* (R. Martin, Trans.). Coral Gables, FL: University of Miami Press. (originally published in 1939)

Kelso, J.A.S. (1984). Phase transitions and critical behavior in human bimanual coordination. *American Journal of Physiology: Regulatory Integrative Comparative Physiology, 15,* R1000-R1004.

Kelso, J.A.S., Holt, K.G., Rubin, P., & Kugler, P.N. (1981). Patterns of human interlimb coordination emerge from the properties of non-linear, limit cycle oscillatory processes: Theory and data. *Journal of Motor Behavior, 13,* 226-261.

Kelso, J.A.S., Putnam, C.A., & Goodman, D. (1983). On the space-time structure of human interlimb co-ordination. *Quarterly Journal of Experimental Psychology, 35A,* 347-375.

Kelso, J.A.S., & Scholz, J.P. (1985). Cooperative phenomena in biological motion. In H. Haken (Ed.), *Complex systems: Operational approaches in neurobiology, physics, and computers* (pp. 124-149). New York: Springer-Verlag.

Kelso, J.A.S., & Schöner, G. (1988). Self-organization of coordinative movement patterns. *Human Movement Science*, **7**, 27-46.

Kelso, J.A.S., Southard, D.L., & Goodman, D. (1979). On the coordination of two-handed movements. *Journal of Experimental Psychology: Human Perception and Performance*, **5**, 229-238.

Klapp, S.T. (1979). Doing two things at once: The role of temporal compatibility. *Memory and Cognition*, **7**, 375-381.

Kugler, P.N., Kelso, J.A.S., & Turvey, M.T. (1980). On the concept of coordinative structures as dissipative structures: I. Theoretical lines of convergence. In G. Stelmach & J. Requin (Eds.), *Tutorials in motor behavior* (pp. 3-47). Amsterdam: North-Holland.

Kugler, P.N., & Turvey, M.T. (1987). *Information, natural law, and the self-assembly of rhythmic movement*. Hillsdale, NJ: Erlbaum.

Marteniuk, R.G., MacKenzie, C.L., & Baba, D.M. (1984). Bimanual movement control: Information processing and interaction effects. *Quarterly Journal of Experimental Psychology*, **36A**, 335-365.

Newell, K.M. (1985). Coordination, control, and skill. In D. Goodman, R.B. Wilberg, & I.M. Franks (Eds.), *Differing perspectives in motor learning, memory, and control*. Amsterdam: North-Holland.

Packard, N.H., Crutchfield, J.P., Farmer, J.D., & Shaw, R.S. (1980). Geometry from a time series. *Physical Review Letters*, **45**, 712-716.

Pollatou, E. (1991). *Interference during bimanual oscillations with frequency ratios of 1:1, 2:1, and 3:1*. Unpublished master's thesis, University of Illinois at Chicago.

Robinson, G.H., & Kavinsky, R.C. (1976). On Fitts' law with two-handed movement. *IEEE Transactions on Systems, Man, and Cybernetics*, **6**, 504-505.

Rosenblum, L.D., & Turvey, M.T. (1988). Maintenance tendency in co-ordinated rhythmic movements: Relative fluctuations and phase. *Neuroscience*, **27**, 289-300.

Schmidt, R.A. (1988). *Motor control and learning*. Champaign, IL: Human Kinetics.

Schmidt, R.A., Zelaznik, H.N., Hawkins, B., Frank, J.S., & Quinn, J.T. (1979). Motor output variability: A theory for the accuracy of rapid motor acts. *Psychological Review*, **86**, 415-451.

Schmidt, R.C., Carello, C., & Turvey, M.T. (1990). Phase transitions and critical fluctuations in the visual coordination of rhythmic movements between people. *Journal of Experimental Psychology: Human Perception and Performance*, **16**, 227-247.

Scholz, J.P., & Kelso, J.A.S. (1989). A quantitative approach to understanding the formation and change of coordinated movement patterns. *Journal of Motor Behavior*, **21**, 122-144.

Schöner, G. (1989). Learning and recall in a dynamic theory of coordination patterns. *Biological Cybernetics*, **62**, 39-54.

Schöner, G. (1990). A dynamic theory of coordination of discrete movement. *Biological Cybernetics*, **63**, 257-270.

Schöner, G., & Kelso, J.A.S. (1988). Dynamic pattern generation in behavioral and neural systems. *Science*, **239**, 1513-1520.

Sherwood, D.E. (1989). The coordination of simultaneous actions. In S.A. Wallace (Ed.), *Perspectives on the coordination of movement*. Amsterdam: North-Holland.

Sherwood, D.E. (1990). Practice and assimilation effects in a multilimb aiming task. *Journal of Motor Behavior*, **22**, 267-291.

Sherwood, D.E., & Canabal, M.Y. (1988). The effect of practice on the control of sequential and simultaneous actions. *Human Performance*, **1**, 237-260.

Stewart, I. (1990). *Does God play dice?* Cambridge, MA: Blackwell.

Swinnen, S.P., De Pooter, A., & Delrue, S. (1991). Moving away from the in phase attractor during bimanual oscillations. In P.J. Beek, R.J. Bootsma, & P.C.W. van Wieringen (Eds.), *Studies in perception and action* (pp. 315-319). Amsterdam: Rodopi.

Swinnen, S.P., & Walter, C.B. (1988). Constraints in coordinating limb movements. In A.M. Colley & J.R. Beech (Eds.), *Cognition and action in skilled behaviour* (pp. 127-143). Amsterdam: North-Holland.

Swinnen, S.P., & Walter, C.B. (1991). Towards a movement dynamics perspective on dual-task performance. *Human Factors*, **33**, 367-387.

Swinnen, S.P., Walter, C.B., Beirinckx, M.B., & Meugens, P.F. (in press). Dissociating the structural and metrical specifications of bimanual movement. *Journal of Motor Behavior*.

Swinnen, S.P., Walter, C.B., Pauwels, J.M., Meugens, P.F., & Beirinckx, M.B. (1990). The dissociation of interlimb constraints. *Human Performance*, **3**, 187-215.

Swinnen, S.P., Walter, C.B., & Shapiro, D.C. (1988). The coordination of limb movements with different kinematic patterns. *Brain and Cognition*, **8**, 326-347.

Swinnen, S.P., Walter, C.B., Vandendriessche, C., & Serrien, D. (1991). The effect of movement speed on upper-limb coupling strength. Manuscript submitted for publication.

Swinnen, S.P., Walter, C.B., & Willekens, V. (1991). Preferred phase relationships in bimanual oscillations with a 2/1 frequency ratio. In P.J. Beek, R.J. Bootsma, & P.C.W. van Wieringen (Eds.), *Studies in perception and action* (pp. 320-324). Amsterdam: Rodopi.

Swinnen, S.P., Young, D.E., Walter, C.B., & Serrien, D.J. (1991). Control of asymmetrical bimanual movements. *Experimental Brain Research*, **85**, 163-173.

Turvey, M.T., Rosenblum, L.D., Schmidt, R.C., & Kugler, P.N. (1986). Fluctuations and phase symmetry in coordinated rhythmic movements. *Journal of Experimental Psychology: Human Perception and Performance*, **12**, 564-583.

Turvey, M.T., Shaw, R.E., & Mace, W. (1978). Issues in the theory of action: Degrees of freedom, coordinative structures and coalitions. In J. Requin (Ed.), *Attention and performance VII* (pp. 557-595). Hillsdale, NJ: Erlbaum.

Walter, C.B., Corcos, D.M., & Swinnen, S.P. (1990). *An experimentally-determined space for the study of multilimb coordination*. Paper presented at the research consortium of the meeting of the American Alliance for Health, Physical Education, Recreation and Dance, New Orleans, LA.

Walter, C.B., & Swinnen, S.P. (1987). The nature of coupling during bimanual actions. In J.M. Flach (Ed.), *Proceedings of the Fourth Annual Mid-Central Human Factors/Ergonomics Conference*, 299-305.

Walter, C.B., & Swinnen, S.P. (1990a). Asymmetric interlimb interference during the performance of a dynamic bimanual task. *Brain and Cognition*, **14**, 185-200.

Walter, C.B., & Swinnen, S.P. (1990b). Kinetic attraction during bimanual coordination. *Journal of Motor Behavior*, **22**, 451-473.

Walter, C.B., & Swinnen, S.P. (in press). Adaptive tuning of interlimb attraction to facilitate bimanual decoupling. *Journal of Motor Behavior*.

Welford, A.T. (1976). *Skilled performance: Perceptual and motor skills*. Glenview, IL: Scott, Foresman.

Yamanishi, J., Kawato, M., & Suzuki, R. (1979). Studies on human finger tapping neural networks by phase transition curves. *Biological Cybernetics*, **33**, 199-208.

Zanone, P.G., & Kelso, J.A.S. (in press). The evolution of behavioral attractors with learning: Nonequilibrium phase transitions. *Journal of Experimental Psychology: Human Perception and Performance*.

Acknowledgments

The authors wish to thank the editors and Richard A. Schmidt for their constructive comments on earlier drafts of this paper. Appreciation is also extended to Hong Yan Pan for help in the preparation of the figures.

Chapter 14

Fluctuations in Interlimb Rhythmic Coordination

Michael T. Turvey

Center for the Ecological Study of Perception and Action
University of Connecticut, Storrs, Connecticut

Haskins Laboratories, New Haven, Connecticut

Richard C. Schmidt

Tulane University, New Orleans, Louisiana

Center for the Ecological Study of Perception and Action
University of Connecticut, Storrs, Connecticut

Peter J. Beek

Free University, Amsterdam, The Netherlands

Center for the Ecological Study of Perception and Action
University of Connecticut, Storrs, Connecticut

Our focus is on the fluctuations evident in the assembling and maintaining of a comparatively uncomplicated and very basic interlimb coordination in which a person moves two limbs, or segments of them, rhythmically at the same frequency. This interlimb coordination pattern typifies the common skills of walking and running, but it is also the cornerstone pattern of uncommon skills such as drumming and juggling. The behavioral simplicity of this coordination pattern of two (or more) body parts moving together at the same tempo is deceptive, as recognized many years ago by the behavioral physiologist von Holst. Investigating the rhythmic fin movements of *Labrus*, a fish that swims with its longitudinal axis immobile, von Holst (1939/1973) observed two archetypal scenarios. In one, the fins maintained a fixed phase relation and oscillated at the same frequency; von Holst referred to this as *absolute coordination*. In the other observed scenario, there was an absence of phase and frequency locking; von Holst referred to this as *relative coordination*. Over a period of observation, examples of both scenarios would be seen intermittently, and whenever one dominated, signs of the other would still be present. An interfin relation that was on the average a strong case of absolute coordination would be peppered by momentary deviations from the strict mode-locked state. Similarly, an interfin relation that was on the average a strong case of relative coordination would be peppered by momentary advances toward phase and frequency locking. On the basis of such observations, von Holst concluded that even when absolute coordination was achieved, competition among rhythmic units (to proceed at one's own pace) remained, and even when relative coordination was occurring, a tendency to cooperate (to proceed at the pace of the others) was still in evidence. He referred to the competitive aspect as the "maintenance tendency" and to the cooperative aspect as the "magnet effect."

A METHODOLOGY FOR INVESTIGATING INTERLIMB COORDINATION

Interlimb coordination in humans exhibits many of the same basic qualities as interfin coordination observed in *Labrus*. They can be demonstrated in a paradigm that permits systematic control over the characteristic frequencies of individual rhythmic units such that a pair of such units can be of identical, similar, or very different eigenfrequencies (Kugler & Turvey, 1987). The paradigm is motivated by a major methodological implication of von Holst's work, namely, that wide-ranging conditions of competing frequencies across to-be-coordinated components are needed to reveal the dynamics of coordination. As depicted in Figure 14.1, a and b, a person is seated and holding a pendulum. The pendulums can vary physically in shaft length and the mass of the attached bob. Because of these physical magnitudes, a person's comfortable swinging of an individual pendulum about an axis in the wrist (with other joints essentially immobile) will exhibit a preferred dynamic. That is, the pendular motions will tend to a particular frequency and a particular amplitude (Kugler & Turvey, 1987). The intrinsic dynamic is not strictly that of a gravitational pendulum, however. The person must use chemical energy in the muscles to sustain the rhythmic movement and must establish a pattern of muscular cocontractions to maintain fairly even periods and amplitudes from cycle to cycle. Nevertheless, it is reasonable to assume that the eigenfrequency of a "wrist-pendulum system" is closely approximated by the angular eigenfrequency of the equivalent simple

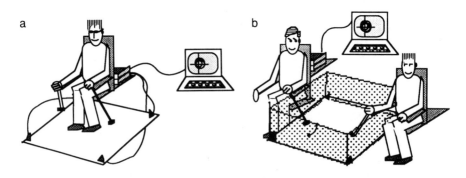

Figure 14.1   The experimental method for investigating interlimb coordination of pendular rhythmic movements. (a) Within-person coordination. (b) Between-persons coordination.

gravitational pendulum, $\omega = 2\pi f = (g/L)^{1/2}$, where L is the simple pendulum length and g is the constant acceleration due to gravity and f is frequency. The quantity L is calculable from the magnitudes of shaft length, added mass, and hand mass, through the standard methods for representing any arbitrary rigid body oscillating about a fixed point as a simple pendulum (den Hartog, 1950; Kugler & Turvey, 1987).

It is evident from the preceding that for bimanual tasks, if the pendulums held in each hand differ in physical dimensions (length, mass), then their eigenfrequencies, their intrinsic dynamics or maintenence tendencies, will not correspond. The component rhythmic units will be in frequency competition.

COARSE-GRAINED
AND FINE-GRAINED FLUCTUATIONS

Consider a subject attempting to swing two pendulums comfortably in antiphase (peaks are separated by 180° or π) at the same frequency, that is, in 1:1 frequency lock. Recordings of the subject's behavior are started from the point in time at which the subject reports having established the requested coordination pattern to his or her own satisfaction. Figure 14.2, a-c, shows, for three subjects (Schmidt, Beek, Treffner, & Turvey, 1991), the continuous phase relation between the two units under three conditions, one in which the two pendulums are nearly identical (left/right ratio Ω of uncoupled eigenfrequencies is nearly unity, $\Omega = 1.03$), and two in which the two pendulums are different, with the eigenfrequency of the left pendulum lower than that of the right pendulum, in the one case ($\Omega = 0.52$), and greater than that of the right pendulum, in the other case ($\Omega = 1.91$). The main features to be noted about Figure 14.2 are that: (a) in all three Ω conditions there is considerable moment-to-moment variation in phase, that is, the moment-to-moment interlimb pattern is primarily that of *relative coordination*; (b) the average deviation from the intended phase relation is larger for $\Omega = 0.52$ and $\Omega = 1.91$ than for $\Omega = 1.03$; and (c) the direction of the deviation from antiphase (180° or π) depends on which rhythmic unit, right or left, had the higher eigenfrequency (compare $\Omega = 0.52$ and $\Omega = 1.91$). There are both fine-grained fluctuations (Feature [a]) and coarse-grained fluctuations (Feature [b]).

The features of the interlimb coordination of pendular rhythmic movements displayed in Figure 14.2 comport with von Holst's observations on interfin coordination. They reinforce two impressions. The first is that mode locking is imperfect, with an attraction to modes (well-defined frequency and phase relations) without a locking into modes. In a phrase, *the coordination is relative with mode attraction, not absolute with mode locking*. The latter may express an important and very general design principle of biological movement systems: Coordination patterns that are very close to being mode locked will be stable; however, to the extent that they are not rigidly mode locked, they will be adaptable. A strategy of gravitating toward mode-locking regimes but operating on the edges of them permits both persistence and change in coordination patterns (Beek, 1989a, 1989b; Beek, Turvey, & Schmidt, 1992; Kelso, DeGuzman, & Holroyd, 1990; Turvey, 1990; Turvey & Beek, 1990). The second impression reinforced by the features evident in Figure 14.2 is that

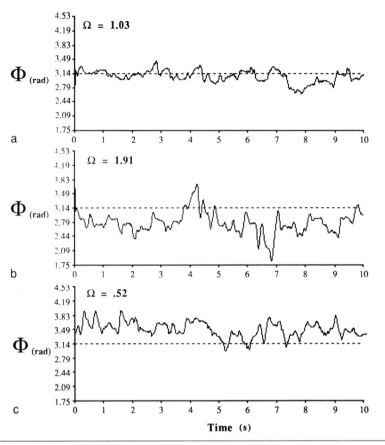

Figure 14.2   (a-c) Examples of relative phase time-series as a function of Ω. (Within-person coordination).

although the phase relation between limbs may be thought of conveniently as a static property characterizing an interlimb coordination, it is in fact a mean or average of a continuously fluctuating process. In general terms, the presence of fluctuations is a consequence of the complexity of a system—multiple components functioning at multiple space and time scales (Chandler, 1987; Soodak & Iberall, 1987). The reinforced impression is that for a biological movement system, as with all complex systems, the observed macroscopic properties reflect statistical laws governing dynamical fluctuations.

The coordination patterns evident in Figure 14.2 can be seen in another behavioral context. Figure 14.3, a and b, shows the continuous relative phase for an interlimb coordination between two people (Schmidt, 1988; Schmidt & Turvey, manuscript submitted for publication). One person swings a pendulum in the right hand, one person swings a pendulum in the left hand, and between the two of them they achieve antiphase 1:1 frequency locking by watching the motions of each other's pendulum (see Figure 14.1b). Comparison of Figures 14.2 and 14.3 suggests that

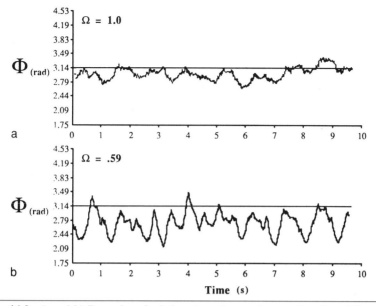

Figure 14.3   (a and b) Examples of relative phase time-series as a function of Ω. (Between-persons coordination).

the description given to interlimb coordination within a person, and the principles used to account for it, may be applicable to interlimb coordination between two persons (Schmidt, Carello, & Turvey, 1990). The dynamics at work in fashioning coordinated states may well generalize over different experimental settings involving very different numbers of microcomponents (e.g., the nervous system of one person vs. the nervous systems of two people) and different perceptual bases for the coordinations (haptic perceptual system vs. visual perceptual system).

OBSERVABLES

The experiments and analyses presented below are directed at understanding the coarse-grained and fine-grained fluctuations of 1:1 frequency locking—the deviation of the average phase from the intended phase, and the time variations in relative phase, respectively. The experiments are connected by a concern for the relations among a common set of measurable quantities, namely, Φ, ω_c, and Ω or $\Delta\omega$. The phase relation Φ between limbs is classified as a *coordination variable*. It can also be described as an order parameter (e.g., Haken, Kelso, & Bunz, 1985; Schmidt et al., 1990), a collective variable that captures the spatiotemporal organization of the component subsystems and changes more slowly than the variables (e.g., velocity, amplitude) characterizing the states of the component subsystems. A slightly more technical paraphrase of the preceding is that an order parameter is a fluctuating variable, the average value of which signifies the order or broken symmetry in a system (Chandler, 1987). The average value of Φ is Φ_{ave}, the intended value is Φ_Ψ (with Ψ symbolizing "intended"), and the variance is σ_Φ^2, as measured by the total power of the spectral density analysis of the Φ time series. The common or coupled frequency ω_c of the limbs is classified as an *energy parameter*. It can also be classified as a control parameter, a variable that is held constant during a given "dynamical run" (Jackson, 1989), and is changed across dynamical runs, and changes in it are associated with changes (bifurcations) in the order parameter at particular critical values. Thus, experiments have shown that as ω_c is increased in steps, an antiphase interlimb coordination will, at some value of ω_c, give way abruptly to an inphase interlimb coordination (Kelso, Scholz, & Schöner, 1986; Schmidt et al., 1990). The ratio Ω of the uncoupled eigenfrequencies (ω_{left}, ω_{right}) of the rhythmic subsystems, or the difference $\Delta\omega$ between them, is classified as a *competition parameter*, in deference to von Holst's intuitions about the nature of interlimb coordination; during a dynamical run the eigenfrequency competition between the rhythmic subsystems will be constant. In the bimanual wrist-pendulum system paradigm depicted in Figure 14.1, the competing eigenfrequencies are calculated, as discussed above, as the frequencies of the corresponding simple gravity pendulums.

The empirically determined relations among the above observables are summarized in two stages, with the first stage focusing on Φ_{ave} and the second stage focusing on σ_Φ^2. With the empirical relations among the observables in place, a dynamical analysis is presented of interlimb rhythmic coordination aimed at providing a theoretical underpinning for the observed dependencies.

OBSERVATIONS ON COARSE-GRAINED FLUCTUATIONS: DEVIATIONS OF Φ_{ave} FROM Φ_Ψ

When a given phase relation is intended, how well is it achieved in the face of the eigenfrequency competition between the component rhythmic units and the tempo at which they are coupled?

Frequency Locking Is Achievable Independently of Ω (or $\Delta\omega$) and Φ_Ψ

It is evident from Figure 14.2 that considerable variation underlies the interlimb coordination of pendular rhythmic movements as indexed by the time series of the coordination variable Φ, especially when there is eigenfrequency competition between the component rhythmic units. In each of the coordination runs depicted in Figure 14.2, the task of the subject was to assemble and maintain a condition of 1:1 frequency locking. Was this goal thwarted by the eigenfrequency competition? Figure 14.4, a-c, shows that over wide variation in Ω (or $\Delta\omega$), subjects were successful in achieving a common frequency for the two wrist-pendulum systems when $\Phi_\Psi = \pi$ and ω_c is the "most comfortable frequency." The mean frequency of the right unit and the mean frequency of the left unit were almost identical; that

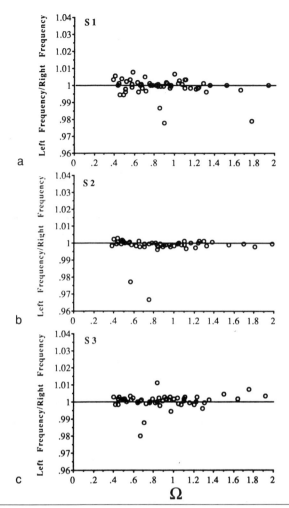

Figure 14.4 (a-c) 1:1 frequency locking as a function of Ω for three subjects under the conditions of $\Phi_\Psi = \pi$ and ω_c = comfort frequency.

is, absolute coordination was achieved in the average state independently of the magnitude of the eigenfrequency competition between the rhythmic units (Rosenblum & Turvey, 1988; Schmidt et al., 1991). In addition, the ability to achieve 1:1 frequency locking when ω_c is the "most comfortable frequency" holds equally for $\Phi_\Psi = 0$ and $\Phi_\Psi = \pi$ (Schmidt, Shaw, & Turvey, in press; Turvey, Rosenblum, Schmidt, & Kugler, 1986).

Φ_{ave} Deviates From Φ_Ψ as a Function of Ω (or $\Delta\omega$)

The preceding makes clear that, although there are fluctuations in Φ during the interlimb coordination of pendular rhythmic movements (as shown in Figures 14.2

and 14.3), the average state of the coordination satisfies 1:1 frequency locking. The subjects had an additional goal in the experiments from which the sample data presented in Figures 14.2 and 14.3 were drawn, namely, to achieve a given Φ_{ave}. Inspection of Figures 14.2 and 14.3 reveals, as noted, that with eigenfrequency competition this goal was not achieved; Φ_{ave} deviates from Φ_ψ. This observation is elaborated in detail in Figures 14.5 and 14.6, a-c. In the experiments of Rosenblum and Turvey (1988; also reported in Schmidt et al., 1991), Φ_{ave} was systematically less than or greater than $\Phi_\psi = \pi$, in accordance with the difference in the uncoupled eigenfrequencies of the two wrist-pendulum systems. In Figure 14.5 the competition parameter is expressed as Ω, and the combined data set of all three subjects in Rosenblum and Turvey (1988) are shown over the full experimental range of Ω. In Figure 14.6, a-c the competition parameter is expressed as $\Delta\omega$, with values limited to the range in which the dependency of the Φ_{ave} deviation from $\Phi_\psi = \pi$ is linear. The data for the individual subjects are shown.

The systematic dependence of Φ_{ave} on Ω (or $\Delta\omega$) seems to be very general. It has been observed at the level of neural oscillatory processes (Stein, 1973) and holds for the interlimb coordination of pendular rhythmic movements achieved visually between two people (Schmidt, 1988; Schmidt & Turvey, manuscript submitted for publication).

Dependency of Φ_{ave} on Ω (or $\Delta\omega$) Is the Same for $\Phi_\psi = 0$ and $\Phi_\psi = \pi$

As noted, there is evidence to suggest that $\Phi_\psi = \pi$ is less stable than $\Phi_\psi = 0$. The evidence comes from the hysteresis of the spontaneous transition induced by ω_c (with increasing ω_c, a transition in Φ_{ave} occurs in the direction $\pi \to 0$ but not in the

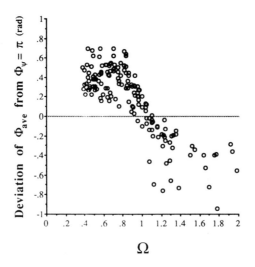

Figure 14.5 Φ_{ave} deviation from $\Phi_\psi = \pi$ as a function of Ω. The data are of the three subjects whose 1:1 frequency-locking behavior is depicted in Figure 14.4.

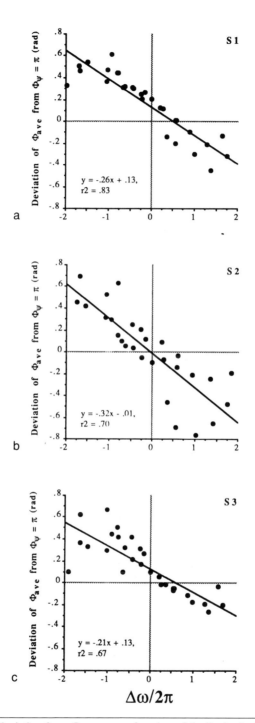

Figure 14.6 Φ_{ave} deviation from $\Phi_\Psi = \pi$ as a function of $\Delta\omega/2\pi$ in the linear range. The data of the three subjects collected in Figure 14.5 are shown individually (a, b, c).

direction $0 \rightarrow \pi$; Haken et al., 1985; Schmidt et al., 1990) and from the observation that "clock variance" (Wing & Kristofferson, 1973) is larger for $\Phi_\Psi = \pi$ (Turvey et al., 1986; see below). The question arises, therefore, of whether or not the relation between Φ_{ave} and Ω (or $\Delta\omega$) is affected differentially by the intended phase mode. The results summarized in Figures 14.5 and 14.6 are for $\Phi_\Psi = \pi$. Are they replicated for $\Phi_\Psi = 0$? Figure 14.7 shows that the deviation of Φ_{ave} from Φ_Ψ as a function of Ω is identical for the two intended modes (Schmidt et al., in press; Turvey et al., 1986).

ω_c and Ω (or $\Delta\omega$) Interact in Determining Φ_{ave} and Do So in the Same Way for Both $\Phi_\Psi = 0$ and $\Phi_\Psi = \pi$

The results summarized in the preceding paragraphs were for the condition in which ω_c is the most comfortable 1:1 frequency lock. But as just noted with increase in ω_c, a coordination satisfying $\Phi_\Psi = \pi$ switches spontaneously to $\Phi_\Psi = 0$. According to the dynamic modeling of this behavioral transition, increases in ω_c bring about a change in the potential function (approximately, the energy landscape) underlying the stability of the two coordination modes of 0 and π, with the local minimum at $\pm\pi$ annihilated at a critical value of ω_c (Haken et al., 1985; Schöner, Haken & Kelso, 1986). It can be expected, therefore, that Φ_{ave} will be affected by ω_c. Figure 14.8, a and b, shows this to be the case. There were three values of ω_c, one at comfort frequency and two higher than comfort frequency. For a given value of Ω, larger ω_c values were associated with larger deviations of Φ_{ave} from Φ_Ψ. Inspection of Figure 14.8 shows that the degree of deviation was the same for both $\Phi_\Psi = 0$ and $\Phi_\Psi = \pi$. Additionally, inspection of Figure 14.8 reveals that the deviation of Φ_{ave}

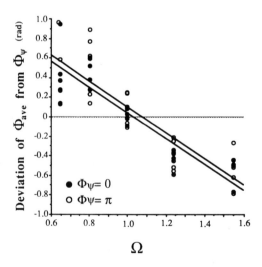

Figure 14.7 Φ_{ave} deviation from $\Phi_\Psi = 0$ and $\Phi_\Psi = \pi$ as a function of Ω. Data points are the mean values of five subjects for each of five values of Ω.

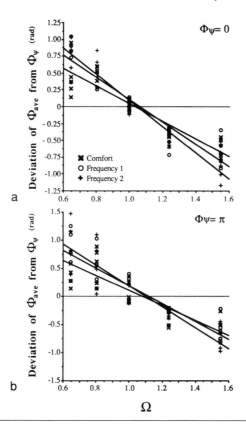

Figure 14.8 (a and b) The interaction between ω_c and Ω in determining the Φ_{ave} deviation from Φ_ψ. Deviations increase with increasing ω_c (the steepest function is for Frequency 2, and the shallowest function is for the comfort frequency). Data points are the mean values of each of five subjects for each combination of ω_c and Ω.

from Φ_ψ induced by ω_c was greater the larger the value of Ω; that is, ω_c and Ω interact in determining the magnitude of Φ_{ave}.

OBSERVATIONS ON
FINE-GRAINED FLUCTUATIONS: σ_ϕ^2

The preceding has underscored that the mean phase departs systematically from intended phase as a function of the eigenfrequency competition and coupled frequency. Does the variance associated with these mean states similarly exhibit a systematic dependence on the control parameters?

σ_ϕ^2 Is a Function of Ω (or $\Delta\omega$)

The measure of σ_ϕ^2 is the total spectral power of Φ. This quantity is the summation of all the fluctuations at all frequency ranges. Figure 14.9 shows the dependence of σ_ϕ^2 on Ω. The data are shown for the three individual subjects in the experiments of Rosenblum and Turvey (1988), additionally analyzed by Schmidt et al. (1991). The conditions were ω_c = comfort frequency and $\Phi_\Psi = \pi$. That is, as the competition between the eigenfrequencies of the component rhythmic units increased, the fluctuations in the coordination variable magnified systematically. It is worth being reminded, through inspection of Figure 14.4, that for all of the competing eigenfrequency conditions depicted in Figure 14.9, 1:1 frequency locking was achieved in the mean.

σ_ϕ^2 Is Greater for $\Phi_\Psi = \pi$ Than for $\Phi_\Psi = 0$, With the Difference Decreasing as Ω Deviates From 1 (or $\Delta\omega$ From 0)

As noted above, available evidence indicates that $\Phi_\Psi = \pi$ is less stable than $\Phi_\Psi = 0$. Accordingly, one might expect to see an influence of Φ_Ψ on σ_ϕ^2. Figure 14.10 shows the dependence of σ_ϕ^2 on Ω for both $\Phi_\Psi = 0$ and $\Phi_\Psi = \pi$ (Schmidt et al., in press). As can be seen, σ_ϕ^2 is larger for $\Phi_\Psi = \pi$. It can also be seen that at the larger deviations of Ω from 1, the σ_ϕ^2 associated with $\Phi_\Psi = \pi$ is less different from the σ_ϕ^2 associated with $\Phi_\Psi = 0$. There is an interaction between Ω and Φ_Ψ in determining σ_ϕ^2 (ANOVA reveals a significant underadditive relation; Schmidt et al., in press).

As Ω Deviates From 1 (or $\Delta\omega$ From 0), Spectral Peaks Present at $\Omega = 1$ Grow, and New Peaks, at Higher Integer Multiples, Are Added

What composes σ_ϕ^2, and how does this composition change with deviations of Ω from $\Omega = 1$? Power spectra analyses of the Φ time series provide an answer (Schmidt

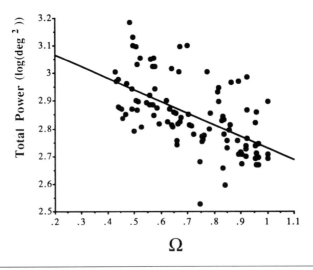

Figure 14.9   Total power of Φ as a function of $\Omega \leq 1$ under the conditions of $\Phi_\Psi = \pi$ and ω_c = comfort frequency (54 data points provided by each of three subjects).

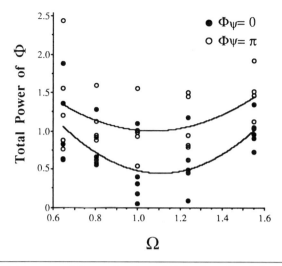

Figure 14.10   Total power of Φ as a function of Ω for $\Phi_\psi = 0$ and $\Phi_\psi = \pi$. Data points are the mean values for each Ω of five subjects.

et al., 1991). When there is no eigenfrequency competition, or very little ($\Omega \approx 1$), there are only two spectral peaks, one at ω_c (referred to as the modal frequency in the subsequent figures) and one at 0.5 ω_c. When the eigenfrequency competition between the component rhythmic units is high ($\Omega \approx 0.5$), there are several spectral peaks—those that were present at $\Omega \approx 1$ and new peaks that are integer multiples of ω_c. Figure 14.11, a and b, gives examples of the power spectra and shows the growth of spectral peaks with Ω.

As Ω Deviates From 1 (or $\Delta\omega$ From 0), Power Becomes More Uniformly Distributed Across Component Frequencies

How should the preceding observation be interpreted? 1:1 frequency locking involves several subtasks—for example, sequential contractions of the flexors and extensors of one limb and sequential contractions of the flexors and extensors of the other limb. Insofar as these latter subtasks are nested within the 1:1 limb pattern, their frequencies are higher. At finer grains we can expect to identify more and more subtasks operating at even shorter time scales. We can also expect subtasks whose time scales are longer than those of the interlimb coordination—for example, a capillary red blood cell flow acting as an oxygen choke limiting the rate of local oxidation in muscle tissue at about 0.01 Hz (Iberall, 1969). For complex systems, a spectral plot shows how the energy of a process is distributed across different kinds of activities at different time scales (Bloch et al., 1971; Iberall, 1977, 1978; Iberall, Soodak, & Hassler, 1978). A prominent observation for such systems is that the power at a given frequency is a function of the inverse of the frequency; that

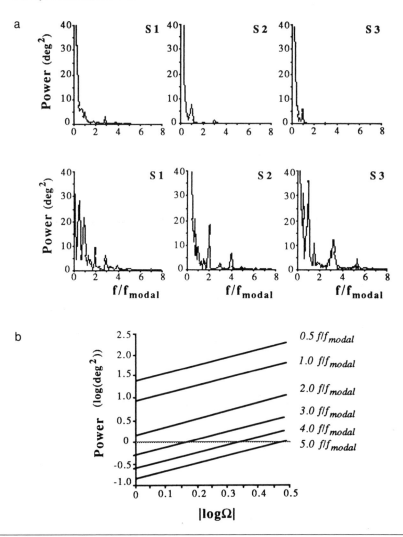

Figure 14.11   (a) Power spectra of Φ for three subjects as a funtion of Ω—upper panel, $\Omega \approx 1$; lower panel, $\Omega \approx 0.5$. (b) Adding of spectral peaks with deviations of Ω from 1. Regression lines are from the data of one subject. The pattern typifies the power spectra dependency on Ω of the three subjects shown in (a).

is, the spectrum can be characterized by a relationship that scales the magnitude of activity to $1/f^\delta$, where f is frequency and δ is a scaling factor. These spectra are called inverse power law spectra and have been reported for a number of biological processes (e.g., Goldberger, Bhargava, West, & Mandell, 1985; Goldberger, Kobalter, & Bhargava, 1986; Koboyashi & Musha, 1982). The scaling factor δ indicates the degree to which the power is concentrated at the lower spectral frequencies. A spectrum that scales power to $1/f^0$ is a flat spectrum of white noise. It represents a random organization of subtasks; fluctuations at any moment are independent of fluctuations at any other moment. Brownian noise, with its frequency spectrum defined by the term $1/f^2$, has fluctuations (and, therefore, subtasks) that are strongly correlated. The $1/f$ spectrum ($\delta = 1$) represents a state of organization such that the power at each frequency is proportional to the inverse of the frequency (power $\propto \tau$). Because power is not localized but distributed across the entire spectrum, fluctuations at any one time scale are only loosely correlated with those of any other time scale. This linear distribution of energy with $1/f$ makes the system adaptive; a perturbation to a process at one time scale will not weaken the system's global integrity (West & Shlesinger, 1990).

The question opening this subsection can be rephrased: How does Ω affect the organization of the rhythmic subtasks producing 1:1 frequency coordination? The spectra of the coordination variable Φ obeys inverse power law behavior, with δ in the neighborhood of 2 but lessening in the direction of 1 as Ω deviates more from 1 (Schmidt et al., 1991). This observation is shown in Figure 14.12, a-c. The figure indicates how the subtasks reacted systematically to Ω, producing varied forms of relative coordination to yield 1:1 frequency locking. The direction of change in δ toward 1 tells us further that the subtask reaction to increasing eigenfrequency competition amounts to a systematic equalizing of activity per time scale. The implication is that, in order to sustain 1:1 frequency locking (the magnet effect) on the average in the face of strong eigenfrequency competition (opposing maintenance tendencies), the movement system assembles coordinations on a fractal plan.

MODELING THE DYNAMICS
OF INTERLIMB RHYTHMIC COORDINATION

In this section, a model will be proposed that captures the coarse (mean deviation of Φ_{ave} from Φ_Ψ) and fine (σ_ϕ^2) fluctuational behavior of the coordination variable Φ as a function of the competition parameter, expressed as Ω or $\Delta\omega$, and the energy parameter, ω_c. Because the behavior to be modeled is the phase locking of biological rhythmic movements, the question can be raised: To what extent can the observed coarse and fine fluctuations in Φ be predicted by dynamical models of coupled oscillators that exhibit phase locking? Implicit in asking this question is the assumption that biological organisms incorporate dynamical regimes in their motor functioning. A dynamical regime is characterized by the "free interplay of forces and mutual influences among the components tending toward equilibrium or steady states" (Kugler, Kelso, & Turvey, 1980, p. 6). Generally speaking, the topology of a regime's equilibrium states or its attractor layout will dictate the observed behavior of the system. Hence, the question to be asked is: To what extent can the attractor layout of a coupled oscillator regime predict (a) the deviation of Φ_{ave} from $\Phi_\Psi = 0$ and $\Phi_\Psi = \pi$, and (b) the magnitude of σ_ϕ^2 as functions of $\Delta\omega$ and ω_c? To anticipate, the presented model will demonstrate that Φ_{ave} is dictated by where the attractor is to be found and σ_ϕ^2 is scaled by the attractor's strength.

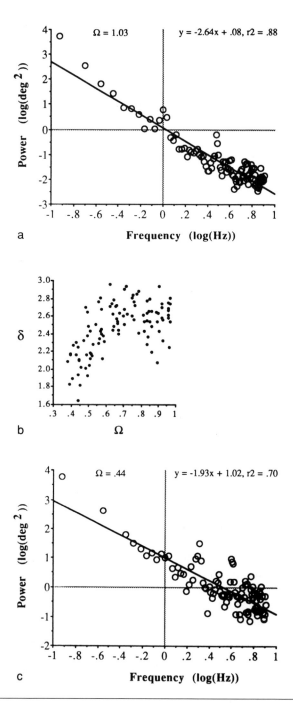

Figure 14.12 Log power versus log frequency as a function of Ω (a and c). The inset (b) shows the flattening in slope of the log power versus log frequency function with deviations of Ω from 1.

Coupled Oscillator Equations

In order to account for observed frequency- and phase-locking phenomena of central pattern generators, Kopell (1988) and Rand, Cohen, and Holmes (1988) have used coupled oscillator equations to model the interaction of neural oscillators (including Stein's 1973 and 1976 crayfish swimmeret data). Under the assumption that the appropriation of these coupled oscillator dynamics is quite general, applying to the microscopic neural scale and also to the macroscopic behavioral scale, we will use the coupled oscillator equations posited for the interactions on the neural level to model the phase-locking phenomena of human interlimb coordination. The model includes two limit-cycle oscillators that are mutually coupled:

$$\dot{x}_1 = F_1(x_1) + G_1(x_1,x_2) \tag{14.1}$$

$$\dot{x}_2 = F_2(x_2) + G_2(x_2,x_1) \tag{14.2}$$

where the x_i are vectors of variables of any dimension for oscillator i, F_i is the limit-cycle component of oscillator i, and G_i is the coupling function that bidirectionally links the two oscillators. Under the assumption that the coupling functions G_i are "weak," the effects of the coupling over each cycle can be averaged to reduce the above complicated set of equations to a simpler set written in terms of the oscillator's phase angles (θ_i):

$$\dot{\theta}_1 = \omega_1 + H_1(\theta_2 - \theta_1) \tag{14.3}$$

$$\dot{\theta}_2 = \omega_2 + H_2(\theta_1 - \theta_2) \tag{14.4}$$

where ω_i is the eigenfrequency of the oscillator. If $\theta_1 - \theta_2 = \Phi$, and we subtract Equation 14.4 from Equation 14.3, then the equation for the change of relative phase Φ is

$$\dot{\Phi} = \omega_1 - \omega_2 + H_1(-\Phi) - H_2(\Phi). \tag{14.5}$$

The equation can be simplified further if the oscillators are considered identical with the same coupling in both directions:

$$\dot{\Phi} = \Delta\omega - 2H(\Phi). \tag{14.6}$$

This equation represents how the relative phase will change as a function of $\Delta\omega$ and the coupling function H. When $\Delta\omega$ and H balance, Equation 14.6 is zero. Assuming that the Φ states of interest are the stable deviations of Φ_{ave} from Φ_Ψ at which Equation 14.6 goes to zero, the equation reduces to the following:

$$\Delta\omega = 2H(\Phi), \tag{14.7}$$

where $\Delta\omega$ and Φ are related analytically.

The Coupling Function

In order to model the behavior of Φ, the coupling function H must be chosen so that the attractor layout has three characteristics. First, the coupling function must have attractive stable points around $\Phi = 0$ and $\Phi = \pi$, regardless of $\Delta\omega$, in order to model the fact that interlimb coordination has two phase modes, symmetric and alternate. If the attractor layout is represented as a potential V where $dV/d\Phi = \dot{\Phi}$, then the attractor's stable point is the bottom of the well where the time rate of change of Φ is 0 (Figure 14.13a). The first condition specifies that the coupling function must have possible potential minima around 0 and π. Further, the position of the attractor's stable point must be a function of $\Delta\omega$. In particular, an increase in $\Delta\omega$ from 0 must cause the stable points to move away from $\Phi_\psi = 0$ and $\Phi_\psi = \pi$ (Figure 14.13a). This condition will ensure that the coupling function models the observed deviations of Φ_{ave} from Φ_ψ with increases in $\Delta\omega$. And lastly, the time to return from perturbation or relaxation time must also be a function of $\Delta\omega$. An increase in the relaxation time means that the potential well becomes broader and

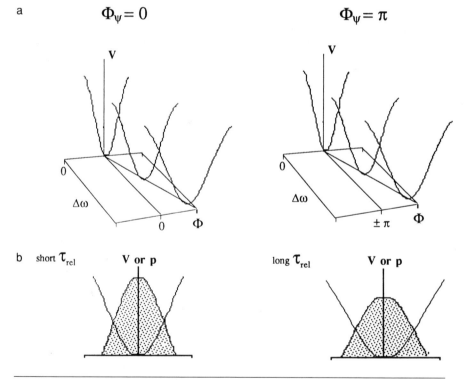

Figure 14.13   (a) The changes in the positioning of the potential minimum and surrounding gradients as a function of $\Delta\omega$ for both $\Phi_\psi = 0$ and $\Phi_\psi = \pi$. (b) The broadening of the potential well with increases in $\Delta\omega$ means a change in the probability distribution of the system's position within the well. Concomitantly, the time to return to the minimum following a perturbation (induced by internal noise) will take longer on the average.

less steep (Figure 14.13a). It also means that the attractor becomes less attractive. If there is a source of noise that is constant in magnitude across perturbations (Schöner et al., 1986), then the broadening of the potential well will increase the magnitude of the fluctuations observed (Figure 14.13b). This condition will ensure that the coupling function models the observed increases in total power of Φ with increases in $\Delta\omega$.

There are two candidate functions for H in the literature that satisfy the above characteristics. Kelso et al. (1990) present a function whose attractor layout contains simultaneously stable points at 0 and π:

$$\dot{\Phi} = \Delta\omega + a\text{Sin}(\Phi) - b\text{Sin}(2\Phi). \tag{14.8}$$

The coupling function of this equation has been used successfully to model the bifurcation that occurs with increasing ω_c from alternate to symmetric phase modes that is seen in quadruped gait transitions and in human interlimb coordination both within-person (Haken et al., 1985) and between-persons (Schmidt et al., 1990). In brief, the magnitudes of a and b are a function of ω_c, and when the ratio of coefficients $b/a = 0.25$, the attractor at $\pm\pi$ disappears, and only the attractor at 0 remains. The other candidate, proposed by Rand et al. (1988), viz.,

$$\dot{\Phi} = \Delta\omega + k\text{Sin}(\Phi), \tag{14.9}$$

has an attractor layout with only a single stable point. That stable point is around 0 for $k > 0$ and around π for $k < 0$. The observed transition from π to 0 phase mode with increasing ω_c can occur if the value of k as a function of ω_c changes from positive to negative. Although both models have the requisite characteristics, the latter model, Equation 14.9, is simpler and fits the data more reliably (see later). Figure 14.14, a and b, demonstrates how this latter model satisfies the deviation in the position of the stable point from $\Phi = 0$ or $\Phi = \pi$ (Figure 14.14a) and the increase in the relaxation time (or strength of the stable point) with changes in $\Delta\omega$ from 0 (Figure 14.14b). Of interest is whether or not the candidate model can predict the fine and coarse fluctuational behavior of Φ, namely, σ_ϕ^2 and the deviation of Φ_{ave} from Φ_Ψ, respectively.

Deviation of Φ_{ave} From Φ_Ψ as a Function of $\Delta\omega$ and ω_c

The success of the model in predicting the deviation of Φ_{ave} from Φ_Ψ can be demonstrated by regressing $\text{Sin}(\Phi)$ on $\Delta\omega/2\pi$. Figure 14.15, a and b, plots Φ_{ave}'s deviation from Φ_Ψ as a function of Ω for the model predictions with data from ω_c = comfort frequency for both $\Phi_\Psi = 0$ (Figure 14.15a) and $\Phi_\Psi = \pi$ (Figure 14.15b).[1] The r^2s of these regressions on comfort mode oscillation Φ_{ave} phase data are significant with the model, predicting 78% of the variance for $\Phi_\Psi = 0$ and 70% of the variance for $\Phi_\Psi = \pi$. The b-weight of the regression indexes the coupling strength

---

[1]The complementary shapes of the plots for the data and the model are due to happenstance. The data are intrinsically nonlinear (as Ω deviates more from 1, the rate of change of Φ_{ave} becomes less), whereas the nonlinearity in the graph for the model arises from plotting the $\Delta\omega_t$ model in Ω coordinates.

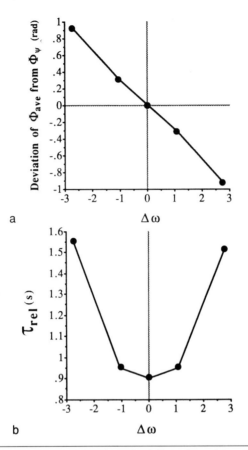

a

b

Figure 14.14   (a) The deviation of Φ_{ave} from Φ_ψ follows from the coupling dynamics represented by Equation 14.9. (b) The change in relaxation time with $\Delta\omega$ follows from the coupling dynamics represented by Equation 14.9.

k. Although the absolute magnitude of k is more for $\Phi_\psi = 0$ (3.76) than for $\Phi_\psi = \pi$ (3.42), the difference is not significant ($p > .05$). When similar regressions were performed on Φ_{ave} data scaled by a metronome to ω_c values of 0.95 Hz or 1.15 Hz, the r^2s were a little higher. The model predicted 80% to 91% of the variance. Further, the absolute magnitude of the coupling strength k decreased as the ω_c increased (0.95 Hz: 3.38 and 3.31; 1.15 Hz: 3.08 and 3.09; for $\Phi_\psi = 0$ and $\Phi_\psi = \pi$, respectively, $p < .05$). Hence, increasing ω_c decreases the strength of the coupling k, and this apparently occurs for both $\Phi_\psi = 0$ and $\Phi_\psi = \pi$.

We could speculate that the rate of change of k with ω_c would become greater for $\Phi_\psi = 0$ than for $\Phi_\psi = \pi$ as ω_c continued to increase, and at some critical frequency would pass through $k = 0$, indicating a transition to $\Phi_\psi = \pi$. In sum, the candidate model, Equation 14.9, does very well in predicting the deviation of Φ_{ave} from $\Phi_\psi = 0$ and $\Phi_\psi = \pi$ observed empirically as a function of $\Delta\omega$ and ω_c. In line with previous accounts (Turvey et al., 1986) that show identical deviations of Φ_{ave} for $\Phi_\psi = 0$ and

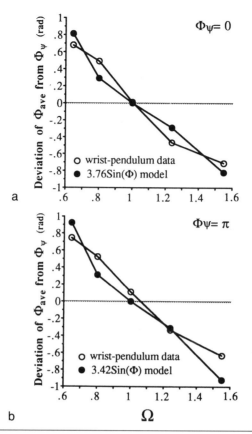

Figure 14.15   Predicted and observed Φ_{ave} deviation from (a) $\Phi_\psi = 0$, and (b) $\Phi_\psi = \pi$, as a function of Ω.

$\Phi_\psi = \pi$ with variation in Ω, the model predicts equal coupling strengths for the two phase modes. However, with larger increases in ω_c, the magnitudes of the two coupling strengths could diverge and lead to a breakdown of $\Phi_\psi = \pi$.

σ_ϕ^2 as a Function of $\Delta\omega$ and Φ_ψ

The model also predicts the magnitude of the variability of Φ around Φ_{ave}, namely, σ_ϕ^2. Figure 14.10 presents how σ_ϕ^2, as measured by log of the total power of Φ, changes as a function of $\Delta\omega$ and Φ_ψ: The total power of $\Phi_\psi = \pi$ is always greater than that of $\Phi_\psi = 0$, and the total power increases for both as the competition between the eigenfrequencies grows. Figure 14.16 replots the total power of Φ as a function of the predicted relaxation times of the candidate model parameterized with the fitted coupling strengths for $\Phi_\psi = 0$ and $\Phi_\psi = \pi$, respectively. The total power of Φ is a function (albeit nonlinear) of the model-predicted relaxation times.

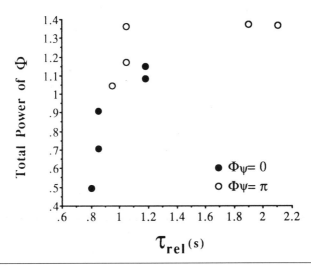

Figure 14.16   The total power of Φ as a function of the predicted relaxation times of the candidate model parameterized with the fitted coupling strengths for $\Phi_\Psi = 0$ and $\Phi_\Psi = \pi$, respectively.

Both the total power and the relaxation times grow larger as the competition $\Delta\omega$ between the rhythmic units increases. Further, both the total power and the relaxation times of $\Phi_\Psi = \pi$ are greater than those of $\Phi_\Psi = 0$. As the relaxation time increases, the strength of the attractor state decreases, with the consequence that microstructural noise of a constant magnitude has a larger effect on the state of the system. The total power indexes this fluctuational state of the system.

In sum, the phase-locked coupled-oscillator model given by Equation 14.9 accommodates a number of the main features of interlimb coordination of rhythmic pendular movements identified in the sections on coarse-grained and fine-grained fluctuations. As yet, it does not address the fluctuational features described when Ω deviates from 1. Other considerations may have to be brought to bear in order to understand the adding of spectral peaks and the gravitation to $1/f$ scaling with increasing deviations from $\Omega = 1$. The section that follows introduces one additional class of considerations that may be required for the fuller account.

CIRCLE MAP DYNAMICS AND FAREY PRINCIPLES

The issue of why σ_ϕ^2 should increase with deviations of Ω from 1 can be looked at from general mode-locking principles. Several qualitative features of the mode-locking structure of driven nonlinear oscillators can be universally described with the aid of the so-called Farey sum (Gonzalez & Piro, 1985). Let $W = K/L$ be the winding number (for K oscillations of one oscillator, the other oscillator completes L oscillations), and let this number be a rational number. Then, Farey summation $\oplus$ is defined as $W_1 \oplus W_2 = (K_1 + K_2)/(L_1 + L_2)$. Two winding numbers that obey the relationship $|K_iL_j - K_jL_i| = 1$ are called unimodular or mod 1 numbers and are said to be adjacent; their Farey sum is called their mediant. Given the preceding, all the rational numbers between 0 and 1 (and, synonymously, all possible phase-locked regions) can be organized hierarchically into a tree in the following manner: Start with the parents 0/1 and 1/1 and form successive layers of mediants. The first level of the tree is defined as $1/2 = 0/1 \oplus 1/1$; the second level is given by $1/3 = 0/1 \oplus 1/2$ and $2/3 = 1/2 \oplus 1/1$, and so on. The Farey tree up to Level 4 is shown in Figure 14.17. Its branching structure summarizes all the possible mode locks that complex dynamical systems may attain and their possible bifurcation routes.

To aid the appreciation of the Farey tree as a representation of mode-locking phenomena, consider a circle map—an iteration of a simple equation that takes one point Θ_n on the circumference of a circle to a second point Θ_{n+1} (with $0 \le \Theta < 1$). A popular version is $\Theta_{n+1} = \Theta_n - [(K/2\pi)\sin2\pi\Theta_n)] + \Omega$, where K represents (as k before) the amplitude of the periodic forcing or coupling, Ω represents (as before) the ratio of uncoupled frequencies, and the sinusoidal term represents the effect of the forcing. In the space of the control parameters K and Ω, the mode-locking regions, or ''Arnold tongues,'' (given by recurring Θ values) are distinct when K < 1, with one region for each rational value of Ω (see Figure 14.18). When forcing is weak to moderate (K < 1), there are distinct regions of stable entrainment or frequency- and phase-locking with quasiperiodicity found in the regions outside the tongues. Of particular importance to the Farey tree representation are the facts that (a) the mode-locked regions differ in width below K < 1, and (b) narrower regions are interspersed with wider regions along the Ω axis. The widest mode-locked regions (1:1, 1:2, 2:3, 1:3, and so on) are the most stable and the most attractive. These regions correspond to the higher Farey levels. With little noise in a system of coupled oscillators, less stable mode-locked states can be maintained; with significant noise, however, the system is likely to be propelled into a neighboring, more stable region (e.g., 3:4 to 2:3).

Let us now return to the issue of why σ_ϕ^2 should increase with deviations of Ω from 1. If Figure 14.18 is expanded, with a focus on W = 1:1, then it can be seen that the conditions for achieving W = 1:1 become more selective as Ω deviates

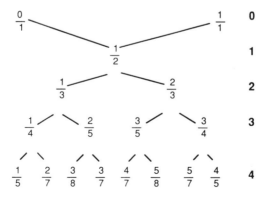

Figure 14.17 The first four levels of the Farey tree.

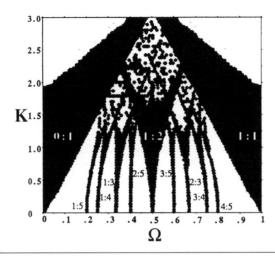

Figure 14.18   Regions of mode locking or Arnold's tongues as a function of Ω and K as generated by the sine circle map equation.
Note. From "Dynamical Substructure of Coordinated Rhythmic Movements" by R.C. Schmidt, P.J. Beek, P. Treffner, and M.T. Turvey, 1991, *Journal of Experimental Psychology: Human Perception and Performance*, **17**(3), pp. 635-651. Copyright 1991 by the American Psychological Association. Adapted by permission of the publisher.

farther and farther from unity (Figure 14.19). The range of permissible K values narrows, and within this narrowing range the magnitude of the most stable regions (largest negative Liapunov exponents), also narrows. In short, as Ω deviates from 1, achieving 1:1 frequency locking becomes more challenging, and the possibility for discovering a stable organization becomes less likely. In sum, an increase in σ_ϕ^2 would be expected if the interlimb coordination of rhythmic pendular movements conforms to a circle map dynamics (Schmidt et al., 1991).

"CLOCK" AND "MOTOR" VARIANCES

To conclude, we consider for completeness a perspective on the fluctuations observed in interlimb rhythmic coordination that attempts to ascribe the fluctuations to independent components of the neural substrate, one functioning as a timekeeper and one implementing the movement pattern. The focus of this perspective, due to Wing and Kristofferson (1973), is not the coordination variable Φ but the cycle time τ of the individual rhythmic units. The cycle-to-cycle variation in τ is analyzed.

The central predictions of Wing and Kristofferson (1973) are that (a) temporally abutting cycles will be negatively correlated and (b) temporally nonabutting cycles, those separated by one cycle or more, will be uncorrelated. The negative sign of the correlation between adjacent cycles means that if one cycle period is by chance shorter than average, the next cycle period will be longer than average, and vice versa (Wing & Kristofferson, 1973; Wing, 1980). The negative dependence does not point to a fed-back influence (in the form of temporal information) from one

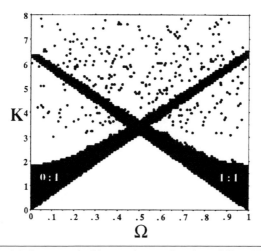

Figure 14.19   The 1:1 frequency-locking mode as a function of Ω and K.
Note. From "Dynamical Substructure of Coordinated Rhythmic Movements" by R.C. Schmidt, P.J. Beek, P. Treffner, and M.T. Turvey, 1991, *Journal of Experimental Psychology: Human Perception and Performance*, **17**(3), pp. 635-651. Copyright 1991 by the American Psychological Association. Adapted by permission of the publisher.

cycle to the next. Rather, it is seen more simply as an inevitable result of the inherent delays in the motor system: A large motor delay in the j-1 cycle will induce a long j-1 cycle period and a short j cycle period. On the average, the predicted negative lag 1 autocorrelation between adjacent cycles is expected to be between 0 (timekeeper variance is very large relative to motor variance) and −0.5 (timekeeper has no variability). Combining equations for the lag 1 autocorrelation and the period variance σ_τ^2 yields measures of the "clock" and "motor" variances:

$$\sigma_c^2 = \sigma_\tau^2 + 2[\sigma_\tau^2 \times \mathrm{cor}(\tau_j \tau_{j-1})] \tag{14.10}$$

and

$$\sigma_{md}^2 = -\sigma_\tau^2 \times \mathrm{cor}(\tau_j \tau_{j-1}), \tag{14.11}$$

respectively.

When people swing two hand-held pendulums in 1:1 frequency locking under the conditions of $\Phi_\psi = \pi$ and ω_c = comfort frequency, the lag 1 autocorrelations are significant, and in the 0 to −0.5 range, whereas the autocorrelations at higher lags are insignificant—a pattern of results complying with the assumption of independent "clock" and "motor" processes (Turvey, Schmidt, & Rosenblum, 1989). Applying Equations 14.10 and 14.11 yields three major observations. Two of the observations are identified in Table 14.1, which shows the results of four multiple regression analyses of the data of Turvey et al. (1989). The analyses are partly new, motivated by the particular quantities that have been the focus of the present chapter. In each analysis, a variance type (left motor, left clock, right motor, right clock) is regressed onto five independent quantities. The table reports the significance level of the independent variable's contribution to the dependent variable. As can be seen, $\Delta\omega$ affects neither the left rhythmic unit's clock variance nor the right rhythmic unit's clock variance; $\Delta\omega$ does, however, affect the motor variance of the right rhythmic unit. Second, the clock variance of one rhythmic unit is markedly dependent on the clock variance of the other rhythmic unit; a similar dependency holds between the motor variances of the two units, but less markedly so. The remaining major observation is from Turvey et al. (1986): Clock variances of both units are larger when 1:1 frequency locking is under $\Phi_\psi = \pi$ than when it is under $\Phi_\psi = 0$; the motor variances, in contrast, are unaffected by Φ_ψ. (From inspection of Table 14.1 it is clear that deviations of Φ from $\Phi_\psi = \pi$ do not further amplify the clock variance.) The three observations together suggest that, in satisfying the task demands of 1:1 frequency locking, (a) the component rhythmic units are organized by a modal task space attractor that acts implicitly as a timekeeping function with greater stability when $\Phi_\psi = 0$ than when $\Phi_\psi = \pi$, and (b) that both units, or at least one (the one on the right), become increasingly irregular in the patterning of muscular activity, with increasing displacements from their preferred tempos, that is, synonymously, with departures from $\Omega = 1$.

Given the interesting implications of the foregoing conclusions for understanding the nature of interlimb coordinations, it would be helpful if future work could place the "clock" and "motor" analysis into a more general dynamics framework. To do so would provide continuity between the differential dependencies of the "clock" and "motor" variances and the coarse-grained and fine-grained fluctuation phenomena previously summarized and interpreted dynamically.

CONCLUDING REMARKS

A cooperative state in a physical system is made up of three levels. For example, the vortex arising in a fluid flowing through a pipe is an in-between level of

Table 14.1   Multiple Regression Analyses of "Clock" and "Motor" Variances

Independent variable	Dependent variable			
	Left motor	Left clock	Right motor	Right clock
$\lvert\Delta\omega\rvert$	ns	ns	.008	ns
ω_c	ns	.01	.002	ns
$\lvert\Phi_{ave} - \Phi_\psi\rvert$	ns	ns	ns	ns
Other-hand motor	.01	ns	.02	ns
Other-hand clock	ns	.0001	ns	.0001
r^2(df = 112)	.10	.47	.45	.41

ns means $p > .05$.

organization that is supported by the interplay between the dynamics of the atomistic (molecular) processes ''below'' and the boundary conditions of the walls of the conduit ''above.'' By analogy, an interlimb coordination of rhythmic pendular movements is three-tiered: The lower level comprises the left and right rhythmic pendular units with their inherent dynamical preferences; the upper level comprises the intention (goal, plan, schema); and the middle level is the interlimb coordination (Kugler & Turvey, 1987; Turvey et al., 1986). All three levels—the (lower) atomistic, the (middle) coordinational, and the (upper) intentional—are characterized dynamically, each governed by one or more attractors. The atomistic level dynamics are expressed in terms of the eigenfrequencies, ω_{left} and ω_{right}. Fluctuations in the periodic timing (σ_τ^2) of a rhythmic pendular unit are least at the eigenfrequency and substantially larger at frequencies other than the eigenfrequency (Rosenblum & Turvey, 1988). More correctly, the atomistic attractors are probably limit cycles with eigenfrequencies close to $\omega = (g/L)^{1/2}$ and amplitudes governed, perhaps, by a constant proportioning of energy to frequency that is scaled to the unit's rotational inertia (Kugler & Turvey, 1987; Kugler, Turvey, Schmidt, & Rosenblum, 1990). Following Schöner and Kelso (1988) and Scholz and Kelso (1990), the intentional level dynamics are expressible through the same collective variables used to characterize the cooperative or coordination level dynamics. Thus, in the experiments reported here, intentional states have been expressed as $\Phi_\Psi = 0$ and $\Phi_\Psi = \pi$. At the intentional level these two phase relations are attractors of equal strength. That is to say, they are describable by potential wells that are of the same depth (identical minimum potential value) and slope (identical gradient descents to minimum value; Schöner & Kelso, 1988). The dynamics of the cooperative or coordination level are expressed through the mean state Φ_{ave}. The analyses of the present chapter have pinpointed the dependence of Φ_{ave} on the atomistic and intentional levels; the collective variable expressing the spatial and temporal order of the middle level (interlimb coordination) was shown to be determined by the interplay of the level above (e.g., $\Phi_\Psi = \pi$) and the level below (e.g., Ω). A major implication of von Holst's (1939/1973) perspective, sketched in the introduction, is that interlimb rhythmic coordinations will be time varying, exhibiting fluctuations in their mean states that are indicative of different attractors competing, both within a level and between levels, for control of the interlimb system's dynamics. Indeed, given that attractor dynamics can be ascribed to all three levels, fluctuations can be expected at each level with the nature of the fluctuations determined, in large part, by the topology of the particular dynamics' stable states.

Focusing, by way of summary, on the cooperative level of interlimb coordination, the experimental results collected above lead to a number of conclusions about movement variability, both specific and general. We highlight two of each, beginning with the specific. In interlimb coordination, the mean state Φ_{ave} identifies, for a given intended stable state Φ_Ψ, the position of the actual stable state in the energy landscape defined by the variation of the potential V with respect to Φ. The manipulation of the atomistic dynamics through Ω results in a repositioning of the attractor in the space of collective variables. Whatever the magnitude of Ω, there are always two attractors (roughly, the symmetric and alternate modes) governing the 1:1 frequency locking of rhythmic units, and they are always separated by π. The strengths of these cooperative or coordinational level attractors, determined conjointly by the dynamics of the upper and lower levels, are revealed by σ_ϕ^2. With the displacement of the attractors due to Ω, the widths and steepnesses of the associated potential wells change, widening and flattening, respectively. The strengths of the two attractors, π distance apart in the potential function, are unequal; the attractor in the vicinity of $\Phi = 0$ is the stronger.

The two general conclusions might be read as variability *principles*, assuming that further research continues to confirm them. The first follows from the observation that the various intra- and interlevel competitions at work in interlimb coordination result in imperfect mode locking: To repeat, interlimb coordination is "relative" with mode attraction, not "absolute" with mode locking. As suggested in the introduction, rhythmic movement patterns might avail themselves of this feature quite generally, given that it provides simultaneously the conditions for stability (resistance to unwanted perturbations) and the conditions for flexibility (docility to intended perturbations). The second follows from the observation that rhythmic movement coordinations are scale free or, synonymously, lack a characteristic time scale. Power spectra of Φ are homogeneous power functions of the form $f^{-\delta}$. This means that changing the frequency scale by any constant factor does not change the frequency dependence of the Φ spectra. The spectra are self-similar, and the underlying processes are statistically self-similar. There are intimations that, as the challenge of assembling a coordination pattern magnifies (greater eigenfrequency competition), the noise progresses from black ($\delta > 2$), through brown ($\delta = 2$), toward pink ($\delta = 1$) (see Figure 14.12b; see Schroeder, 1991, for a detailed discussion of the "colors" of noise). That is, the underlying processes progress toward a condition in which the power is distributed evenly across the entire spectrum—each band of frequencies 1 octave wide contains the same energy. This arrangement would endow the movement system, with its multiple degrees of freedom at multiple process scales, with an optimal balance of stability and flexibility. Movement systems might apply this fractal principle whenever and wherever the challenges of coordination require it. As is apparent, the two putatively general principles identified in the foregoing achieve the same consequence. Future research on movement variability might be expected to provide other insights into the smart ways in which coordinated movement patterns manage their stability and adaptability demands.

REFERENCES

Beek, P.J. (1989a). *Juggling dynamics*. Amsterdam: Free University Press.

Beek, P.J. (1989b). Timing and phase locking in cascade juggling. *Ecological Psychology*, **1**, 55-96.

Beek, P.J., Turvey, M.T., & Schmidt, R.C. (1992). Autonomous and nonautonomous dynamics in coordinated rhythmic movements. *Ecological Psychology*, **4**, 65-96.

Bloch, E.S., Cardon, S., Iberall, A., Jacobowitz, D., Kornacker, K., Lipetz, L., McCulloch, W., Urquhart, J., Weinberg, M., & Yates, F. (1971). *Introduction to a biological systems science* (NASA Report No. CR-1720). Springfield, VA: National Technical Information Service.

Chandler, D. (1987). *Introduction to modern statistical mechanics*. Oxford: Oxford University Press.

Den Hartog, J.P. (1948). *Mechanics*. New York: Dover.

Goldberger, A.L., Bhargava, V., West, B.J., & Mandell, A.J. (1985). On a mechanism of cardiac electrical stability: The fractal hypothesis. *Biophysics Journal*, **48**, 525-528.

Goldberger, A.L., Kobalter, K., & Bhargava, V. (1986). 1/f-like scaling in normal neutrophil dynamics: Implications for hematologic monitoring. *IEEE Transactions on Biomedical Engineering*, **33**, 874-876.

Gonzalez, D.L., & Piro, O. (1985). Symmetric kicked self-oscillators: Iterated maps, strange attractors, and symmetry of the phase locking Farey hierarchy. *Physical Review Letters*, **55**, 17-20.

Haken, H., Kelso, J.A.S., & Bunz, H. (1985). A theoretical model of phase transitions in human hand movements. *Biological Cybernetics*, **51**, 347-356.

Holst, E. von (1973). Relative coordination as a phenomenon and as a method of analysis of central nervous system function. In R. Martin (Ed. and Trans.), *The collected papers of Erich von Holst. Vol. 1. The behavioral physiology of animal and man*. Coral Gables, FL: University of Miami Press. (Paper originally published in 1939)

Iberall, A.S. (1969). A personal overview, and new thoughts in biocontrol. In C. Waddington (Ed.), *Towards a theoretical biology. 2. Sketches*. Chicago: Aldine.

Iberall, A.S. (1977). A field and circuit thermodynamics for integrative physiology. I. Introduction to the general notions. *American Journal of Physiology*, **233**, R171-R180.

Iberall, A.S. (1978). A field and circuit thermodynamics for integrative physiology. III. Keeping the books—a general experimental method. *American Journal of Physiology*, **234**, R85-R97.

Iberall, A.S., Soodak, H., & Hassler, F. (1978). A field circuit thermodynamics for integrative physiology. II. Power and communicational spectroscopy in biology. *American Journal of Physiology*, **234**, R3-R19.

Jackson, E.A. (1989). *Perspectives of nonlinear dynamics*. Cambridge, England: Cambridge University Press.

Kelso, J.A.S., DeGuzman, G.C., & Holroyd, T. (1990). The self organized phase attractive dynamics of coordination. In A. Babloyantz (Ed.), *Self organization, emerging properties, and learning*. New York: Plenum Press.

Kelso, J.A.S., Scholz, J.P., & Schöner, G. (1986). Nonequilibrium phase transitions in coordinated biological motion: Critical fluctuations. *Physics Letters*, **118**, 279-284.

Koboyashi, M., & Musha, T. (1982). 1/f fluctuation of heartbeat period. *IEEE Transaction on Biomedical Engineering*, **29**, 456-457.

Kopell, N. (1988). Toward a theory of modelling central pattern generators. In A.H. Cohen, S. Rossignol, & S. Grillner (Eds.), *Neural control of rhythmic movements in vertebrates*. New York: Wiley.

Kugler, P.N., Kelso, J.A.S., & Turvey, M.T. (1980). On the concept of coordinative structures as dissipative structures: I. Theoretical lines of convergence. In G.E. Stelmach & J. Requin (Eds.), *Tutorials in motor behavior* (pp. 3-47). Amsterdam: North-Holland.

Kugler, P.N., & Turvey, M.T. (1987). *Information, natural law, and the self-assembly of rhythmic movement.* Hillsdale, NJ: Erlbaum.

Kugler, P.N., Turvey, M.T., Schmidt, R.C., & Rosenblum, L.D. (1990). Investigating a nonconservative invariant of motion. *Ecological Psychology,* **2,** 151-189.

Rand, R.H., Cohen, A.H., & Holmes, P.J. (1988). Systems of coupled oscillators as models of central pattern generators. In A.H. Cohen, S. Rossignol, & S. Grillner (Eds.), *Neural control of rhythmic movements in vertebrates.* New York: Wiley.

Rosenblum, L.D., & Turvey, M.T. (1988). Maintenance tendency in coordinated rhythmic movements: Relative fluctuations and phase. *Neuroscience,* **27,** 289-300.

Schmidt, R.C. (1988). *Dynamical constraints on the coordination of rhythmic limb movements between two people.* Unpublished doctoral dissertation, University of Connecticut, Storrs.

Schmidt, R.C., & Turvey, M.T. (1992). Phase entrainment dynamics of visually coupled rhythmic movements. Manuscript submitted for publication.

Schmidt, R.C., Beek, P.J., Treffner, P., & Turvey, M.T. (1991). Dynamical substructure of coordinated rhythmic movements. *Journal of Experimental Psychology: Human Perception and Performance,* **17,** 635-651.

Schmidt, R.C., Carello, C., & Turvey, M.T. (1990). Phase transitions and critical fluctuations in the visual coordination of rhythmic movements between people. *Journal of Experimental Psychology: Human Perception and Performance,* **16,** 227-247.

Schmidt, R.C., Shaw, B., & Turvey, M.T. (in press). Coupling dynamics in interlimb coordination. *Journal of Experimental Psychology: Human Perception and Performance.*

Scholz, J.P., & Kelso, J.A.S. (1990). Intentional switching between patterns of bimanual coordination depends on the intrinsic dynamics of the patterns. *Journal of Motor Behavior,* **22,** 98-124.

Schöner, G., Haken, H., & Kelso, J.A.S. (1986). A stochastic theory of phase transitions in human hand movements. *Biological Cybernetics,* **53,** 247-257.

Schöner, G., & Kelso, J.A.S. (1988). A dynamic pattern theory of behavioral change. *Journal of Theoretical Biology,* **135,** 501-524.

Schroeder, M. (1991). *Fractals, chaos, power laws.* New York: Freeman.

Soodak, H., & Iberall, A.S. (1987). Thermodynamics and complex systems. In F.E. Yates (Ed.), *Self-organizing systems: The emergence of order* (pp. 459-470). New York: Plenum Press.

Stein, P.S.G. (1973). The relationship of interlimb phase to oscillator activity gradients in crayfish. In R.B. Stein, K.G. Pearson, R.S. Smith, & J.B. Redford (Eds.), *Control of posture and locomotion* (pp. 621-623). New York: Plenum Press.

Stein, P.S.G. (1976). Mechanisms of interlimb coordination. In R.M. Herman, S. Grillner, P.S.G. Stein, & D.G. Stuart (Eds.), *Neural control of locomotion* (pp. 465-487). New York: Plenum Press.

Turvey, M.T. (1990). Coordination. *American Psychologist,* **45,** 938-953.

Turvey, M.T., & Beek, P.J. (1990). Invariants of perception and action. *Proceedings of the Sixth Yale Workshop on Adaptive and Learning Systems* (pp. 201-205). New Haven, CT: Yale University.

Turvey, M.T., Rosenblum, L.D., Schmidt, R.C., & Kugler, P.N. (1986). Fluctuations and phase symmetry in coordinated rhythmic movements. *Journal of Experimental Psychology: Human Perception and Performance*, **12**, 564-583.

Turvey, M.T., Schmidt, R.C., & Rosenblum, L.D. (1989). ''Clock'' and ''motor'' components in absolute coordination of rhythmic movements. *Neuroscience*, **33**, 1-10.

West, B.J., & Shlesinger, M.F. (1990). The noise in natural phenomena. *American Scientist*, **78**, 40-45.

Wing, A.M. (1980). The long and short of timing in response sequences. In G.E. Stelmach & J. Requin (Eds.), *Tutorials in motor behavior* (pp. 469-485). Amsterdam: North-Holland.

Wing, A.M., & Kristofferson, A.B. (1973). Response delays amd the timing of discrete motor responses. *Perception & Psychophysics*, **14**, 5-12.

Acknowledgments

Preparation of this manuscript and the research reported were supported by NSF grant BNS-8811510. Correspondence concerning this article should be addressed to M.T. Turvey, CESPA, U-20, 406 Babbidge Road, University of Connecticut, Storrs, CT 06269.

PART IV

VARIABILITY AND MOVEMENT DISORDERS

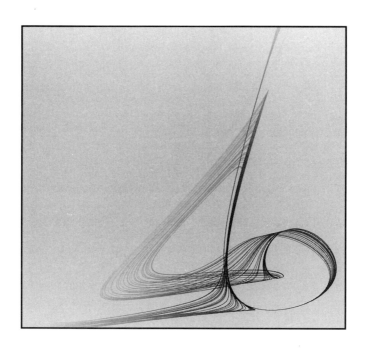

Chapter 15

Slicing the Variability Pie: Component Analysis of Coordination and Motor Dysfunction

Richard Ivry
University of California at Berkeley

Daniel M. Corcos
University of Illinois at Chicago

As emphasized by the chapters in this volume, the study of variability is essential for understanding coordination. We can choose to focus on the competence of the performer, identifying his or her capabilities and skills. But in doing so, we are acknowledging the limitations of this performer. An important source of constraint underlying these limitations is that the individual performances of a particular task are variable. Differences between individuals may result from differences in variability between individuals on a particular task. On the other hand, individual differences may exist in terms of general abilities required for coordinated movement.

TASK-SPECIFIC APPROACH TO VARIABILITY

One approach to studying variability is to examine constraints imposed by a particular task. For instance, Kelso (1984; Kelso & Ding, this volume, chap. 11) has identified conditions conducive to stable performance on repetitive bimanual rhythmic movements and conditions yielding unstable, variable performance on this same task. Although the same approach may be applied to other tasks, a basic goal is to describe the coordinative structures specific for that task. Research on individual differences has also emphasized the need for examining task-specific constraints. For example, from a large set of correlational studies, Fleishman (1966) concluded that, although there may be a number of basic component abilities shared across tasks, extended practice increased the importance of task-specific sources of variation.

COMPONENT-ANALYSIS APPROACH
TO VARIABILITY

In our research, we have taken a different approach to the study of variability. Like Fleishman, we start with the premise that there are many sources of variability in the performance of any task. Some of these, including certain biomechanical constraints, may be idiosyncratic to that particular task. Others arise because of variability in the operation of component mental processes. Of course, these hypothetical mental processes cannot be directly observed; we must define an appropriate dependent variable that is observable or derived from an overt action—and in doing so, a new problem arises. Our observable measure will likely include many sources of variability, making it difficult to tease out the hypothetical basic component sources.

Our solution to this dilemma has been to devise model tasks that can provide experimental tests of putative mental operations involved in the production of coordinated actions. In designing these tasks, we attempt to isolate one operation in each task, or at least structure that task so that the operation under consideration would be expected to contribute greatly to the total variability. Two such tasks are a repetitive tapping task and a force control task (Keele, Ivry, & Pokorny, 1987). The former was chosen to measure variability of an internal timing system. In contrast, the latter task was chosen to measure variability of processes involved in the regulation of force output.

The tapping task is based on a paradigm introduced by Wing and Kristofferson (1973). On each trial, a computer presents a series of evenly spaced tones. In Keele et al. (1987), the tones were separated by an interval of 400 ms. When ready, the subject begins tapping on a response key, attempting to synchronize his or her responses with the pacing tones. After 12 paced responses, the tones cease, and the subject continues tapping for 30 more intervals. Subjects are generally able to respond at a mean rate close to the target interval. Our primary measure of the consistency of the internal timing system is the standard deviation of the intertap intervals during the unpaced portion of the trials.

Each trial in the force control task begins with the presentation of a horizontally oriented line on a computer monitor. The vertical placement of the line indicates the target force for that trial: The higher the line, the greater the target force. The computer then plays a tone, after which the subject makes an isometric force pulse on a strain gauge. Feedback is presented graphically to indicate if the produced force is greater or less than the target force. Six responses with feedback are made in this manner. Then the subject makes six more responses to the same target without feedback. Each response is initiated only after the computer plays a stimulus tone. The consistency of force control is assessed as the standard deviation of the forces produced without feedback.

Thus, in both the timing and the force control tasks, we primarily measure the consistency of the subjects' responses when external information about performance is absent. Although forces were measured during the tapping task, subjects were not given any explicit instructions about regulating force. Similarly, we attempted to minimize timing requirements in the force control task by randomly spacing the stimulus tones.

In separate blocks, the timing and force control tasks were performed with movements of either the index finger or the forearm (Keele et al., 1987). The correlation matrix of the standard deviation scores is shown in Table 15.1. When calculated by task, the correlations were high, reaching .90 on the timing task and .76 on the force task. Correlations by effector were much smaller, ranging from .18 to .34. These results suggest that the two tasks were dependent on different component

Table 15.1   Correlation Matrix Between Timing and Force Control

| | Timing | | Force |
	Finger	Arm	Finger
Timing			
Arm	.90		
Force			
Finger	.30	.34	
Arm	.18	.21	.76

operations. We attributed the timing correlation to the operation of an internal timing system (see also Keele, Pokorny, Corcos, & Ivry, 1985) and the force correlation to the operation of an internal process regulating force output or a variable correlated with force control. We argued that previous difficulties in accounting for individual differences in coordination may have arisen in part because the selected tasks involved substantial contributions from both the timing and the force control systems.

STUDIES OF TIMING VARIABILITY

In this section, we focus on timing variability. This work is presented in two parts. First, we review our studies of patients with neurological impairments. This work provides converging evidence supporting the hypothesis that timing can be considered one component of motor control. Moreover, the evidence implicates the cerebellum as playing a critical role in the operation of an internal timing system. Deficits in this neural system are associated with increases in timing variability in both movement and perception tasks. The latter finding is perhaps the strongest piece of evidence that a common timing module is exploited across multiple task domains.

Second, we will present some new data exploring an alternative way to assess timing variability. These data complement the correlational work with healthy subjects and the patient research in pointing toward an internal timing system that spans motor and perceptual domains. The method also has the potential for a logical extension of the component approach. Our earlier work has primarily focused on separating variance associated with timing from variance associated with other components of motor control such as force control. One goal of this new approach is to begin a component analysis of the clock itself. That is, in order to understand the operation of an internal clock at a mechanistic level, we believe it will be necessary to develop a model of the component processes that form the clock.

Timing Variability in Patients With Neurological Lesions

There recently has been a burgeoning interest among cognitive psychologists in the study of patients with neurological disorders. Although there are many reasons for this, both theoretical and technological, three are of particular interest for our purposes. First, the logic underlying the study of patient populations is similar to that developed in the individual difference studies with healthy populations. The individual difference studies assume that there are consistent differences in the operation of a given process. A range in performance will be achieved because of this variation: For example, in some individuals, an internal timing process is more consistent than in other individuals. In the patient studies, the range of differences may be amplified as a consequence of the neurological impairment. If the internal timing process is dependent on a particular neural system, then lesions of this system are expected to produce increased timing variability. Note that for both methodologies, the evidence is essentially correlational. In the studies with healthy subjects, the correlations are based on individual variation found in a sample drawn from a homogeneous population. In the patient studies, the correlations are based on group differences that arise as a result of samples being drawn from heterogeneous populations. These populations are developed by categorizing patients according to the neural system(s) affected by their lesions.

Second, the study of neurological patients can provide converging evidence for the utility of a cognitive model. Based on studies with healthy subjects, we have argued that force and timing can be considered to be two relatively independent components of coordination. Although this model does not provide any a priori constraints concerning physiological mechanisms, a reasonable conjecture would be that different neural systems are involved in the operation of these two components. If this were so, then lesions of one neural system should produce a deficit in timing control, whereas lesions of a different neural system should produce a deficit in force control. Double dissociations of this sort are generally interpreted as strong evidence for the existence of two processes in neuropsychological research.

Third, linkage of a particular mental operation with a given neural system is of interest for localization theories of the brain. Many students of behavior have kept an eye on the relation of mind and matter and view neuropsychological research as a valuable tool for providing insight on this issue. Thus, the study of patients can prove useful for further developing our cognitive model as well as for identifying the crucial neural systems required for timing and force control.

Repetitive Tapping Task

Patients with cortical and subcortical motor disorders were tested on the repetitive tapping task (Ivry & Keele, 1989). Patients were assigned to one of three groups, depending on whether their lesions were centered in the cerebellum, basal ganglia, or frontal cerebral cortex. Classification criteria were based on a clinical examination and neuroradiographic data.

The cerebellar group ($n = 27$) included patients with either focal or diffuse lesions. The focal lesions ($n = 11$) were the result of tumor or stroke, and the resultant motor deficits were restricted to the hand ipsilateral to the lesion. The diffuse lesions ($n = 16$) were the result of atrophic processes. In some of the patients with a degenerative disorder, there was evidence of extracerebellar involvement, especially of the pons and olive.

For studying the effects of basal ganglia lesions, we recruited 28 patients with Parkinson's disease. All of these patients were taking L-dopa medication at the time of testing. This treatment ameliorates some Parkinson symptoms, but there were still obvious motor problems in these patients.

The third patient group ($n = 7$) had focal lesions from strokes, centered in the posterior region of the frontal cortex. All of these subjects presented some degree of hemiparesis in the hand contralateral to the lesion, indicating that the lesions included upper limb areas of motor cortex. However, we selected only subjects for whom the deficit was not so severe as to prevent performance of the tapping task.

A fourth group of elderly control subjects ($n = 21$) consisted of healthy people with no history of neurological disease or disturbance. The mean age of this group, 67 years old, was slightly older than the mean ages of the patient groups (range from 51 to 63 years old).

As described previously, each trial of the continuation tapping task required the production of 12 paced and 30 unpaced intervals. The target intertap interval was 550 ms in the patient study. This pace was chosen so that the subjects would not be performing near their maximal rate of tapping. The subjects completed at least 12 tapping trials, grouped into blocks of 6 trials each. As in our studies with healthy subjects (Keele et al., 1985, 1987), the primary measure of interest was the variability of the unpaced intertap intervals.

Figure 15.1 shows the mean standard deviations for the four groups. Two results stand out. First, there was no difference between the Parkinson patients and the control subjects. This result was striking, considering that these patients showed the usual array of Parkinson symptoms, including bradykinesia and rigidity. Despite these deficits, the Parkinson patients were as consistent as the age-matched control subjects. Second, the mean standard deviation for both the cerebellar and the cortical groups was significantly higher than for either the control subjects or the Parkinson patients. There was no difference between the mean standard deviations for the cerebellar and the cortical patients.

The data presented in Figure 15.1 look at overall variability on the tapping task. There are many reasons that a person could have trouble with this task. One reason may be because of inconsistency in an internal timing process. Alternatively, the clock may operate properly, indicating the appropriate time at which a response should be made, but the motor system may have difficulty in executing that response. That is, the clock may correctly determine when a series of responses should occur, but the motor apparatus may introduce variability in implementing those responses. We next turn to a finer grained analysis of the patients' variability on the continuation tapping task.

Analysis of Tapping Data With the Wing-Kristofferson Model

Wing and Kristofferson (1973) proposed a formal model for decomposing the total variability on the repetitive tapping task into two independent sources. The model is described in detail in their paper (see also Wing, 1980; Ivry, Keele, & Diener, 1988). Briefly, the key assumptions of the model are as follows. Each interval is assumed to be the sum of three events. Two of the events are attributed to the implementation system—namely, the time required to implement the key press that

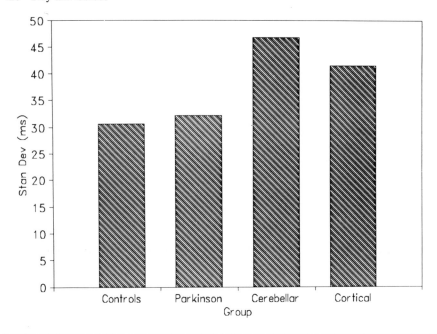

Figure 15.1   Mean standard deviations of the intertap intervals for the four groups on the tapping task.
Note. From "Timing Functions of the Cerebellum" by R. Ivry and S. Keele, 1989, *Journal of Cognitive Neuroscience*, **1**, pp. 136-152. Adapted by permission.

initiates the interval and the time required to implement the key press that terminates the interval. The third event is the interval metered out by a central clock. Wing and Kristofferson treat the clock and implementation processes as two random variables with normal variances. The mean of the clock is set (by the pacing signal) to the target interval, and the mean of the implementation durations, referred to as motor delays, is an unknown constant. Because the two processes are assumed to be independent, the total variability is simply the sum of the variances of the component parts. That is,

$$\sigma_I^2 = \sigma_C^2 + 2\sigma_{MD}^2 \tag{15.1}$$

where c and md stand for clock and motor delay (implementation), respectively.

A critical assumption of the Wing-Kristofferson model is that all of the component events occur independently. Each output from the clock process is assumed to be independent of preceding clock outputs, each motor delay is assumed to be independent of other motor delays, and, as stated previously, all of the clock outputs and motor delays are assumed to be independent of one another. In other words, the model assumes that the task is performed in an open-loop mode. From this assumption, Wing and Kristofferson have shown that an estimate of the variance associated with the implementation process is a function of the covariance between successive intervals, or

$$\sigma^2_{MD} = -\text{autocov}(1) \tag{15.2}$$

A graphic depiction of this formalization is given in Ivry et al. (1988).

The variance associated with the clock can now be estimated by subtraction. The overall variance of the intertap intervals is obtained directly from the data. An estimate of motor delay variance is obtained from the covariance analysis. Subtracting this value from the overall variance will yield an estimate of clock variance.

The Wing-Kristofferson model has received empirical support in a number of studies with healthy subjects (reviewed in Wing, 1980). An alternative test of the model can be made using patients with peripheral neuropathies (Ivry & Keele, 1989; Keele & Ivry, 1987). Given that timing variability correlates across effectors (Keele et al., 1985, 1987), we assume that the clock is a central process, one that is accessible to all effectors. Thus, the model should attribute any increase in variability in peripheral patients to the implementation system. We would not expect lesions of the peripheral nervous system to affect the timing process.

Four patients with peripheral neuropathies were tested (Ivry & Keele, 1989). Their etiologies varied: Two had ulnar nerve damage, one median nerve damage, and one had suffered an entrapped nerve at the shoulder. The patients were selected not so much for a specific lesion, but rather because their coordination problems were the result of a peripheral neuropathy.

One important feature in testing these patients is that their deficits were unilateral. The patients were only impaired when using the hand ipsilateral to the lesion. Thus, a within-subject design can be employed in which performance with an impaired effector is compared to performance with an unimpaired effector. This generally involved comparing tapping performance with the index finger on the left and the right hand. In one case the comparison was between two fingers on the same hand.

Overall, the mean standard deviation when tapping with the unimpaired effector was 28 ms. When tapping with the impaired effector, the mean rose to 34 ms, about a 20% increase. The clock and motor delay estimates are shown in Figure 15.2. There is a negligible increase in the clock estimate for tapping with the impaired hand. In contrast, the motor delay estimate is over 40% larger for the impaired hand. Although the percentage increase varied from 15% to 75% across patients, the impaired hand yielded a higher motor delay estimate on almost 95% of the block-by-block comparisons. Thus, as predicted, the model attributed increased variability in patients with peripheral neuropathies to the implementation process.

Fortified by this neuropsychological validation of the Wing-Kristofferson model, we then performed a series of within-subject comparisons using patients with lesions of the central nervous system (Ivry & Keele, 1989; Ivry et al., 1988). Seven Parkinson patients were recruited to perform the tapping task under two medication conditions. In the *on* condition, the patients were tested while following their normal medication regimen. In the *off* condition, the patients skipped their morning medication period(s) prior to testing. Clinically, this *on-off* manipulation produced a striking effect; all of the patients became much more rigid and bradykinetic off medication. Despite these changes, there was no effect in their performance on the tapping task. The overall variability as well as the estimates of the clock and implementation components were essentially identical under both conditions. As in the group comparison,

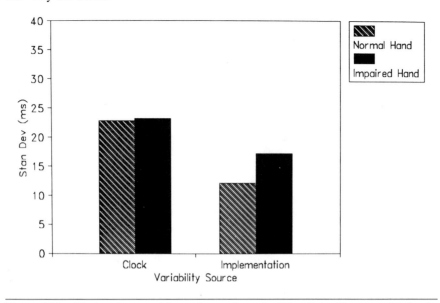

Figure 15.2 Clock and motor delay estimates from the Wing-Kristofferson model for the patients with peripheral neuropathies.

Note. From ''Timing Functions of the Cerebellum'' by R. Ivry and S. Keele, 1989, *Journal of Cognitive Neuroscience*, **1**, pp. 136-152. Adapted by permission.

the within-subject experiment confirmed that lesions of the basal ganglia do not affect tapping consistency or the operation of a central timing process.

Within-subject comparisons were made for seven patients with cortical lesions and eight patients with focal cerebellar lesions (Ivry & Keele, 1989). In all of these cases, the lesions were unilateral, allowing a comparison to be made between an impaired effector and unimpaired effector on the tapping task. The mean clock and motor delay estimates for these two groups are shown in Figure 15.3, a and b. Averaging across the patients within the two groups, the Wing-Kristofferson model attributed the increased variability to both the clock and the implementation components.

However, by averaging within each group, we may have obscured individual deficits that can be identified by the within-subject comparisons. We were unable to identify any such differences by further analysis of the tapping data for patients with cortical lesions. Moreover, the data for some of the cortical patients showed consistent violations of the Wing-Kristofferson model.

A clearer picture emerged in an extended analysis of a group of cerebellar patients with focal lesions (Ivry et al., 1988). Seven patients were tested. Each subject produced a minimum of eight six-trial blocks, four blocks with each hand. The patients were separated into two subgroups, based on neuroradiographic and clinical criterion. For four of the subjects, the lesion foci were lateral, encompassing portions of the cerebellar hemisphere on the impaired side. These patients all presented symptoms associated with hemispheric lesions, notably dysmetria in voluntary movements. The lesions were more medial for the other three patients, and their primary

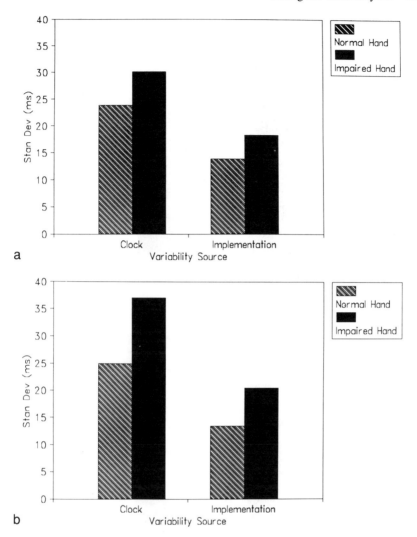

Figure 15.3   Clock and motor delay estimates from the Wing-Kristofferson model for patients with (a) unilateral cortical or (b) cerebellar lesions.
Note. From "Timing Functions of the Cerebellum" by R. Ivry and S. Keele, 1989, *Journal of Cognitive Neuroscience*, **1**, pp. 136-152. Adapted by permission.

symptoms, disturbances of balance and gait, were more typical of these types of lesions.

The results from the Wing-Kristofferson analysis are shown in Figure 15.4, a and b. The increased tapping variability for the patients with lateral lesions was attributed to the clock process. In contrast, the increased tapping variability was attributed to the implementation process for the patients with medial lesions. This result was extremely consistent over tapping blocks: The seven patients produced

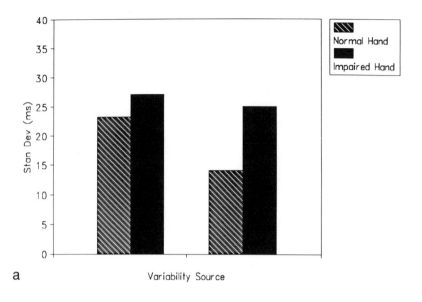

a

Variability Source

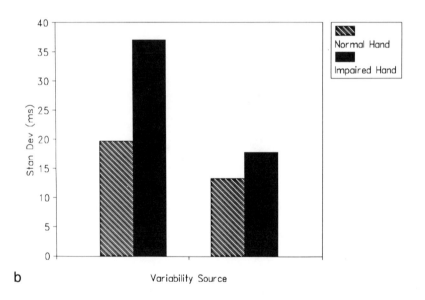

b

Variability Source

Figure 15.4 Clock and motor delay estimates from the Wing-Kristofferson model for patients with lesions centered in either (a) the medial cerebellum or (b) the lateral cerebellum.

Note. From "Dissociation of the Lateral and Medial Cerebellum in Movement Timing and Movement Execution" by R. Ivry, S. Keele, and H. Diener, 1988, *Experimental Brain Research*, **73**, pp. 167-180. Adapted by permission.

a total of 100 six-trial blocks, or 50 impaired-unimpaired comparisons. The pattern described above was reversed in only 1 of the 50 comparisons.

These data indicate that an internal timing system is disrupted following lesions of the lateral regions of the cerebellum. Medial cerebellar lesions may increase tapping variability, but this increase appears to be the result of added noise in processes associated with implementing a response. This double dissociation is in accord with functional models of the cerebellum derived from the different patterns of neural connectivity of the lateral and medial regions (e.g., Allen & Tsukahara, 1974). Much of the output from the lateral regions projects to motor and premotor cortical areas via the ventrolateral thalamus (Asanuma, Thach, & Jones, 1983a; Goldberg, 1985; Schell & Strick, 1984). Medial cerebellar regions generally innervate descending pathways in the brainstem and spinal cord (Asanuma, Thach, & Jones, 1983b; Wilson, Uchino, Maunz, Susswein, & Fukushima, 1978). A crude distinction can be made between neural pathways that ascend and those that descend. The former could presumably contribute to motor planning and programming, whereas the latter would be expected to contribute to movement implementation (Allen & Tsukahara, 1974). As argued previously, setting the temporal requirements of the movement may be one component of motor programming, at least in tasks where timing control is part of the explicit movement goal. Our data suggest that the lateral cerebellum plays a critical role in this process.

Time Perception in Patients With Neurological Lesions

One problem with the tapping data is that the results were quite similar for patients with cerebellar and cortical lesions. Both groups were more variable than the control subjects and the Wing-Kristofferson estimates were problematic for the cortical patients. It is possible that both the cortex and the cerebellum are part of an internal timing system. That is, timing may be a distributed process that involves pathways spanning a number of neural structures. On the other hand, one of these neural areas may be most critical for timing. Lesions of the other area may appear as a timing deficit because of limitations in the Wing-Kristofferson model. Note that the model partitions total variability into two components, labeled the clock and the implementation systems. The implementation estimate is theoretically derived; once this component is subtracted out, the remaining variance is attributed to the clock component. However, variability in other central (i.e., nonimplementation) processes will also be contained in the remainder (Ivry & Keele, 1989). Although Wing and Kristofferson referred to the two components as clock and implementation, a more accurate dichotomy would be central and peripheral components. The clock may be just one part of the central component.

Given these limitations, it is important to consider other tasks that require accurate timing. One task we have employed involves the perception of temporal intervals. On each task, the subject hears four 50-ms tones, grouped into two pairs of two tones each. The first two tones are separated by a fixed interval. In most of our experiments, this interval was set at 400 ms. Then, after a 1-s pause, the second pair of tones is presented. The interval between this pair is varied, and the subject's task is to judge whether the second, comparison interval is shorter or longer than the first, standard interval. Based on the correctness of the subject's response, the duration of the comparison interval is adjusted. After a number of trials, an estimate of perceptual acuity is obtained. This estimate is given as a standard deviation, corresponding to a threshold at which the subject's performance is approximately 75% correct (Pentland, 1980).

If there exists a task-independent internal timing system, then we would expect to find a positive correlation between performance on the motor tapping task and performance on the perception-of-duration task. Keele et al. (1985) obtained a significant correlation of .53 (.60 following a reliability correction) between these two tasks in a study with 32 healthy college students. This result, coupled with the cross-effector tapping correlations, formed the cornerstone for the hypothesis that one component of coordination was an internal timing process.

The perception task can also be used in the patient research. If a particular neural system is part of an internal timing system, then patients with lesions of this system should be more variable in making duration judgments. Indeed, the perception-of-duration task has a major advantage over the tapping task, in that there are no motor requirements. For the tapping task, we selected patients with disorders of movement, but we were constrained in that we could not test patients with the most severe problems, because they were unable to complete the task. The perception task is not similarly constrained; the only requirement is that the patients be able to understand the directions.

Eight patients with cortical lesions, 28 Parkinson patients, 27 patients with cerebellar lesions, and 21 elderly control subjects were recruited (Ivry & Keele, 1989). Most of the subjects had also been tested in the tapping study. The subjects were tested on two perception tasks: the perception-of-duration task and a control task in which they compared the loudness of auditory stimuli. As in the duration perception task, each trial for the control task consisted of two pairs of two tones each. The interval between both pairs was always 400 ms. The volume of the second pair was either more or less intense than the volume of the first pair. The same psychophysical procedure was used to obtain loudness thresholds. This task was included to ensure that any deficit obtained on the perception-of-duration task could not be attributed to a generalized problem with auditory tasks or psychophysical testing procedures.

The results for the two tasks are shown in Figure 15.5, a and b (Ivry & Keele, 1989). Statistical analyses revealed a second double dissociation implicating the cerebellum in timing control. The cerebellar patients were significantly more variable than the control subjects on the perception-of-duration task. The cortical patients performed approximately as well as the control subjects on this task. However, these subjects were significantly more variable on the perception-of-loudness task. Although it was not predicted, we believe this latter result arose because some of the cortical lesions extended into the auditory cortex. Nonetheless, this finding strengthens the perception-of-duration results in two ways. First, it demonstrates that both perception tasks were sensitive enough to identify potential deficits. Second, it emphasizes that the cerebellar deficit on the perception-of-duration task is specific and not the result of a generalized impairment.

Role of the Cerebellum in Other Tasks Requiring Timing

The finding of a deficit on a purely perceptual task following lesions of the cerebellum was exciting. This neural system has traditionally been associated with motor

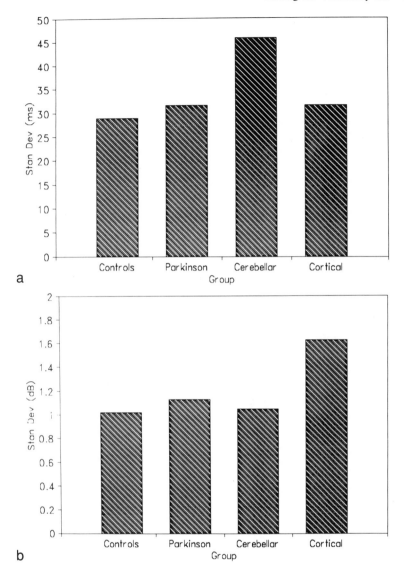

Figure 15.5   Mean standard deviation on (a) the perception of duration and (b) the perception of loudness tasks.
Note. From ''Timing Functions of the Cerebellum'' by R. Ivry and S. Keele, 1989, *Journal of Cognitive Neuroscience*, **1**, pp. 136-152. Adapted by permission.

functions. Our results do not dispute this belief: They simply provide a more precise hypothesis regarding one computational role of the cerebellum. Many movements, especially skilled movements, require precise coordination of both spatial and temporal events. We have hypothesized that the cerebellum can be viewed as an internal timing system. At least one function of this neural structure is to perform the component operations needed to produce the temporal aspects of coordinated movements. Moreover, we believe this computational capability extends beyond the motor domain. The results on the perception-of-duration task indicate that a common internal timing system is invoked across a variety of tasks if those tasks have similar computational requirements.

We have replicated the perception-of-duration results with both auditory and visual stimuli (Ivry & Gopal, in press). We have also found that patients with cerebellar lesions are impaired on a perceptual task in which judgments are made about the velocity of a moving stimulus (Ivry & Diener, 1991). This latter task was selected for two reasons. First, velocity, by definition, involves a computation that occurs over time. Second, lesions of the cerebellum have been associated with eye movement disorders (Aschoff & Cohen, 1971; Ritchie, 1976), and we were interested in whether these disorders might, in part, reflect impaired perception of a to-be-tracked stimulus.

Moreover, we have argued that the cerebellar timing hypothesis can account for a number of disparate functions associated with the cerebellum (Keele & Ivry, in press). Some of these are summarized in Table 15.2. Together with our empirical results, we believe that a compelling argument can be made that timing control can be viewed as a component operation of coordination.

Slope Analysis of Timing Variability

The Wing-Kristofferson model provides one way to partition variance on the repetitive tapping task. Wing and Kristofferson have typically referred to the two subcomponents as clock and implementation (or motor delay). However, as discussed earlier, only the implementation component is theoretically derived; once this is

Table 15.2   Generalization of the Timing Hypothesis to Other Functions Associated With the Cerebellum

Deficit	Timing interpretation and selected reference
Hypermetria in rapid movements	Loss of ability to temporally coordinate agonist/antagonist activity, especially antagonist onset (Hallett, Shahani, & Young, 1985)
Locomotion ataxia	Loss of ability to coordinate phase-phase relations between different limbs (Arshavsky, Gelfand, & Orlovsky, 1983)
Abolition of conditioned learning	Loss of ability to represent temporal relationship of conditioned stimulus to unconditioned stimulus necessary for making conditioned response adaptive (Thompson, 1986)
Efference copy	Loss of ability to anticipate afferent information (Gellman, Gibson, & Houk, 1985)
Cerebellar dysarthria	Deficit in temporally coordinating interarticulatory actions (Ivry & Gopal, in press)

estimated, the remaining variance is, by default, attributed to the clock. We now turn to an alternative approach to partitioning variance on timing tasks.

Logic of Approach and Background/Previous Work

We call this approach a slope analysis. The basic idea is quite simple and has been employed by others (e.g., Getty, 1975; Killeen & Weiss, 1987; Wing, 1980). The variability of a timing system is assumed to increase with the duration of the interval being timed. Thus, if timing variability is measured as a function of interval duration, the slope of this function provides an estimate of clock variability. This hypothesis rests on one critical assumption, namely, that the only duration-dependent process involved in a timing task is the clock. As in the Wing-Kristofferson model, the observed variability is assumed to be the sum of the contributions of a number of independent processes. One of these processes is a timing system, another process may be involved in implementing responses, and there may be other, unknown processes. However, for the slope analysis to be valid, we must postulate that all of the processes other than the timing system are independent of interval duration. If the subject is tapping repetitive intervals of 350 ms or 550 ms, the variability due to the implementation component is invariant.

This assumption is also an essential part of the Wing-Kristofferson model. The model assumes independence of clock and implementation outputs in order to estimate the implementation variability. Wing (1980) reports one test of the model in which clock and motor delay estimates were derived for tapping at 10 different target intervals over the range 220 to 490 ms. As predicted, the motor delay estimate was essentially constant over the different intervals. In contrast, the estimate of clock variance was highly dependent on the produced interval, ranging from a low of approximately 95 ms^2, when the intertap interval was 220 ms, to a high of approximately 465 ms^2, when the intertap interval was 490 ms. These data are one of the strongest sources of support for the Wing-Kristofferson model.

While the basic assumptions of the Wing-Kristofferson model predict that only the clock estimate will vary with interval duration, the exact form of this relationship is dependent on additional assumptions about the mechanisms of timing (see Killeen & Weiss, 1987). Two-process clock-counter models generally predict a linear relationship between variance and interval duration (Abel, 1972; Creelman, 1962). Alternative models in which the time-dependent variability is attributed to a single process such as variable activation times (e.g., Grossberg & Schmajuk, 1989; also Rosenbaum, 1990) predict that the linear relationship will be between the standard deviation and interval duration (general form of Weber's law, see Getty, 1975). Wing (1980) plotted the estimate of clock variance as a function of interval duration and observed a significant linear fit. Over 96% of the variance was accounted for by the linear component. However, a linear relationship appears to be just as strong when the data are replotted with standard deviation depicted on the ordinate (Figure 15.6, a and b). Indeed, the proportion of clock variance accounted for by a linear component actually increases slightly.

Obviously, from a regression analysis it is unclear whether the linear relationship of timing variability as a function of interval duration is in terms of variance or standard deviation. However, the second parameter of the regression analysis, the

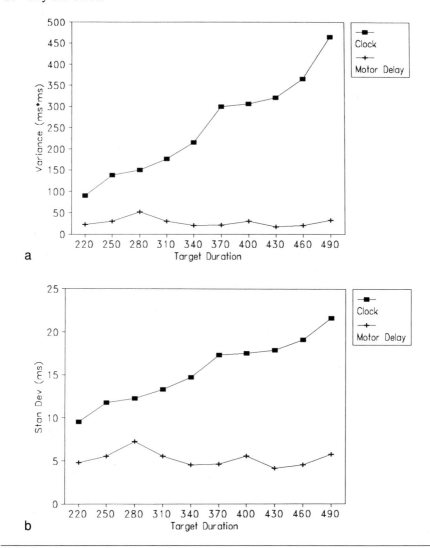

Figure 15.6   Clock and motor delay estimates as a function of target interval. Data are estimated from Wing (1980). Variance is plotted on the ordinate in (a), and standard deviation is plotted on the ordinate in (b).

Note. From "The Long and Short of Timing in Response Sequences" by A.M. Wing. In *Tutorials in Motor Behavior* by G. Stelmach and J. Requin (Eds.), 1980, New York: Elsevier Science Publishers. Adapted by permission.

intercept, may be informative. In the Wing (1980) data, the intercept for the clock variance function is -155 ms^2. In contrast, the intercept for the clock standard-deviation function is essentially 0 ms. On intuitive grounds, we would expect the intercept to be zero: as the interval to be timed approaches zero, the variance should also become negligible. The large negative value yielded by the regression equation for the variance function is problematic, especially considering that variances by definition must be positive. Thus, our working hypothesis is that the standard deviation function is more accurate.

The variance–standard deviation debate is secondary to our current interest. Regardless of the outcome, the basic point is that, assuming linearity, the slope analysis provides an alternative way to estimate clock variability. This approach has a number of strengths in comparison to the Wing-Kristofferson model. First, although Wing (1980) has plotted the clock estimate, there is no need to perform the decomposition into clock and motor delay. The slope analysis can be performed directly on the observed data. This bypasses error that will be introduced by the estimation procedure. This is especially important given that the clock and motor delay components are not estimated independently; in the Wing-Kristofferson model, the clock estimate is made by subtraction. Any error in estimating the motor delay component will also introduce error in the clock estimate. The slope analysis eliminates the decomposition stage.

Second, the slope analysis provides an alternative method for comparing motor timing with perceptual timing. The strength of the Wing-Kristofferson model for motor timing was that it provided an analytic tool for separating central and peripheral sources of variability. The perception task used in our patient studies does not allow for a similar decomposition: All of the variability is treated as a lump whole. However, it is reasonable to assume that there is also peripheral, or nonclock, variability contributing to performance on this task. For example, there may be variability in the perceived onset of the tones. The slope analysis can provide an analytic method for decomposing performance on either motor or perceptual timing tasks. The slope provides an estimate of the clock component. If a common internal clock is involved in both tasks, then the slope values should be comparable. The following sections contain a preliminary report of two experiments testing this prediction.

Slope Experiment 1

In the first experiment, subjects performed the repetitive tapping task and the perception task at four different target durations, 325 ms, 400 ms, 475 ms, and 550 ms. To improve the stability of the data, each subject completed four sessions composed of eight blocks each. Four blocks involved tapping, one block at each duration. Each block consisted of a practice trial and six test trials. The other four blocks tested time perception in which the four durations served as the standard on different blocks. The order of blocks was counterbalanced with the constraint that tapping and perception for a given duration were paired, with the tapping block always preceding the perception block. To date, five subjects have completed this task.

Figure 15.7, a and b, presents the variability data plotted as a function of duration. Variance is plotted on the ordinate in Figure 15.7a, and standard deviation is plotted on the ordinate in Figure 15.7b. Separate functions are shown for the overall variability on the tapping task as well as estimated clock and motor delay components. Note that the current data replicate Wing (1980) and provide strong support for the basic slope prediction derived from the Wing-Kristofferson model. The motor delay estimate varies minimally with duration, whereas the clock estimate increases monotonically.

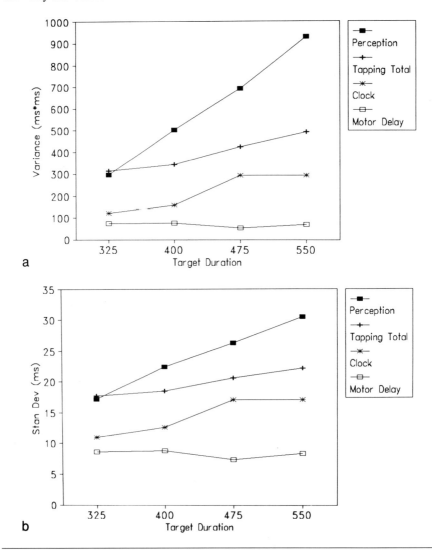

a

b

Figure 15.7   Results of Slope Experiment 1. Data from the repetitive tapping task are plotted in terms of overall variability (a), and standard deviation (b), with estimates of clock and motor delay components from the Wing-Kristofferson model.

Table 15.3 summarizes the regression analyses. A linear component accounts for over 85% of the variance for all of the functions except the motor delay estimates. As in our reanalysis of the Wing (1980) data, the intercepts indicate that the standard deviation functions are more meaningful. Large negative intercepts are obtained for both the clock and the perception variance functions. In contrast, the intercept values for the standard deviation functions are in agreement with a couple of different predictions. First, the clock intercept is close to zero. Second, the intercept for the

Table 15.3   Regression Analysis for Slope Experiment 1

Measure	Slope (ms)	Intercept (ms)	r^2
Standard deviation			
Tapping total	0.0213	10.6	0.97
Clock	0.0311	1.1	0.88
Motor delay	−0.0032	9.6	0.21
Perception	0.0584	−1.4	0.99
Variance			
Tapping total	0.8483	28.4	0.96
Clock	0.8824	−163.2	0.87
Motor delay	−0.0527	90.9	0.22
Perception	2.7893	−615.3	0.99

overall variability data on the tapping task is slightly larger, 10.6 ms, than the mean motor delay estimate of 8.2 ms. This is predicted because this intercept should include not only the motor delay component, but also other central sources of variability that are not duration dependent. One troubling aspect of the intercept scores, however, is that the perception intercept is negative, −1.4 ms. This intercept was predicted to estimate nonclock variability on the perception task. We assume that the negative intercept reflects measurement error. It is also possible that other sources of variability on the perception task are negligible.

Given the intercept results, we focus on the slopes for the standard deviation functions.[1] Contrary to our prediction, the mean slope estimates are different for the tapping and the perception functions. Even though only five subjects have completed this task, the difference is marginally significant, $t(4) = 2.35$, $p < .10$, two-tailed test. The perception slope is almost three times as large as the tapping slope. The difference is reduced if the clock slope for the tapping task is substituted for the overall scores. However, the perception slope is still 88% larger. From these data, the slope analysis does not provide converging evidence that a common timing system is invoked in the repetitive-tapping and the perception tasks.

Slope Experiment 2

There are a number of possible explanations for the differences in slope. First, although each subject completed four sessions, the data may still not be stable enough for this type of analysis; the slope values are heavily weighted by the fastest

---

[1]The intercept of the function for overall variance on the tapping is positive, suggesting that variance functions may be viable. However, not only is the variance function for the perception task negative, but, by our approximations, the variance function in Wing (1980) yields a negative intercept of −122 ms^2. The intercept for the standard deviation function from Wing's data is approximately 4.9 ms.

and slowest target durations, and error in estimating these data points would distort the functions. Second, the central assumption underlying the slope analysis may not be correct. Sources of variability other than the clock may contribute to the slope values. Third, the timing demands in the tapping and perception tasks may not be comparable. In particular, each trial in the tapping task requires the production of a series of 42 consecutive intervals, 12 with a pacing signal and 30 unpaced. In contrast, each trial on the perception task requires a comparison between 2 isolated intervals, the standard interval and the comparison interval. It is possible that the repetitive aspect of the tapping task serves to stabilize the operation of an internal timing system. This would produce a decrease in variability on the tapping task.

We thus modified the procedures in a second experiment to make the two tasks more comparable. The modified tapping task began with a paced phase in which the computer generated a single interval marked by two 50-ms tones. The word *tap* then appeared on the screen, and the subject made two key presses, attempting to reproduce the target interval. This procedure was repeated until the subject had produced 12 isolated intervals following the presentation of the target interval. Following this, the tones were eliminated, and the subject produced 30 more intervals, each individually initiated after the word *tap* was displayed on the computer. The response-stimulus interval was randomly varied to prevent subjects from adopting a rhythmic mode of responding. After producing 30 unpaced intervals, the subject was provided with feedback. A block consisted of one practice and six test trials. Each subject completed four blocks, one at each of the four target durations.

After completing a block of tapping, the subjects were tested on a modified version of the perception task. On each trial, only a single test interval was presented. The subject judged whether the interval was shorter or longer than an implicit standard. To help the subjects establish an implicit standard, the first 10 trials of a block involved relatively easy comparisons. For example, if the target interval was 400 ms, the durations used in the first 10 trials were either less than 325 ms or greater than 475 ms. Subjects rarely made errors with these values when performing the perception task with a standard interval. In addition, the preceding tapping trials were expected to establish an appropriate standard interval, because the subjects had just completed a set of tapping trials at that duration.

To summarize, in the second slope experiment, the intervals were generated individually rather than repetitively. Correspondingly, the perceptual judgments were made on isolated intervals. The same five subjects were tested. One subject completed the 2nd slope experiment prior to the 1st slope experiment.

The variability functions and regression analyses are summarized in Figure 15.8, a and b, and Table 15.4, respectively. Over 88% of the variance is accounted for by a linear function for both functions when plotted by variance (Figure 15.8a) and when plotted by standard deviation (Figure 15.8b). Of primary interest, the slope values are essentially identical for the tapping and perception functions, $t(4) = 0.18$. These results support the hypothesis that a common clock is used for both tasks. This finding is in accord with our correlational studies with normal subjects (Keele et al., 1985, 1987) and patient research (Ivry & Keele, 1989). The earlier studies, however, had assessed performance only at a single interval. In the current approach, the properties of the clock lead to the expectation that the variability of the timing process should increase in a systematic manner as the target interval is lengthened.

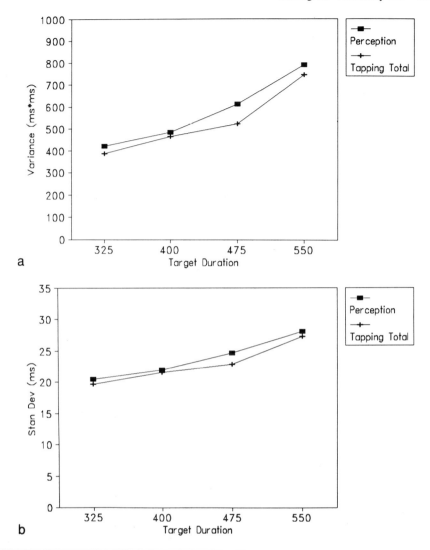

Figure 15.8 Results of Slope Experiment 2 in terms of variance (a), and standard deviation (b) functions. Only overall tapping variability is depicted because tasks involved the production of isolated intervals.

Moreover, we expected the increase to be the same for both movement and perception if a common timing system was involved. These predictions were confirmed.

Clock and motor delay functions are not shown in Figure 15.8. They cannot be estimated, because the Wing-Kristofferson model requires the production of consecutive intervals. However, as stated previously, an estimate of variability sources that are not duration dependent (i.e., nonclock variability) should be obtained from the intercept values. Once again, the data favor models predicting linear fits

Table 15.4   Regression Analysis for Slope Experiment 2

Measure	Slope (ms)	Intercept (ms)	r^2
Standard deviation			
Tapping	0.0387	5.7	0.91
Perception	0.0340	9.0	0.97
Variance			
Tapping	1.8088	−271.9	0.89
Perception	1.6480	−145.0	0.96

when standard deviation is plotted on the ordinate. Both the tapping and the perception values are positive and of seemingly reasonable magnitude. In contrast, the variance functions yield large negative intercepts.

Summary of the Different Slicings of the Timing Pie

Figure 15.9, a and b, summarizes two ways of slicing up the variability pie on timing tasks. The Wing-Kristofferson model (Figure 15.9a) slices off the implementation component and attributes the remaining variability to central processes, most notably a clock. The slope analysis (Figure 15.9b) slices off the clock component and attributes the remaining variability to all other processes involved in the tasks. Taken together, the two methods provide converging operations to analyze timing variability.

Moreover, the methods may facilitate a more finely grained analysis of the different components involved in timing tasks. For example, on the repetitive tapping task, an estimate of central processes that are not part of the clock component can be obtained by examining performance on both tasks. This component can be inferred from the disjunctive set (see Figure 15.9c), the region of the pie that is not directly estimated by either procedure. If timing is attributed to the clock component, and the implementation of the response is attributed to the motor delay component, what computational process might generate this remaining source of variability? One possibility is that a timing process can be conceived of as having more than one component, only one of which is concerned with the actual timing. For example, clock-counter models postulate a second component that keeps track of the number of outputs of the clock. An alternative two-component model assumes variability in a process that accesses the timer. Consider an analogy in which a foot race is to be timed with a stopwatch. An erroneous time could be attributed to either a problem with the stopwatch or a problem with starting or stopping the stopwatch. Indeed, in track and field, the latter type of error is sufficiently great that hand-held timings are generally not accepted as official.

The preceding statements begin a natural extension of our component analysis of coordination. The 1st step led to a decomposition of variability into components

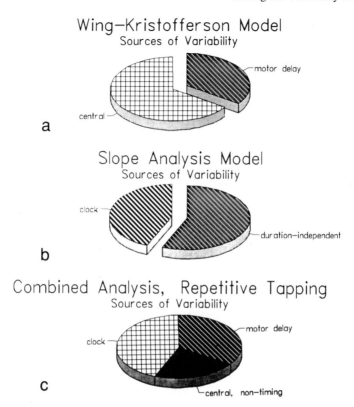

Figure 15.9   Three different slicings of the variability pie on timing tasks.

such as force and timing. A 2nd step, based on the Wing-Kristofferson model, decomposes timing variability into central and peripheral components. The slope analysis provides a 3rd step for further decomposing the central system itself. For example, manipulations that only affect the intercept can be used to investigate timing components that are duration independent. Accessing the timer is predicted to be such a process. On the other hand, manipulations that affect the slope presumably reflect properties of the duration-dependent component of the timing system. To date, we have tested patients only at a single duration. If a set of durations were used, we predict that steeper slopes would be obtained for patients with lesions of the lateral cerebellum than for control subjects or other patients with coordination deficits. In this manner, we anticipate that the slope analysis should prove useful for identifying the components of an internal timing system and for understanding the different sources of variability manifest in the operation of this system.

STUDIES OF FORCE VARIABILITY

As reviewed earlier, a strong case can be made that an internal timing system, common to both perception and production, involves the cerebellum. We now present a brief summary of preliminary research exploring the neural systems involved in force control. The evidence suggests that the basal ganglia play a crucial role in this component operation. The nature of this computation, however, is unclear. We do not postulate that the basal ganglia control the recruitment of motor units. Rather, we expect that the contribution of the basal ganglia in force regulation is less direct. One possibility is that the basal ganglia computation is more related to shifts in motor set that may precede or trigger motor unit recruitment (Mink & Thach, 1991; Wing, 1988).

Force Control in Parkinson's Disease

A large body of research with both animals and humans has investigated the effects of basal ganglia dysfunction on movement kinematics and kinetics. The most consistent finding is that lesions of the substantia nigra or globus pallidus, two of the basal ganglia nuclei, reduce the speed at which movements occur. For example, Horak and Anderson (1984) found that kainic acid injections into the globus pallidus in monkeys led to a slowing of movement time and that this deficit became more pronounced as the pathological consequences of the injections advanced. This finding matches the clinical observations of Parkinson's disease in humans. This disease, in which extensive cell death in the substantia nigra is observed, is characterized by a slowness in movement and rigidity. This slowness, or bradykinesia, has been documented in numerous studies (e.g., Benecke, Rothwell, Dick, Day, & Marsden, 1987; Hallett & Khoshbin, 1980; Stelmach, Teasdale, Phillips, & Worringham, 1989).

Superficially, bradykinesia might be interpreted as a timing deficit, because the movements are abnormally slow. Such an interpretation would appear to be at odds with our findings on the tapping and time perception tasks in which Parkinson patients performed as well as age-matched healthy subjects (Ivry & Keele, 1989). It may be necessary to make a distinction between tasks that require explicit timing control and those in which temporal properties arise as the result of an interaction of many processes, only one of which may be the operation of an internal clock. However, examining the variability of the movements produced by Parkinson patients is informative. Teasdale, Phillips, and Stelmach (1990) measured movement time and movement time variability in a group of Parkinson patients. Although the patients moved more slowly than the control subjects, the patients' movement times were not more variable once the differences in absolute movement time were taken into account. This result meshes with our null findings on the tapping task. Indeed, we chose a relatively slow tapping rate (ITI = 550 ms) to ensure that the Parkinson patients were able to keep up with the pace. Given this allowance, the temporal characteristics of the patients' movements were as consistent as for healthy subjects. Teasdale et al. (1990) obtain the same result in a unidirectional movement task.

Force variability in Parkinson patients has been examined in two recent studies. Stelmach and Worringham (1988) first measured the maximum isometric force capability for Parkinson and control subjects. Then each subject was asked to produce force pulses to match targets that were either 25%, 50%, or 75% of that person's maximum capability. Three different accuracy conditions were tested, and the subjects' responses were scored as to whether they fell within or outside the target region. Although not significant, the maximum force produced by the Parkinson patients was about 25% lower than the maximum force produced by the controls. Most interesting, the Parkinson patients were as accurate as the control subjects.

Rather than simply determine if the produced force fell within or outside the target area, Stelmach et al. (1989) used a quantitative measure of force variability in which they recorded the actual forces produced for each target. When expressed as a coefficient of variation (standard deviation of peak force/peak force), no differences were observed between the Parkinson patients and the control subjects. The coefficient of variation for both groups averaged about 10% across the range of forces tested.

We have conducted a similar experiment using our force control task. Seven Parkinson patients and 11 age-matched control subjects were tested. As in the tapping study, the Parkinson patients were tested under *on* and *off* medication conditions, the latter being when the patients skipped their morning medication cycles. Thus, this design provided a between-subject comparison (controls vs. Parkinsons) and a within-subject comparison for the patients (*on* vs. *off*).

As described in the section on the task-specific approach to variability, all forces were produced by isometric contractions of the index finger. The target forces were the same for all subjects, corresponding to 1, 3, 5, and 7 N. These force levels were well below maximum force capability for all of the subjects. Each trial consisted of six force pulses with feedback and six force pulses without feedback. Subjects completed six trials at each of the four targets.

The mean forces produced by the control subjects and the Parkinson patients in the *on* and *off* medication conditions approximated the target forces. The Parkinson patients under both conditions tended to produce slightly less force than the control subjects, and this effect was more pronounced when the patients were off medication.

The variability data are presented in Figure 15.10, a and b. The standard deviation for the three groups at each target level is plotted in Figure 15.10a. These data are replotted in Figure 15.10b as coefficient-of-variation measures, which look at variability as a percentage of the target. In accord with previous findings (Carlton & Newell, this volume, chap. 2), the coefficient-of-variation functions are downward sloping and concave. Most important for our present concern is the finding that the coefficient-of-variation functions are essentially identical for all three groups. As in the Stelmach studies, these results would suggest that the Parkinson patients are no more variable than controls at regulating force.

However, Figure 15.10 does not reflect one important aspect of the results. The Parkinson patients were much slower in generating the force pulses than were the control subjects. Moreover, this effect was greatly magnified by the medication manipulation. The mean contraction time for the control subjects was 407 ms. The mean contraction times for the Parkinson subjects were 483 ms and 742 ms in the *on* and *off* conditions, respectively. A number of researchers have attempted to theoretically derive the relationship of movement variability to kinematic, kinetic, and temporal properties of the movement (e.g., Carlton & Newell, this volume, chap. 2; Meyer, Smith, & Wright, 1982; Newell & Carlton, 1988; Schmidt, Zelaznik, & Frank, 1978). In each of these models, variability is expected to increase

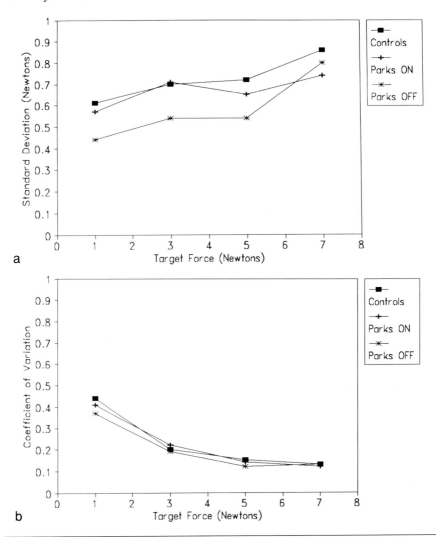

a

b

Figure 15.10   Results of Force Control Experiment 1. Mean standard deviation of peak force is plotted as a function of target force (a). The data are transformed into coefficient of variation measures (SD/Force) in (b).

with force (or distance), as was found in the current experiment. In addition, all of these models predict that variability should be inversely related to impulse duration (or movement time). In other words, there should be a speed-accuracy trade-off.

Assuming that this trade-off was operative, the data in Figure 15.10 can be reinterpreted as demonstrating a force control deficit in the Parkinson patients. In the *off* condition, the Parkinson patients took over 80% longer to complete their force pulses in comparison to the control subjects. Despite this increase in impulse duration, the variability of the two groups was essentially identical. It is possible that increased force variability in the patients is offset by a decrease in force variability resulting from the slower generation of the force pulses. An even more powerful demonstration of this finding is seen in the within-subject comparison. A 50% increase in impulse duration when the task was performed in the *off* condition led to no observed decrease in variability. Again, we assume that the increased variability caused by the drug manipulation is obscured by the reduction in variability resulting from the longer impulse durations.

The preceding argument is admittedly post hoc: We had originally expected that a force variability deficit could be identified by examining the variability functions. However, when the temporal differences were considered, it became clear that this analysis was too simplistic. The argument of a force deficit in Parkinson patients only emerges when the data are analyzed in light of current models of force production. It is possible, however, that the Parkinson patients do not have a force deficit, but rather that, for some unknown reason, their movements do not conform to a normal speed-accuracy trade-off.

This possibility was tested in a second experiment. Four Parkinson patients, all of whom had been in the preceding experiment, were tested on the force control task under two different instructions. In both conditions, the patients were instructed to make single, smooth force pulses, trying to match the target force. In the *fast* condition, the patients were instructed to generate rapid force pulses. In the *slow* condition, the patients were instructed to move more slowly while still making a single pulse. Subjects completing the *slow* condition first were instructed to move about twice as fast in the *fast* condition; subjects completing the *fast* condition first were instructed to move about twice as slow in the *slow* condition.

The subjects were able to follow the instructions. The mean impulse durations for the *slow* and *fast* conditions were 588 ms and 252 ms, respectively. The variability data are shown in Figure 15.11, a and b. The Parkinson patients demonstrate a strong dependency of variability on impulse duration. Both the absolute measure of variability (Figure 15.11a) and the relative measure of variability (Figure 15.11b) reveal greater variability for the faster impulses. These results indicate that the performance of Parkinson patients does follow a general speed-accuracy trade-off. Thus, the results of this experiment strengthen the interpretation that the basal ganglia lesions of Parkinson's disease impair force control.

In summary, the evidence reviewed in this section suggests that the loss of coordination in Parkinson's disease may reflect an increase in variability of a different component operation than that observed in the studies with the cerebellar patients. It should be reemphasized that to call the deficit a force control problem may be misleading. A more accurate description of the Parkinson deficit may be in terms of regulating the force-time profile of an isometric contraction. We do not think the Parkinson patients adopt a strategy of moving slower as a way of compensating for increased variability in force control. Rather, it appears that the disease impairs their ability to produce normal force-time pulses. As noted previously, the computation performed by the basal ganglia may be related only indirectly to force control. The computational process may be one of implementing transitions between different states of muscular activity. Note that even in this model the proposed deficit is not attributed to a process involved in explicitly

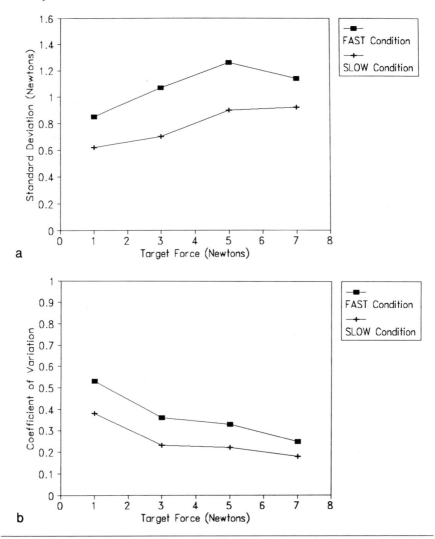

Figure 15.11 Results of Force Control Experiment 2 in terms of mean standard deviation of peak force plotted as a function of target force in (a), and coefficient of variance measures (SD/Force) in (b).

regulating temporal aspects of the contraction, a process we would expect to involve the cerebellum.

Force and Timing Deficits in Clumsy Children

We have recently applied a component analysis to a developmental issue (Lundy-Ekman, Ivry, Keele, & Woollacott, 1991); namely, Why are some children clumsy, and can the motor variability in clumsy children be attributed to specific component processes? Different subtypes of clumsiness have been noted (e.g., Dare & Gordon, 1970), although an etiological framework for these differences has not been articulated. One hypothesis concerning developmental motor problems is that the problems may reflect mild forms of brain dysfunction (Tupper, 1987). Although structural abnormalities are not found (Henderson, 1987), the child may present a set of symptoms that are similar to the problems seen in patients with focal lesions. Thus, the term *soft neurological signs* has been employed to describe these problems in contrast to the ''hard signs'' that can be ascribed to specific diseases or lesions.

Lundy-Ekman et al. (1991) identified a group of clumsy children and examined them for the presence of soft neurological signs. From a group of 60 seven- and eight-year-old children who showed some element of clumsiness, 25 were selected for further study. Fourteen of these children demonstrated soft signs consistent with cerebellar dysfunction, such as dysmetria and intentional tremor. The other 11 demonstrated soft signs consistent with basal ganglia dysfunction. Assessment of basal ganglia signs included choreiform, athetoid, and synkinesis. Note that these signs are not associated with Parkinson's disease, but rather with another basal ganglia disorder, Huntington's disease. The signs observed in the 35 children excluded from the study were either mixed or inconsistent.

The 25 clumsy children and 10 normal children were tested on the force control, tapping, perception-of-duration, and perception-of-loudness tasks (Figure 15.12, a and b). The results provided a striking double dissociation. On the tapping task, the children with soft cerebellar signs were more variable than either the control group or the children with basal ganglia signs (Figure 15.12a). This result was also obtained on the perception-of-duration task, but not on the control perception experiment involving loudness discrimination. In contrast, the children with soft basal ganglia signs were more variable on the force control task than were the other two groups (Figure 15.12b). The force control deficit for the children with soft basal ganglia signs can be assessed more directly in this experiment than in our Parkinson studies, because both groups of clumsy children produced force pulses of approximately the same duration.

SUMMARY OF COMPONENT ANALYSIS OF VARIABILITY

The research reviewed in this chapter demonstrates the strength of a component analysis. We began with a set of mental operations that were hypothesized to be involved in the performance of coordinated action. Correlational studies with normal subjects were conducted to assess the validity of these putative operations. These studies were then followed by neuropsychological investigations. Our goal in conducting these studies was twofold: First, we sought converging evidence for the existence of the operations identified in the correlational work; second, we wished to explore the neural basis of the operations. The patient research has provided strong evidence for the role of the cerebellum in timing functions. More recent work has implicated the basal ganglia in the regulation of force output, or a variable correlated with force such as shifts in motor set (e.g., equilibrium point).

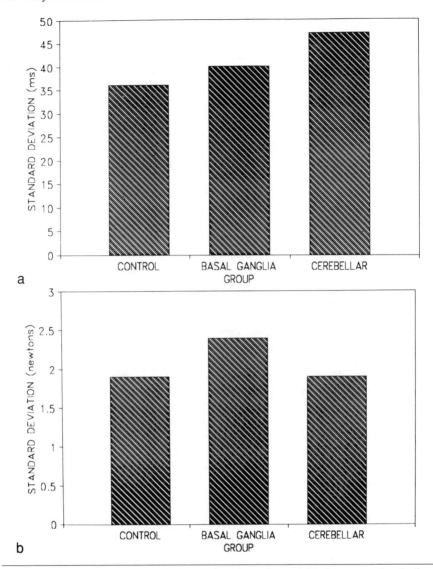

Figure 15.12   Mean standard deviation on the (a) tapping and (b) force control tasks for clumsy children.
Note. From "Timing and Force Control Deficits in Clumsy Children" by L. Lundy-Ekman, R. Ivry, S. Keele, and M. Woollacott, 1991, *Journal of Cognitive Neuroscience,* **3,** pp. 368-377. Adapted by permission.

The dissociation between the contributions of the cerebellum and the basal ganglia in timing and force aspects of movement emphasizes the usefulness of a component analysis for understanding variability. As shown in neuroscience textbooks (e.g., Ghez & Fahne, 1985), these structures are part of two primary subcortical pathways of the motor system. It is noteworthy that there is little interaction between these two pathways, at least prior to motor and premotor cortex (see Goldberg, 1985). Assuming the computations performed within these pathways are nonredundant, it would be expected that variability that arises within each pathway would be independent. However, if the observable behavior requires the successful operation of both of these pathways (in addition to other pathways), then the variability in the behavior cannot be attributed to a single source. A component analysis is a critical prerequisite for identifying the appropriate pieces of the variability pie.

REFERENCES

Abel, S. (1972). Discrimination of temporal gaps. *Journal of the Acoustical Society of America*, **52**, 519-524.

Allen, G., & Tsukahara, N. (1974). Cerebrocerebellar communication systems. *Physiological Reviews*, **54**, 957-1006.

Arshavsky, Y., Gelfand, I., & Orlovsky, G. (1983). The cerebellum and control of rhythmic movements. *Trends in Neuroscience*, **6**, 417-422.

Asanuma, C., Thach, W., & Jones, E. (1983a). Distribution of cerebellar terminations and their relation to other afferent terminations in the ventral lateral thalamic region of the monkey. *Brain Research Reviews*, **5**, 237-265.

Asanuma, C., Thach, W., & Jones, E. (1983b). Brainstem and spinal projections of the deep cerebellar nuclei in the monkey with observations on the brainstem projections of the dorsal column nuclei. *Brain Research Reviews*, **5**, 299-322.

Aschoff, J., & Cohen, B. (1971). Changes in saccadic eye movements produced by cerebellar cortical lesions. *Experimental Neurology*, **32**, 123-132.

Benecke, R., Rothwell, J., Dick, J., Day, B., & Marsden, C. (1987). Simple and complex movements off and on treatment in patients with Parkinson's disease. *Journal of Neurology, Neurosurgery, and Psychiatry*, **50**, 296-303.

Creelman, C. (1962). Human discrimination of auditory durations. *Journal of the Acoustical Society of America*, **34**, 582-593.

Dare, M., & Gordon, N. (1970). Clumsy children: A disorder of perception and motor organization. *Developmental Medicine and Child Neurology*, **12**, 178-185.

Fleishman, E. (1966). Human abilities and the acquisition of skill. In E.A. Bilodeau (Ed.), *Acquisition of skill*. New York: Academic Press.

Gellman, R., Gibson, A., & Houk, J. (1985). Inferior olivary neurons in the awake cat: Detection of contact and passive body displacement. *Journal of Neurophysiology*, **54**, 40-60.

Getty, D. (1975). Discrimination of short temporal intervals: A comparison of two models. *Perception and Psychophysics*, **18**, 1-8.

Ghez, C., & Fahne, S. (1985). The cerebellum. In E. Kandel & J. Schwartz (Eds.), *Principles of neural science* (2nd ed.). New York: Elsevier.

Goldberg, G. (1985). Supplementary motor areas structure and function. Review and hypothesis. *Behavioral and Brain Sciences*, **8**, 567-616.

Grossberg, S., & Schmajuk, N. (1989). Neural dynamics of adaptive timing and temporal discrimination during associative learning. *Neural Networks*, **2**, 79-102.

Hallett, M., & Khoshbin, S. (1980). A physiological mechanism of bradykinesia. *Brain*, **103**, 301-314.

Hallett, M., Shahani, B., & Young, R. (1975). EMG analysis of patients with cerebellar lesions. *Journal of Neurology, Neurosurgery, and Psychiatry*, **38**, 1163-1169.

Henderson, S.E. (1987). The assessment of ''clumsy'' children: Old and new approaches. *Journal of Child Psychology and Psychiatry*, **28**, 511-527.

Horak, F.B., & Anderson, M.E. (1984). Influence of globus pallidus on arm movements in monkeys. I. Effects of kainic acid–induced lesions. *Journal of Neurophysiology*, **52**, 290-304.

Ivry, R., & Diener, H.C. (1991). Impaired velocity perception in patients with lesions of the cerebellum. *Journal of Cognitive Neuroscience*, **3**, 355-366.

Ivry, R., & Gopal, H. (in press). Speech perception and production in patients with cerebellar lesions. In D. Meyer & S. Kornblum (Eds.), *Attention and Performance* (Vol. 14).

Ivry, R., & Keele, S. (1989). Timing functions of the cerebellum. *Journal of Cognitive Neuroscience*, **1**, 136-152.

Ivry, R., Keele, S., & Diener, H. (1988). Dissociation of the lateral and medial cerebellum in movement timing and movement execution. *Experimental Brain Research*, **73**, 167-180.

Keele, S., & Ivry, R. (in press). Does the cerebellum provide a common computation for diverse tasks? A timing hypothesis. In A. Diamond (Ed.), *Developmental and neural basis of higher cognitive function*.

Keele, S., & Ivry, R. (1987). Modular analysis of timing in motor skill. In G. Bower (Ed.), *The psychology of learning and motivation: Advances in research and theory*. New York: Academic Press.

Keele, S., Ivry, R., & Pokorny, R. (1987). Force control and its relation to timing. *Journal of Motor Behavior*, **19**, 96-114.

Keele, S., Pokorny, R., Corcos, D., & Ivry, R. (1985). Do perception and motor production share common timing mechanisms: A correlational analysis. *Acta Psychologia*, **60**, 173-191.

Kelso, J. (1984). Phase transitions and critical behavior in human bimanual coordination. *American Journal of Physiology: Regulatory, Integrative, and Comparative Physiology*, **15**, R1000-R1004.

Killeen, P., & Weiss, N. (1987). Optimal timing and the Weber function. *Psychological Review*, **94**, 455-468.

Lundy-Ekman, L., Ivry, R., Keele, S., & Woollacott, M. (1991). Timing and force control deficits in clumsy children. *Journal of Cognitive Neuroscience*, **3**, 370-377.

Meyer, D., Smith, J., & Wright, C. (1982). Models for the speed and accuracy of aimed movements. *Psychological Review*, **89**, 449-482.

Mink, J., & Thach, W. (1991). Basal ganglia motor control: III. Pallidal ablation: Normal reaction time, muscle cocontraction, and slow movement. *Journal of Neurophysiology*, **65**, 330-351.

Newell, K.M., & Carlton, L.G. (1988). Force variability in isometric responses. *Journal of Experimental Psychology*, **14**, 37-44.

Pentland, A. (1980). Maximum likelihood estimation: The best PEST. *Perception and Psychophysics*, **28**, 377-379.

Ritchie, L. (1976). Effects of cerebellar lesions on saccadic eye movements. *Journal of Neurophysiology*, **39**, 1246-1256.

Rosenbaum, D. (1990). *Broadcast theory of sequencing and timing.* Paper presented at the 31st annual meeting of the Psychonomic Society, New Orleans, LA.

Schell, G., & Strick, P. (1984). The origin of thalamic inputs to the arcuate premotor and supplementary motor areas. *Journal of Neuroscience*, **4**, 539-560.

Schmidt, R.A., Zelaznik, H.N., & Frank, J.S. (1978). Sources of inaccuracy in rapid movement. In G. Stelmach (Ed.), *Information processing in motor control and learning* (pp. 183-203). New York: Academic Press.

Stelmach, G.E., Teasdale, N., Phillips, J., & Worringham, C.J. (1989). Force production characteristics in Parkinson's disease. *Experimental Brain Research*, **76**, 165-172.

Stelmach, G.E., & Worringham, C.J. (1988). The preparation and production of isometric force in Parkinson's disease. *Neuropsychologia*, **26**, 93-103.

Teasdale, N., Phillips, J., & Stelmach, G. (1990). Temporal movement control in patients with Parkinson's disease. *Journal of Neurology, Neurosurgery, and Psychiatry*, **53**, 862-868.

Tupper, D.E. (1987). The issues with "soft signs." In D.E. Tupper (Ed.), *Soft neurological signs* (pp. 1-16). Orlando, FL: Grune & Stratton.

Thompson, R. (1986). The neurobiology of learning and memory. *Science*, **233**, 941-947.

Wilson, V., Uchino, Y., Maunz, R., Susswein, A., & Fukushima, K. (1978). Properties and connections of cat fastigiospinal neurons. *Experimental Brain Research*, **32**, 1-17.

Wing, A. (1980). The long and short of timing in response sequences. In G. Stelmach & J. Requin (Eds.), *Tutorials in motor behavior*. New York: North-Holland.

Wing, A. (1988). A comparison of the rate of pinch grip force increases and decreases in Parkinsonian bradykinesia. *Neuropsychologia*, **26**, 479-482.

Wing, A., & Kristofferson, A. (1973). Response delays and the timing of discrete motor responses. *Perception and Psychophysics*, **14**, 5-12.

Acknowledgments

The first author was supported by an NINDS grant (NS-30256) and a Sloan Fellowship in Neuroscience. The second author was supported by an NIH grant (NS-23593). The authors are grateful to Steve Keele, Alan Wing, David Rosenbaum, and Seth Roberts for their comments.

Chapter 16

Movement Variability in Limb Gesturing: Implications for Understanding Apraxia

Eric A. Roy
University of Waterloo, Waterloo, Ontario, Canada
Toronto Western Hospital, Toronto, Ontario, Canada

Liana Brown, Michael Hardie
University of Waterloo, Waterloo, Ontario, Canada

Gestural movements such as those involved in a wave when greeting a friend have been of interest in a number of contexts. One focus has been on limb gestures in communication. In nonverbal communication, a number of studies have catalogued the hand and arm postures and movements associated with particular symbols or signs (e.g., Argyle, 1988; Johnson, Ekman, & Friesen, 1975). Cross-cultural differences in the posture and movements signifying a particular symbol have received considerable attention (e.g., Argyle, 1988; Efron, 1941; Ramsey, 1984). One aspect of nonverbal communication that has enjoyed particular scrutiny is the movements involved in sign language. Both movement notation (Stokoe, 1972) and kinematic analyses (Poizner, Klima, & Bellugi, 1987) have been applied to the study of these movements.

Another line of research emerging out of studies linking the control of limb movements involved in gesturing and those involved in speech has investigated the gestural movements that accompany speech (Argyle, 1988; Jarvella & Klein, 1982; Kendon, 1980; McNeil, 1985). This work has found that, in right-handers, movements with the right hand more frequently accompany speech than those with the left hand (e.g., Kimura, 1976). Because, in most right-handers, speech and language functions are controlled in the left hemisphere, activation of this hemisphere during speech would more likely activate the right hand, due to the closer proximity in cerebral space of the control center for this hand to the activated speech and language areas. This activation of the right hand might then precipitate an increased incidence of right-hand movements during speaking.

Another dimension of gestural movements involves the disruption to gestural movements associated with brain damage. This work relates to the aforementioned studies linking speech and right-hand movements, which point to the potential role of the left hemisphere in the control of limb and speech movements (e.g., Kimura, 1982). In concert with this interpretation, these brain damage studies have frequently linked impairments in speech and language function and impairments in limb gesturing in patients with damage to the left hemisphere (e.g., Kertesz, 1985; Square-Storer, Roy, & Hogg, 1990). One gestural disorder of particular importance in this discussion, termed apraxia, is a movement disorder resulting more frequently from left-hemisphere damage in which the patient is unable to pantomime or imitate gestures upon request (Roy, 1985).

In characterizing apraxia, a number of studies have looked at analyses of errors in performance (e.g., Haaland & Flaherty, 1984; Rothi, Mack, Verfaellie, Brown, & Heilman, 1988; Roy, Square, Adams, & Friesen, 1985; Roy, Square-Storer, Adams, & Friesen, 1987) as well as detailed kinematic analyses of the gestural movements of the affected patients (e.g., Poizner, Mack, Verfaellie, Rothi, & Heilman, 1990). These studies have provided some interesting insights into the nature of apraxia. Some of our recent work, however, indicates that there may be considerable variability in gestural performance, suggesting that it may be difficult to clearly define what is an error in the performance of brain-damaged patients without having some idea of the variability in normals. The purpose of this chapter is to describe our explorations into ways of describing this variability in gestural performance.

Our chapter will begin with a brief review of the impairments to limb gesturing seen in apraxia, focusing particularly upon error-notation and kinematic analyses used in describing this disorder. We will then turn to a review of our studies, which have attempted to characterize variability in gestural performance using these types of analysis. With these findings in mind, we will conclude the presentation with some considerations for future research into analyses of variability in gesturing.

ANALYSES OF LIMB APRAXIA

Limb apraxia represents an inability to perform gestures such as demonstrating how to stir coffee or showing how to do a military salute. Although the affected patients are unable to perform these gestures in a clinical examination, they are frequently able to perform them when in the appropriate environmental context—for example, when making a cup of tea at breakfast. As with any disorders arising from brain damage, apraxia is defined by both exclusion and inclusion. The exclusionary definition indicates that the disorder is defined as apraxia only if visual recognition disorders, basic motor control impairments, verbal comprehension disorders, and generalized impairments to cognitive-behavioral function (dementia) can be ruled out. The definition by inclusion places particular emphasis on the types of errors that characterize apraxia. These errors will be considered in detail shortly.

Apraxia has proven difficult to understand, as the problem appears to be not with movement per se, but rather with the processes involved with planning or programming the movements involved in performing the gesture. Much of the research on apraxia, then, has been directed toward the study of task demands that affect these processes. Such dimensions as input modality (e.g., verbal command vs. imitation), movement complexity (e.g., number of movement elements), and the movement system used (e.g., oral vs. limb movements) have been of interest (for details, see DeRenzi, 1985; Roy & Hall, 1992; and Roy & Square-Storer, 1990).

A recent study by Roy (Roy, Square-Storer, Hogg, & Adams, 1991) examined the simultaneous effects of several task demands in a battery of gestural tests: the movement system (limb vs. axial movements), input modality (command vs. imitation), movement complexity (single gestures vs. a sequence of gestures), type of limb gesture (transitive vs. intransitive), and the representational nature of the gestures (representational vs. nonrepresentational gestures). On this battery, it was apparent that gestural performance was selectively and dramatically impaired with damage to the left hemisphere. The left-hemisphere-damaged (LHD) patients were impaired relative to right-hemisphere-damaged (RHD) patients in performing all of the gestures. This rather pervasive nature of the apractic deficit was consistent with work reported by Poeck (Poeck, Lehmkuhl, & Willmes, 1982), who also observed apractic deficits in a wide variety of gestures involving arm, leg, and axial movements. Although the LHD patients were impaired across all task dimensions, several task demands exhibited greater effects on performance than others, and the pattern of these different effects provided some insight into the nature of the apractic deficit.

First, the representational or symbolic nature of the gestures did not seem to be an important factor, in that the LHD patients were equally impaired on the representational and nonrepresentational gestures, a finding that concurred with a number of other studies (e.g., DeRenzi, 1985; Kimura & Archibald, 1974). Considering the movement system involved in gesturing, the LHD patients exhibited an equivalent impairment in performing limb and axial gestures. This finding speaks against the notion that apraxia is a disorder affecting only the pyramidal motor system and is consistent with other work by Poeck et al. (1982) showing that apraxia may affect upper and lower limb as well as axial gestures.

Turning next to the factor of movement complexity, the LHD patients were impaired on all the gestures, including the simple and the more complex ones. Nevertheless, the more complex gestures (gestures with two or three movement components) were performed less accurately than the single gestures, suggesting that some movement complexity factor related to the number of movement elements may be an important basis for the impairment in apraxia. Although complexity of the gesture seemed to increase the degree of impairment with left-hemisphere damage, with right-hemisphere damage complexity seemed to be an important factor in eliciting the gestural impairment, in that the RHD patients were impaired only on the more complex gestural sequences involving three movement elements. This finding concurred with work by Kolb and Milner (1981), who found an impairment in their RHD patients only for the multiple-movement gestures.

ERROR-NOTATION AND KINEMATIC DESCRIPTIONS OF APRACTIC PERFORMANCE

Given this review of the work on apraxia, it is clear that this movement disorder is a difficult one to understand. One approach that has been used to gain some insight involves manipulating various task dimensions in an attempt to observe their effect on the patient's performance. This approach, which is well represented in the work we have just considered, can lead to identifying impairments in cognitive and motor control processes that may underlie apraxia (e.g., Jason, 1983a, 1983b, 1985,

1986, 1990; Roy & Hall, 1992). An alternative approach, which is of particular interest in this chapter, focuses more upon describing the characteristics of the movements observed in apraxia. One type of analysis here involves examining the frequency of various types of error in performance. The other is concerned more with detailed kinematic analyses of movement.

Kaplan (1968) was one of the first to provide a system for classifying gestural performance into categories based on the characteristics of the movements observed. The system involved 16 categories that described changes in gestural performance in normal children over development. One of the characteristic errors observed in young children was a hand posture error referred to as body part as object (BPO). One form of this BPO error involved the hand shape taking on characteristics of the object in such a way that the hand virtually became the object. For example, in demonstrating how to brush teeth, the child extended the index finger and used it to rub against the teeth as if the finger were the toothbrush. Another form of this error is seen later in development. In this case, the hand posture reflects the appropriate grasp to hold the object, but the hand is held too close to the surface on which it is acting. For example, in pantomiming brushing teeth, the child's hand is held up against the teeth without taking into account the size of the object. Although Kaplan (1968) described a number of other behavioral categories, this BPO error in particular has been adopted and observed frequently in apraxia.

Haaland and Flaherty (1984) and Rothi et al. (1988) incorporated these BPO errors into their error systems for describing gestural performance in adult apraxia. Along with these hand posture errors a number of other categories were developed that reflected spatial (e.g., hand position and orientation), temporal (e.g., delay in initiating movement), and action (e.g., augmented or distorted movements) components. Roy (Roy et al., 1985) derived an error-notation system based on categories involved in describing the movements in American sign language (Stokoe, 1972). This system involved 10 categories, which included temporal (delay, Category 8), spatial (location, plane of movement, orientation; Categories 6, 5, and 2), form (posture, Category 4), action (Category 3), and added movement (target effector or noneffector movements, Categories 7 and 1) elements. As well, there were categories indicating that no movement was attempted (Category 0) or the correct gesture was performed correctly (Category 9). Further qualification to these categories was provided through a system of diacritical markers, which describe the error (e.g., added non-target-effector movement) more specifically (e.g., the other hand). Using these markers, a more precise description of an error category was afforded. For example, a non-target-effector movement (i.e., a part of the body other than the one that is to be used in the gesture) can, in the case of a limb gesture, involve movements of the head, the face, or the other arm or a verbalization. The diacritics provide a means of notating just which of these "other" movements occurred.

This system was used to examine the gestural performance of a group of LHD patients (Roy et al., 1987, 1991). Interrater reliability of this system ranged from .78 to .92, depending on the category examined. Comparisons of the relative frequency of various error categories between the apractic and the nonapractic patients revealed a significantly higher incidence of errors in the hand posture, location, added target effector movement, and action categories.

These analyses of errors reveal something of the nature of apraxia. Generally, apractics exhibit errors of hand posture, location, and action. Although the errors suggest that apraxia might be characterized as a problem affecting the spatial, postural, and dynamic action components of gesture, the observed errors must be viewed to some extent as products of an impaired control system (cf. Roy, 1990). Some insight into how these errors might arise in performance requires a more detailed examination of the movement dynamics. This type of kinematic analysis of apractic performance was recently completed by Poizner et al. (1990). The apractic patients in this study exhibited marked differences from the normal adult in their performance of two transitive gestures (roll down a car window, carve a turkey with a knife). One particular finding of interest here was that the apractics made movements having components that were out of the principal plane of movement exhibited in the normal adult.

KINEMATIC DESCRIPTIONS OF GESTURAL PERFORMANCE IN NORMAL ADULTS

Roy and Brown (1991) have begun to examine gestural performance using three-dimensional kinematic analyses of movement. Gestural performance in normal adults is being studied in this work and is designed to provide a normative base against which to compare the performance of brain-damaged adults.

Method

In this study, four gestures, two transitive (hammer a nail and bring a spoon to the mouth) and two intransitive (wave goodbye and salute), were examined. Four normal female right-handed adults ranging in age from 20 to 50 years participated as subjects. The movements were analyzed using the WATSMART three-dimensional movement analysis system. Movements of the subject's right arm were reflected in the trajectories of four light-emitting diodes (IREDs), one placed on the dorsum of the hand and three on the arm—one on the forearm midway between the elbow and the head of the radius, one on the elbow, and one midway on the upper arm between the elbow and the shoulder. An IRED was also placed on the right side of the head above the tip of the ear midway between the ear and the crown of the head. A small metal washer was taped on the lateral aspect of the hand, and the subject began each trial with this washer placed on a metal plate at his or her midline. The break in contact between the washer on the hand and the metal plate on the table arising when the subject began the gesture signaled the beginning of the movement. The experimenter signaled the end of the movement by depressing a microswitch when the subject had achieved the target gesture. These contact breakpoints signifying the start and end of the movement were recorded as voltage changes on an A/D system (WATSCOPE). These contact breakpoints were used to define the beginning and end of the movement trajectories recorded by the WATSMART system. The subjects performed five trials of each of the four gestures, with the order of gestures counterbalanced across the four subjects.

Data Analysis

Initial analyses of the limb movements examined only the trajectory of the IRED on the hand. Several measures were examined. The trajectory of the hand in each of the three axes was analyzed. The complete trajectory for the spoon and salute gesture was examined, whereas the trajectory of the hand from movement initiation to the onset of the reciprocal part of the gesture was examined in the wave and hammer gestures. The resultant displacement trajectory of the hand, reflecting movement in all three axes, was differentiated to provide a velocity profile. From this profile the movement time, peak velocity, and time to peak velocity were examined. The velocity profile was normalized to equate for differences in movement time, and the percent time after peak velocity was examined.

Results

Representative velocity profiles for one subject performing the body-centered gestures are depicted in Figure 16.1, a and b. Kinematic analyses of the movement of the IRED on the hand (Table 16.1) revealed that gestures directed toward the body (body-centered gestures: spoon and salute) involved significantly greater movement time and time spent after peak velocity than the gestures directed away from the body (allocentric gestures: hammer and wave). Although these gestures also exhibited lower peak velocities, only the spoon gesture was significantly different from the allocentric gestures.

Why might these gestures exhibit different kinematic patterns? Possibly the body-centered gestures are controlled somewhat differently than those directed away from the body. The lower peak velocity in these gestures may indicate a smaller amplitude of movement. Much work (e.g., Nelson, 1983) has shown a strong relationship between peak velocity and movement amplitude. This explanation is unlikely, as attempts were made to equate the gestures on movement amplitude in terms of the linear distance from the start position to the "target" area for the hand.

A second possibility is that this velocity-deceleration time pattern may reflect the fact that, in these gestures, the hand makes contact with the body. This pattern, then, may serve to prevent the hand from impacting with the head on its approach.

The third hypothesis is that this velocity-deceleration time pattern may arise from the increased spatial precision demands for the body-centered gestures. These gestures are directed toward rather well-defined spatial targets (i.e., the mouth or the side of the forehead), whereas the allocentric gestures place no such demands on spatial precision. Work on reaching by Soechting (1984) and Marteniuk and MacKenzie (e.g., Marteniuk, MacKenzie, Jeannerod, Athenes, & Dugas, 1987) has demonstrated that as the spatial demands of the task increase (e.g., with decreasing target size), the velocity profile of the reaching movement becomes increasingly asymmetrical, with greater time spent after peak velocity in deceleration.

Comparison of the two body-centered gestures seems to favor this latter precision demand hypothesis. The salute gesture differs significantly from the allocentric gestures only with respect to the time in deceleration, whereas the spoon gesture differs in both peak velocity and time in deceleration. This difference in kinematic

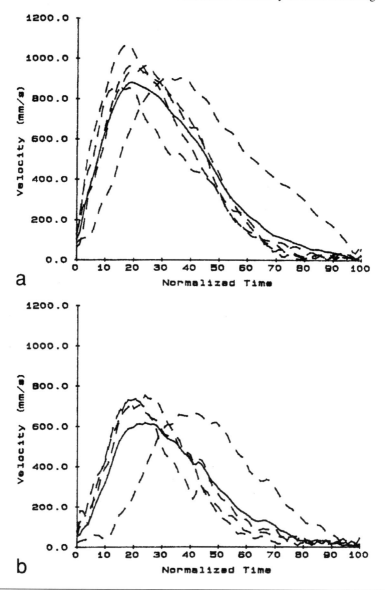

Figure 16.1 Velocity profiles for subject LZ performing the (a) salute and (b) spoon gestures. (The mean profile is the solid line.)

pattern would not likely be predicted by the second (impact) hypothesis, because in both gestures the hand makes contact with the head. Rather, it would seem to favor the precision demand hypothesis. Indeed, in examination of the two gestures, the spoon gesture would seem to demand more spatial precision, as the hand is directed to a smaller target (the mouth) than that (side of the forehead) in the salute gesture. This tendency to slow down the movement (lower peak velocity) *and* to

Table 16.1   Movement Time, Peak Velocity, and Time to and After Peak Velocity for Each Gesture

Gesture	MT	PV	Time to PV	Time after PV	Percent time to PV
Wave	628	1,269	269	359	42.8
Hammer	597	1,410	317	281	53.0
Spoon	1,367	886	465	902	33.9
Salute	1,127	1,296	287	839	25.5

spend more time in deceleration when performing the spoon gesture, then, may reflect the constraints imposed by the smaller size of the target toward which the hand is directed.

In order to identify the nature of the coordination involved in performing these gestures, the spoon and salute gestures were examined in more detail. To analyze the range of movement at each joint, the distance through which the IREDs moved in the resultant trajectory was examined. Movements of the IRED on the hand reflect movements at both the elbow joint and the shoulder joint. Movements of the IRED at the elbow reflect primarily movement at the shoulder; changes in the angle subtended by IREDs at the forearm, elbow, and upper arm reflect movements at the elbow.

Analyses of movement distance for the IREDs on the hand and elbow indicated significantly greater movement of the hand than of the elbow (see Figure 16.2). Comparison between the salute and the spoon gestures revealed a gesture by IRED

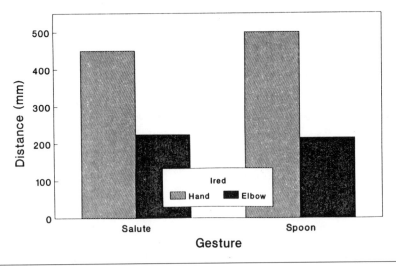

Figure 16.2   Total distance moved by the hand and elbow IRED in the salute and spoon gestures.

interaction, such that the gestures differed only in the degree of movement at the hand, with the spoon gesture exhibiting significantly greater movement.

The finding that movement of the hand IRED (movement at the elbow and the shoulder) is significantly greater than that of the elbow (movement at the shoulder) suggests that movement of the hand is only partly described by movement at the shoulder. A major contribution for hand movement must also be made by movement at the elbow joint. The finding that the spoon gesture differs from the salute gesture primarily in terms of degree of movement at the hand suggests that the contribution of movement at the elbow must be greater for the spoon gesture than for the salute gesture. Indeed, an examination of range of movement at the elbow revealed significantly greater angular displacement for the spoon gesture (see Figure 16.3).

VARIABILITY IN GESTURAL PERFORMANCE

So far these analyses reveal something of the nature of control involved in common limb gestures and the types of errors (i.e., hand posture, hand location, and action) that appear in the performance of apractic adults. One aspect that these descriptions do not capture, however, is the variability in performance. Because one of our ultimate goals is to use these gestural tasks as diagnostic tools to identify movement disorders such as apraxia, one type of variability that is of particular interest reflects individual differences in performance. This notion of variability in performance has only recently been incorporated into areas of clinical medicine such as neurology. Until the development of behavioral neurology, the focus in detecting and classifying neuropathology had largely been on identifying pathognomonic signs (e.g., tremor, bradykinesia, hypotonia) and making cross-body comparisons (e.g., relative strength in the left vs. the right hand). In contrast, almost from its inception clinical

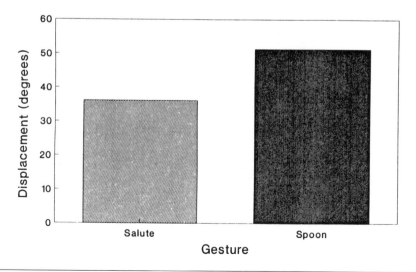

Figure 16.3   Angular displacement at the elbow in the salute and spoon gestures.

neuropsychology adopted the psychometric tradition from educational and clinical psychology, which recognized variability as the cornerstone of clinical diagnosis (e.g., Kolb & Whishaw, 1990). Measures of the variability in normal individuals matched on age and other relevant characteristics (e.g., sex, education) to the person being assessed are of particular import in diagnosis.

The other type of variability focuses on differences in performance within subjects across a series of trials. This measure, reflecting performance consistency, has received much less attention in the clinical sciences but has a rich tradition in the movement sciences (e.g., Darling & Cooke, 1987). This type of variability may also be important in the diagnosis of limb apraxia, as studies of verbal apraxia indicate that variability in performance across multiple attempts at uttering a target word is characteristic of the disorder (e.g., Roy & Square-Storer, 1990).

Variability Revealed Through Movement Notation Analyses

With the former measure of variability (i.e., individual differences) in mind, we began to look more closely at the performance of the normal adults we had used in our studies (Roy et al., 1987, 1989, 1991). An initial informal examination of the videotaped performance of these people revealed considerable variability in performance. This finding was somewhat surprising to us, because the gestures used in our work are well-known, conventional gestures in our culture that others (e.g., Johnson et al., 1975) have suggested are performed with relatively little inter-individual variation. Further, in a preliminary normative study these gestures were selected from a corpus of gestures because of the apparently consistent way they were performed across a sample of normal adults.

In an attempt to capture this variability we embarked on more detailed analyses using the system described earlier to notate various types of error. You will recall that this system reflects several dimensions of gestural performance, temporal, spatial (e.g., location) and action dimensions, as well as noting any added movements observed. Borrowing from a number of different movement notation systems (e.g., Argyle, 1988) we developed a means for notating more precisely the characteristics of the movement relevant to each dimension. As an example here let us focus on the location and action dimensions.

The location of the hand in body space for the target gesture at completion (e.g., the location of the hand when waving) is represented in the cells of a 6-by-6 grid reflecting positions in the Y-Z (coronal) plane (see Figure 16.4). The grid was designed using anatomical landmarks. The grid lines in the Y (horizontal) axis are defined by the shoulders, sides of the head, and nose; those for the Z axis are defined by the chest, shoulders, chin, nose, and top of the head. With regard to the action dimension, the action used in performing the gesture is reflected by notating at which joints movement was occurring (see Figure 16.5). The fingers, wrist, elbow, and shoulder are focused on here.

Using this approach we examined the videotaped performance of 30 normal adults and 30 LHD patients matched on age and handedness (right). Performance with the left hand was examined, because the majority of the LHD patients were unable to use the right hand due to the left-hemisphere stroke. Two observers independently notated the gestural performance of each subject. Examination of interrater consistency using the point-to-point method revealed reliabilities exceeding .80 across the various dimensions of performance. The data presented here, then, reflect the average frequencies for the two observers and focus on the wave gesture performed to verbal command.

Looking first at the location dimension (Figure 16.4), the majority of both the normal adults (40%) and the LHD patients (33.3%) fell in the 3 (Z), 5 (Y) cell above the left shoulder next to the face below the level of the nose. The remaining 60% of the normal adults placed their hands in cells immediately adjacent to this

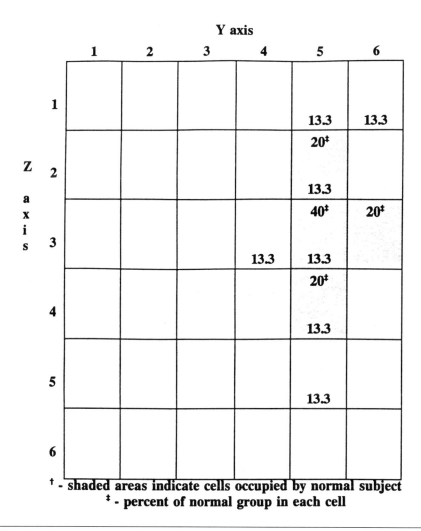

Figure 16.4 Distribution of hand locations in the wave gesture for the normal adults and left-hemisphere-damaged patients.

cell. Approximately 27% of the LHD patients also located their hands in these adjacent cells. The remaining approximately 40% of the LHD patients located their hands outside the cells occupied by the normal adults. These 8 LHD patients would be considered to have made a location error.

Looking next at the action pattern, it appears (see Figure 16.5) that the pattern for the normal adults involves movement at the wrist and shoulder, at the wrist and elbow, or at the wrist alone. Of the LHD patients, 64.4% exhibited one of these patterns. The remaining 36.6% exhibited different patterns, involving movement at the shoulder alone, at the wrist, elbow, and shoulder, or at the fingers alone.

In comparing across these dimensions to determine how many patients exhibited one or both errors, it was apparent that 7 patients (23.3%) exhibited both errors, 3 (10%) exhibited a location error alone, and 4 (13.3%) displayed an action error alone. One might argue that the patients displaying multiple errors are more impaired than those demonstrating only one type of error. This notion of multiple errors signifying greater impairment promises to provide a means of making a discrimination among various degrees of impairment in gestural performance.

As we discussed earlier, this error-notation approach provides a rather low-level analysis of gestural performance. As well, it essentially focuses on the gestural product (e.g., a location error) rather than the gestural process (e.g., kinematic details of movement execution), which might describe how the product emerged out of the way the subject performed the gesture (e.g., Roy, 1990). The other approach we have been using, which we described in the previous section, involves 3-D kinematic analyses of movement and has the potential to provide information about both of these aspects of performance. Further, although the focus in the error-notation analyses is largely on variability due to individual differences, in these kinematic analyses it is possible to examine variability due both to individual differences and

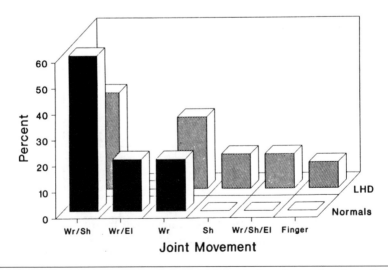

Figure 16.5   Percentage of normal adults and left-hemisphere-damaged patients displaying movement at the shoulder, elbow, wrist, and/or fingers in the wave gesture.

to the consistency of performance within subjects across trials. As an example of these analyses let us look more closely at the body-centered gestures, salute and spoon.

Movement Variability Revealed Through Kinematic Analyses

Our earlier analyses suggested that the normalized resultant velocity profiles for movement of the hand provided some insight into the strategies used to control movement. In comparison to the two allocentric gestures (wave and hammer), the spoon gesture exhibited both a lower peak velocity and a greater percentage of time in deceleration, whereas the salute gesture displayed only a greater percentage of time in deceleration. These gestures then, on the average, differ in peak velocity but not in percent time in deceleration. Does each subject demonstrate this pattern? How consistent are these effects within individuals across trials?

Representative velocity profiles for one subject for both gestures are depicted in Figure 16.1. The means and standard deviations for peak velocity and percent time in deceleration (Table 16.2) reveal considerable variability between subjects. LL demonstrated the highest peak velocity, and LZ the lowest; EL demonstrated the least time in deceleration, and LZ the most. Despite these differences it is clear that all subjects demonstrated the expected pattern: a substantial decrease in peak velocity for the spoon gesture, with percent time in deceleration remaining unchanged. To the extent that the shape of the velocity curve (relative time in acceleration and deceleration) reflects something of each subject's control strategy, it is interesting to see that the relative ranking of subjects in terms of the time spent in deceleration is the same for the two gestures, that is, for EL, AN, LL, LZ. This finding suggests

Table 16.2   Peak Velocity and Percent Time to Peak Velocity for Each Subject

Subject	Gesture	Peak velocity[a]		Percent time after peak velocity	
		M	SD	M	SD
EL	Salute	1,030.67	126.62	54.20	3.37
	Spoon	975.17	71.06	47.50	8.26
LL	Salute	1,608.87	63.76	67.00	1.67
	Spoon	894.63	145.73	72.80	8.26
AN	Salute	1,325.78	71.16	69.20	5.49
	Spoon	1,069.45	205.96	66.60	8.43
LZ	Salute	952.84	69.22	78.60	6.09
	Spoon	718.61	33.45	73.75	9.34

[a]mm/s.

that subjects adopted a consistent strategy across the two gestures, although these strategies differed between subjects.

These findings suggest that the normalized velocity profile may provide a rather reliable measure of normal performance of a gesture against which the performance of neurological patients might be compared. One approach here might be to compare a patient's profile to the normal profile to determine if it falls within the one- or two-standard-deviations bandwidth around the average profile. Such a "normal" profile for the salute gesture, based on the performance of the four subjects, is depicted in Figure 16.6.

The notion that subjects seem to adopt a consistent strategy (e.g., time spent in deceleration) across the gestures is supported when one examines performance across trials. Even with so few trials there is a remarkable degree of consistency in the velocity profiles within individuals. The coefficients of variation range from .03 to .19 for peak velocity and from .07 to .35 for percent time after peak velocity. Although there are considerable interindividual differences, then, it appears that there is consistency within individuals in the strategy used to control movement within a gesture over trials.

Although there is considerable consistency across trials for each subject, examination of the velocity profiles reveals differences between subjects within a gesture and differences within subjects across the two gestures in the degree to which this consistency is exhibited. This observation points to the importance of measuring

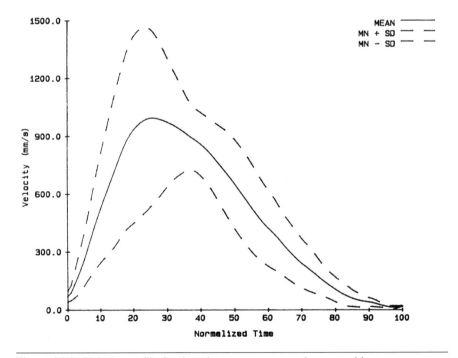

Figure 16.6   Velocity profile for the salute gesture averaged across subjects.

within-subject variability more precisely. As well it appears that one of the characteristic features of apraxia is the high degree of within-subject variability observed
over successive attempts at performing a gesture (Roy & Square-Storer, 1990). How
might within-subject variability be measured? One might begin by presenting a
visual representation that exhibits the one-standard-deviation bandwidth around the
mean profile for each subject. Examination of the profiles for two of the subjects
(EL and LZ) performing the salute gesture (Figure 16.7, a and b) reveals a larger
standard deviation bandwidth for LZ.

A more precise reflection of this variability can be derived by calculating a
coefficient of variation (CV) that reflects the summation over the entire movement
of the ratio of the standard deviation to the mean velocity at each sampling frame.
In essence this measure represents the root mean square width of the standard
deviation band expressed as a percentage of the magnitude of the velocity. To
examine whether a particular patient exhibits greater than normal within-subject
variability, her or his CV measure would be compared to a distribution based on
CV measures for all the normal subjects to determine whether it falls within the
bounds of normal. This approach, introduced by Winter (1984, 1989) in his analyses
of gait, has the potential to answer more penetrating questions than the rather simple
diagnostic one we alluded to. One might ask, for example, whether certain task
demands have more impact than others on within-subject variability and whether
these effects are similar across different types of neuropathology (e.g., stroke,
Parkinson's disease).

Although the relative consistency of the normalized velocity profiles has interesting implications for clinical diagnosis, this measure reflects resultant velocity that
is a composite measure of movements in the three spatial planes. Another approach
to examining the performance of these gestures is afforded through analyses of the
movements within each of these planes. Poizner et al. (1990) have suggested that
movement in the gestures they examined in one normal adult tended to be focused
in one of the three spatial planes. In examining the performance of two apractic
patients, they found that a significant proportion of patients' movements occurred
outside the principal plane observed in the normal adult. The notion that these out-
of-principal-plane movements are characteristic of apraxia rests on the assumption
that there is some consistency in normal adults in the plane in which movement is
focused.

In order to look at this aspect of performance, we have selected two of the subjects
who seem to reflect quite different velocity profiles (see Table 16.2). EL consistently
spends the least time in deceleration, whereas LZ spends the most time in deceleration. Are these differences in the velocity profiles for the hand reflected in any way
in the movements observed in each of the spatial planes? To look at this question,
we focused on the salute gesture. We first examined the total resultant distance in
each plane (Table 16.3). LZ generally moved through a greater distance in each
plane, but the spatial plane involving the most movement differed between the two
subjects. LZ moved the greatest distance in the Y-Z (coronal) plane. EL, however,
moved the greatest distance in the X-Z (sagittal) plane.

We also examined the variability of movement within each plane. This analysis
was done using the ellipse method introduced by Georgeopoulos, Kalaska, & Massey
(1981). The variability over trials is calculated as a series of ellipses, one at each

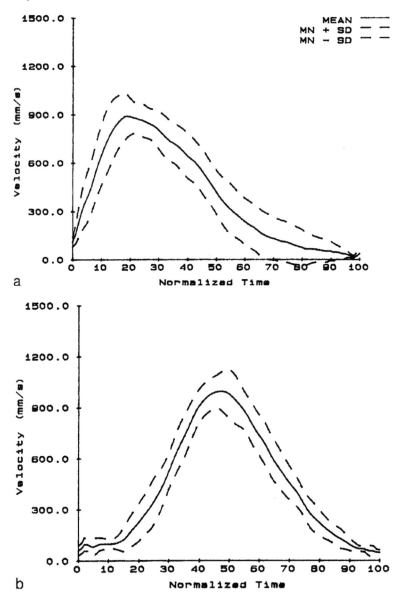

Figure 16.7   Velocity profile for the salute gesture averaged across trials for (a) subject LZ and (b) subject EL.

frame. Each ellipse has an area based on the standard deviation over trials in each axis of the plane. A representative example of the variability for LZ in the Y-Z (coronal) plane is depicted in Figure 16.8. In examining this variability (Table 16.3), we found that the pattern was consistent with that observed for the distance moved in each plane. LZ tended to demonstrate the greatest variability, and, for both

Table 16.3   Total Resultant Distance and Variability for Hand Movement in Each Plane for Salute Gesture

Subject	Measure	Plane		
		XY	XZ	YZ
EL	SD[a]	2,324.67	5,197.99	4,035.43
	Distance	147.59	237.26	229.21
LZ	SD[b]	3,540.39	5,117.73	7,809.55
	Distance	243.30	266.18	346.92

[a]mm². [b]mm.

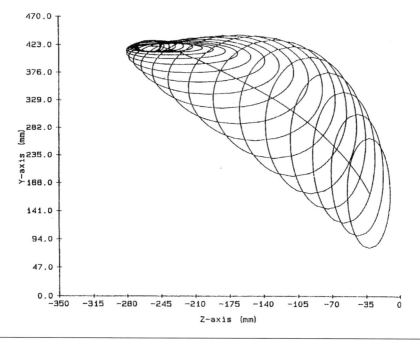

Figure 16.8   Spatial variability of movement trajectory in the coronal plane for subject LZ performing the salute gesture.

subjects, the plane exhibiting the greatest movement also exhibited the greatest variability.

It is difficult to ascertain exactly what the source of these different movement patterns might be, but it is possible that one source is the way the gesture is performed in terms of movements at each joint. Recall from our earlier analyses (Figure 16.2) that movement of the IRED on the elbow reflects movement at the shoulder, whereas changes in the angle subtended by the second (forearm), third (elbow), and fourth (upper arm) IREDs reflects movement at the elbow. When we compared the total distance moved at each joint for each subject (Table 16.4), we found greater movement for LZ at each joint, although particularly at the shoulder joint. This finding suggests that for LZ much more proximal movement at the shoulder is used in control of the arm. Because the shoulder has many more degrees of freedom than the elbow joint, it is reasonable that gestural hand movements involving more movement at the shoulder would be more variable. Further, because the predominant movement at the shoulder in this salute gesture is adduction, reflected in greater movement of the elbow IRED in the Y-Z plane for each subject, one might very well expect that the hand would make a greater movement and exhibit greater variability of movement in this plane, if the subject (LZ) controlled the gesture more at the shoulder.

The findings to this point suggest that these subjects have organized this gesture quite differently. LZ seems to control the gesture more proximally and appears to use a more complex movement, as greater movement was exhibited than for EL at both the elbow and the shoulder. These differences seem reflected in both the distance moved and the variability of movement in each plane, with LZ exhibiting greater movement and greater variability in all planes. Are these differences in the way the gesture is organized in any way related to the differences observed in the resultant velocity profiles alluded to above?

To examine this question we normalized the trajectories in time in order to remove differences between the subjects in movement time and then examined the variability in three-dimensional space using a derivation of the ellipse analysis described above. In this case the variability over trials is calculated as a series of spheres, one at each frame. Each sphere has a volume based on the standard deviation over trials in each axis of three-dimensional space. Looking first at the average variability for each subject, we see that LZ exhibits a greater volume (i.e., variability, 63790.92 mm^3) than EL (7680.09 mm^3), in accord with the previous analyses, which revealed greater variability for LZ in each plane.

The question of what relationship this variability might have with the observed velocity profile was examined by comparing the position in time of the peak variability with the relative time of the peak velocity (see Figure 16.9, a-d). For EL, percent time to peak variability was 30%, whereas her peak velocity occurred at 45%. For LZ these percentages were 37% and 21.4%, respectively. Given these data, there does not appear to be any clear relationship between peak variability in the movement

Table 16.4   Distance Moved at Elbow and Shoulder for Salute Gesture

Subject	Elbow[a]	Shoulder[b]
EL	27.50	66.18
LZ	50.99	151.72

[a]degrees. [b]mm.

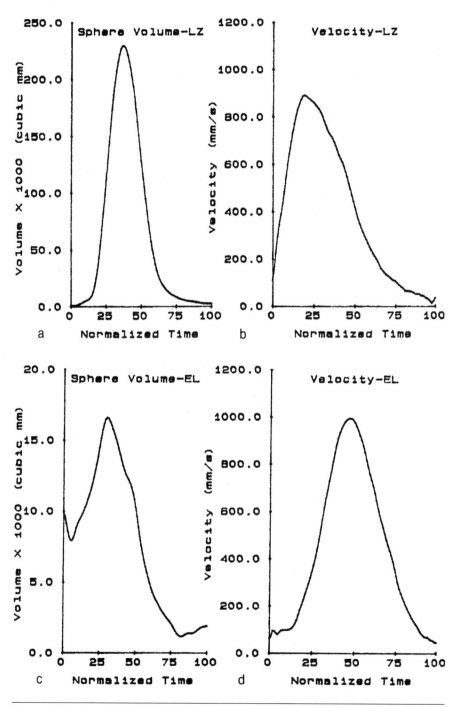

Figure 16.9 Spatial variability profiles (a, c) and velocity profiles (b, d) for the salute gesture for subjects LZ (a, b) and EL (c, d).

trajectory and peak velocity. Indeed, peak variability occurs before peak velocity for EL but after peak velocity for LZ. Possibly this difference in pattern relates to the different movement organization exhibited by these two subjects. For example, if we envisage that subjects are controlling velocity, it is possible that a person such as LZ, who appears to use a more complex coordination between the elbow and shoulder joints to execute the movement, might achieve peak velocity for the hand early in the trajectory prior to peak variability in order to have control of the movement on approaching the body.

Taken together these findings suggest that the way in which the subject executes the gesture may influence the movement strategy, as revealed through the normalized velocity profile. The velocity profile, then, may reflect both the demands of the task, as we discussed earlier with regard to differences between body-centered and allocentric gestures in terms of spatial precision, as well as the way in which the subject organizes the movement.

Another implication of these findings is that although more movement does appear to occur in one plane in this gesture, as Poizner et al. (1990) found in the gestures they examined, this plane is not consistently the same across subjects. Given these individual differences it is possible that Poizner's finding of greater out-of-principal-plane movements in his apractic patients may, rather than being a characteristic of apractic performance, reflect, to some degree, normal variation in gestural performance.

Considering that Poizner et al. (1990) used transitive gestures (pantomimed object-use gestures, cutting bread, and opening a car window with a rotary handle) and the gesture described here is intransitive (pantomimed gesture not involving objects), however, it is difficult to conclude how our findings relate to those of Poizner et al. (1990). It may be, for example, that the degree to which individual differences are apparent depends on the type of gesture. Transitive gestures may exhibit much less individual variation in which plane includes the greatest movement, because these gestures are meant to simulate the use of an object. Indeed, in actual use of the object (e.g., a knife to cut bread), movement is constrained by the environment (e.g., the location of the bread), by the object (e.g., the size of the handle), and by the task objective (e.g., cut a slice of bread vertically) such that effective performance necessitates that movement be focused in one plane (e.g., the sagittal). Intransitive gestures, on the other hand, are used more for interpersonal communication. Because a gestural message (e.g., a salute or a wave goodbye) can be communicated with a variety of movements (e.g., one can wave with movement originating from the wrist, elbow, and/or shoulder; see Figure 16.7), it is possible that considerable interindividual variation might be exhibited in pantomiming these gestures. Clearly, much more work needs to be done on examining these potential sources of individual variation in gestural performance in normal adults before we can begin to adequately describe the nature of apraxia.

IMPLICATIONS FOR STUDYING APRAXIA

In reviewing these findings it seems possible to capture a description of gestural performance at two levels of analysis. One, a categorical notation system, proffers a rather low-level description reflected in dimensions such as the location of the hand in space. At this level, variability representing primarily individual differences indicates the distribution of normal performance that is central to defining errors in performance. This approach potentially has important implications for the clinical assessment of apraxia, in that the presence of an error in a particular performance dimension is empirically determined based on the distribution of the performance of normals in each dimension.

Kinematic analyses of performance, the other level of analysis, affords a description that provides much more insight into how the gesture is organized and controlled in both time and space. At this level, variability reflects both performance consistency within individuals over trials as well as individual differences. Here, both sources of variability permit an examination of the degree to which the patient's performance falls within the bounds of normality. As we considered in our discussion of variability, it is possible to determine whether a patient's velocity profile is within a standard deviation band encompassing the normal profile. Again, as with the error-notation approach, we have an objective means of defining performance errors, an important feature in clinical assessment.

Unlike error-notation analyses, however, kinematic analyses afford a much richer source of information about aberrant performance in brain-damaged patients. In fact one might view the velocity profile comparisons as only the first step in analyzing a patient's performance. Subsequent steps might focus on the components of the velocity profile, asking whether the patient's profile differs from the normal in peak velocity and/or time after peak velocity. Analyses examining the variability in the spatial trajectory of the hand movement within each plane and others investigating the organization of the gesture in joint space may provide progessively more detailed accounts of how the patient's performance is impaired. Because our findings here point to the sensitivity of the velocity profile to task demands (e.g., spatial precision), this type of serial approach to analyzing the patient's performance may yield important insights into differential effects of task demands in various neuropathologies (e.g., stroke, neurodegenerative diseases), thus potentially affording a means for identifying the contribution of different brain regions (e.g., frontal lobe, basal ganglia) to gestural performance (cf. Roy et al., 1991).

The error-notation and kinematic analyses described herein are only first approximations to the types of analyses necessary to understand more clearly the nature of apraxia. One of the most important directions is to develop a better method for examining the organization of movement in joint space. Analyses of the relationships among angular displacements at each joint will afford a more lucid description of intersegmental coordination. Kinetic analyses of the joint moments will reveal where the forces are being generated, permitting one to discern which joint is contributing the most to the movement.

As for the application of this work to clinical practice, most clinical sites would not have the capability to carry out kinematic analyses. Error-notation approaches, on the other hand, are commonly used in many fields of clinical practice, such as speech pathology (e.g., Square-Storer, Qualizza, & Roy, 1989). It is important, however, that the dimensions of performance in the notation system used and the distribution of normal performance in each dimension be valid descriptors of gestural performance. One way to ensure this is to derive in some sense the notation system from the kinematic analyses. Work in finite state-space models of motor control (e.g., Tomovic, Anastasijevic, Vuco, & Tepavac, 1990) provides some clues as to how one might make such derivations. Considering the action dimension in the notation system, for example, the state of each joint (e.g., flexion/extension) can be derived from the joint displacement curves revealed in the kinematic analysis. This state-space depiction of the gesture, providing an account of which joints are active during performance, may then define what is the "normal" joint action pattern. When the observational techniques used in the notation system are applied, the action pattern observed in a patient can be compared to this normal pattern to ascertain whether it is different. When derived in this way the error-notation system, then, could be used as a screening tool to determine which patients might reasonably be examined using more fine-grained kinematic analyses. This type of progression through levels of analysis is employed in general medical diagnosis and other areas of clinical neuroscience (e.g., speech pathology) and, so, may be useful in developing more complete descriptions of apractic performance.

REFERENCES

Argyle, M. (1988). *Bodily communication.* London: Methuen.

Atkeson, C.G., & Hollerbach, J.M. (1985). Kinematic features of unrestrained arm movements. *Journal of Neuroscience,* **5,** 2318-2330.

Charlton, J., Roy, E.A., Marteniuk, R.G., MacKenzie, C.L., & Square-Storer, P.A. (1988). Disruptions to reaching in apraxia. *Society for Neuroscience Abstracts,* **14,** 1234.

Darling, W.G., & Cooke, J.D. (1987). Changes in the variability of movement trajectories with practice. *Journal of Motor Behavior,* **19,** 291-309.

DeRenzi, E. (1985). Methods of limb apraxia examination and their bearing on the interpretation of the disorder. In E.A. Roy (Ed.), *Advances in Psychology. Volume 23. Neuropsychological studies of apraxia and related disorders,* (pp. 45-64). Amsterdam: North-Holland.

Efron, D. (1941). *Gesture and environment.* New York: King's Crown Press.

Friesen, H., Roy, E.A., Square-Storer, P.A., & Adams, S. (1987). *Apraxia: Interrater reliability of a new error notation system for limb apraxia.* Poster presentation at the annual meeting of the North American Society for the Psychology of Sport and Physical Activity, Vancouver, B.C.

Georgeopoulos, A.P., Kalaska, J.F., & Massey, J.T. (1981). Spatial trajectories and reaction times of aimed movements: Effects of practice, uncertainty and change in target location. *Journal of Physiology,* **46,** 725-743.

Haaland, K.Y., & Flaherty, D. (1984). The different types of limb apraxia errors made by patients with left or right hemisphere damage. *Brain and Cognition,* **3,** 370-384.

Jarvella, R.J., & Klein, W. (1982). *Speech, place, and action.* Chichester, England: Wiley.

Jason, G. (1983a). Hemispheric asymmetries in motor function. I. Left hemisphere specialization for memory but not performance. *Neuropsychologia,* **21,** 35-46.

Jason, G. (1983b). Hemispheric asymmetries in motor function. II. Ordering does not contribute to left hemisphere specialization. *Neuropsychologia*, **21**, 47-58.

Jason, G. (1985). Manual sequence learning after focal cortical lesions. *Neuropsychologia*, **23**, 35-46.

Jason, G. (1986). Performance of manual copying tasks after focal cortical lesions. *Neuropsychologia*, **23**, 41-78.

Jason, G. (1990). Disorders of motor function following cortical lesions: Review and theoretical considerations. In G. Hammond (Ed.), *Cerebral control of speech and limb movements*. Amsterdam: Elsevier.

Johnson, H.G., Ekman, P., & Friesen, W.V. (1975). Communicative body movements: American emblems. *Semiotica*, **15**, 335-353.

Kaplan, E. (1968). *The development of gesture*. Unpublished doctoral dissertation, Clark University, Worcester, MA.

Kendon, A. (1980). Gesticulation and speech: Two aspects of the process of utterance. In M.R. Key (Ed.), *The relationship of verbal and nonverbal communication*. New York: Mouton.

Kertesz, A. (1985). Apraxia and aphasia: Anatomical and clinical relationship. In E.A. Roy (Ed.), *Advances in psychology. Volume 23. Neuropsychological studies of apraxia and related disorders* (pp. 163-178). Amsterdam: North-Holland.

Kertesz, A., & Hooper, P. (1982). Praxis and language. The extent and variety of apraxia in aphasia. *Neurophysiologia*, **20**. 275-286.

Kimura, D., & Archibald, Y. (1974). Motor functions of the left hemisphere. *Brain*, **97**, 337-350.

Kimura, D. (1976). The neural basis of language *qua* gesture. In H. Whitaker & H.A. Whitaker (Eds.), *Studies in neurolinguistics* (Vol. 2). New York: Academic Press.

Kimura, D. (1982). Left-hemisphere control of oral and brachial movements and their relationship to communication. *Philosophical Transactions of the Royal Society of London*, **B298**, pp. 135-149.

Kolb, B., & Milner, B. (1981). Performance of complex arm and facial movements after focal brain lesions. *Neurophysiologia*, **14**, 491-503.

Kolb, B., & Whishaw, I. (1990). *Fundamentals of human neuropsychology* (3rd ed.). New York: W.H. Freeman.

Liepmann, H. (1920). *Ergenbnisse der Gesamten Medizin*, **1**, 516-543.

Marteniuk, R.G., MacKenzie, C.L., Jeannerod, M., Athenes, S., & Dugas, C. (1987). Constraints on human arm movement trajectories. *Canadian Journal of Psychology*, **41**, 365-378.

McNeil, D. (1985). So you think gestures are nonverbal? *Psychological Review*, **92**, 350-371.

Nelson, W.L. (1983). Physical principles for economies of skilled movements. *Biological Cybernetics*, **46**, 135-147.

Poeck, K. (1983). Ideational apraxia. *Journal of Neurology*, **230**, 1-5.

Poeck, K., Lehmkuhl, G., & Willmes, K. (1982). Axial movements in ideomotor apraxia. *Journal of Neurology, Neurosurgery, and Psychiatry*, **45**, 1125-1129.

Poizner, H., Klima, E.S., & Bellugi, U. (1987). *What the hands reveal about the brain*. Cambridge, MA: MIT Press.

Poizner, H., Mack, L., Verfaellie, M., Rothi, L.J.G., & Heilman, M. (1990). Three-dimensional computergraphic analysis of apraxia: Neural representations of learned movement. *Brain*, **113**, 85-101.

Ramsey, S. (1984). Double vision: Nonverbal behavior East and West. In A. Wolfgang (Ed.), *Nonverbal behavior*. Lewiston, NY: Hofgrefe.

Rothi, L.J.G., Mack, L., Verfaellie, M., Brown, P., & Heilman, K.M. (1988). Ideomotor apraxia: Error pattern analysis. *Aphasiology*, **2**, 381-387.

Roy, E.A. (1982). Action and performance. In A. Ellis (Ed.), *Normality and pathology in cognitive function* (pp. 265-298). New York: Academic Press.

Roy, E.A. (Ed.). (1985). *Advances in psychology. Volume 23. Neuropsychological studies of apraxia and related disorders*. Amsterdam: North-Holland.

Roy, E.A. (1990). The interface between normality and pathology in understanding motor function. In G. Reid (Ed.), *Problems in motor control* (pp. 3-30). Amsterdam: Elsevier.

Roy, E.A., & Hall, C. (1992). Limb apraxia: A process approach. In D. Elliott & L. Proteau (Eds.), *Vision and motor control* (pp. 261-282). Amsterdam: Elsevier.

Roy, E.A., & Brown, L. (1990). *Kinematic analyses of gesturing*. Unpublished bachelor of science thesis, Kinesiology Department, University of Waterloo, Ontario.

Roy, E.A., Friesen, H., Square-Storer, P.A., & Adams, S. (1987, June). Apraxia: Interrater reliability of a new error notation system for limb apraxia. Poster presentation at the annual meeting of the North American Society for the Psychology of Sport and Physical Activity, Vancouver, British Columbia, Canada.

Roy, E.A., Square, P.A., Adams, S., & Friesen, H. (1985). Error/movement notation systems in apraxia. *Recherches Semiotiques/Semiotics Inquiry*, **5**, 402-412.

Roy, E.A., & Square-Storer, P.A. (1990). Evidence for common expressions of apraxia. In G. Hammond (Ed.), *Cerebral control of speech and limb movements*. Amsterdam: Elsevier.

Roy, E.A., Square-Storer, P.A., Adams, S., & Friesen, H. (1989). Disruptions to central programming of sequences. *Canadian Psychology*, **30**, 423.

Roy, E.A., Square-Storer, P.A., Hogg, S., & Adams, S. (1991). Analysis of task demands in apraxia. *International Journal of Neuroscience*, **56**, 177-186.

Soechting, J.F. (1984). Effect of target size on spatial and temporal characteristics of a pointing movement in man. *Experimental Brain Research*, **54**, 121-132.

Square-Storer, P., Qualizza, L., & Roy, E.A. (1989). Isolated and sequenced oral posture production under different input modalities by left-hemisphere damaged adults. *Cortex*, **25**, 371-386.

Square-Storer, P.A., Roy, E.A., & Hogg, S. (1990). The dissociation of aphasia from apraxia of speech, ideomotor limb and bucchofacial apraxia. In G. Hammond (Ed.), *Cerebral control of speech and limb movements*. Amsterdam: Elsevier.

Stokoe, W.C. (1972). *Semiotics and human sign languages*. The Hague: Mouton.

Tomovic, R., Anastasijevic, R., Vuco, J., & Tepavac, D. (1990). The study of locomotion with finite state models. *Biological Cybernetics*, **63**, 271-276.

Winter, D.A. (1984). Kinematic and kinetic patterns in human gait: Variability and compensating effects. *Human Movement Science*, **3**, 51-76.

Winter, D.A. (1989). Coordination of motor tasks in human gait. In S.A. Wallace (Ed.), *Perspectives on the coordination of movement* (pp. 330-363). Amsterdam: Elsevier.

Acknowledgments

The authors wish to thank the editors for their constructive comments on earlier drafts of this paper. Appreciation is also extended to Tammi Winchester for help in the preparation of the figures.

Chapter 17

Stereotypy and Variability

Karl M. Newell, Richard E.A. van Emmerik, and Robert L. Sprague
University of Illinois at Urbana-Champaign

The concepts of variability and stereotypy in motor control are in many respects antithetical. On the one hand, the notion of variability traditionally focuses on variation in movement sequences and their outcomes, with the primary theoretical perspective being that the existence of variability, or increased variability relative to some a priori standard, reflects a problem of control or instability in the sensorimotor system. On the other hand, the notion of stereotypy usually refers to the lack of variation in movement sequences, and this condition in turn is also interpreted theoretically as reflecting a problem for system control. Variation in the movement sequence and its outcome has been, therefore, a common index for considering instability and stereotypy in motor control.

Although many definitions of stereotypic behavior exist, stereotypic acts are usually taken to reflect superfluous patterns of coordination. They involve repetitive movements or postural sequences that appear unrelated to a specific action goal or, at least most certainly, to an externally defined action goal. These stereotypic movements can occur during the execution of a particular task-oriented action, or they may be isolated as the execution of the stereotypic activity itself. Stereotypic activities have been studied from a variety of theoretical and practical perspectives, including their relation to normal infant motor development (Thelen, 1980; Wolff, 1968); rearing experiences of the individual (Mason & Berkson, 1975); institutional stereotypies of the developmentally disabled (Baumeister & Forehand, 1973; Berkson, 1967; Lewis & Baumeister, 1982); oscillatory properties of the neuromotor system (Pohl, 1977); stimulus characteristics that induce the repetitious behaviors (Berkson, 1983; Berkson & Gallagher, 1986); and the neurotransmitter properties of the brain (Waddington, Molloy, O'Boyle, & Pugh, 1990).

The measurement of stereotypy has relied primarily on natural observation and checklist categorization of the movement sequences examined. In contrast, studies of movement variability usually focus on the within-subject standard deviation of a given movement parameter (outcome, some discrete kinematic or kinetic property), such as in work on the movement speed-accuracy trade-off, isometric force variability, and joint variability that relates to movement outcome (e.g., Hancock & Newell, 1985; Newell, Carlton, & Hancock, 1984). Thus, not only have the concepts of stereotypy and variability been examined with different operational movement assessment techniques, but they have also focused on different properties of movement coordination patterns. The literature on movement stereotypies has been primarily concerned (albeit implicitly) with the qualitative "form" properties of movement sequences, whereas the movement variability literature has focused on the quantitative properties of certain discrete movement parameters and their outcomes. And viewed from another standpoint, the stereotypies domain has focused on the invariance in movement sequences, whereas the variability domain has focused on the variance of discrete movement properties.

In this chapter we examine the variability of the qualitative and quantitative properties of stereotypic movement sequences. The focus of our discussion of stereotypy and variability will be the stereotypic movements that arise as an emergent side effect from prolonged neuroleptic medication—a syndrome known as tardive dyskinesia (APA, 1980; Kalachnik, 1984; Sprague & Newell, 1987). We suggest that the recognized invariance of stereotypic movements is in the qualitative "form" properties of the movement coordination patterns, rather than the quantitative properties of a given discrete dynamic variable.

A number of theoretical viewpoints have been advanced in relation to stereotypies (cf. Cooper & Dourish, 1990). All of these theoretical accounts suggest that a description of the movement sequence is a necessary but not sufficient component to understanding stereotypy. It is our position that the invariance in stereotypic movements appears to reflect problems in the adaptability of the sensorimotor system. We interpret this stereotypic invariance in terms of a dynamic systems account of motor control, in which variability is viewed within bounds as a facilitory factor to system control, rather than the traditional problematic interpretation that arises from the direct association of movement variability and system noise.

STEREOTYPY AND SKILL

To highlight the difficulty of distinguishing the unique features of stereotypic movements, we begin this chapter by briefly contrasting the concepts of stereotypy and skill. The terms *stereotypy* and *skill* in movement are usually inferred to be at opposite ends of a continuum on some poorly defined, or even undefined, dimension. Stereotypy and skill are rarely integrated into a unified theoretical perspective in motor control, in part because the two concepts are typically confined to separate domains of empirical investigation. However, in some respects these concepts hold common ground, whereas in others they are remarkably different. A key factor in understanding these two action concepts is the dimension of movement variability, which, as we have already indicated, is often invoked to define, partially or in total, both stereotypy and skill. In this section we explore stereotypy and skill with respect to this variability dimension and propose that movement variability is not a sufficient basis on which to distinguish these concepts.

The common ground in stereotypy and skill can be found in both concepts' having reduced variability of some aspect of the movement sequence as a characteristic feature. As outlined earlier, a hallmark feature of stereotypic actions is reduced variability (Berkson, 1983), a property that even dictionary definitions of stereotypy often emphasize. However, most definitions of stereotypy are vague with respect to the exact movement properties that exhibit reduced variability. Similarly, most definitions of skill incorporate either directly or indirectly reduced variability as an inherent feature. For example, Guthrie (1935) suggested that the bringing about of some predetermined outcome with *maximal certainty* is an integral part of skilled performance. Thus, both concepts, at least on the surface, are defined by the common index of movement variation, with reduced variability reflecting both enhanced stereotypy and higher levels of skill—a schematic of which is shown in Figure 17.1. There are, however, a number of caveats to the relation between stereotypy and skill as defined on a variability dimension.

One limitation to considering variability as a sufficient basis for distinguishing stereotypy and skill is revealed through an examination of the boundaries to the variability evident in movement sequences. For example, if we consider movement variability on a ratio scale dimension, then in principle one could record no variability in certain properties of movement. Would no variability on a given movement dimension (or dimensions) be the ultimate expression of skill in movement control? This is perhaps the literal implication from most definitions of skill, but such an extreme interpretation is almost certainly unwarranted, as it is counter to another hallmark characteristic of skill, namely, adaptability.

It seems that there are some bounds at the lower end of the variability dimension for an action to be inferred to be skilled, beyond which one moves closer to the common interpretation of the concept of stereotypy. This variability threshold for determining skill and stereotypy is most certainly task dependent. Furthermore, at the upper end of the movement variability continuum, high variability would probably lead to negating an interpretation of stereotypy, whereas high variability might be invoked as reflecting a low level of skill. The discrete movement parameter selected to estimate skill level is also a factor, as skilled performers may minimize variability on a particular dimension (often the task criterion) but have higher joint motion variability than their less skilled counterparts (Arutyunyan, Gurfinkel, & Mirskii, 1968, 1969). Thus, for a number of reasons, the degree of variability is a relative issue, and the concepts of stereotypy and skill cannot be considered as

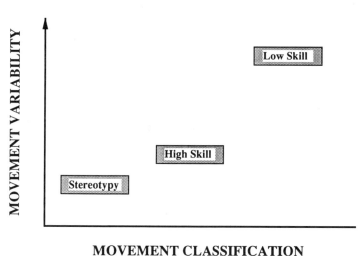

Figure 17.1   Schematic of the relation between stereotypy, skill, and variability.

opposites on a variability dimension without identifying the relevant movement parameters.

Another important qualifier that needs to be invoked in considering the role that variability plays in distinguishing stereotypy and skill is what dimension (or dimensions) of movement or action the variability is being considered. Typically, there is a significant difference in the dimension on which movement variability is assessed in the stereotypy and skill domains. The stereotypy literature usually focuses on the invariance of the form of the movement sequence itself, whereas the skill literature usually focuses on the level of variability of the movement outcome. This distinctive emphasis on variability of movement form and outcome is more a matter of tradition in each domain, as a consequence of the tasks that have been used to examine performance, than it is a reflection of any fundamental requirement of the definitions of stereotypy and skill (Newell, 1985).

It is also worth noting, by way of closing this introductory section, that stereotypic and skilled movements hold commonality on dimensions other than variability. For example, many stereotypic actions appear very smooth and efficient, again qualitative features of movement that are characteristic of what is usually termed a skilled performance. This apparent similarity on an efficiency dimension is not surprising, because stereotypic actions are in many instances the products of an extensive practice regimen involving the performance of many trials over a considerable period of time. Thus, even Guthrie's (1935) complete criteria for skill, namely, the bringing about of some predetermined outcome with maximal certainty and minimal outlay of time and energy, cannot be considered a sufficient basis to distinguish stereotypy and skill. It is also not beyond the bounds of reasonableness to suggest that some stereotypic actions even exhibit aesthetic qualities in the movement sequence.

In summary, it seems evident that one may need to look beyond the dimensions of variability, efficiency, and aesthetics to distinguish the concepts of stereotypy and skill. Descriptions of movement sequences on these dimensions do not appear to be *sufficient* to characterize the nature of actions. It is proposed here that it is the adaptability of the sensorimotor system in relation to the goal of the task and the changing constraints to action that is an important dimension for distinguishing stereotypy and skill. Adaptive control is a dimension that has not been studied directly in relation to stereotypies.

ANALYSIS OF STEREOTYPIC MOVEMENTS

Natural observation of stereotypic actions leads intuitively to the categorization of behaviors such as body rocking, hand waving, facial grimacing, tongue thrusting, and so on. The rating scale categorization of stereotypic movements provides a useful approach to the screening and identification of many stereotypic actions, and it can also provide a basis for the development of movement disorder screening tests that can be administered by nursing and allied health personnel in a number of clinical settings (e.g., Kalachnik, Sprague, & Slaw, 1988; Sprague & Kalachnik, 1991). However, the rating scales do not provide any direct measures of the motor control characteristics of the stereotypic actions, and the observer-quantified approach is probably not the most sensitive technique for discriminating among various populations or pathological conditions (Newell & Sprague, 1990).

It would seem that only a formal dynamic analysis of stereotypic actions will open the window to an understanding of the motor control properties of these abnormal movements. This approach would be particularly useful if we are to distinguish differences between, for example, stereotypies induced by prolonged medication regimens and those that arise from isolation. One instance of this category distinction is the stereotypic movements of the developmentally disabled who are diagnosed as having tardive dyskinesia and those stereotypic actions of the institutionalized developmentally disabled who have not been on medication regimens. The movement control characteristics of the developmentally disabled have received only minor theoretical consideration (Berkson, 1983; Lewis & Baumeister, 1982; Pohl, 1977), in part because of the rather simplistic approaches to characterizing the dynamic (kinematic and kinetic) properties of the stereotyped actions. And on the occasions when the dynamic characteristics of tardive dyskinesia or stereotypic behaviors have been reported (e.g., Gardos, Cole, & La Brie, 1977; Lewis, MacLean, Johnson, & Baumeister, 1981), the experimental work has been so limited that neither the description of the movement sequence nor the theoretical link to the movement control domain has been adequately examined. The Eshkol-Wachman Movement Notation System is used now to understand the movement components of stereotypic actions of drug-treated animals (cf. Teitelbaum, Pellis, & DeVietti, 1990) and the phylogenetic stereotypic sequences that appear to be a part of certain aspects of animal behavior (Golani, 1976).

It is evident, therefore, that there are no data available that bear directly on the variability of the qualitative and quantitative properties of stereotypic movements, in spite of the centrality of movement variability to the notion of stereotypic behavior. In a recent paper, one of us outlined an approach to operationally distinguishing the qualitative and quantitative properties of movement that are usually associated with the action characteristics of coordination, control, and skill (Newell, 1985). In this framework, the relational or qualitative properties of motion characterize coordination, whereas the quantitative properties of discrete variables of movement dynamics characterize control. One can observe variability in both the qualitative and the quantitative properties of movement, a feature that has particular utility for attempts to understand stereotypic behavior (Newell, 1986a). In short, this movement skill-oriented framework provides an operational approach to characterizing what Berkson and Gallagher (1986) called the topographic structure of stereotypic movements.

STEREOTYPIES AND TARDIVE DYSKINESIA

The stereotypic motions that arise from prolonged use of neuroleptic medication can be so severe that these movement abnormalities can interfere with the conduct of activities of daily living, including the execution of vocational activities. The time course influences of medication on the development of stereotypies is still not well understood in humans, in part because of the difficulty of monitoring the behavior of individuals over the full time-span of drug intake and drug withdrawal periods. The onset of observable movement disorders often takes many months from the beginning of the medication regimen, and the development of tardive dyskinesia stereotypies usually becomes progressively more prevalent and observable over time. The stereotypies become even more noticeable on the immediate withdrawal of the medication and then dissipate to some degree over prolonged periods of drug withdrawal (Kalachnik et al., 1984; Sprague, 1983). There is still controversy and debate about the reversibility of tardive dyskinesia, in part because the long-term drug withdrawal time-series data are not available to make the appropriate determination.

There seems to be a higher degree of facial stereotypies in tardive dyskinesia than in other movement disorder syndromes that exhibit stereotypic behavior, although comprehensive syndrome-by-stereotypy-category data are not available. Certainly, the stereotypies of tardive dyskinesia are not confined to the face and can be observed in all effector units of the action system. It is generally recognized, however, that tardive dyskinesia has an emphasis on stereotypies in the more peripheral effector units of face, fingers, and toes (Kalachnik, 1984; Sprague & Newell, 1987).

There are two general action categories in which stereotypies are observed. The most prevalent category is the natural stereotypies produced in the absence of any specific external goal or stimulation. The other category where stereotypies are observed is during the execution of a given externally defined and task-oriented action. This latter action category for stereotypies has been less studied than the former, although in both cases there are very few data available that relate directly to the variability characteristics of the movement dynamics of stereotypies.

In the remainder of this section of the paper we present some data from our ongoing work that speak to a number of aspects of the variability of stereotypic movements of tardive dyskinesia, including both natural stereotypies and those that emerge during the execution of an externally driven task-oriented action. Some of the kinematic data have been obtained through standard videocamera movement recording techniques. It should be self-evident that stereotypies can be recorded only when the subject naturally exhibits the behavior—by definition, therefore, they cannot be invoked or recorded on demand by the experimenter. As a consequence, the operational control available in the natural videotape recording of these motions is reduced when contrasted to the typical ideal operational methods for analysis of movement dynamics. This is because it is counterproductive to constrain the subject and influence the stereotypic behavior by imposing markers on the body or by limiting the tardive dyskinetic individual to movement in a particular plane of motion. Nevertheless, it is our estimation that the kinematic analyses produced under these natural recording circumstances are adequate to infer both the qualitative and the relative quantitative characteristics of stereotypic and other rhythmical behaviors.

Facial Stereotypies

The data reported here are drawn from a study of 11 adult subjects that were diagnosed by both clinical judgment and rating scale methodology to have tardive dyskinesia (van Emmerik, Sprague, Slobonouv, & Newell, 1992). The exact medication history of these individuals is not available, as is often the case with this category of subjects. All subjects, however, had been on neuroleptic medication for some years, were diagnosed as tardive dyskinetic by the supervising medical personnel, and readily produced clear and observable movement stereotypies.

All data were derived from videotapes of the subjects' movements that were recorded while they were being examined clinically for abnormal tardive dyskinetic movements. Most of the stereotypic actions observed were recorded during either the waiting period or the testing phase of the rating scale examination for tardive dyskinesia (DISCUS; Sprague, Kalachnik, & Slaw, 1989). For the facial stereotypies recorded, the horizontal and vertical movements of the tongue, the vertical movements of the jaw, and the vertical movements of the upper lip and lower lip were analyzed. As we could record the stereotypies only as they occurred, not all subjects produced all categories of the facial movements studied, and each subject produced a different number of cycle durations and number of trials of each stereotypy. Thus, the data inevitably have a certain individual flavor to them, although where possible the data are grouped to provide the basis for more general assessments. The facial stereotypic movement patterns were evaluated according to different classes of dependent variables that are related to the qualitative and quantitative aspects of the movement sequences.

The quantitative aspects refer to the discrete measures used to quantify the variability in the stereotypic movement patterns. These measures include mean cycle duration, mean maximum position, mean minimum position, mean peak-to-peak distance and duration, as well as the standard deviation for all these measures. The variability of the stereotypic movement patterns was assessed by using the individual subject coefficient of variation of these discrete measures. The coefficient of variation (the standard deviation divided by the mean) is a relative dimensionless measure and allows the assessment of movement variability in each effector system independently of knowing the absolute spatial measures on some given externally defined dimension for each subject's movement profile.

Figure 17.2, a and b, displays position-time movement profiles from the upper and lower lips for two subjects during the execution of behavioral sequences that were defined by observers of the natural action as stereotypic. These profiles reveal a high degree of coupling between the upper and lower lips. Indeed, the cross-correlations between the positions of the upper and lower lip in the two examples were $r = .86$ and $r = .92$ for Figures 17.2a and 17.2b, respectively. These kinematic profiles are generally reflective of the lip stereotypies observed in our subjects, although some subjects demonstrate out-of-phase lip relations, rather than the inphase lip relations shown in Figure 17.2.

The form of the stereotypic motions was assessed through pattern recognition techniques and assessment of the degree of coupling between pairs of body parts, such as the upper and lower lips. To examine the cycle-to-cycle stability in the form of the stereotypic movement sequence, relative movement patterns were created for each cycle in the kinematic time series. These relative movement patterns were obtained by plotting (a) the data in configuration space, for example, lower lip vertical displacement versus upper lip vertical displacement; and (b) the phase plane, which represents the plot of the displacement of one segment or axis of motion

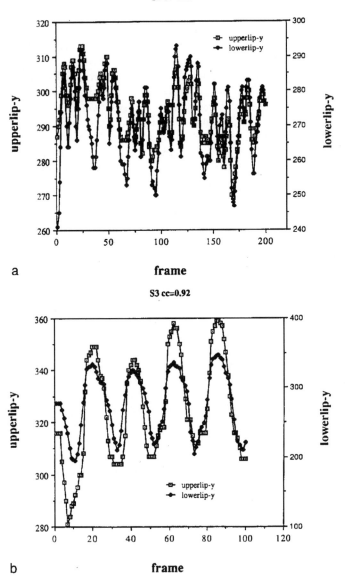

Figure 17.2 Sample upper and lower lip position over-time stereotypic profiles for two different tardive dyskinetic subjects. Video frame rate is 30 Hz, hence each frame = 33.33 ms. Lip motion unit is pixels.

versus the velocity in that same segment or axis of motion. The degree of coherence in the movement form between cycles was obtained by means of pattern recognition analysis (Sparrow, Donovan, van Emmerik, & Barry, 1987). This analysis yields a recognition coefficient with a value of 1.0, reflecting absolute similarity of movement form, and conversely a coefficient of 0.0, reflecting no coherence between the kinematics of the lip cycles.

The recognition coefficients for the stereotypic lip motion cycles ranged from .5 to .9 across the set of trials examined for our subjects. Thus, there were considerable individual differences in the stability of the movement form of the lip motions, with some subjects on some trials exhibiting tight consistency in the form of the cyclical behavior and other subjects exhibiting low- to medium-level consistency. Even in the trials that yielded a low stability measure, it is important to note that natural observation of the movement sequence on the videotape would still determine that the stereotypy was being exhibited. In other words, the variability in kinematic form as shown by our relatively fine-grained kinematic analysis was not so severe that an observer of the natural stereotypy would infer that the stereotypic movement had changed. This finding is consistent with the perceptual constancy literature that shows that considerable variability is affordable for the perception of a given form in a variety of dimensions.

The lip motion profiles shown in Figure 17.2, a and b, also reveal that there is considerable variability in the discrete cyclical properties of a given lip position–time trace. That is, there is variability in the amplitude and frequency of the lip profile from cycle to cycle. The variability of the lip position–time cycles was analyzed and expressed as a coefficient of variation of either space or time. Figure 17.3, a and b, shows the coefficient of variation for several upper lip and lower lip kinematic parameters over trial cycles for four of the tardive dyskinetic subjects.

Again, there are considerable differences between subjects and between movement parameters. Some subjects have low coefficients of variation on certain movement properties of the lip cycle but not on others. The patterns of the coefficients of variations tend, however, to show two regularities—namely, that there is a similar relative subject variability across the coefficients of variation for the different dependent variables both within *and* between the lips. This finding supports the notion that the form of the movement sequence is being retained over the stereotypic cycle even though the variation in a given discrete kinematic movement parameter can be quite considerable. It should be noted that the absolute levels of the coefficients of variation are considerably higher than those found in the analysis of movement outcome variability (cf. Newell, Carlton, & Hancock, 1984).

To make a direct comparison of the variability of these stereotypic motions to the variability of similar movements in normal subjects, we have analyzed the form and quantitative variability profiles of normal adult subjects producing rhythmical preferred motions in the effector units that are common to tardive dyskinesia. Thus, in relation to the preceding lip motion analysis of subjects diagnosed as having tardive dyskinesia, we asked six normal adult subjects to produce two 15-s duration trials with a preferred rhythmical lip motion. The form and variability of the lip cycles were determined using videotape methods and analysis techniques similar to those reported earlier. The coefficients of variation for the six subjects varied as a function of the movement parameter under consideration, but in general the

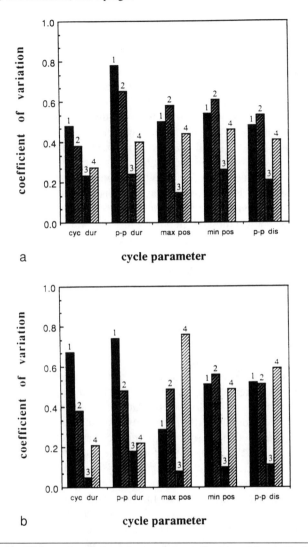

Figure 17.3   Coefficient of variation of several upper lip (a) and lower lip (b) motion parameters during mouth stereotypies for four subjects. The cycle parameter (dependent variable) abbreviations: cyc dur = cycle duration; p-p dur = peak-position-to-peak-position duration; max pos = maximum peak position; min pos = minimum peak position; and p-p dis = peak-to-peak distance (movement amplitude).

coefficients of variation for a respective kinematic parameter were always lower than those produced by the tardive dyskinetic subjects. The coefficients of variation for the peak amplitudes and frequencies for the upper and lower lip motions ranged from .008 to .039. The mean of the peak amplitude coefficient of variation across subjects and trials was .018. A similar trend of lower coefficients of variation in the normal subjects was evident in the analysis of other lip kinematic parameters.

These lower coefficients of variation in the normal subjects were also obtained on what turned out in the main to be longer duration time series than those produced by the tardive dyskinetic subjects.

Thus, the variability of individuals without tardive dyskinesia producing rhythmical movement forms that were similar to the facial stereotypies was lower than that produced by the tardive dyskinetic subjects while executing their stereotypic motions. It should be remembered that the normal subjects were not well practiced at producing the rhythmical lip motion and that this was probably the first occasion on which they produced these movements in a systematic fashion. These data from the normal subjects reveal that the variability of the stereotypies produced by the tardive dyskinetic subjects was, in the main, higher than that evident in normal healthy adult subjects producing similar lip motions. These preliminary findings on the variability of stereotypic movements suggest limitations to considering variability of motion as a defining feature of stereotypy.

Similar findings occurred with the analysis of the vertical and horizontal tongue motions. In Figure 17.4 we show the coefficient of variation for the vertical displacement of the tongue on five tongue cycle kinematic parameters for four tardive dyskinetic subjects. Again, there was considerable individual difference in the degree of variability, with the overall coefficient of variation being high in three of the subjects. The coefficient of variation on the same parameter for the six normal subjects oscillating their tongues in the vertical plane at a preferred motion ranged from .012 to .043, with the mean coefficient of variation across subjects and trials being .0215. Thus, as with the lip motions, the coefficent of variation of the individual

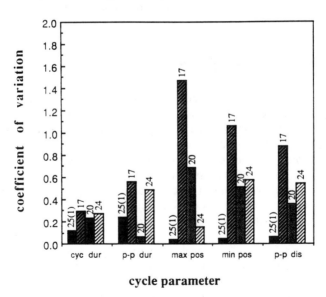

Figure 17.4   Coefficients of variation for several vertical tongue parameters during mouth stereotypies for four subjects: cyc dur = cycle duration; p-p dur = peak-position-to-peak-position duration; max pos = maximum peak position; min pos = minimum peak position; and p-p dis = peak-to-peak distance (movement amplitude).

stereotypic motions is as high as, and often higher than, that of preferred rhythmical motions of similar form in normal healthy subjects.

We have also analyzed the rhythmical properties of limb and torso motions of subjects exhibiting stereotypic movements in a manner similar to that described above for facial stereotypies. There are common forms of stereotypies such as body rocking and pill rolling, but in addition there is a range of individual specific movement forms exhibited by subjects showing stereotypic behavior. The limb and torso stereotypies that were analyzed were from adult subjects that were diagnosed as developmentally disabled (profoundly and severely mentally retarded) and tardive dyskinetic (Sprague, Slobonouv, & Newell, 1992). Again, a normal healthy adult control group of subjects were asked to produce preferred rhythmical motions of the torso and arms that were similar in form to the stereotypies produced. These control motions had the subjects oscillating at the hip in the sagittal plane with a preferred motion that was deemed to be comfortable and efficient. The control arm rhythmical movements were with the arms held out straight with the oscillations occurring in the vertical plane. In general, the findings matched those from the lip and tongue analyses reported earlier in showing that the variability of normal rhythmical motion was generally less than that of the tardive dyskinetic stereotypies.

In summary, our preliminary kinematic analyses of natural stereotypies suggest that the dimension of variability is not sufficient to distinguish stereotypies from the rhythmical motions that are produced by normal healthy individuals. This is the case with respect to both the qualitative and the quantitative kinematic properties of the stereotypies. We now turn to converging operational approaches that provide complementary data that are consistent with this general proposition.

Drug Withdrawal and Stereotypies

The evidence available suggests that the withdrawal of medication from subjects diagnosed as tardive dyskinetic produces an immediate and prolonged activation of the abnormal movements (Kalachnik et al., 1984; Sprague, 1983). In other words, withdrawing the agent that induced the movement disorder actually enhances the movement problem, as observed from typical clinical recording techniques. There are no long-term studies of the movement dynamics of human tardive dyskinetic stereotypies. This kind of examination of stereotypies is part of our ongoing work on the study of the movement disorders associated with tardive dyskinesia.

The influence of medication withdrawal on the dynamics of tardive dyskinesia may be reflected in a case study that we have conducted of a single 70-year-old male adult individual. The individual in question had a long history of tardive dyskinesia and produced observable stereotypies in a variety of effector units. Here we report the motion of the upper and lower lips of the individual while he was sitting at rest in an examination room. The individual was videotaped on medication withdrawal, immediately on medication reinstatement, after a 12-month period of medication reinstatement, and finally during engagement in the task-oriented action of handwriting. Approximately 10-s segments of lip motion were analyzed from the videotape recordings.

A sample position-over-time trace of the upper and lower lips at each of the testing conditions is shown in Figure 17.5, a-d. These samples are representative of the lip motion at each testing condition. The positions of the upper and lower lips were cross-correlated over the duration of each analysis segment to provide an index of the coupling between the two lips. The coupling reflects the coordination in the motion of the upper and lower lips or the degree to which the lip segments move cooperatively. During medication withdrawal there was a relatively high degree of coupling of the lips, as revealed by the correlation of $r = .76$. On medication reinstatement, the coupling between lips was not present, as indexed by the low correlation of $r = .12$. The independence of the lip motions was still present after another 12 months of the medication regimen, as shown by the correlation of $r = .07$. Finally, when the subject was engaged in a task-oriented action of handwriting while under medication, the correlation returned to a higher level of coupling, with a correlation of $r = .75$.

The generalizability of single-subject data is always problematic. However, the trends suggested by the case study of lip motions as a function of medication manipulation in this adult subject are consistent with theoretical expectations. The drug withdrawal exacerbates the problem of motor control by constraining the degrees of freedom to act as a unit. This coupling or constraint on the system is reduced or even eliminated on medication reinstatement, and this demonstration of independent lip control is consistent over a medication regimen of a year. The primary task demands of handwriting lead to the reintroduction of the lip coupling, a feature consistent with the effects of stress on exacerbating stereotypic behavior.

Thus the medication manipulation influenced both the form of the lip coordination pattern as well as the scaling of this coordination function. The medication influence, therefore, is not simply one of greater or lesser amplitude of movement. It influences the structural properties of the coordination function as reflected in the more con-strained relations between lip degrees of freedom during the execution of a movement sequence. In short, the prolonged neuroleptic medication regimen reduces the flexibility in the organization of the qualitative properties of the motor system.

Stereotypies in Movement Tasks

Stereotypies are usually observed and analyzed in non-task-oriented situations—namely, in situations where there is no externally driven or discernible goal that specifies the production of the stereotypy. However, stereotypies also occur during the execution of a given task-oriented action, and in many instances the stereotypy interferes with the realization of the externally defined goal. One can view this situation as the production of two unrelated actions or the production of a single action that is encompassed by two typically individual action components.

In our studies of the postural control of adults diagnosed as developmentally disabled (profoundly and severely mentally retarded) with tardive dyskinesia, we have recorded the production of what would classically be labeled as a postural stereotypy during the execution of certain postural tasks (Ko, van Emmerik, Sprague, & Newell, in press). In this study 31 adult subjects were examined standing on a force platform while performing certain postural tasks that included (a) standing

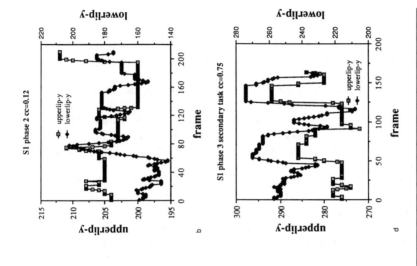

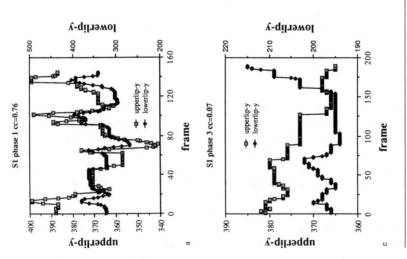

Figure 17.5 Sample upper and lower lip position over-time profiles of a single adult tardive dyskinetic subject as a function of medication conditions and primary task demands. (a) Medication withdrawal. (b) Medication reinstatement. (c) Twelve-month retest session on medication. (d) During primary task of handwriting. Video frame rate is 30 Hz, hence each frame = 33.33 ms. Lip motion unit is pixels.

still with arms at side; (b) standing still with one arm (right and left arm separately) up in front of the body, still and parallel to the ground; or (c) both arms up in front of the body, still and parallel to the ground. Finally, in an arm-swinging task, subjects were asked to swing both arms rhythmically back and forth in the sagittal plane at a preferred rhythm. Only the center-of-pressure data derived from the platform force and moment components during each postural task will be discussed here. There were two 10-s trials at each postural task for each subject, and most of the tardive dyskinetic subjects completed all of the trials.

In normal adult posture there is what is usually interpreted as an unstructured pattern to the forces exerted at the feet (e.g., Goldie, Bach, & Evans, 1989; Murray, Wood, & Sepic, 1975). There is more motion in the anterior-posterior direction than the lateral direction, but the pattern of the center of pressure is typically viewed as unstructured, random, and noiselike. A prototypical normal adult center-of-pressure profile is shown in Figure 17.6a. Abnormal movements are usually associated with increased variability of this unstructured center-of-pressure pattern (e.g., Diener, Dichgans, Bacher, & Gompf, 1984; Lucy & Hayes, 1985). However, in 38% of the trials recorded of our developmentally disabled with tardive dyskinesia group there was an observable systematic rhythmical structure to the center-of-pressure profile. A sample rhythmical postural trial is shown in Figure 17.6b. There were three forms to the rhythmical center-of-pressure profiles, in the anterior-posterior, lateral, and diagonal directions. These rhythmical center-of-pressure movement forms would usually be interpreted as a stereotypic behavior that was unrelated to the successful execution of the postural tasks.

The rhythmical center-of-pressure profiles can also be interpreted, however, as coordination solutions, different from that found in normal stance, to each of the respective postural tasks. The rhythmical center-of-pressure forms are a viable solution to the postural task, in that the subjects remained standing during each of the trials. And although the standard center-of-pressure dependent variables, namely, the mean and standard deviation of the length and area of the center of pressure, were usually greater in this developmentally disabled tardive dyskinetic group than in the normal adult control group, this comparison depends on the plane of motion in which the measurement occurs. For example, in the trial shown in Figure 17.6b, the variability (standard deviation) of the center of pressure in the anterior-posterior direction is very small and within the range of (and sometimes lower than) that of normal subjects.

The above individual trial examples raise the question whether the standard measures of center-of-pressure variability are *sufficient* indices of postural stability. It is typical in the postural literature to equate measures of the variability of the center of pressure with the stability of posture. Namely, the more variable the center of pressure, as measured by the standard deviation in the respective plane of motion, the less stable the postural support, and vice versa. Our study of postural control in tardive dyskinesia suggests, however, that the variability of postural center of pressure may be a necessary, but not sufficient, measure to adequately invoke the concept of stability.

Assessments of postural stability also need to take into account the nature of the dynamical organization, or the type of ''attractor'' that is organizing the postural control. Variability on a given postural dimension needs to be considered in relation

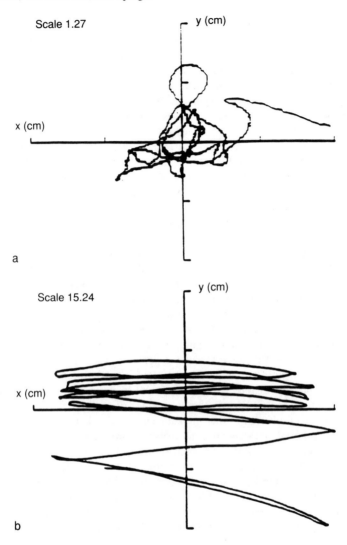

Scale 1.27

y (cm)

x (cm)

a

Scale 15.24

y (cm)

x (cm)

b

Figure 17.6 Sample center-of-pressure profiles over 10-s trials for two subjects. (a) Normal subject. (b) Tardive dyskinetic subject.

to the attractor type supporting the observed dynamics (Newell, Kugler, van Emmerik, & McDonald, 1989). A given absolute amount of center-of-pressure variability in a given plane of motion will probably reflect different degrees of stability in, for example, a point-versus-limit-cycle attractor organization. Intuitively, it appears that the attractor space of the rhythmical center-of-pressure forms is a limit cycle, whereas the normal adult data seem to reflect a higher dimensional (fractal) attractor structure as the organizing dynamic. Indeed, our preliminary dimensional analysis of the rhythmical and nonrhythmical trials confirms this intuition

(Newell, van Emmerik, & Sprague, 1992). If our estimate of a different attractor type organizing posture is correct, then postural variability on a given plane of motion cannot be directly linked to absolute estimates of stability, without considering the nature of the attractor supporting the action. If the postural control dynamics were, for example, organized in the form of a limit-cycle type of organization, then these dynamics would be highly stable, although the absolute deviations in the excursions of the center of pressure could be relatively large. In short, our data from tardive dyskinetic subjects suggest that the traditional measures of the variability of center of pressure are not necessarily sufficient indices of postural stability and, in many instances, not a distinguishing feature of tardive dyskinetic versus normal posture.

It is also worth mentioning at this juncture that there are other considerations that undermine the equating of postural variability with postural stability. These relate to the idea that stability must be considered in a subject-relevant intrinsic frame of reference rather than an extrinsic subject-irrelevant (or merely experimenter-relevant) frame of reference. Stability of postural control is defined over the inter-action of organismic, environmental, and task constraints (Newell, 1986b; Riccio & Stoffregen, 1988). Although organismic factors such as body size may be presumed to be equated through randomization procedures across groups, the general assess-ment (and control) of the subject-relevant frame of reference for action may be more difficult with so-called special populations than it is with normal subjects. The nonlinearity of the perceptual-motor system suggests that more attention needs to be paid to individual subject and trial measurement in this and related movement variability domains.

VARIABILITY AND ADAPTABILITY

The lines of evidence we have discussed from different operational approaches to the study of stereotypic movements in tardive dyskinesia all suggest that variability is not a sufficient index of stereotypy. The variability of both the qualitative and the quantitative properties of the stereotypic motions is as high as, and usually higher than, that produced by normal healthy subjects producing preferred rhythmical motions of similar form. The withdrawal of medication also seems to have a profound effect on the qualitative and quantitative properties of the organization of the motor system.

Berkson and Gallagher (1986) concluded, without the support of adequate direct measurement, that many aspects of the kinematics of stereotypies in normal and severely mentally retarded infants were similar, although they differed in certain ways, including the movement form. To show that different populations can produce different stereotypies or movement forms is not counter to the proposal that different populations can produce, under certain conditions, the same movement forms. Our emphasis here has been on the relative variability of rhythmical motions in normal and tardive dyskinetic subjects, and we show that low variability is not a defining feature of the stereotypies associated with tardive dyskinesia.

All biological systems at all levels of analysis demonstrate some degree of variabil-ity in their rhythmicity. This variability is traditionally interpreted as reflective of system noise, but more recent interpretations through chaotic dynamic systems suggest a functional role for variability (Glass & Mackey, 1988; Mpitsos, 1990; Winfree, 1990), the significance of which we are only just beginning to understand. Ironically, the traditional noise interpretation of variability in the motor system, when taken in conjunction with the traditional minimal variability view of stereotypy, would lead logically to stereotypy being defined from a variability perspective as a desirable state! This line of reasoning brings us full circle to the earlier discussion of the similarities and dissimilarities between the concepts of skill and stereotypy.

As suggested in the introduction to this chapter, the movement organization of stereotypies reflects many of the qualities of skilled behavior. Thus, to observe the dynamic organization of natural stereotypies as we have done may not be sufficient to unravel the nature of the dynamical disease (Glass & Mackey, 1988). The nature of the dynamical disease may be realized only by manipulating the informational input to the natural dynamic of the stereotypy to examine the influence of small changes in the initial conditions. Berkson and Gallagher (1986) have concluded that self-stimulation is a key component in initiating and maintaining the production of stereotypies. This view of stereotypy stresses the functional role of stereotypies to the individual.

A common interpretation of stereotypic behavior is that it is purposeless or nonadaptive (e.g., Teitelbaum et al., 1990). The assessment of nonadaptiveness is usually post hoc, in the sense that the state of purposelessness is ascribed from observations of the behavior. For example, many instinctive behaviors, such as animal courting displays (Golani, 1976), are seen to be stereotyped but are also viewed to be adaptive on the larger time scale of evolutionary considerations. Thus not all stereotyped behaviors are nonadaptive or purposeless. A major problem in understanding the functional role of stereotypies seems to arise from the multitude of interpretations of the term *adaptive*.

The confusion about the significance of adaptation in sensorimotor control arises, in part, from the different frames of reference or levels of analysis used to interpret the term. For example, variability in a system parameter may be viewed as adaptive at one level of analysis but not at another. Conversely, a lack of variability can be viewed as nonadaptive at one level of analysis but not at another.

In our studies of the variability of stereotypic behavior we have confined our observations to the kinematics of the movement. We have made no inferences as to the goal of the stereotypic act or the long-term adaptive influence of the stereotypic behavior on the individual. Within our frame of reference to stereotypies we have shown that the tardive dyskinetic movements are no less variable than similar form-movements of normal healthy adults. Furthermore, variability can be construed as functional at this level of analysis, given that movement variability affords information about the dynamics supporting the action.

CONCLUDING REMARKS

In summary, we propose that stereotypic movements are defined in part by the form of the movement sequence, in just the same way as other so-called normal movement patterns are perceived and categorized by observers. The form of movements is characterized by the topological characteristics of the kinematics, that is, those characteristics that remain invariant to transformations of scale on some given movement dimension, such as movement time or amplitude (Newell, 1985). Thus, when stereotypic movements are defined or referred to as invariant or lacking in variability, the definition is referring to the variability of the topological movement properties.

Our studies show that perceptual constancy affords a considerable range of variation to exist in the absolute movement properties that define a given movement form or stereotypy. Thus, the variability of the discrete quantitative kinematic variables of stereotypies is considerably greater than would be expected by traditional intepretations of stereotypy. Indeed, examinations of discrete kinematic movement properties of stereotypic movement sequences reveal that they are no more invariant or lacking in variability than those same movement parameters in the movement sequences of normal individuals producing rhythmical movement forms similar to common stereotypies. Indeed, in many instances the variability of discrete kinematic measures in certain stereotypies of certain individuals is considerably higher than that of normal individuals. Thus, expressed another way, stereotypic movements are no less variable than voluntary skilled movements, in spite of the general dogma on stereotypic behavior.

Our studies suggest that one of the consequences of prolonged neuroleptic medication is that the movement system is overconstrained in its organization of the available degrees of freedom of the effector system. Thus, tardive dyskinesia is characterized not only by the exhibition of certain stereotypies, particularly those in the face, fingers, and toes, but also by a change in the organization of the motor output that offers reduced adaptability to changing environmental demands. At this point in time, direct tests of this hypothesis have not been conducted, but our preliminary data analysis of natural stereotypies, and those produced during the execution of task-defined actions, suggests that adaptability of the sensorimotor system is the dimension on which to consider stereotypic movement sequences, rather than the often-promoted dimension of variability. Variability is not a *sufficient* index of stereotypy.

REFERENCES

American Psychiatric Association. (1980). *Task force report: Tardive dyskinesia.* Washington, DC: Author.

Arutyunyan, R.H., Gurfinkel, V.S., & Mirskii, M.L. (1968). Investigation of aiming at a target. *Biophysics,* **13,** 536-538.

Arutyunyan, R.H., Gurfinkel, V.S., & Mirskii, M.L. (1969). Organization of movements on execution by man of an exact postural task. *Biophysics,* **14,** 1162-1167.

Baumeister, A.A., & Forehand, R. (1973). Stereotyped acts. In N.R. Ellis (Ed.), *International review of research in mental retardation* (Vol. 6). New York: Academic Press.

Berkson, G. (1967). Abnormal stereotyped motor acts. In J. Zubin & H.F. Hunt (Eds.), *Comparative psychopathology—animal and human* (pp. 76-94). New York: Grune & Stratton.

Berkson, G. (1983). Repetitive stereotyped behaviors. *American Journal of Mental Deficiency*, **88**, 239-246.

Berkson, G., & Gallagher, R.J. (1986). Control of feedback from abnormal stereotyped behaviors. I. In M.G. Wade (Ed.), *The development of coordination, control, and skill in the mentally handicapped* (pp. 7-24). Amsterdam: North-Holland.

Cooper, S.J., & Dourish, C.T. (Eds.) (1990). *Neurobiology of stereotyped behavior.* Oxford: Clarendon Press.

Diener, H.C., Dichgans, J., Bacher, M., & Gompf, B. (1984). Quantification of postural sway in normals and patients with cerebellar diseases. *Electroencephalography and Clinical Neurophysiology*, **57**, 134-142.

Emmerik, R.E.A. van, Sprague, R.L., Slobonouv, S., & Newell, K.M. (1992). *Stereotypic actions and tardive dyskinesia: Facial movements.* Manuscript submitted for publication.

Gardos, G., Cole, J.O., & La Brie, R.L. (1977). The assessment of tardive dyskinesia. *Archives of General Psychiatry*, **34**, 1206-1212.

Glass, L., & Mackey, M.C. (1988). *From clocks to chaos: The rhythms of life.* Princeton, NJ: Princeton University Press.

Golani, I. (1976). Homeostatic motor processes in mammalian interaction: A choreography of display. In P.T.G. Bateson & P.H. Klopher (Eds.), *Perspectives in ethology* (pp. 69-134). New York: Plenum Press.

Goldie, P.A., Bach, T.M., & Evans, O.M. (1989). Force platform measures for evaluating postural control: Reliability and validity. *Archives of Physical Medicine and Rehabilitation*, **70**, 510-517.

Guthrie, E.R. (1935). *The psychology of learning.* New York: Harper.

Hancock, P.A., & Newell, K.M. (1985). The movement speed-accuracy relationship in space-time. In H. Heuer, U. Kleinbeck, & K.H. Schmidt (Eds.), *Motor behavior: Programming, control, and acquisition* (pp. 153-188). Berlin: Springer-Verlag.

Kalachnik, J.E. (1984). Tardive dyskinesia and the mentally retarded: A review. In S.E. Bruening (Ed.), *Advances in mental retardation and developmental disabilities* (Vol. 2, pp. 329-356). Greenwich, CT: JAI Press.

Kalachnik, J.E., Harder, S.R., Kidd-Nielson, P., Errickson, E., Doebler, M., & Sprague, R.L. (1984). Persistent tardive dyskinesia in randomly assigned neuroleptic reduction, neuroleptic nonreduction, and no neuroleptic history groups: Preliminary results. *Psychopharmacology Bulletin*, **20**, 27-32.

Kalachnik, J.E., Sprague, R.L, & Slaw, K.M. (1988). Training clinical personnel to assess tardive dyskinesia. *Progress in NeuroPsychopharmacology and Biological Psychiatry*, **12**, 749-762.

Ko, Y.G., van Emmerik, R.E.A., Sprague, R.L., & Newell, K.M. (in press). Postural stability, tardive dyskinesia, and developmental disability. *Journal of Mental Deficiency Research*.

Lewis, M.H., & Baumeister, A.A. (1982). Stereotyped mannerisms in mentally retarded persons: Animal models and theoretical analyses. In N.R. Ellis (Ed.), *International review of research in mental retardation* (Vol. 11). New York: Academic Press.

Lewis, M.H., MacLean, W.E., Johnson, W.L., & Baumeister, A.A. (1981). Ultradian rhythms in stereotyped and self-injurious behavior. *American Journal of Mental Deficiency, 85*, 601-610.

Lucy, S.D., & Hayes, K.C. (1985). Postural sway profiles: Normal subjects and subjects with cerebellar ataxia. *Physiotherapy Canada, 37*, 140-148.

Mason, W.A., & Berkson, G. (1975). Effects of maternal mobility on the development of rocking and other behaviors on rhesus monkeys: A study with artificial mothers. *Developmental Psychobiology, 8*, 197-211.

Mpitsos, G.J. (1990). Chaos in brain function and the problem of nonstationarity: A commentary. In E. Basar (Ed.), *Chaos in brain function.* Berlin: Springer-Verlag.

Murray, M.P., Wood, A.A.S., & Sepic, S.B. (1975). Normal postural stability and steadiness: Quantitative assessment. *Journal of Bone and Joint Surgery, 57A*, 510-515.

Newell, K.M. (1985). Coordination, control, and skill. In D. Goodman, I. Franks, & R. Wilberg (Eds.), *Differing perspectives in motor control* (pp. 295-318). Amsterdam: North-Holland.

Newell, K.M. (1986a). Comments on coordination, control, and skill papers. In M.G. Wade (Ed.), *Motor skill acquisition of the mentally handicapped* (pp. 101-112). Amsterdam: North-Holland.

Newell, K.M. (1986b). Constraints on the development of coordination. In M.G. Wade & H.T.A. Whiting (Eds.), *Motor development in children: Aspects of coordination and control* (pp. 341-360). Dordrecht: Martinus Nijhoff.

Newell, K.M., Carlton, L.G., & Hancock, P.A. (1984). Kinetic analysis of response variability. *Psychological Bulletin, 96*, 133-151.

Newell, K.M., Kugler, P.N., van Emmerik, R.E.A., & McDonald, P.V. (1989). Search strategies and the acquisition of coordination. In S.A. Wallace (Ed.), *Perspectives on the coordination of movement* (pp. 85-122). Amsterdam: North-Holland.

Newell, K.M., & Sprague, R.L. (1990). Early diagnosis of tardive dyskinesia. In A. Vermeer (Ed.), *Motor development, adapted physical activities, and mental retardation* (pp. 30-46). Amsterdam: North-Holland.

Newell, K.M., van Emmerik, R.E.A., & Sprague, R.L. (1992). *On postural stability and variability.* Manuscript under review.

Pohl, P. (1977). Voluntary control of stereotyped behavior by mentally retarded children: Preliminary experimental findings. *Developmental Medicine and Child Neurology, 19*, 811-817.

Riccio, G.E., & Stoffregen, T.A. (1988). Affordances as constraints on the control of stance. *Human Movement Science, 7*, 265-300.

Sparrow, W.A., Donovan, E., van Emmerik, R.E.A., & Barry, E.B. (1987). Using relative motion plots to measure change in intra-limb and inter-limb coordination. *Journal of Motor Behavior, 19*, 115-129.

Sprague, R.L. (1983). *Rated abnormal movements in a study of retarded subjects randomly assigned to psychotropic medication withdrawal and control groups.* Paper presented at the meeting of the American College of Neuropsychopharmacology, San Juan.

Sprague, R.L., & Kalachnik, J.E. (1991). Reliability, validity, and a total score cutoff for the Dyskinesia Identification System: Condensed User Scale (DISCUS) with mentally ill and mentally retarded populations. *Psychopharmacology Bulletin*, **27**, 51-58.

Sprague, R.L., Kalachnik, J.E., & Slaw, K.M. (1989). Psychometric properties of the Dyskinesia Identification System: Condensed User Scale (DISCUS). *Mental Retardation*, **27**, 141-148.

Sprague, R.L., & Newell, K.M. (1987). Toward a movement control perspective of tardive dyskinesia. In H.Y. Meltzer (Ed.), *Psychopharmacology: The third generation of progress* (pp. 1233-1238). New York: Raven Press.

Sprague, R.L., Slobonouv, S., & Newell, K.M. (1992). *Stereotypic actions and tardive dyskinesia: Torso and limb movements.* Manuscript in preparation.

Sprague, R.L., White, D.M., Ullmann, R., & Kalachnik, J.E. (1984). Methods for selecting items in a tardive dyskinesia rating scale. *Psychopharmacology Bulletin*, **20**, 339-345.

Teitelbaum, P., Pellis, S.M., & DeVietti, T.L. (1990). Disintegration into stereotypy induced by drugs or brain damage: A microdescriptive behavioral analysis. In S.J. Cooper & C.T. Dourish (Eds.), *Neurobiology of stereotyped behavior* (pp. 200-231). Oxford: Clarendon Press.

Thelen, E. (1980). Determinants of amounts of stereotyped behavior in normal human infants. *Ethology and Sociobiology*, **1**, 141-150.

Waddington, J.L., Molloy, A.G., O'Boyle, K.M., & Pugh, M.T. (1990). Aspects of stereotyped and non-stereotyped behavior in relation to dopamine receptor subtypes. In S.J. Cooper & C.T. Dourish (Eds.), *Neurobiology of stereotyped behavior* (pp. 64-90). Oxford: Clarendon.

Winfree, A.T. (1990). *The geometry of biological time.* Berlin: Springer-Verlag.

Wolff, P.H. (1968). Stereotypic behavior and development. *Canadian Psychologist*, **9**, 474-484.

Acknowledgments

The preparation of this paper was supported in part by grant HD-21212 from the National Institutes of Health. Alex Antoniou, Young Ko, and Sam Slobonouv helped with some of the data analysis reported here. Requests for reprints should be addressed to K.M. Newell, Department of Kinesiology, University of Illinois at Urbana-Champaign, Freer Hall, 906 S. Goodwin Avenue, Urbana, IL 61801. Richard van Emmerik is now at the Free University, the Netherlands.

Credits

Developmental Editor: Judy Patterson Wright, PhD
Assistant Editors: Dawn Roselund, John Wentworth, Moyra Knight, and Julie Swadener
Copyeditor: Wendy Nelson
Indexer: Theresa Schaefer
Production Director: Ernie Noa
Typesetter and Text Layout: Sandra Meier
Text Design: Keith Blomberg
Paste Up: Denise Lowry
Printer: Braun-Brumfield

Cover Description

The cover illustrates the dynamics of a chaotic regime obtained from a computer simulation of a small catalytic network. Dynamical principles inherent in such networks have been implicated in a wide variety of biological systems from prebiotic evolution, to population dynamics and neural networks. They may be useful in gaining an understanding of memory storage in neural systems and in the generation of patterns of activity that they produce. The three-dimensional phase portrait shown here was constructed by plotting one of the four variables in this network on the horizontal axis. The second variable was plotted on the vertical axis. The color flow from deep red toward white represents the range of values in the third variable. Deep red projects into the plane of the cover away from the reader. As the red color diminishes, the flow of values is out of the plane of the cover toward the reader.

This illustration was prepared by George J. Mpitsos and H. Clayton Creech following Andrade, M.A., Nuño, J.C., Morán, F., Montero, F., & Mpitsos, G.J. (1993). Complex dynamics of a catalytic network having faulty replication into an error species. *Physica D*. In press. Work was supported by AFOSR-92J0140.

Index

A

Abbs, J.H., 66, 176
Abrams, R.A., 92, 99, 121-122, 132
Absolute coordination, 297-304, 381-382, 408
Adaptive control
    and postural behavior, 322-323, 328
    and stereotypy, 476
Adaptive mechanisms
    definitions of, 227n, 492
    stereotypies as, 492
Adey, R.W., 249-250
Advantage, mechanical, 201-204
Agarwal, Gyan C., 117-155
Aimed-hand movements
    description of task, 93
    kinematic properties of, 101-103
    in speed-accuracy trade-off research, 92-93, 101-103
Aiming movements, spatial variability in, 53-62. *See also* Aimed-hand movements
Akazawa, K., 197
Allard, R., 99
Alpha-motoneurons (α-MNs)
    in FLETE model, 189-201, 202, 204, 217
    in speed-accuracy trade-off, 133, 134, 162, 168
American sign language, 452
Anderson, M.E., 438
Andrade, M.A., 237, 242, 245
Aplysia, 247, 252, 254, 260-265
Apraxia. *See* Limb apraxia
Aristotle, 228, 229
Arutyunyan, G.H., 8, 66
Atomistic level of coordination, 407
Attractors and attracting states. *See also*
    Chaotic attractors; Limit-cycle attractors
    in bimanual actions, 362, 364, 365, 366-371, 374-375
    in chaotic dynamical systems, 304-305
    and coordination dynamics, 293, 294, 300
    definitions of, 362
    in distributed function, 237-239, 244
    in interlimb rhythmic coordination, 395, 398-399, 407-408
    and neural networks in behavior, 237-239, 244, 273-274, 276
    and postural behavior, 323, 332, 333, 346, 349, 490-491
    and stability in dynamical systems, 362-363
    and stereotypies, 490-491

B

Backpropagation method, 274
Badminton strategy (r/τ-strategy), 177-178
Bailey, M.A., 338-340, 342
Bak, P., 239, 268
Balance dynamics, 323-324, 326, 327
BCNs (buccal-cerebral neurons), 231, 236-237, 240, 247, 272
Becchi, M.P., 243
Beek, Peter J., 5, 293, 381-411
Behavior. *See also* Postural behavior
    behavioral choice, 234-236
    behavioral hierarchy and reflexes, 235
    and breaking symmetry, 299
    defined by Mpitsos & Soinila, 229, 245-246
    and impulse variability models, 97-99
    and parallel force unit model, 51
    of *Pleurobranchaea*, general description of, 229-230
    and skill, 359
Behavioral choice, 234-236
Bellman, Kirsti L., 235
Berkson, G., 479, 491, 492
Bernstein, Nicholai A., 8, 160, 163
Bias compensation in FLETE model, 198-199
Bifurcation
    in computer simulations, 270-273
    and converging/diverging connections, 243
    in coordination dynamics, 300, 301
    in interlimb rhythmic coordination, 403
    and variability in general, 239-243, 244, 248-249
Bimanual actions. *See also* Coordination
    dynamics; Interlimb rhythmic
    coordination
    attractors in, 362, 364, 365, 366-371, 374-375
    collective variables for, 362
    continuous, 363-365
    continuous vs. discrete processes in, 374
    coordinated state in, 366-367, 371
    discrete, 365-373
    effect of practice on, 365, 371-373, 374, 375-376
    movement speed in, 367, 370
    as a source of variability, 361, 363, 374
    spatial variability in, 371, 373
    stability of, 363-373
    symmetry/asymmetry in, 366, 369, 371, 373, 374-375